Handbook of Soil Fertility

Handbook of Soil Fertility

Editor: Donald Cronin

R CALLISTO REFERENCE

www.callistoreference.com

Callisto Reference,
118-35 Queens Blvd., Suite 400,
Forest Hills, NY 11375, USA

Visit us on the World Wide Web at:
www.callistoreference.com

ISBN: 978-1-63239-978-6 (Hardback)

Cataloging-in-Publication Data

Handbook of soil fertility / edited by Donald Cronin.
 p. cm.
Includes bibliographical references and index.
ISBN 978-1-63239-978-6
1. Soil fertility. 2. Soil productivity. 3. Agriculture. I. Cronin, Donald.
S596.7 .H36 2018
631.422--dc23

Table of Contents

Permissions

List of Contributors

Index

Preface

This book has been an outcome of determined endeavour from a group of educationists in the field. The primary objective was to involve a broad spectrum of professionals from diverse cultural background involved in the field for developing new researches. The book not only targets students but also scholars pursuing higher research for further enhancement of the theoretical and practical applications of the subject.

Soil fertility is an integral part of agriculture and this book unravels the recent studies in this field. Key characteristics of soil like porosity, aeration, texture and drainage determine the quality of the soil. The presence of organic matter, nutrition and soil biota are also essential for the growth of crops. Methods that increase soil nutrition are also important topics of this area of study. As this field is emerging at a rapid pace, the contents of this book will help the readers understand the modern concepts and applications of the subject. Students, researchers, experts and all associated with soil science and agriculture will benefit alike from this book.

It was an honour to edit such a profound book and also a challenging task to compile and examine all the relevant data for accuracy and originality. I wish to acknowledge the efforts of the contributors for submitting such brilliant and diverse chapters in the field and for endlessly working for the completion of the book. Last, but not the least; I thank my family for being a constant source of support in all my research endeavours.

Editor

Environmental Modeling and Exposure Assessment of Sediment-Associated Pyrethroids in an Agricultural Watershed

Yuzhou Luo[1,2]*, Minghua Zhang[1,2]*

1 Wenzhou Medical College, Wenzhou, China, 2 Department of Land, Air, and Water Resources, University of California Davis, Davis, California, United States of America

Abstract

Synthetic pyrethroid insecticides have generated public concerns due to their increasing use and potential effects on aquatic ecosystems. A modeling system was developed in this study for simulating the transport processes and associated sediment toxicity of pyrethroids at coupled field/watershed scales. The model was tested in the Orestimba Creek watershed, an agriculturally intensive area in California' Central Valley. Model predictions were satisfactory when compared with measured suspended solid concentration ($R^2 = 0.536$), pyrethroid toxic unit (0.576), and cumulative mortality of *Hyalella azteca* (0.570). The results indicated that sediment toxicity in the study area was strongly related to the concentration of pyrethroids in bed sediment. Bifenthrin was identified as the dominant contributor to the sediment toxicity in recent years, accounting for 50–85% of predicted toxicity units. In addition, more than 90% of the variation on the annual maximum toxic unit of pyrethroids was attributed to precipitation and prior application of bifenthrin in the late irrigation season. As one of the first studies simulating the dynamics and spatial variability of pyrethroids in fields and instreams, the modeling results provided useful information on new policies to be considered with respect to pyrethroid regulation. This study suggested two potential measures to efficiently reduce sediment toxicity by pyrethroids in the study area: [1] limiting bifenthrin use immediately before rainfall season; and [2] implementing conservation practices to retain soil on cropland.

Editor: Stephen J. Johnson, University of Kansas, United States of America

Funding: The authors would like to acknowledge funding support from the Coalition for Urban/Rural Environmental Stewardship and California State Water Quality Control Board. This work was also supported by the 2005 to 2006 Consolidate Grants-Proposition 50 Agricultural Water Quality Grant Program, State of California, as well as the support from Zhejiang and Wenzhou (2008C03009, 20082780125). The funders had no role in study design, data collection and analysis, decision to publish, or preparation of the manuscript.

Competing Interests: The authors have declared that no competing interests exist.

* E-mail: yzluo@ucdavis.edu (YZ); mhzhang@ucdavis.edu (MZ)

Introduction

Use of pesticides in crop production has been an important practice in modern agriculture, especially in the Central Valley of California, the most dynamic agricultural region in the world. Pesticide use can lead to severe environmental problems due to their toxicity to humans and many ecosystem organisms. Synthetic pyrethroids have become increasingly popular following outright bans or limitations on the use of cholinesterase-inhibiting insecticides, such as organophosphates (OPs). Previous studies have indicated that the decrease in OP use in California was related to the substitution with pyrethroids [1]. Pyrethroid insecticides are associated with selective potency in insects and relatively low potency in mammals. However, results of exposure monitoring and pesticide illness surveillance suggested that field residues of pyrethroids can cause irritant respiratory symptoms, nausea and headache [2]. Furthermore, pyrethroids are very acutely toxic to fish and invertebrates, with the 10-day LC50 values ranging from 2–140 ng/L in water (*Americamysis bahia* and *Ceriodaphnia dubia*) and 4–110 ng/g in sediment (*Hyalella azteca*) [3]. Surface water monitoring indicated widespread presence of pyrethroids and associated toxicity in agricultural and urban waterways in California [4,5,6]. Identifying the distribution of pyrethroids in surface waters and their effects on aquatic

organisms is very important in pesticide regulation and water management of pyrethroids.

Monitoring data are usually insufficient to characterize the spatial distribution and the main sources of pesticide residues. Therefore, mathematical models are used to simulate the effects of pesticide use, management practices, and environmental factors on pesticide fate and distribution. In addition, the regulatory burden has evolved to currently consider negative impacts of pesticides on aquatic organisms. Detailed information on pesticide residues, such as the magnitude, timing and frequency of peak concentrations, are required to examine the overall ecosystem exposure by the use of pesticides. Therefore, continuous modeling at the field scale is essential for decision making processes to adequately meet regulatory requirements and improve management practices.

Recent developments in GIS technology enable the application of field-scale models on a large landscape by incorporating spatially distributed simulations of water and chemical movement in river networks. The integrated systems with field-scale models routing algorithms have been successfully applied to simulate pesticide fate and behaviors in streams. Most of those models were originally designed for the simulation of pesticides in the dissolved phase, indicating appropriate model applications on pesticides with lower adsorption coefficients. With octanol-water partition coefficients (K_{OW}) values of 10^5–10^7, pyrethroids tend to adsorb to soil and sediment rather than remain in the dissolved phase [7]. Therefore, accurate prediction of pyrethroid fate and transport must incorporate the simulations of soil erosion, in-stream sediment transport, and pyrethroid partitioning. Due to inadequate representation of the above hydrologic and transport processes, most of the existing field-scale models and in-stream routing models are not appropriate for predicting environmental behaviors of pyrethroids. For example, RZWQM (Root zone water quality model) were developed for water flow and solute transport, and thus do not simulate soil erosion and adsorbed pesticide removal. Consequently, edge-of-field pesticide fluxes are underestimated, especially for those with strong sediment/soil sorption [8]. In the GLEAMS (Groundwater loading effects of agricultural management system) model, solid-bound pesticide concentration in eroded soil is determined based on a prescribed soil mass per unit runoff volume regardless of the actual soil erosion rate [9]. In addition, many popular routing models are not able to sufficiently capture the dynamics of pesticide partitioning and transport. They either assume a steady state hydraulics (e.g., River and stream water quality model, QUAL2K [10]), or utilize prescribed suspended solid concentrations (e.g., River water quality model, RIVWQ [11]).

PRZM (Pesticide root zone model) estimates soil erosion based on a modified Universal Soil Loss Equation. In our previous study [12], PRZM model was coupled with a linear routing model for assessing pesticide dynamics and distribution in crop fields and stream networks. The coupled system provided a suitable modeling platform for determining environmental concentration and toxicity of pyrethroids. However, some model improvements were required. For example, a minor deficiency has been identified in the PRZM algorithm for adsorbed pesticide removal [13]. This paper presents an improved modeling system based on our previous study [12] for simulating the environmental fate and dispersion of pyrethroid insecticides. The specific purposes for the proposed model were to: (a) account for pyrethroid entry into surface water via soil erosion, (b) predict dynamics and distribution of pyrethroids in channel flow and bed sediment, and (c) characterize the toxicity by pyrethroids to sediment-dwelling organisms. This is one of the first studies on the dynamic modeling of pyrethroids at watershed scale, responding to the emerging research need for pyrethroid reevaluation and watershed management planning. Simulation capability of the developed model was demonstrated by applying it to the Orestimba Creek Watershed (Figure 1), an agriculturally dominated watershed in the California's Central Valley, with four pyrethroids of bifenthrin, λ -cyhalothrin, esfenvalerate, and permethrin as test agents.

Results and Discussion

Evaluation of the improved algorithm

The improved algorithm for adsorbed pesticide removal in PRZM was evaluated in a melon crop field with historical bifenthrin applications in the Orestimba Creek watershed. This field is close to the monitoring site OCER (Orestimba Creek at Eastin Road, Figure 1 and Table 1), with an average annual bifenthrin use of 0.08 kg per treated hectare. The concentration of eroded bifenthrin was reported as the total amount of eroded bifenthrin divided by the total amount of eroded soil during the simulation period. Results of the improved PRZM in this study were generally invariant with the depth of soil compartment used in the numerical analysis (Figure 2). This confirmed that the improved algorithm removed adsorbed pesticide from all compartments within the soil-interaction depth (D_E, see Materials and Methods), thus the resulting removal was not dependent with the depth of each compartment. As discussed in Materials and Methods, the original PRZM considers only the top-

most soil compartment for adsorbed pesticide removal; therefore, the results were very sensitive to the depth. The original PRZM generated similar results as the improved one when the depth of soil compartment was close to D_E (1 cm in this case study). Figure 2 demonstrated PRZM simulations for soil erosion and associated pesticide removal with depth of soil compartment up to 1 cm. In the real PRZM modeling, however, small depth was required for accurate numerical simulation of water and chemical movement in the soil. For example, all crop scenarios for PRZM require depth of 0.1 cm or less for the top soil horizon developed by U.S. Environmental Protection Agency (USEPA) [14]. With small depth of soil compartment, the original PRZM significantly underestimated the adsorbed pesticide removal. The improved PRZM should be applied for consistence estimations of adsorbed pesticide release from the applied field.

Sediment Loadings

Due to their high adsorption coefficients (K_{OW}), pyrethroids are typically adsorbed to soil particles and transported with suspended solids in surface runoff and stream flows. Therefore, a reasonable estimation of sediment concentration in a stream is the first necessary step in simulating pyrethroid partitioning between dissolved and particulate phases. Figure 3 shows the flow-weighted suspended solid concentration on a monthly basis observed and predicted at site OCRR (Orestimba Creek at River Road, close to the watershed outlet, Figure 1 and Table 1). The temporal trend of predictions followed the measured data ($R^2 = 0.536$), indicating a satisfactory simulation of suspended solid transport processes based on the model evaluation guidelines by Moriasi et al. [15]. High concentrations of suspended solids were observed during the irrigation season, especially in July.

According to the USGS sampling results, in-stream concentrations of particulate organic carbon (OC) and suspended solids were strongly correlated ($r = 0.90$, $p < 0.001$). Therefore, agreement between the predicted and observed concentrations of suspended solids also indicated a reasonable simulation for the particulate OC concentrations in the study area.

Pyrethroid Toxicity

Significant correlation ($r = 0.72$, $p = 0.004$) was observed between the predicted and measured toxicity units (TU) of pyrethroids in 15 samples at the three monitoring sites during 2007 and 2008 (Figure 4a). This indicated that the model generally captured the spatial variability and seasonality of the pyrethroid distribution in bed sediment. Based on the model prediction and field measurements, the OCRR site was generally associated with low pyrethroid TU relative to the other sites. Samples with undetected pyrethroids (plotted at 0.01 for measured TU in the figure) were all collected at the OCRR site during dormant seasons or early irrigation seasons. Located at the outlet of the Orestimba Creek, OCRR has larger drainage areas and a longer transport path for pesticides compared to other two sites. Pesticide residues have been largely decayed and diluted before reaching the water-sediment system at this site. For those undetected samples in the OCRR site, corresponding model results yielded TU values of 0.16– 0.21, suggesting that the actual toxicity level was undetectable by the analytical methods applied in the sampling projects. This might also be the reason why the model overestimated measured data with low TUs. In the range of higher toxic levels (TU $> = 0.3$), the model had better agreement with the measurements, and the predicted and measured TUs approached the 1:1 line on the plot.

Figure 4b compares the predicted TUs based on the simulated pyrethroids to the measured cumulative mortality of Hyalella azteca. For mortality values <40%, the predicted TUs are significantly correlated to the measured cumulative mortality ($r = 0.75$,

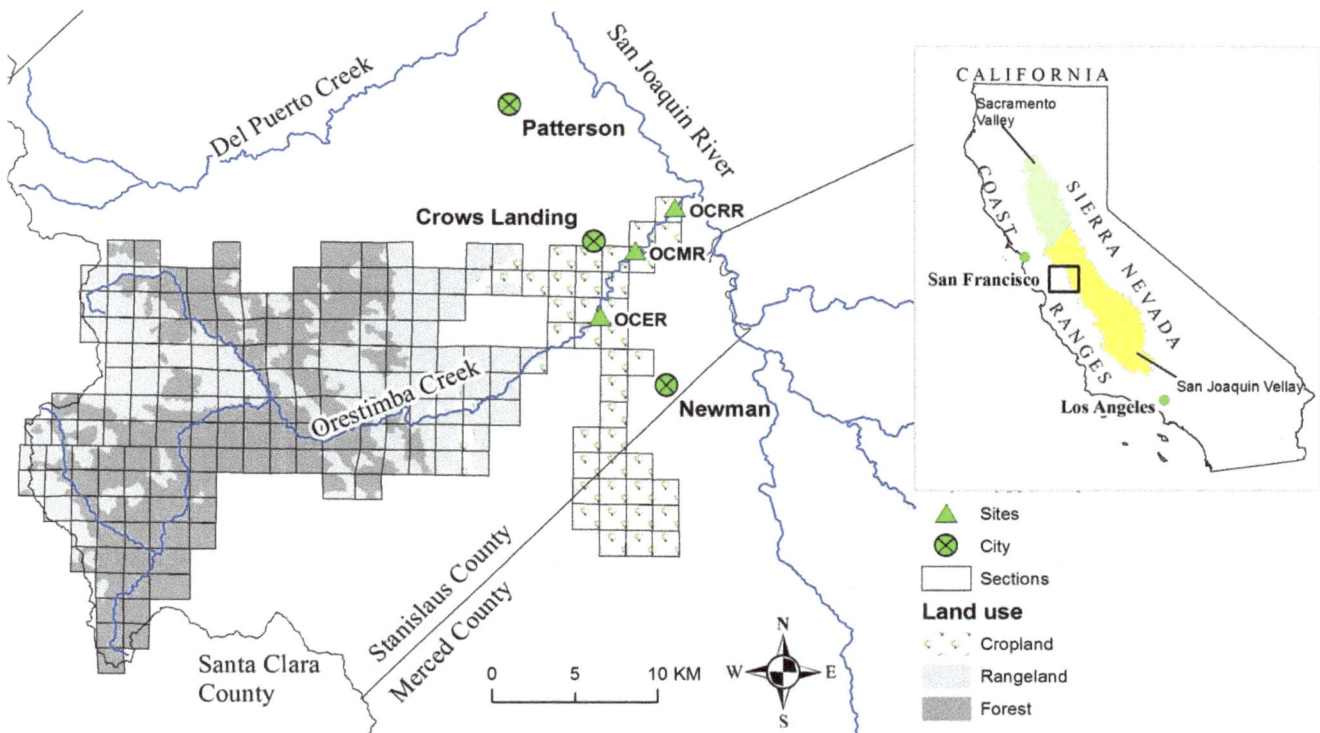

Figure 1. Orestimba Creek watershed and the sampling sites.

p = 0.005). By fitting a toxicity-mortality curve in logistic form [16], the resultant R^2 was 0.578, suggesting the model satisfactorily captured the dose-expose relationship in the evaluated sampling site. This also supported the hypothesis that sediment toxicity in the study area was mainly associated with pyrethroid concentrations. However, the correlation was not as strong among the samples with higher mortality values of >40%. For those samples, predicted TUs were about 0.2 based on the modeled pyrethroid concentrations in this study. This deficiency may have arisen due to the substantial contributions of other pesticides or other toxic compounds which were not modeled in this study to the measured sediment toxicity. This possibility was confirmed by the fact that, for those samples with high mortality, the measured TUs in sediment were also low, ranging from non-detected to 0.1. Another issue was that some samples with relatively high measured TUs were associated with low mortality. For instance, the sample at site OCMR in September 2007 had a measured TU of 0.4 and mortality of 6%. In this case, the respective model prediction of TU = 0.23 gave a more reasonable match to the observed mortality.

Table 1. Sampling sites for model evaluation.

Name	Site ID	USGS ID	Latitude	Longitude
Orestimba Creek at Eastin Road	OCER	-	37.35	−121.07
Orestimba Creek at Morris Road	OCMR	-	37.39	−121.04
Orestimba Creek at River Road	OCRR	11274538	37.41	−121.02

Characterization of Pyrethroid Exposures

In the Orestimba Creek watershed, there was a general increasing trend of total pyrethroid use during 1990–2004 (Figure 5). This increase was mainly attributed to esfenvalerate for years before 2000, and to bifenthrin and λ-cyhalothrin after 2000. After 2004, the amount of pyrethroids used has decreased, except for 2007 when reported permethrin use was very high. Figure 6 shows the predicted TUs at the OCRR site on a monthly basis, presenting sub-chronic risks of benthic organisms to pyrethroid exposures. Similar temporal trends of the predicted pyrethroid TUs were shared at the three sites in this study. Before 2000, the predicted sediment toxicity in the study was low, with maximum monthly TUs less than 0.5. Esfenvalerate was the major contributor to TUs during this period. The use of bifenthrin was started from July 1992 and explained a significant portion of the predicted TU. However, there was a general decreasing trend for both bifenthrin use (from 23.4 to 3.2 kg/year) and predicted TUs during 1992–1999. After 2000, bifenthrin use was increased again and predicted TU in sediment was substantially elevated and peaked in 2002–2004. Another potential reason for the elevated TUs was the change in application timing of bifenthrin. While it was mainly applied in July and August before 2000, significant amounts of bifenthrin were also applied during the late irrigation months of September and October and subject to the significant runoff events induced by winter precipitation. Consequently, bifenthrin became the major contributor to sediment toxicity for the last 10 years, accounting for 50–85% TU in sediment. Esfenvalerate and λ-cyhalothrin were also important contributors, especially during the irrigation season when they explained up to 50% of TU in sediment. During the recent years of 2004–2008, λ-cyhalothrin accounted for 38% of total pyrethroid used in the study area, followed by permethrin (32%), esfenvalerate (17%), and bifenthrin (10%). It is noteworthy that the

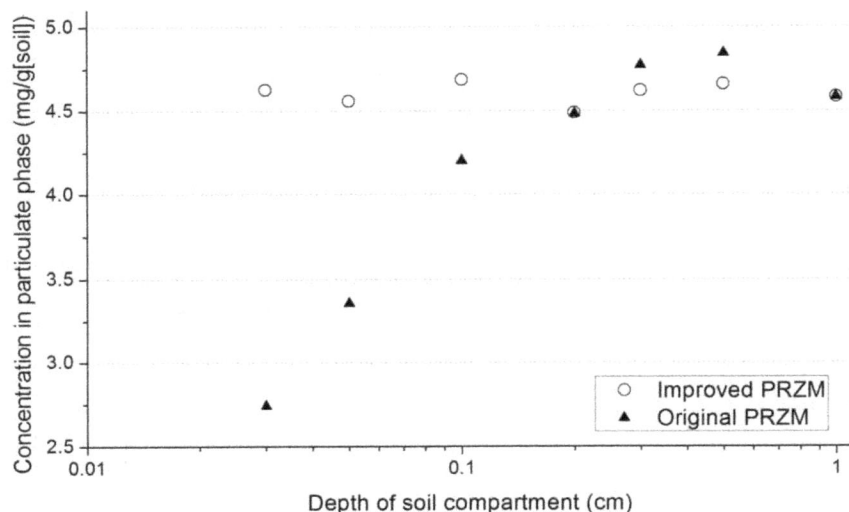

Figure 2. PRZM-simulated removals of absorbed pesticide with depths of soil compartment, tested with historical bifenthrin applications at a melon crop field in the study area.

use of λ-cyhalothrin was first reported in this area in 1998 and its use has been significantly increased since 2004, with an annual rate of about 180 kg in recent years. However, λ-cyhalothrin has a relatively short half-life in sediment (12 days), limiting its persistence and toxicity in aquatic ecosystems.

During the study period, about 50% of annual pyrethroids are applied in July and August, and 70% during June to September. However, high concentrations and TU values of pyrethroids in sediment were predicted during rainfall seasons. Predicted TUs from November to January were significantly higher than the annual average. Therefore, no linear relationship was confirmed for pyrethroid uses and predicted TUs on either monthly or annual bases. In previous studies, however, such correlations were reported for organophosphate pesticides [12,17]. With relatively high adsorption coefficients, the off-site movement of pyrethroids is mainly associated with soil erosion from agricultural fields and suspended solid transport in channels. Bifenthrin is highly persistent in soil and sediment with half-lives in aerobic soils of 85 days and in aquatic sediments of 251 days used in the modeling (Table 2). The applied pyrethroids might persist in soil and sediment long enough, and be available for the subsequent winter storms. Therefore, it is possible that predicted TUs in sediment reflected pyrethroid uses in the previous growing season. Significant time-lagged correlations were detected between the annual maximum TU and total prior bifenthrin use during the late irrigation season of September and October ($r = 0.74$, $p = 0.022$) for years when bifenthrin usage in September and October were observed (Table 3). Monthly precipitation corresponding to the maximum TU was identified as the second most important factor. Further analysis indicated that the precipitation and September+October bifenthrin use explained

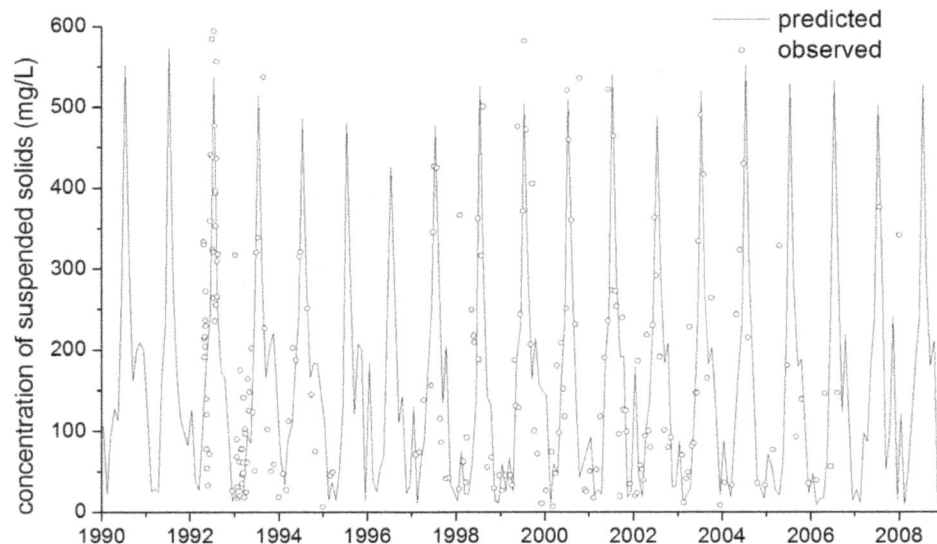

Figure 3. Observed and predicted monthly flow-weighted concentrations of suspended solids at the site OCRR (Orestimba Creek at River Road).

[a]

[b]

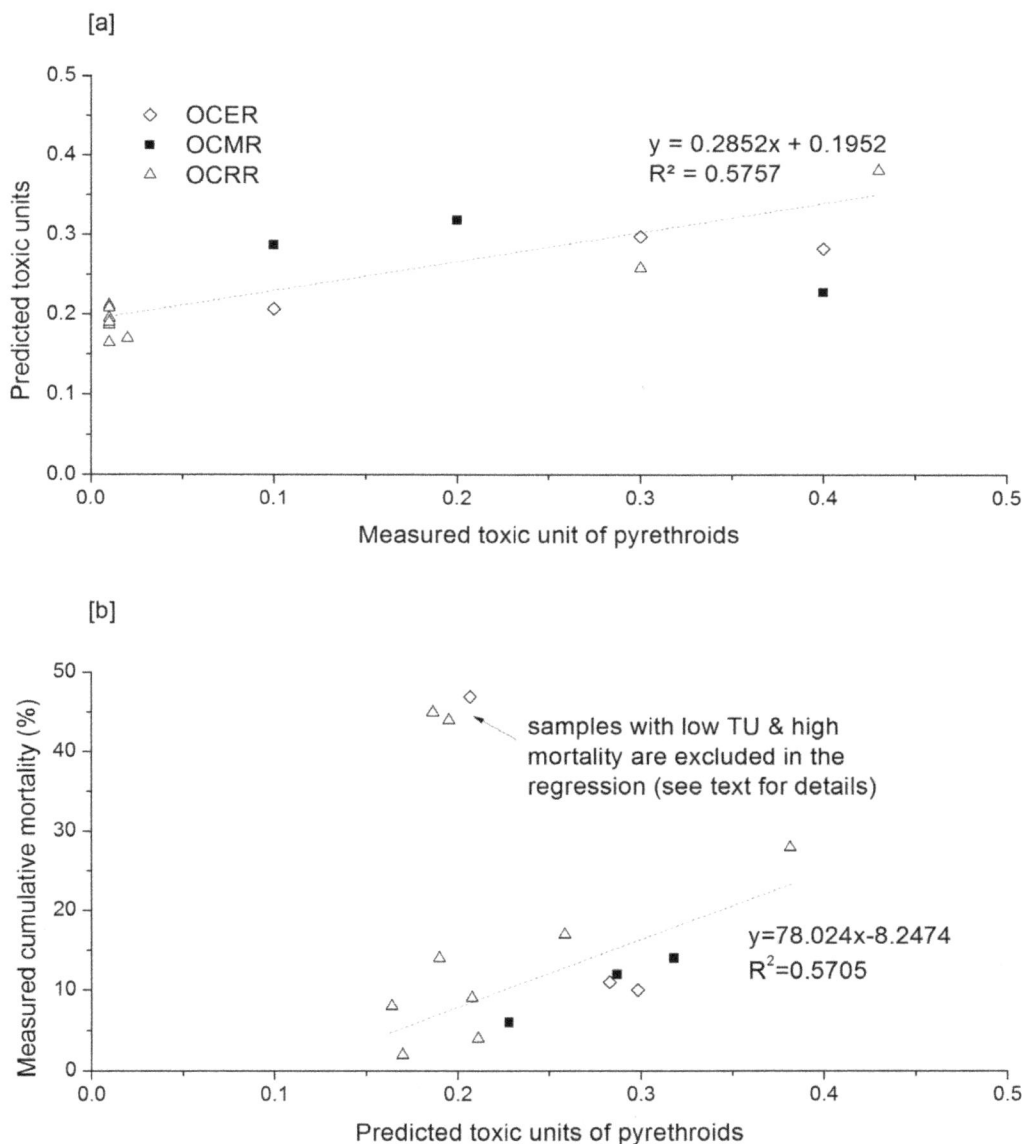

Figure 4. Predicted toxic units (TU) of the simulated pyrethroids in this study, in comparisons with [a] measured TUs (samples without detected pyrethroids are arbitrarily plotted on the figure at 0.01 TU), and [b] measured cumulative mortality.

90.2% of the variance of the annual maximum TUs, among which 54.9% was contributed by the bifenthrin use and 35.3% by the precipitation. This finding suggested two potential measures to efficiently reduce sediment toxicity by pyrethroids in the study area: [1] limiting bifenthrin use immediately before rainfall season; and [2] implementing conservation practices to retain soil on cropland, which would mitigate suspended soild transport to surface water bodies during early rainfall season.

Materials and Methods

PRZM application at watershed scale

A geo-referenced modeling system has been developed in our previous study [12] for tracking pesticide transport from its field application to the receiving waters. Pesticide discharges from the soil-canopy system were simulated by PRZM model. PRZM [18] is a one-dimensional dynamic model, primarily designed to predict the influence of climate, land/soil properties, and agricultural management on the physical and biochemical dynamics of pesticides in the environment. PRZM was selected based on its ability to simulate relevant governing processes of pesticide transport and its preferential use by the USEPA for pesticide-associated risk assessment [19]. Pre-calibrated PRZM parameters were recommended in the USEPA Standard Tier 2 scenarios for the major crops throughout the United States [14]. GIS technology was used to extend the PRZM capability for geo-referenced parameterization and application at a watershed scale.

Based on a linear routing model, edge-of-field fluxes of water and pesticides predicted by PRZM were routed through stream channels to a downstream location, e.g., a monitoring site. For water transport, stream flows at the routing destination were calculated as the summation of convolutions between PRZM-predicted runoff and corresponding watershed unit hydrograph in each simulation zone. The hydrologic response was presented by a flow-path redistribution

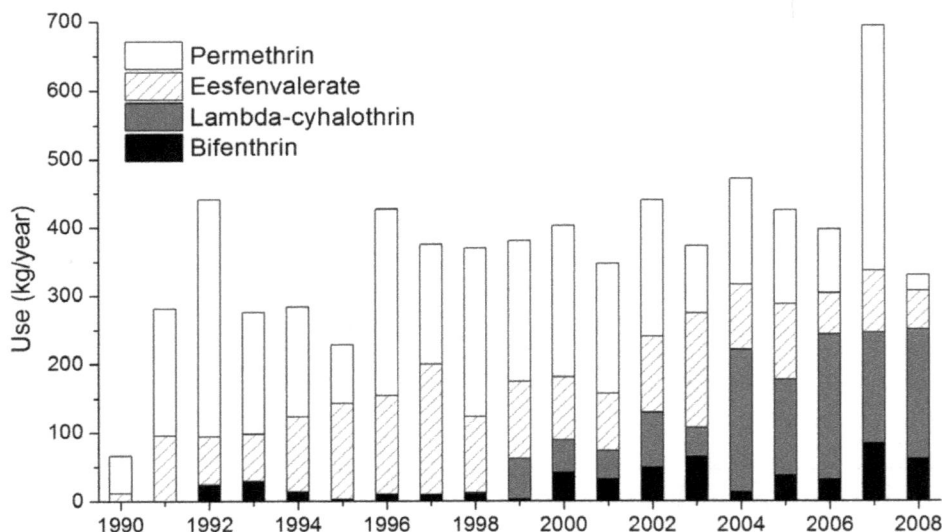

Figure 5. Annual uses of the simulated pyrethroids in the study area.

function (U) based on the first passage time distribution [20,21],

$$U_i(t) = \frac{1}{2t\sqrt{\pi(t/T_i)\varDelta_i}} \exp\left\{ - \frac{[1-(t/T_i)]^2}{4(t/T_i)\varDelta_i} \right\} K_i \qquad (1)$$

where i is a running index for simulation zone, T (s) is the lag time in the flow-path, $\varDelta_i$ (dimensionless) represents the shear and storage effects on the flow, and K_i (dimensionless) is the loss factor accounting for evaporation and transmission losses.

The same flow-path redistribution functions were also applied in the transport simulation of dissolved pesticides. A pesticide dispersion coefficient was determined as the sum of molecular diffusivity and flow diffusivity for the corresponding flow paths, and

pesticide decay rate was calculated from its aquatic half-life. Modeling nodes in the channel network were selected to correspond with tributary/drainage junctions and monitoring locations. The developed model was applied to the Orestimba Creek watershed during 1990 through 2006, with diazinon and chlorpyrifos as test agents. The model yields reasonable agreements with measured data for the stream flow and dissolved pesticide loads [12].

Pesticide transport with eroded soil

PRZM is known to inadequately predict pesticide transport associated with soil erosion [13]. In PRZM, soil column was divided into compartments according to user-defined numerical simulation interval of soil depth. Adsorbed pesticide is only removed from the top-most compartment. Therefore, the removal of pesticide in

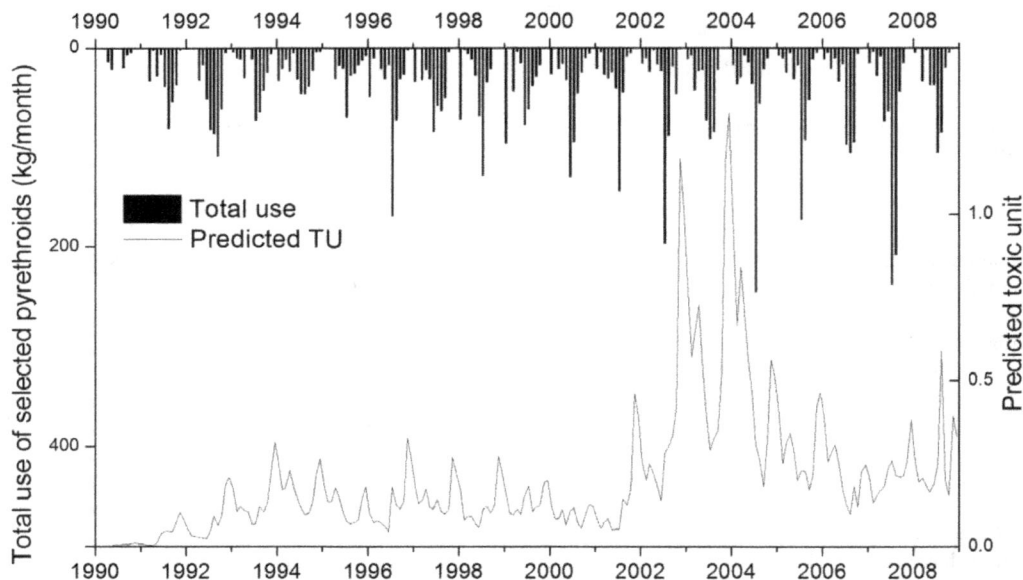

Figure 6. Predicted monthly toxic units (TU) of pyrethroids in bed sediment at the site OCRR (Orestimba Creek at River Road) and corresponding monthly pyrethroid uses in the drainage area.

Table 2. Chemical and toxic data for the simulated pyrethroids.

Parameter (unit)	Bifenthrin	λ-cyhalothrin	Esfenvalerate	Permethrin
Molecular weight (g/mol)	422.9	449.85	419.9	391.3
Octanol-water partition coefficient (K_{OW}, L/kg)	2.00E+07	7.94E+06	1.74E+06	1.26E+06
Vapor pressure (Pa)	1.78E−05	2.00E−07	1.20E−09	2.00E−06
Henry's law constant (Pa*m^3/mol)	7.74E−05	2.00E−02	4.90E−04	1.89E−01
Organic-carbon normalized partition coefficient (L/kg)	2.37E+05	1.57E+05	5300	1.00E+05
Half-life (day)				
in soil	85	25	44	42
in water	8	8	30	23
in sediment	251	12	71	40
Method detection limit (ng/g)				
in Domagalski et al. [29]	2.2	2.4	2.1	1.0
in Ensminger et al. [30]	0.5	2.0	1.0	2.0
10-day LC50 (ng/g)	5.2	4.5	15.4	108

Notes:
[1] LC50 = median lethal concentration for *Hyalella azteca* in sediment containing 1% organic carbon.
[2] Data sources: physicochemical properties and reaction half-lives were retrieved from FOOTPRINT pesticide properties database [35]. Method detection limits were taken from the respective studies. LC50 values were compiled by Domagalski et al. [29] from the literature.

adsorbed phase is primarily a function of soil compartment depth. Small depths of soil compartments, which are likely to be applied by model users to improve the numerical calculations, will result in

Table 3. Precipitation, bifenthrin use, and predicted max TUs at the site OCRR during 1990–2008.

Year	Max TU (rainfall season)	Precipitation (cm) Nov & Dec	annual	Bifenthrin use (kg) Sep & Oct	annual
1990	0.0117	1.72	17.87	0.00	0.00
1991	0.1045	2.4	23.05	0.00	0.00
1992	0.2072	2.52	22.9	0.00	24.05
1993	0.3128	0.89	19.16	0.00	29.41
1994	0.2629	0.71	6.37	0.00	13.66
1995	0.1798	3.73	29.54	1.34	2.77
1996	0.3257	12.12	41.76	0.00	10.42
1997	0.2682	12.3	26.03	0.00	9.65
1998	0.2715	3.96	47.44	0.00	10.24
1999	0.1976	2.24	28.2	0.00	3.15
2000	0.1286	0.65	19.34	8.14	41.10
2001	0.4581	9.2	26.03	0.00	31.93
2002	1.1647	11.44	24.53	13.34	48.73
2003	1.3027	7.23	16.9	18.73	64.19
2004	0.5627	8.49	25.67	1.34	13.52
2005	0.4622	3.92	25.67	6.38	36.93
2006	0.2462	2.8	19.01	9.60	31.01
2007	0.3816	3.47	12.91	4.02	83.24
2008	0.3923	3.26	12.95	1.76	61.13

significantly less pesticide mass removed by erosion, especially for high-sorbing compounds such as pyrethroids.

In this study, we improved the PRZM simulation algorithm by introducing a soil-interaction depth (D_E). It is assumed that all soil layers from the ground to the depth of D_E were subjected to the soil erosion process. This concept is similar to the extraction model used in transport model for estimating dissolved chemicals in surface runoff. For example, PRZM estimates the amount of dissolved pesticide runoff based on the average concentration of dissolved pesticide concentration weighted by an exponential curve for all compartments from the surface to a depth of 2 cm. The depth D_E, which could be initialized and calibrated by users, is independent from the compartment size for numerical calculation, and remains a fixed value during each PRZM simulation run.

Weighted average concentration of pesticide adsorbed on soil particles subject to erosion ($C_{S,E}$, g/g) for all compartments within the depth of D_E was first determined as:

$$C_{S,E} = \sum_{j=1}^{N_E} C_s(j) \cdot w(j) \qquad (2)$$

where j is a running index for compartments, N_E is total compartments within D_E, C_s (g/g) is the concentration of soil-bound pesticides, and w is a return-to-unit weighting function, i.e., sum(w) = 1. The amount of adsorbed pesticide transported out of the field (J_{ER}, g/day) is calculated by:

$$J_{ER} = p X_e r_{om} C_{S,E} \qquad (3)$$

This equation was the same as Eq. (6.13) in the PRZM manual [18], with X_e (ton/day) as the erosion sediment loss, r_{om} as the enrichment ratio for organic matter, and p as a units conversion factor (g/ton). To implement the above equation, source codes of PRZM were modified and the new procedure for determining pesticide removal with eroded soil was:

[1] initialize the soil-interaction depth (D_E), and define a weighting curve as a function of depth (w);

[2] determine the affected compartments within D_E, and calculate weighting factors for each compartment;

[3] adjust "B" term in the PRZM numerical solution as:

$$B(j) = B(j) + ELTERM * w(j) * DELT, j = 1, \ldots N_E \quad (4)$$

where B (day^{-1}) is the diagonal element in the tri-diagonal matrix solution (Thomas algorithm) utilized by the PRZM code for the governing equations of pesticide transport, ELTERM (day^{-1}) is the erosion loss term for pesticide balance, and DELT (day) is the simulation time step.

Sediment and pesticide transport in stream network

Chemical partitioning and degradation in channel transport have been formulated in the linear routing model as described in our previous study [12]. In this study, improvements were made mainly for sediment routing and partitioning/transport of pesticide associated with suspended solids and bed sediment. A concept of sediment transport capacity was applied in this study to predict sediment deposition in the channels. By following the algorithm of Soil and water assessment tool (SWAT) [22], the maximum sediment concentration ($C_{ss,max}$, kg/m^3) that can be transported from a reach segment is calculated as:

$$C_{ss,max} = SPCON \cdot V^{SPEXP} \quad (5)$$

where V (m/s) is the peak channel velocity, and SPCON and SPEXP are coefficients to be determined. Sedimentation flux was determined by comparing the initial concentration of suspended solids in a reach at a time step ($C_{ss,0}$, kg/m^3) to $C_{ss,max}$. For instance, if $C_{ss,0} > C_{ss,max}$, the exceeding amount of suspended solids and associated pesticides in the adsorbed phase would be transported into bed sediment. The resulting sedimentation flux was used to adjust the initial concentration of suspended solids. A similar methodology was applied in the calculation of resuspension fluxes of bed sediment and pesticides by introducing factors for channel erodibility and channel cover [22]. The predicted concentrations of suspended solids were applied to determine pesticide partitioning between dissolved and adsorbed phases in the water column:

$$F_{wd} = \frac{1}{1 + K_d C_{ss}} \quad (6)$$

where F_{wd} (dimensionless) is the fraction of total pesticide of the water column in dissolved phase, K_d (L/kg) is the pesticide partition coefficient, and C_{ss} (kg/m^3) is the predicted concentration of suspended solids.

Pesticide simulation in bed sediment was only conducted for the active sediment layer with user-defined depth. Based on the solid-liquid partitioning, the fraction (F_{dd}, dimensionless) of total sediment pesticide in the dissolved phase was calculated as:

$$F_{dd} = \frac{1}{\phi + (1 - \phi)\rho_s K_d} \quad (7)$$

with Φ (dimensionless) denoting sediment porosity and ρ_s (g/m^3) particle density. Pesticide decay and burial in bed sediment were combined and simulated according to first-order kinetics. Pesticide transport flux by sedimentation was calculated as the product of previously determined sedimentation flux of suspended solids and the pesticide concentration in suspended solids. Similarly, pesticide resuspension flux was based on the sediment resuspension flux and the concentration of sediment-bound pesticide. Therefore, sedimentation and resuspension processes for both suspended solids and solid-bound pesticides were simulated dynamically rather than being prescribed.

Pesticide diffusion flux (J_{diff}, $kg/m^2/day$) between the water and bed sediment was formulated using a multimedia environmental fate modeling approach [23]:

$$J_{diff} = D_{wd} \left(\frac{C_w}{Z_w} - \frac{C_d}{Z_d} \right)$$
$$D_{wd} = \left(\frac{\delta_{wd\_w}}{Z_w K_w} + \frac{\delta_{wd\_d}}{Z_d K_d} \right)^{-1} \quad (8)$$

where subscripts w and d are for water compartment and sediment compartment, respectively, D_{wd} (kg/Pa/day) is the Mackay-type mass transfer coefficient, the Z's ($mol/Pa/m^3$) are the fugacity capacity of pesticide, δ's (m) are boundary layer depths at the water-sediment interface, and K's (m^2/day) are pesticide diffusivities. As suggested by the CalTox model [24], the boundary layer thickness in water side (δ_{wd_w}) was set as 0.02 m, while that in sediment side was estimated as $318 K_d^{0.683}$.

Site Description

The modeling system newly developed in this study was applied to the field conditions of the Orestimba Creek watershed of California (Figure 1). Located in western Stanislaus County, the creek originates in the mountainous areas of the Coast Range, and discharges into the San Joaquin River. Characterized by heavier textured soils and greater slopes relative to eastside watersheds of the San Joaquin River, the Orestimba Creek watershed represents a worse-than-average condition for pesticide contamination in surface water. Climate and landscape characteristics for the studied watershed were summarized in the previous studies [12,25].

The lower reach of the Orestimba Creek flows through agricultural lands in California's Central Valley, the most dynamic agricultural region in the world. Pyrethroids are applied to control a myriad of pests, and in this study the most important crops receiving these insecticides are orchards and row crops. Sediment from this creek was also found to be toxic to sediment-dwelling organisms, most likely because of high levels of pyrethroids [26]. Based on sediment sampling in the Orestimba Creek during 2004 irrigation season, high sediment concentrations of bifenthrin and λ-cyhalothrin were reported with acute toxicity to sensitive aquatic species [27]. In the 2010 Clean Water Act 303(d) report of California, the Orestimba Creek was listed for sediment toxicity, however the source pollutants were not yet fully identified in the report [28].

Pesticide Data Acquisition

The case study was based on two monitoring studies of pyrethroid concentrations and aquatic toxicity in streambed sediments of the Orestimba Creek watershed. Our previous monitoring study included 20 sampling sites throughout the San Joaquin River Valley, with 3 sites situated within the Orestimba Creek watershed [29] (Figure 1 and Table 1). Field measurements were conducted for 9 pyrethroids during the irrigation season of 2007. In an associated study, Ensminger et al. [30] collected monthly water and sediment samples at the site OCRR from December 2007 through June 2008, to determine concentrations of organophosphate and pyrethroid insecticides. In addition to chemical analyses, both studies conducted sediment toxicity tests with *Hyalella azteca*, following the standard USEPA protocols [31]. More details on experimental design, analytical and sediment toxicity methods, and monitoring results for the two studies can be found in Domagalski et al. [29] and Ensminger et al. [30].

Among all analyzed pyrethroids, only those detected at least twice in the two sampling studies, including bifenthrin, λ -cyhalothrin, esfenvalerate, and permethrin, were selected for model application in this study. Table 2 lists the physicochemical properties, reaction half-lives, and toxicity benchmark (as 10-d median lethal concentration, LC50, for *Hyalella azteca* in sediment) of the simulated pyrethroids. Pesticide application data were retrieved from the Pesticide Use Reporting (PUR) database maintained by California Department of Pesticide Regulation [32]. The PUR database records daily pesticide use by active ingredient and crops for each Meridian-Township-Range-Section (MTRS, or section) following the United States Land Survey System.

Simulation Design

The improved PRZM and routing simulations were performed to simulate water, sediment, and pesticide transport processes in the Orestimba Creek watershed at a daily time step for the period 1990–2008. The watershed was delineated into sections for the convenience of incorporating pesticide use data from the PUR database. Multiple fields were simulated in each section, based on the contemporary land use mapping in the study area [33]. A soil interaction depth of 1 cm and uniform weighting factors for each soil compartment were used in the case study, as suggested by SWAT documentation [22]. Individual conservation practices were not included in the model configuration. Instead, the model was calibrated based on the field measurements of water flow, sediment loading, and pesticide concentrations. Therefore, the model parameterization and simulation results reflected the overall reduction of pesticide use due to various best management practices (BMPs) implemented in the study area.

Channel parameters, including Manning's roughness coefficient, flow diffusivity, coefficients for sediment transport capacity, and depth for active sediment layers, were taken from previous studies in the Orestimba Creek watershed [17,25]. Automatic calibration was conducted for the USLE crop factor (USLE_C) to match the measured suspended solid concentrations at the watershed outlet. The calibrated model was assumed to establish a reliable hydrologic framework for the study area, and applied to the dynamic simulation of pyrethroids.

Model Evaluation

The modeling system has been validated in our previous study for its simulation capacity for stream flow and organophosphate pesticides in the dissolved phase [12]. In this study, therefore, model evaluation was emphasized on transport simulation of the suspended solids and absorbed pesticides. The location of the site OCRR is also gauged by a USGS station (#11274538) for stream flow, suspended solids, and organic carbon in suspended solids [34]. Flow-weighted concentrations of suspended solids and associated organic carbon on a monthly basis were calculated from the measurements at sampling days.

In the chemical analysis of pyrethroids, reported results were associated with different method detection limits (MDLs) (Table 2), and chemicals with concentrations lower than MDLs were reported as zeros. In addition, most of the reported concentrations of detectable pyrethroids in the study area were below the nominal reporting limit [29]. Therefore, it is not appropriate to directly compare predicted and observed concentrations for individual pyrethroids. In this study, predicted and observed pyrethroid concentrations were first converted into toxic units (TU), which is based on the assumption of toxicity additivity and is widely used as an estimate for aquatic toxicity. For each sample, the TU value was calculated as a summation of concentrations normalized by the corresponding sediment LC50 on an organic carbon (OC) basis. When no pyrethroids were detectable, the TU value was set as 0.01 for plotting convenience. To evaluate the model efficiency in predicting pyrethroid transport, the predicted TU values were compared with the measured values for each monitoring day at the three sites. It is important to note that the predicted TU values were calculated based on daily average predictions of pyrethroid concentrations, while measured values were from instantaneous samples. The model evaluation also compared the predicted TUs and cumulative mortality of *Hyalella azteca* to collected bed sediment samples. The cumulative mortality reflected the actual sediment toxicity by all chemicals, including those not analyzed or not detected, in the bed sediment. Thus, comparisons between predicted TUs and mortality were anticipated to provide useful information on the toxicity identification evaluations.

Acknowledgments

The authors acknowledge the two anonymous reviewers who have helped improve the present article with their most appropriate suggestions.

Author Contributions

Conceived and designed the experiments: YL MZ. Performed the experiments: YL. Analyzed the data: YL. Contributed reagents/materials/analysis tools: YL MZ. Wrote the paper: YL MZ.

References

1. Epstein L, Bassein S (2003) Patterns of pesticide use in California and the implications for strategies for reduction of pesticides. Annual Review of Phytopathology 41: 351–375.
2. Spencer J, O'Malley M (2006) Pyrethroid Illnesses in California, 1996–2002. Reviews of Environmental Contamination and Toxicology 186: 57–72.
3. Hladik ML, Kuivila KM (2009) Assessing the Occurrence and Distribution of Pyrethroids in Water and Suspended Sediments. Journal of Agricultural and Food Chemistry 57: 9079–9085.
4. Weston DP, You J, Lydy MJ (2004) Distribution and Toxicity of Sediment-Associated Pesticides in Agriculture-Dominated Water Bodies of California's Central Valley. Environmental Science & Technology 38: 2752–2759.
5. Amweg EL, Weston DP, Ureda NM (2005) Use and toxicity of pyrethroid pesticides in the Central Valley, California, USA. Environmental Toxicology and Chemistry 24: 966–972.
6. Weston DP, Zhang M, Lydy MJ (2008) Identifying the cause and source of sediment toxicity in an agriculture-influenced creek. Environmental Toxicology and Chemistry 27: 953–962.
7. Laskowski DA (2002) Physical and chemical properties of pyrethroids. Reviews of environmental contamination and toxicology 174: 49–170.
8. Ma Q, Don Wauchope R, Ma L, Rojas KW, Malone RW, et al. (2004) Test of the Root Zone Water Quality Model RZWQM for predicting runoff of atrazine, alachlor and fenamiphos species from conventional-tillage corn mesoplots. Pest Management Science 60: 267–276.
9. Leonard RA, Knisel WG, Still DA (1987) GLEAMS: Groundwater Loading Effects of Agricultural Management Systems. Transactions of the ASAE (American Society of Agricultural Engineers) 30: 1403–1418.
10. Chapra SC, Pelletier GJ, Tao H (2008) QUAL2K: A Modeling Framework for Simulating River and Stream Water Quality, Version 2.11: Documentation and User's Manual. Medford, MA: Civil and Environmental Engineering Department, Tufts University.
11. Williams WM, Zdinak CE, Ritter AM, Cheplick JM, Singh P (2004) RIVWQ Chemical transport model for riverine environments, user's manual and program documentation, version 2.02. Leesburg, VA: Waterborne Environmental, Inc.

12. Luo Y, Zhang M (2009) A geo-referenced modeling environment for ecosystem risk assessment: organophosphate pesticides in an agriculturally dominated watershed. Journal of Environmental Quality: 38(32): 664–674.

13. USEPA (2010) Pesticide Root Zone Model (PRZM) release notes (http://www.epa.gov/ceampubl/gwater/przm3/prz3reln.html, accessed 10/2010). Athens, GA: U.S. Environmental Protection Agency, Center for Exposure Assessment Modeling.

14. USEPA (2008) USEPA Tier 2 crop scenarios for PRZM/EXAMS Shell (http://www.epa.gov/oppefed1/models/water/index.htm, accessed 09/2010). Washington, DC: U.S. Environmental Protection Agency, Office of Pesticide Programs.

15. Moriasi DN, Arnold JG, Liew MWV, Bingner RL, Harmel RD, et al. (2007) Model evaluation guidelines for systematic quantification of accuracy in watershed simulations. Transaction of the American Society of Agricultural and Biological Engineers (ASABE) 50: 885–900.

16. Scholze M, Boedeker W, Faust M, Backhaus T, Altenburger R, et al. (2001) A general best-fit method for concentration-response curves and the estimation of low-effect concentrations. Environmental Toxicology & Chemistry 20: 448–457.

17. Luo Y, Zhang X, Liu X, Ficklin D, Zhang M (2008) Dynamic modeling of organophosphate pesticide load in surface water in the northern San Joaquin Valley watershed of California. Environmental Pollution 156: 1171–1181.

18. USEPA (2006) PRZM-3, a model for predicting pesticide and nitrogen fate in the crop root and unsaturated soil zones: users manual for release 3.12.2. Center for Exposure Assessment Modeling, U.S. Environmental Protection Agency. EPA/600/R-05/111 EPA/600/R-05/111.

19. USEPA (2006) Organophosphate pesticides: revised cumulative risk assessment (http://www.epa.gov/pesticides/cumulative/rra-op/, accessed 10/2010). Washington, DC: U.S. Environmental Protection Agency.

20. Olivera F, Maidment DR, Charbeneau RJ (1996) Spatially distributed modeling of storm runoff and non-point source pollution using Geographic Information Systems (GIS). Center for Research in Water Resources, the University of Texas at Austin (http://ceprofs.tamu.edu/folivera/UTexas/disstn/header.htm. Accessed 09/2008). CRWR Report 96-4 CRWR Report 96-4.

21. Olivera F, Maidment D (1999) Geographic information systems (GIS)-based spatially distributed model for runoff routing. Water Resources Research 35: 1155–1164.

22. Neitsch SL, Arnold JG, Kiniry JR, Williams JR (2005) Soil and Water Assessment Tool theoretical documentation, Version 2005. College Station, TX: Agricultural Research Service and Blackland Research Centre, Texas A&M University.

23. Luo Y, Gao Q, Yang X (2007) Dynamic modeling of chemical fate and transport in multimedia environments at watershed scale–I: Theoretical considerations and model implementation. Journal of Environmental Management 83: 44–55.

24. McKone TE, Maddalena RL, Bennett DH (2003) CalTOX 4.0 a multimedia total exposure model. Berkeley, CA: Lawrence Berkeley National Laboratory.

25. Luo Y, Zhang M (2009) Management-oriented sensitivity analysis for pesticide transport in watershed-scale water quality modeling. Water Research 157: 3370–3378.

26. Meadows R (2005) Pyrethroids in Central Valley stream sediments toxic to bottom-dwellers. California Agriculture 59: 5–6.

27. CVRWQCB (2005) Conditional Waiver for Irrigated Agriculture Monitoring Program Phase II, Quarterly Report, Activities from July 1, 2004-September 30, 2004. Sacramento, CA: Central Valley Regional Water Quality Control Board.

28. CEPA (2010) Proposed 2010 Integrated Report (Clean Water Act Section 303(d) List/305(b) Report), http://www.waterboards.ca.gov/water_issues/programs/tmdl/integrated2010.shtml (verified 06/2010). Sacramento, CA: California Environmental Protection Agency, State Water Resources Control Board.

29. Domagalski JL, Weston DP, Zhang M, Hladik M (2010) Pyrethroid insecticide concentrations and toxicity in streambed sediments and loads in surface waters of the San Joaquin Valley, California, USA. Environmental Toxicology and Chemistry 29: 813–823.

30. Ensminger M, Bergin R, Spurlock F (2009) Pesticide Concentrations in Water and Sediment and Associated Invertebrate Toxicity in Del Puerto and Orestimba Creeks, California (http://www.cdpr.ca.gov/docs/emon/pubs/eha-preps.htm). Sacramento, CA: California Environmental Protection Agency, Department of Pesticide Regulation.

31. USEPA (2000) Methods for measuring the toxicity and bioaccumulation of sediment-associated contaminants with freshwater invertebrates (EPA/600/R-99/064). Washington DC: U.S. Environmental Protection Agency, Office of Research and Development.

32. CEPA (2010) Pesticide Use Reporting (PUR), http://www.cdpr.ca.gov/docs/pur/purmain.htm (accessed 09/2010). Sacramento, CA.: California Environmental Protection Agency, Department of Pesticide Regulation.

33. CDWR (2010) California land and water use: survey data access (http://www.landwateruse.water.ca.gov/, accessed 09/2010). Sacramento, CA.: California Department of Water Resources.

34. USGS (2010) National Water Information System: Web Interface. United States Geographical Survey, http://waterdata.usgs.gov/nwis (accessed 09/2010).

35. FOOTPRINT (2010) The FOOTPRINT Pesticide Properties Database (http://sitem.herts.ac.uk/aeru/footprint/en/index.htm, verified 05/2010). Hatfield, Herts, UK: The Agriculture & Environment Research Unit (AERU) at the University of Hertfordshire.

2

Effect of Cry1Ab Protein on Rhizobacterial Communities of Bt-Maize over a Four-Year Cultivation Period

Jorge Barriuso, José R. Valverde, Rafael P. Mellado*

Centro Nacional de Biotecnología (CSIC), Campus de la Universidad Autónoma, Cantoblanco, Madrid, Spain

Abstract

Background: Bt-maize is a transgenic variety of maize expressing the Cry toxin from *Bacillus turingiensis*. The potential accumulation of the relative effect of the transgenic modification and the cry toxin on the rhizobacterial communities of Bt-maize has been monitored over a period of four years.

Methodology/Principal Findings: The accumulative effects of the cultivation of this transgenic plant have been monitored by means of high throughput DNA pyrosequencing of the bacterial DNA coding for the 16S rRNA hypervariable V6 region from rhizobacterial communities. The obtained sequences were subjected to taxonomic, phylogenetic and taxonomic-independent diversity studies. The results obtained were consistent, indicating that variations detected in the rhizobacterial community structure were possibly due to climatic factors rather than to the presence of the Bt-gene. No variations were observed in the diversity estimates between non-Bt and Bt-maize.

Conclusions/Significance: The cultivation of Bt-maize during the four-year period did not change the maize rhizobacterial communities when compared to those of the non-Bt maize. This is the first study to be conducted with Bt-maize during such a long cultivation period and the first evaluation of rhizobacterial communities to be performed in this transgenic plant using Next Generation Sequencing.

Editor: Mark R. Liles, Auburn University, United States of America

Funding: The Spanish Ministry of the Environment and Rural and Marine affairs has commissioned and supported this research (Grant No. EGO22008). The funders had no role in study design, data collection and analysis, decision to publish, or preparation of the manuscript.

Competing Interests: The authors have declared that no competing interests exist.

* E-mail: rpmellado@cnb.csic.es

Introduction

Rhizobacterial communities are a key element of soil quality and fertility [1]. The cultivation of genetically modified plants can alter these soil bacterial communities, hence endangering cultivar sustainability. Among these genetically modified maize, the so-called Bt-maize, harbours the *Bacillus thuringiensis* gene, which encodes the Cry protein, rendering the plant resistant to the attack of corn borer Lepidopterae.

The presence of the Bt protein potentially modifies the composition of root exudates of the transgenic plants and additionally may exerts a direct effect on non-target species of soil microorganisms [2,3,4,5].

Many techniques have been used to analyse soil bacterial communities, including classic approaches based on the cultivation of viable bacteria, metabolic profiling studies and nucleic acid-based methods [6]. Most of the existing studies to date suggest that the cultivation of transgenic Bt-maize plants has minor effects or no effects at all on the rhizobacterial communities. However, these studies have not employed the Next Generation Sequencing (NGS) technique, which, upon being used for the sequencing of the SSU rRNA hypervariable regions, has proven to be very useful for the diversity studies of bacterial communities in many habitats, including soil [7,8,9,10,11]. Moreover, few studies are available for long-term periods of Bt-maize cultivation. We monitored the potential accumulation of the relative effects of Bt-maize on the

rhizobacterial communities when compared to non-Bt maize by using NGS over a four-year period of cultivation. To this end, we have examined the structure of the rhizobacterial communities using a taxonomic approach [12,13] conducted phylogenetic distances analyses [14,15,16], and bacterial diversity estimations. It has recently been reported that differences in diversity estimation may depend on the method used [17,18,19], therefore, we employed two commonly used methods to calculate diversity [12,19], which have proven to be appropriate for the number and length of the sequences obtained [20].

To our knowledge, this is the first study conducted on rhizobacterial communities of Bt-maize during a four-year period using NGS, and the results obtained may be relevant as to a potential renewal of the authorisation to cultivate Bt-maize in the European Union.

Results

Sequences obtained from the Bt-maize and its non-transgenic counterpart at two different sampling times were compared with the sequences present in the RDP database and grouped using the MEGAN program and NCBI taxonomy to generate the taxonomic trees. Figure 1 shows the taxonomic breakdown as a result of sequencing the hypervariable V6 region of the 16S rDNA genes, and the corresponding taxonomic trees are included in

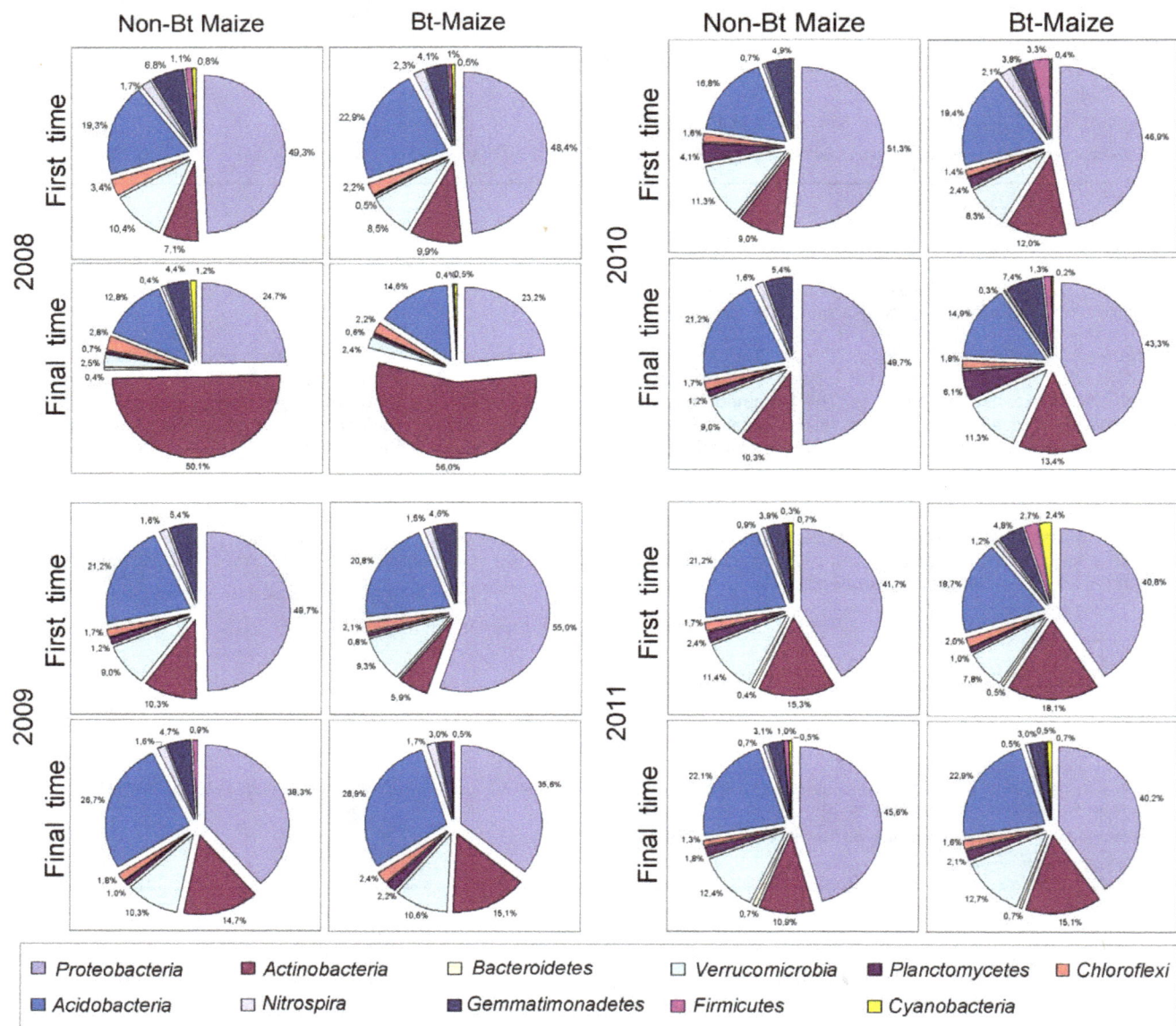

Figure 1. Taxonomic breakdown of the more relevant phyla. The percentages of *Proteobacteria*, *Actinobacteria*, *Bacteroidetes*, *Verrucomicrobia*, *Planctomycetes*, *Chloroflexi*, *Acidobacteria*, *Nitrospira*, *Gemmatimonadetes*, *Firmicutes* and *Cyanobacteria* are indicated and do not include the unassigned sequences. Unclassified sequences were not included as they were of no taxonomic use.

figure S1 as Supplementary Information. The prominent phyla in all the samples were *Proteobacteria*, *Actinobacteria* and *Acidobacteria*.

A total of 1959 sequences were analysed from the rhizosphere of non-Bt maize at the first sampling time in the first year; 22.6% of these remained unassigned. From the assigned sequences, 49.3% belonged to the *Proteobacteria* phylum, 19.3% were *Acidobacteria*, 7.1% *Actinobacteria* and 23.7% belonged to other taxa. A very similar distribution of taxa (48.4% *Proteobacteria*, 22.9% *Acidobacteria*, 9.9% *Actinobacteria*, and 18.3% other taxa) was found for the 2445 total sequences from the Bt-maize rhizosphere at an equivalent sampling time, where 24.4% of sequences remained unassigned. At the final sampling time, *Proteobacteria* and *Acidobacteria* were again prominent among the assigned sequences: 24.7% and 12.8% respectively in the non-Bt maize (2680 total sequences), and 23.2% and 14.6% respectively in the Bt maize

(2534 total sequences), while the presence of *Actinobacteria* markedly increased to 50.1% in the non-Bt maize and 56.0% in Bt maize. 23.6% and 21.2% were unassigned sequences respectively in the non-Bt and Bt-maize.

In the second year, the distribution of taxa in the assigned sequences of the non-Bt and Bt-maize was similar to that of the previous year at the first sampling time; *Proteobacteria*, *Actinobacteria* and *Acidobacteria* were predominant (Fig. 1). At the first sampling time 24.3% and 24.4% of the sequences remained unassigned for the non-Bt and Bt-maize, yielding a total of 2334 and 2777 sequences respectively, while at the second sampling time 22.1% and 22.6% of sequences remained unassigned, with 2177 and 2374 total sequences for the non-Bt and Bt-maize respectively.

The relative presence of the three major phyla was similar in the samples from the third year (Fig. 1), where 22.9% of the sequences were unassigned at the first sampling time for the non-Bt maize (2916 total sequences) and 19.2% for the Bt-maize (3344 total sequences); 21.0% of the sequences were unassigned for the non-Bt maize at the final sampling time, and 22.9% for the Bt-maize, with 4236 and 4523 total sequences respectively.

In the fourth year, an equivalent presence of the predominant phyla was once again found at the first and final sampling time in the non-Bt or Bt-maize, where the relative abundance of *Proteobacteria*, *Actinobacteria* and *Acidobacteria* remained comparatively similar in all cases (Fig. 1). The total number of sequences was 13332 for the non-Bt maize and 24084 for the Bt-maize at the first sampling time, with 15.8% and 16.5% of unassigned sequences respectively. The unassigned sequences were 16.2% for the non-Bt maize and 13.9% for Bt-maize at the final sampling time, with 36314 and 27202 total sequences for the non-Bt and Bt-maize respectively.

The hierarchical clustering tree of samples based on the UniFrac metric (Fig. 2) shows that the rhizobacterial communities from each particular year are grouped in a separate branch of the tree, except for the year 2008, where samples from the first sampling time were grouped with samples from 2011, and samples from the final sampling time appeared as an independent branch of the tree, as they differ in taxonomic composition (Fig. 1). Samples are also grouped within each year depending on the sampling time. The UniFrac significance test showed no statistical differences between the non-Bt and Bt-maize within each year at any sampling time.

When estimating bacterial diversity, we have previously shown that ESPRIT as well as the web-based workflow, RDP pyrosequencing pipeline, produced the more accurate equivalent results as to the size and length of the hypervariable V6 region of the 16S rDNA sequences obtained from the non-Bt and Bt-maize rhizospheres when compared to other methodologies of analysis [20], therefore, we have used both of them in a comparative manner. Table 1 shows a comparison of the OTUs analyses performed using the RDP pyrosequencing pipeline compared to using the ESPRIT package, at three different dissimilarity levels. Species richness was determined using the Chao1 and ACE estimators for the ESPRIT software, and only Chao1 for the RDP software, since it does not use the ACE estimator (Table 1). No significant differences were detected among the bacterial communities from the non-Bt or Bt-maize, or from one sampling time to another or even from one year to another. The number of OTUs estimated when using the ESPRIT package was always slightly lower, as previously reported [19], however the Chao1 estimator was slightly higher when calculated with RDP, probably due to its own implementation of the algorithm. In all cases the results were consistent irrespective of the methodology used.

Discussion

In this study the potentially cumulative effect of Bt-maize cultivation on rhizobacterial communities has been monitored over a four-year continuous cultivation of MON810 maize. The effect of Bt-maize is not only exerted by the exudation of the Cry1Ab toxin through the roots [4], as it is constitutively produced within the whole plant and Cry proteins may also confer unexpected properties to the plant, for example, by increasing lignin content [21] or provoking gene instability at or around the insertion site of the Bt coding gene in the plant genome [22].

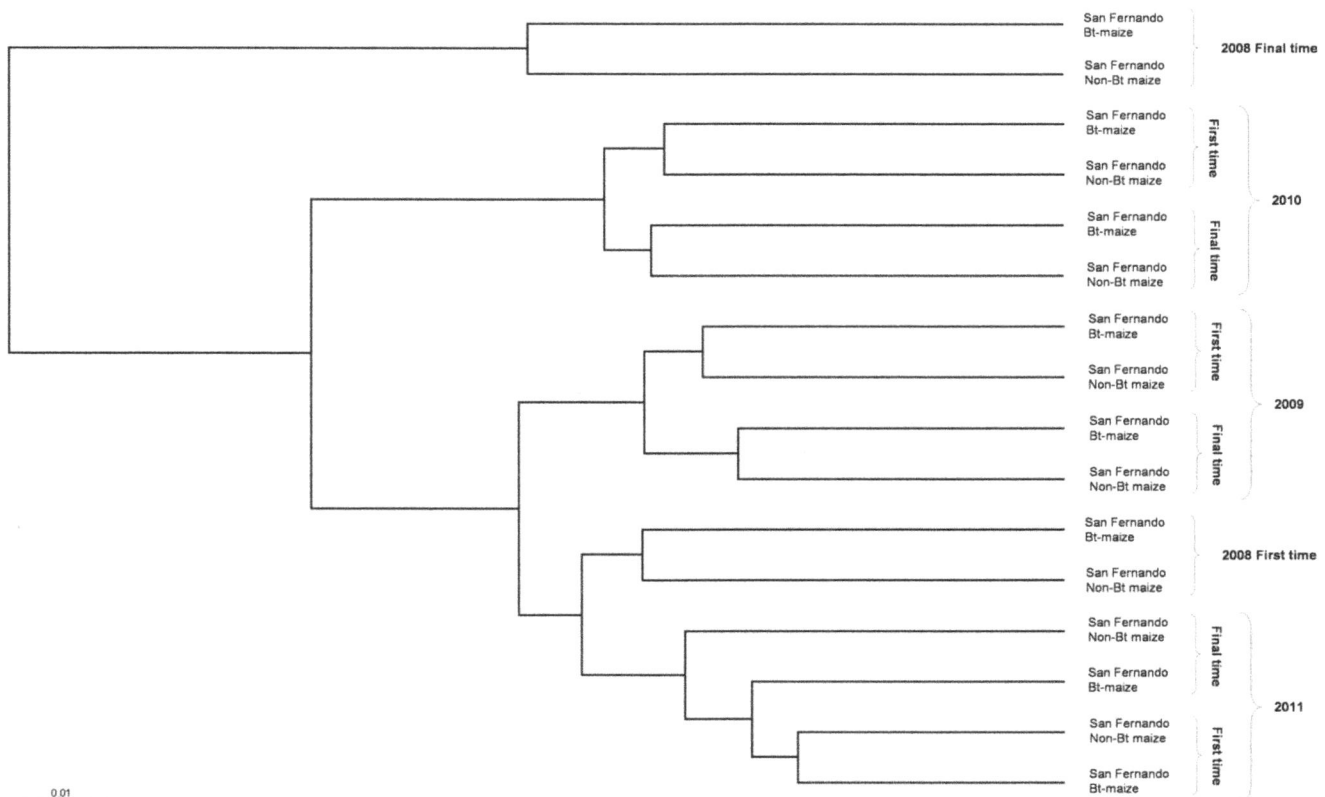

Figure 2. Fast UniFrac hierarchical clustering tree. Analysis of the different soil samples was carried out using normalised abundance weights.

Table 1. Similarity-based OTUs and species richness estimates at 3%, 5% and 10% dissimilarity levels.

					OTUs	ACE	Chao1						OTUs	ACE	Chao1
2008	First time	non-Bt	RDP	0.03	1148	NC	2480±279	2010	First time	non-Bt	RDP	0.03	952	NC	2126±266
				0.05	846	NC	1608±164					0.05	765	NC	1429±172
				0.1	527	NC	670±49					0.1	465	NC	668±77
			ESPRIT	0.03	1031	1842.63	1895±194				ESPRIT	0.03	916	2117.53	1894±227
				0.05	728	1110.54	1065±98					0.05	745	1531.48	1430±182
				0.1	316	341.97	333±14					0.1	433	568.97	534±44
		Bt	RDP	0.03	1135	NC	2469±272			Bt	RDP	0.03	920	NC	2199±297
				0.05	918	NC	1528±173					0.05	711	NC	1289±156
				0.1	541	NC	745±81					0.1	425	NC	592±69
			ESPRIT	0.03	1035	2075	2009±215				ESPRIT	0.03	886	2228.54	1970±259
				0.05	795	1309.08	1269±126					0.05	692	1315.88	1179±136
				0.1	447	549.7	540±42					0.1	406	543.15	529±54
	Final time	non-Bt	RDP	0.03	1053	NC	2820±406		Final time	non-Bt	RDP	0.03	1011	NC	2414±311
				0.05	797	NC	1687±211					0.05	815	NC	1595±196
				0.1	498	NC	703±50					0.1	537	NC	807±94
			ESPRIT	0.03	982	1951.57	2087±260				ESPRIT	0.03	1010	2508.72	2201±260
				0.05	755	1247.91	1303±153					0.05	805	1556.53	1384±146
				0.1	415	502.03	506±44					0.1	469	642.18	630±65
		Bt	RDP	0.03	1216	NC	2230±258			Bt	RDP	0.03	999	NC	2420±317
				0.05	963	NC	1474±179					0.05	819	NC	1783±243
				0.1	568	NC	679±71					0.1	493	NC	688±73
			ESPRIT	0.03	1029	1992.92	1947±204				ESPRIT	0.03	1015	2679.5	2341±293
				0.05	726	1095.15	1013±84					0.05	821	1692.52	1596±196
				0.1	326	352.58	342±14					0.1	485	632.34	609±51
2009	First time	non-Bt	RDP	0.03	1307	NC	2565±350	2011	First time	non-Bt	RDP	0.03	898	NC	2138±296
				0.05	1033	NC	1661±204					0.05	691	NC	1300±171
				0.1	608	NC	663±85					0.1	429	NC	606±71
			ESPRIT	0.03	1105	2395.43	2287±249				ESPRIT	0.03	847	1916.25	1774±228
				0.05	865	1583.68	1602±186					0.05	680	1265.76	1204±149
				0.1	473	602.51	593±51					0.1	398	523.2	517±53
		Bt	RDP	0.03	1184	NC	2454±253			Bt	RDP	0.03	884	NC	2253±332
				0.05	910	NC	1650±178					0.05	690	NC	1430±208
				0.1	503	NC	654±60					0.1	407	NC	540±58
			ESPRIT	0.03	1104	2315.27	2303±255				ESPRIT	0.03	852	1990.86	2068±609
				0.05	847	1445.27	1448±154					0.05	651	1178.58	1194±160
				0.1	470	574.01	556±39					0.1	360	441.06	429±35
	Final time	non-Bt	RDP	0.03	1252	NC	3070±395		Final time	non-Bt	RDP	0.03	902	NC	2068±273
				0.05	1010	NC	1897±241					0.05	717	NC	1318±161
				0.1	596	NC	819±76					0.1	411	NC	512±53
			ESPRIT	0.03	1063	2108.05	1899±179				ESPRIT	0.03	853	1961.8	1730±213
				0.05	770	1266.42	1234±128					0.05	675	1235.8	1095±119
				0.1	362	397.08	387±18					0.1	370	460.93	433±32
		Bt	RDP	0.03	1204	NC	2730±301			Bt	RDP	0.03	868	NC	2184±326
				0.05	904	NC	1636±176					0.05	665	NC	1208±155
				0.1	539	NC	728±71					0.1	380	NC	529±65
			ESPRIT	0.03	996	1808.36	1810±186				ESPRIT	0.03	828	1925.67	1793±242
				0.05	716	1112.28	1062±101					0.05	648	1211	1153±149

Table 1. Cont.

	OTUs	ACE	Chao1		OTUs	ACE	Chao1
0.1	338	367.01	354±13	**0.1**	357	456.86	444±43

The species richness estimates were determined using the RDP pyrosequencing pipeline or the ESPRIT program, as described in Materials and Methods. NC = non computable.

Furthermore, transgene rearrangements may occur with a potential to produce changes in plant gene expression and phenotype [23]. The fate and effects of insect-resistant toxin in soil ecosystems have been reviewed [2], and the Cry1Ab protein (event MON810) has been detected to remain in soils after four years of Bt-maize cultivation in field conditions, whereas other Cry proteins as Cry3Bb1 (event MON863) were not [24]. Cry proteins are rapidly adsorbed to clay minerals, which render the proteins resistant to biodegradation in soil, thus facilitating a potential longer exposure of non-target organisms to the toxin [2,25,26]. The effect of Cry root exudates on different soil non-target organisms (namely earthworms, nematodes and protozoa) has been investigated extensively with apparently little significance or none all, while infective fungal mycorrhizae could colonize Bt maize roots more efficiently than non-Bt ones, and so the persistence of Cry proteins in soil may also be related to the decrease of some microbial activity [2].

There has been an increasing concern in recent years with respect to altering the composition of soil microbial communities and the studies performed differ in their objective and methodology. This four-year study was conducted to get a further insight into the possible harmful effect that the continued presence of Bt-toxin may exert on the structure of maize rhizobacterial communities.

In our study we found three predominant phyla in all the rhizospheres: *Proteobacteria*, *Acidobacteria* and *Actinobacteria*. The *Proteobacteria* are of great importance to global carbon, nitrogen and sulfur cycling [27]; *Actinobacteria* are an important component of soil communities playing a major role in organic matter turnover in soils, due to their ability to decompose organic materials [28]; *Acidobacteria* are commonly detected in soils [29] and it has been suggested that members of this phylum are likely to play a relevant role in conducting processes in terrestrial ecosystems [30].

The overall distribution of these three phyla did not change from the non-Bt to the Bt-maize at any sampling time. The reason why the structure of the rhizobacterial community changed at the final sampling time in the first year is still unknown, however, a particular climate change comprising a heavier period of rainfall between the first and final sampling times was registered, when almost three times more rain accumulated than during the same period in subsequent years (http://clima.meteored.com). At this particular sampling time the *Actinobacteria* became very predominant, which is compensated mostly by a reduction in *Proteobacteria*, and to a lesser extent in *Acidobacteria* (Fig. 1). It is not infrequent to see that climate changes alter the structure of rhizobacterial communities, and this has also been reported in maize long-term cultivation studies [31]. It is also remarkable to observe the natural resilience of the rhizobacterial communities, which at the beginning of the second year had recovered completely from the climatic episode.

No consistent statistically significant differences have been reported in the number of different groups of microorganisms, enzyme activities or pH between non-Bt and Bt-maize rhizo-spheres [2], or in soil improved with Bt-maize biomass [32]. Other short-term experiments using techniques such as Phospholipid Fatty Acid profiles (PLFA) [33], polymerase chain reaction–

denaturing gradient gel electrophoresis (PCR–DGGE) [34], automated ribosomal intergenic spacer analysis (ARISA) [35] or genome-wide commercially available DNA microarrays [36] concluded that the effects of transgenic Bt-maize on the bacterial community structure are minimal, and that the growth stage of plant or environmental factors may exert a more noticeable effect on the microbial community. The UniFrac statistical phylogenetic analysis did not show significant differences between the non-Bt maize and Bt- maize rhizospheres at any sampling time, and grouped the samples according to the sampling time, confirming the trends observed in the taxonomic analysis (Fig. 2).

The bacterial diversity analysis showed no differences in species abundance or richness between non-Bt and Bt-maize at any sampling point estimated with the RDP pyrosequencing pipeline or the ESPRIT software (Table 1). Moreover, no differences were found throughout the four-year experiment despite changes in the community structure. Although the two methods used to analyse bacterial diversity showed some minor differences, both of them offered consistent results when estimating the diversity of the bacterial communities. These two methods produced qualitatively equivalent results and have been shown to be an adequate choice, both being equally suitable for assessing the diversity of rhizobacterial communities.

Some studies have suggested that the repeated cultivation of transgenic Bt-plants may lead to the accumulation and persistence of Bt proteins in soil [26,32,37]. Taking together the results obtained we can conclude that no effects may be attributed to the transgenic Bt-maize when compared to its respective isogenic counterpart over a four-year period of seasonal cultivation. The differences perceived in the composition of the rhizobacterial communities were most likely due to the fluctuations in climate, which affected the non-Bt and Bt-maize almost equally. Nevertheless, further studies concerning soil microbial communities functioning should contribute to a better understanding of the relationship of the bacterial communities with the plant through-out its development.

Materials and Methods

Plant Materials and Sampling

Bt-maize MON810, variety DKC6451YG, expressing the Cry1Ab protein (Monsanto Agricultura, Spain) and its isogenic non-Bt line DKC6450 (Monsanto Agricultura, Spain) were grown in experimental maize fields located in San Fernando de Henares, Madrid, Spain (N40° 25′ 14″ W3° 29′ 30″). Current agricultural practises were maintained throughout the four years of cultivation period and crop residues were removed after each vegetative cycle. The surface of each experimental field was 40 m^2; both fields were annexed to each other and separated by a four meters wide path. Four extra rows of non-Bt maize surrounded both fields. The non-Bt and Bt-maize plants were harvested at two different growth stages: about 90 days after seeding (first sampling time), when the plants had around 8 leaves, and just before crop harvesting at final growth (final sampling time). The four extra rows were not

Table 2. Multiplex identifiers (MIDs).

Year	Samplig time	MID Bt-maize	MID Non Bt-maize
2008	First	MID1 (5'-ACGAGTGCGT-3')	MID2 (5'- ACGCTCGACA -3')
	Final	MID3 (5'- AGACGCACTC -3')	MID4 (5'- AGCACTGTAG -3')
2009	First	MID5 (5'-ATCAGACACG-3')	MID6 (5'- CGTGTCTCTA -3')
	Final	MID7 (5'- CGTGTCTCTA -3')	MID8 (5'- CTCGCGTGTC -3')
2010	First	MID1 (5'-ACGAGTGCGT-3')	MID2 (5'- ACGCTCGACA -3')
	Final	MID3 (5'- AGACGCACTC -3')	MID4 (5'- AGCACTGTAG -3')
2011	First	MID1 (5'-ACGAGTGCGT-3')	MID2 (5'- ACGCTCGACA -3')
	Final	MID3 (5'- AGACGCACTC -3')	MID4 (5'- AGCACTGTAG -3')

The Multiplex identifiers (MIDs) used for pyrosequencing the different samples are shown.

sampled. Roots and adhered soil measuring approximately 2 mm or less in diameter were separated from the bulk soil by gently shaking the root system. The term "rhizosphere" describes the carefully separated soil adhered to these roots. Given the small size of the maize fields, these were divided into subplots and 3 samples were taken from each subplot at the time of collection. A total of 9 subplots were collected from each maize field at every collection and an equal amount of soil from each of the 27 samples was pooled in all cases. The experiments were performed during development throughout 2008, 2009, 2010 and 2011. The performed studies did not involved human or animal participants and, in this regard, did not required specific permits.

Texture and Chemical Properties of the Soils

The texture of the soil appears to be loamy containing 16.5% clay, 50% silt and 33.5% sand, as determined by Agriquem S.L. (Seville, Spain). Average organic matter (OM) content was 3.03% with a Cationic Exchange Capacity of 4.7 meq/100 g. Total rainfall registered in the experimental field over the four-year study was collected from http://clima.meteored.com.

DNA Extraction, PCR Amplification and Pyrosequencing

Rhizospheres from each of the different collection times were pooled and the soil was subjected to three independent DNA extractions using the PowerMax Soil DNA kits (MO Bio Laboratories Inc., USA) following instructions from the supplier. Soil DNA from each of the three independent extractions was used as template for PCR amplification of the V6 hypervariable region of the 16S rRNA gene. The oligonucleotide design included 454 Life Science's Titanium A or B sequencing adaptors fused to the 5' end of primer 967F (5'- CAACGCGAAGAACCTTACC –3') and 1046R (5'- CGACAGCCATGCANCACCT –3'), where a MID (Multiplex Identifier) was included immediately preceding the V6 specific primer, allowing the samples to be analysed in a single lane of the 454 pyrosequencer first sampling time. A different MID was included for each sample as indicated in Table 2.

PCR amplification was performed by incubation at 95°C for 5 min, followed by 30 cycles of incubation at 95°C (30 sec), 63°C (45 sec) and 72°C (1 min), with a final extension cycle of 5 min at

72°C. The amplified DNA resulting from the three independent PCR reactions for each DNA template preparation was pooled and cleaned (Illustra GFX PCR DNA purification kit, GE Healthcare), checked with Bioanalyser 2100 (Agilent technologies), quantified with Quant-IT-picogreen (Invitrogen) and used to make the single strands on beads, as required for 454 Titanium pyrosequencing [38]. The obtained sequences were deposited in the NCBI sequence reads archives (accession number SRA009281).

For taxonomical purposes, the 454 reads for the V6 regions of each soil were filtered to eliminate the short sequences that account for 50% of all pyrosequencing errors [17] and then compared with the RDP database version 10 [12] using BLASTN. Files containing the 25 best matches for each of the 454 determined sequences were used as input to generate the corresponding taxonomic trees by means of the MEGAN 2.0 program [13].

Fast UniFrac (http://bmf2.colorado.edu/fastunifrac) [14,15,16] was used to perform a hierarchical clustering of the samples based on their phylogenetic distances. The analysis was conducted with the Greengenes core as a reference phylogenetic tree, Jackknife supporting values, and calculated using normalized abundance weights. The Fast UniFrac significance test was used to assess the existence of statistically significant differences among the 16S rDNA sequences from each soil sample, based on their phylogenetic distances.

To assess taxonomic-independent diversity, an equal number of sequences (1959) were randomly selected from each soil, and the selected pools of sequences were procesed with the RDP pyrosequencing pipelines (http://pyro.cme.msu.edu), which build an initial MSA using the Infernal tool [39] and then proceed directly to perform a complete linkage clustering to cluster OTUs (operational taxonomic units), and calculate species richness with the Chao1 estimator. The obtained results were compared with those generated by the ESPRIT software (using the –f parameter), which avoids the initial multiple sequence alignment step by applying an efficient k-tuple based distance filter, subsequently aligning the sequences using the Needleman-Wunsch method and computing pairwise distances using the quickdist algorithm [19]. Hence, a comparison is made of the efficiency of both methods and species diversity estimation. Regarding species richness estimation, confidence intervals were calculated at a 95% confidence level for all Chao1 data.

Supporting Information

Figure S1 Taxonomic trees. Taxonomic trees resulting from pyrosequencing the V6 region of the 16S rDNA extracted from each field at the indicated sampling times are shown. The size of the dots reflects the relative amount of taxa assigned to each particular node.

Acknowledgments

We wish to thank S. Marín for her technical help.

Author Contributions

Conceived and designed the experiments: RPM. Performed the experiments: JB. Analyzed the data: JB JRV. Contributed reagents/materials/analysis tools: JRV JB. Wrote the paper: RPM JB.

References

1. Jangid K, Williams MA, Franzluebbers AJ, Sanderlin JS, Reeves JH, et al. (2008) Relative impacts of land-use, management intensity and fertilization upon soil microbial community structure in agricultural systems. Soil Biol Biochem 40: 2843–2853.

2. Icoz I, Stotzky G (2008) Fate and effects of insect-resistant Bt crops in soil ecosystems. Soil Biol Biochem 40: 559–586.

3. Liu B, Zeng Q, Yan FM, Xu HG, Xu CR (2005) Effects of transgenic plants on soil microorganisms. Plant Soil 271: 1–13.

4. Saxena D, Stotzky G (2000) Insecticidal toxin from *Bacillus thuringiensis* is released from roots of transgenic Bt corn *in vitro* and *in situ*. FEMS Microbiol Ecol 33: 35–39.
5. Priestley AL, Brownbridge M (2009) Field trials to evaluate effects of Bt-transgenic silage corn expressing the Cry1Ab insecticidal toxin on non-target soil arthropods in northern New England, USA. Transgenic Res 18: 425–43.
6. Amann RI, Ludwig W, Schleifer KH (1995) Phylogenetic identification and in situ detection of individual microbial cells without cultivation. Microbiol Rev 59: 143–169.
7. Acosta-Martinez V, Dowb SE, Sun Y, Wester D, Allen V (2010) Pyrosequencing analysis for characterization of soil bacterial populations as affected by an integrated livestock-cotton production system. App Soil Ecol 45: 3–25.
8. Barriuso J, Marin S, Mellado RP (2010) Effect of the herbicide glyphosate on glyphosate-tolerant maize rhizobacterial communities: a comparison with pre-emergency applied herbicide consisting of a combination of acetochlor and terbuthylazine. Environ Microbiol 12: 1021–1030.
9. Roesch LFW, Fulthorpe RR, Riva A, Casella G, Hadwin AKM, et al. (2007) Pyrosequencing enumerates and contrast soil microbial diversity. ISME J 1: 283–290.
10. Sogin ML, Morison HG, Huber JL, Welch D, Huse SM, et al. (2006) Microbial diversity in the deep sea and the unexplored "rare biosphere". Proc Nat Acad Sci USA 103: 12115–12120.
11. Barriuso J, Valverde JM, Mellado RP (2011) Effect of the herbicide glyphosate on the culturable fraction of the glyphosate-tolerant maize rhizobacterial communities using two different growth media. Microbes Environ 26: 332–338.
12. Cole JR, Wang Q, Cardenas E, Fish J, Chai B, et al. (2009) The Ribosomal Database Project: improved alignments and new tools for rRNA analysis. Nucleic Acid Res 37: D141.
13. Huson DH, Richter DC, Mitra S, Auch AF, Schuster SC (2009) Methods for comparative metagenomics. BMC Bioinformatics 10(Suppl 1): S12.
14. Hamady M, Lozupone C, Knight R (2010) Fast UniFrac: facilitating high-throughputphylogenetic analyses of microbial communities including analysis of pyrosequencing and PhyloChip data. ISME J 4: 17–27.
15. Lozupone C, Hamady M, Knight R (2006) UniFrac – an online tool for comparing microbial community diversity in a phylogenetic context. BMC Bioinformatics 7: 371.
16. Lozupone C, Knight R (2005) UniFrac: a new phylogenetic method for comparing microbial communities. Appl Environ Microbiol 71: 8228–8235.
17. Huse SM, Welch DM, Morrison HG, Sogin ML (2010) Ironing out the wrinkles in the rare biosphere through improved OTU clustering. Environ Microbiol 12: 1889–98.
18. Schloss PD (2010) The Effects of alignment quality, distance calculation method, sequence filtering, and region on the analysis of 16S rRNA gene-based studies. PLoS Comput Biol 6(7): e1000844.
19. Sun Y, Cai Y, Liu L, Yu F, Farrell ML, et al. (2009) ESPRIT: estimating species richness using large collections of 16S rRNA pyrosequences. Nucleic Acid Res 37: e76.
20. Barriuso J, Valverde JR, Mellado RP (2011) Estimation of bacterial diversity using Next Generation Sequencing of 16S rDNA: a comparison of different workflows. BMC Bioinformatics 12: 473.
21. Saxena D, Stotzki G (2001) Bt corn has a higher lignin content than non-Bt corn. Am J Botany 88: 1704–1706.
22. van Leeuwen W, Ruttnik T, Borst Vrenssen AWM, van der Plas LHW, van der Krol AR (2001) Characterisation of position induced spatial and temporal regulation of transgene promoter activity in plants. J Exp Bot 52: 949–959.
23. Windels P, Taverniers I, Depicker A, Van Bockstaele E, De Loose M (2001) Characterisation of the Roundup Ready soybean insert. Eur Food Res Technol. 213: 107–112.
24. Icoz I, Saxena D, Andow DA, Zwahlen C, Stotzky G (2008) Microbial populations and enzyme activities in soil in situ under transgenic corn expressing cry proteins from *Bacillus thuringiensis*. J Environ Qual 37: 647–62.
25. Koskella J, Stotzky G (1997) Microbial utilization of free and claybound insecticidal toxins from *Bacillus thuringiensis* and their retention of insecticidal activity after incubation with microbes. App Environ Microbiol 63: 3561–3568.
26. Stotzky G (2004) Persistence and biological activity in soil of the insecticidal proteins from *Bacillus thuringiensis*, especially from transgenic plants. Plant Soil 266: 77–89.
27. Kersters K, De Vos P, Gillis M, Swings J, Vandamme P, et al. (2006) Introduction to the Proteobacteria. In: Dwarkin M, Falkow S, Rosenberg E, Schleifer K-H, Stackebrandt E (eds). The Prokaryotes, 3rd edn, vol. 5. Springer: New York, 3–37.
28. Hodgson DA (2000) Primary metabolism and its control in *Streptomycetes* a most unusual group of bacteria. Ad Microb Physiol 42: 47–238.
29. Kielak AM, Pijl AS, van Veen JA, Kowalchuk GA (2009) Phylogenetic diversity of Acidobacteria in a former agricultural soil. ISME J 3: 378–382.
30. Kielak AM, van Veen JA, Kowalchuk GA (2010) Comparative Analysis of Acidobacterial Genomic Fragments from Terrestrial and Aquatic Metagenomic Libraries, with Emphasis on Acidobacteria Subdivision 6. Appl Environ Microbiol 76: 6769–6777.
31. Barriuso J, Marin S, Mellado RP (2011) Potential accumulative effect of the herbicide glyphosate on glyphosate-tolerant maize rhizobacterial communities over a three-year cultivation period. PLoS ONE 6(11): e278558.
32. Saxena D, Stotzky G (2001) *Bacillus thuringiensis* (Bt) toxin released from root exudates and biomass of Bt corn has no apparent effect on earthworms, nematodes, protozoa, bacteria, and fungi in soil. Soil Biol Biochem 33: 1225–1230.
33. Blackwood CB, Buyer JS (2004) Soil microbial communities associated with Bt and non-Bt corn in three soils. J Environ Qual 33: 832–836.
34. Tan FX, Wang JW, Feng YJ, Chi GL, Kong HL, et al. (2010) Bt corn plants and their straw have no apparent impact on soil microbial communities. Plant Soil 329: 349–364.
35. Brusetti L, Francia P, Bertolini C, Pagliuca A, Borin S, et al. (2004) Bacterial communities associated with the rhizosphere of transgenic Bt 176 maize (*Zea mays*) and its non transgenic counterpart. Plant Soil 266: 11–21.
36. Val G, Marin S, Mellado RP (2009) A sensitive method to monitor *Bacillus subtilis* and *Streptomyces coelicolor*-related bacteria in maize rhizobacterial communities: The use of genome-wide microarrays. Micro Ecol 58: 108–115.
37. Muchaonyerwa P, Waladde S, Nyamugafata P, Mpepereki S, Ristori GG (2004) Persistence and impact on microorganisms of *Bacillus thuringiensis* proteins in some Zimbabwean soils. Plant Soil 266: 41–46.
38. Margulies M, Egholm M, Altman WE, Attiya S, Bader JS, et al. (2005) Genome sequencing in microfabricated high-density picolitre reactors. Nature 437: 376–380.
39. Nawrocki EP, Kolbe DL, Eddy SR (2009) Infernal 1.0: inference of RNA alignments. Bioinformatics 25: 1713–1713.

3

Identification of Autotoxic Compounds in Fibrous Roots of Rehmannia (*Rehmannia glutinosa* Libosch.)

Zhen-Fang Li, Yan-Qiu Yang, Dong-Feng Xie, Lan-Fang Zhu, Zi-Guan Zhang, Wen-Xiong Lin*

Agroecological Institute, Fujian Agriculture and Forestry University, Fuzhou, China

Abstract

Rehmannia is a medicinal plant in China. Autotoxicity has been reported to be one of the major problems hindering the consecutive monoculture of Rehmannia. However, potential autotoxins produced by the fibrous roots are less known. In this study, the autotoxicity of these fibrous roots was investigated. Four groups of autotoxic compounds from the aqueous extracts of the fibrous roots were isolated and characterized. The ethyl acetate extracts of these water-soluble compounds were further analyzed and separated into five fractions. Among them, the most autotoxic fraction (Fr 3) was subjected to GC/MS analysis, resulting in 32 identified compounds. Based on literature, nine compounds were selected for testing their autotoxic effects on radicle growth. Seven out of the nine compounds were phenolic, which significantly reduced radicle growth in a concentration-dependent manner. The other two were aliphatic compounds that showed a moderate inhibition effect at three concentrations. Concentration of these compounds in soil samples was determined by HPLC. Furthermore, the autotoxic compounds were also found in the top soil of the commercially cultivated Rehmannia fields. It appears that a close link exists between the autotoxic effects on the seedlings and the compounds extracted from fibrous roots of Rehmannia.

Editor: Alejandra Bravo, Universidad Nacional Autonoma de Mexico, Instituto de Biotecnologia, Mexico

Funding: This work was supported by The National Natural Science Foundation of China (No. 30772729), Provincial Natural Science Foundation of Fujian, China (No. 2008J0051), and the key discipline program of Fujian Province, China (grant No. 20020F012). The funders had no role in study design, data collection and analysis, decision to publish, or preparation of the manuscript.

Competing Interests: The authors have declared that no competing interests exist.

* E-mail: wenxiong181@163.com

Introduction

Rehmannia (*Rehmannia glutinosa* Libosch) is in the Scrophulariaceae family and is one of the most common and important medicinal herbal plants in China. It is perennial and its fresh or dried tuberous roots are used as a high demand traditional Chinese medicinal ingredient for hematologic conditions, sedation, insomnia and diabetes [1,2]. Its commercial cultivation has been practiced for almost 1500 years in China. However, the consecutively monocultured plants are prone to severe diseases resulting in reduced biomass, especially the tuberous products. To maintain the cultivation, the farmers commonly limited the cultivation on a same plot once every eight years. Therefore, less desirable areas outside Jiaozuo had to be used for the planting with decreased tuber yields and lower product quality [3].

The autotoxicity issue has attracted much attention [4,5]. Autotoxicity is the phenomenon whereby mature plants inhibit the growth of their own seedlings through the release of autotoxic chemicals. It has been found to exist in various crops [6,7], such as greenhouse crops [8,9], fruits [10,11], forage [12,13], horticultural and medicinal plants [4,5,14,15,16]. Several groups of chemicals have been implicated in autotoxicity, including terpenoids, phenolics, steroids, alkaloids, and cyanogenic glycosides. Recently, autotoxicity in Rehmannia has been reported [15,16,17] especially in relation to the compounds derived from the root exudates. However, to date, the degradation of fibrous roots and its products had not been studied, and the mechanism of autotoxicity in Rehmannia remains unknown.

This study aims to identify substances that contribute directly to Rehmannia autotoxicity. A number of potentially autotoxic compounds from the fibrous roots were isolated and characterized. The inhibitory effect of these compounds on seedling growth was observed. Furthermore, the concentration of these bioactive compounds in the top soil collected from one-year cultivated and two-year consecutively moncultured Rehmannia fields was determined.

Materials and Methods

Sample collection and autotoxic compound extraction

Water extraction. The fields were located in Jiao-zuo County (113°21′E, 35°24′N), He-nan province of China, which is the optimal production areas of Rehmannia. The samples were collected in October 2008 (Figure S1).

Fibrous roots of one-year cultivated Rehmannia plants at the mature stage were collected. The air-dried roots (500 g), passed 2 mm sieve, were soaked in 1000 mL distilled water at 25–30°C for 48 h. The extract was filtered, and the extraction was repeated three times. The aqueous extracts from the three extractions were combined and concentrated to 20 mL under vacuum at 50°C, then freeze-dried under liquid N_2 at $-180°C$ Approximately 530 mg of the dried material were obtained from the 500 g of air-dried fibrous roots.

Top soil samples (20 cm depth) were collected from both one-year cultivated and two-year consecutively moncultured Rehmannia fields in Jiaozuo county at harvest time. A soil sample from an adjacent uncultivated field was collected as a control. Potential autotoxic compounds were extracted from the soil samples using the same method for the fibrous roots. Approximately 400 mg of dried material were obtained from the 500 g air-dried soil samples.

Ethanol extraction and partitioning. Air-dried fibrous roots (2 kg) were extracted with 95% ethanol (5 L) at room temperature for 5 d. This process was repeated once. The extract was concentrated by evaporation to 200 mL at 50°C under vacuum followed by freeze-drying under liquid N_2 ($-180°C$), and then dissolved in 200 mL distilled water.

The aqueous solution was consecutively partitioned with petroleum ether, chloroform, ethyl acetate, and n-butanol, as shown in Figure 1. The solvent extractions were performed by shaking in separation funnels for 10 min, followed by evaporation under vacuum.

Bioassays and statistical analysis

To examine the functional effect of the potential autotoxic compounds on Rehmannia growth, measurements of Rehmannia radicles in petri dishes was performed. The extracts of petroleum ether, chloroform, ethyl acetate, n-butanol and the ethyl acetate-extracted fractions (1–5), of fibrous roots, monocultured soils for 1 year and 2 years, and uncultivated soil, were diluted with distilled water into 2, 5, 10, 20 and 50 mg L^{-1}. 5 mL of each of the diluted solutions were transferred into the 10-cm diameter petri dishes containing double-layered filter paper (Whatman No. 42). In addition, individual compounds, including 7 aliphatic and 2 phenolics

as identified in Fr3, were dissolved in a small volume of methanol and transferred onto the double-layered filter paper in the petri dishes. The solvents were evaporated in a draft chamber for 1 h. The filter paper containing these compounds was moistened with 5 mL distilled water. The final concentrations of the compounds in water were 2, 10, and 50 mg L^{-1}, while distilled water was used as control. Twenty Rehmannia seeds were placed on the filter paper in the petri dishes. All dishes were maintained in a tissue culture room at 26°C with fluorescent lights for 11 h (8:00–20:00) as described previously [7]. The fluorescent light intensity was $4.17 \pm 0.18 \times 103$ lux. Germinated seeds with >1 mm radius were recorded and the radicle lengths were measured after incubation for 5 d.

The radicle lengths of the treated ones in comparison with the controls were used as an index of the inhibition rates (IRs). IR was calculated as: IR = (control - treatment)/control×100%. When IR is greater than 0, it indicates an inhibition; conversely when IR is less than 0, it indicates promoting growth [18]. All data were subject to an analysis of variance using the Statistical Analysis System Program (SPSS). Each value was expressed as the mean of three replicates±standard error (SE).

Identification of autotoxic chemicals

The ethyl acetate extract was further analyzed. About 8 g of the extract was subjected to column chromatography (CC), packed with silica gel (200–300 mesh) and eluted with chloroform ($CHCl_3$)-methanol (MeOH), followed by a gradient solvent system (0–100 MeOH) to yield five fractions (Fr1-5). The fractions were

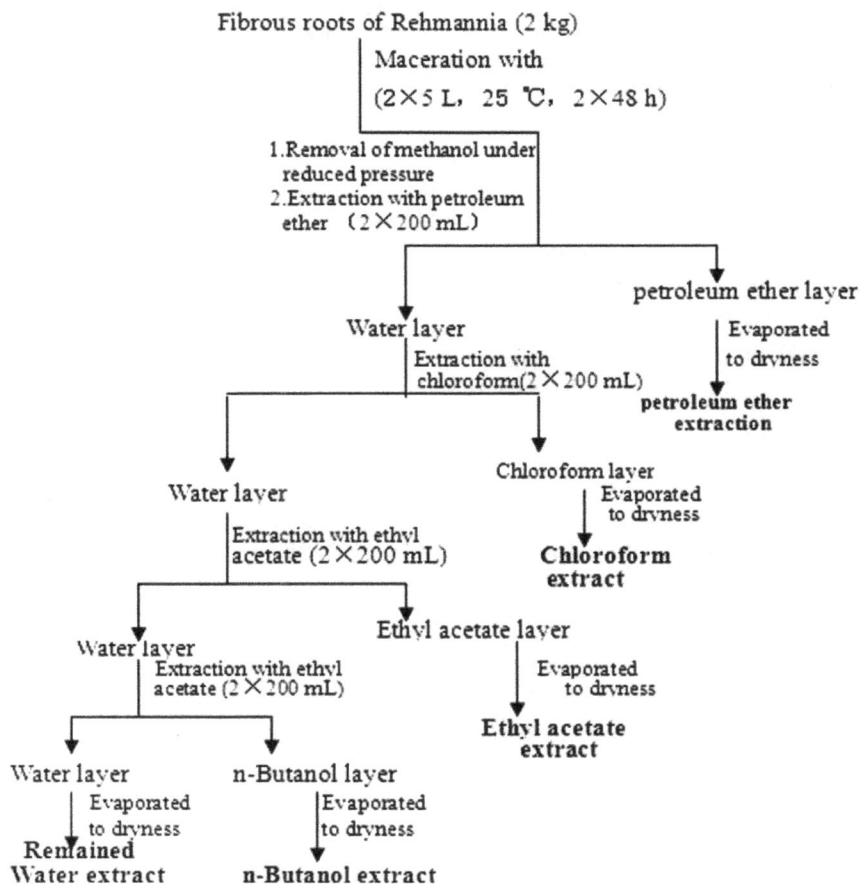

Figure 1. Flow chart of extraction procedures.

Table 1. Inhibition effect of aqueous extracts from soil and fibrous roots on Rehmannia radicles growth.

Concentration	Uncultivated (control soil) Radicle length (mm)	IR (%)	Monocultured soil (1-year) Radicle length (mm)	IR (%)	Monocultured soil (2-year) Radicle length (mm)	IR (%)	Fibrous roots Radicle length (mm)	IR (%)
2 mg·L^{-1}	9.67±0.23a	-	9.44±0.09b	−3.09	9.22±0.18c	−5.15	8.68±0.14d	−10.31
5 mg·L^{-1}	10.12±0.07a	-	8.37±0.21b	−16.83	8.04±0.16c	−20.79	7.46±0.25d	−25.74
10 mg·L^{-1}	9.33±0.27a	-	6.67±0.08b	−27.96	5.14±0.16c	−45.16	4.92±0.28c	−47.31
20 mg·L^{-1}	9.07±0.23a	-	5.45±0.24b	−39.56	4.31±0.13c	−52.75	3.23±0.15c	−64.84
50 mg·L^{-1}	9.92±0.30a	-	2.86±0.14b	−70.71	2.53±0.21b	−74.75	2.35±0.24b	−75.76

Note: Values on the same row followed by the same lowercase letters are not statistically different at P = 0.05 by Duncan's test, the same below.

analyzed by thin-layer chromatography (TLC) on the same Rf value. Preliminary bioassay study indicated that fraction 3 (Fr3) exhibited the greatest inhibitory activity on radicle growth of Rehmannia. Thus, Fr3 was further purified using the reversed phase multi-purpose GC column chromatography (Varian, California, U.S.A., VF-5 ms, 30 m×0.25 mm×0.25 μm).

The GC/MS fingerprints of Fr3 were obtained with a GC/MS (Autospec-240 MS Ion Trap mass spectroscopy, Varian, California, U.S.A.). Fraction 3 was subsequently dissolved into 2 mL redistilled MeOH. One μL aliquot of this solution was evaporated under a stream of helium to remove residual water. A mixture of 10 μL redistilled MeOH and 10 μL N, O-Bis (trimethylsilyl) trifluoro acetamide (BSTFA) + 1% trimethylchlorosilane (TMCS) was added to the residue to produce trimethylsilyl derivatives by heating at 100°C for 1 h. The solution was filtered through a 0.45 μm filter before injected into the GC-MS system.

The injection volume was 1 μL. Helium was used as the carrier gas, and the flow rate was adjusted to 10 mL min^{-1}. The oven temperature was initially programmed at 50°C and ramped to 270°C at a rate of 5°C·min^{-1}, where it remained for 10 min. MS data were acquired in the negative ionization mode. The full scan mass covered a range from m/z 50 to 1500. The m/z values for standard autotoxins and the compounds as available in the literature were used to match with those obtained in the spectra for the study.

HPLC analysis

To determine whether the autotoxic compounds identified in Rehmannia root extracts also existed in ground soil, we tested the soil samples.

The concentration of autotoxic compounds in soil samples was determined using a Waters HPLC system (Alliance HPLC system Massachusetts, U.S.A.).The chromatographic system consisting of a Waters 2695 HPLC system with a reversed-phase column Zorbax Eclipse XDB –C18 (250 mm×4.6 mm, 5 μm column) was

used at a flow rate of 1.0 mL min^{-1}. The solvent system was a linear gradient of solvent A (methanol) and solvent B (0.5 mol L^{-1} acetic acid in water): from 1% to 25% A in 0–10.0 min, and hold at 25% A for 5 min;from 25% to 80% A in 15.0–25.0 min, and hold at 80% A for 5 min;from 80% to 25% A in 30.0–40.0 min, and hold at 25% A for 10 min. The injected volume was 10 mL of a water solution of the extracts (10 mg mL^{-1}). The column temperature was maintained at 35°C. The UV detector was performed at 280 nm. The chromatographic data were recorded and processed with a Waters empower workstation.

HPLC grade acetonitrile, acetic acid (Merck, Darmstadt, Germany), and filtered bi-distilled water, were used for HPLC analysis. The methanol used for extraction was from "Honeywell International" (New Jersey, U.S.A.). Standards of phenolic acids, including (gallic acid, 4-hydroxybenzoic acid, vanillic acid, protocatechuic acid, ferulic acid, benzoic acid, and salicylic acid,) were purchased from Sigma Chemicals Co., U.S.A.. Solvents and standards of phenolic acids were chromatographic grade. The specific recovery rates (%) for the standards were: gallic acid, 96.24±4.42;4-Hydroxybenzoic acid, 92.26±5.35; vanillic acid, 92.15±3.22; protocatechuic acid, 92.44±5.06; ferulic acid, 93.25±5.31; benzoic acid, 90.01±4.38; and salicylic acid, 90.06±3.21.

The dry materials from the water extract were dissolved into 100 mL methanol and passed through a 0.22 μm glass fiber sieve. 7 Phenolic acids compounds found in the samples extracts were dentified by matching the retention time and their spectral characteristics against those of standards. The separation procedures were repeated six times for each standard compound, and data were presented as mean±SE.

Results

Bioassay of fibrous root extracts

When compared with the root exudates and the extracts of three different soils (i.e. control soil, one-year monocultured and

Table 2. Inhibitory effect on the growth of Rehmannia radicles when exposed to fibrous root extracts.

Fraction	Radicle length of Rehmannia (mm) 2 mg·L^{-1}	5 mg·L^{-1}	10 mg·L^{-1}	20 mg·L^{-1}	50 mg·L^{-1}
Distilled water (control)	9.89±0.16b	9.89±0.16b	9.89±0.16a	9.89±0.16a	9.89±0.16a
Petroleum ether extract	10.06±0.14a	10.14±0.21a	9.19±0.18b	9.06±0.24c	9.03±0.12b
Chloroform extract	10.13±0.12a	10.14±0.16a	10.12±0.13a	9.68±0.12b	9.32±0.16b
Ethyl acetate extract	8.67±0.09d	7.85±0.24d	4.79±0.23d	4.32±0.16e	3.54±0.14d
n-Butanol extract	9.35±0.25c	9.11±0.15c	8.17±0.16c	8.31±0.23d	7.34±0.17c

Table 3. Effect of Fractions 1–5 from Rehmannia fibrous root's ethyl acetate extract on radicle growth of Rehmannia.

| Fraction | Radicle Length of Rehmannia (mm) | | | | |
	2 mg·L^{-1}	5 mg·L^{-1}	10 mg·L^{-1}	20 mg·L^{-1}	50 mg·L^{-1}
Fr 1	9.37±0.12a	10.11±0.25a	9.34±0.09a	9.22±0.13a	9.13±0.17b
Fr 2	7.06±0.09c	6.10±0.14d	6.53±0.12c	6.06±0.15c	6.03±0.15d
Fr 3	4.37±0.23e	3.15±0.27e	2.79±0.18d	0 d	0 f
Fr 4	6.15±0.16d	6.62±0.18c	6.13±0.14c	5.65±0.18c	5.33±0.14e
Fr 5	8.39±0.13b	8.31±0.15b	8.17±0.21b	8.08±0.20b	7.96±0.22c
Sterilized water (control)	9.67±0.23a	9.67±0.23a	9.67±0.23a	9.67±0.23a	9.67±0.23a

two-year monocultured soils), the autotoxic compounds extracted from the fibrous roots showed greater inhibitive effects on the radicle growth of Rehmannia (Table 1). The inhibition rate increased with higher concentrations of the extracts. The IR peaked at 50 mg·L^{-1} with 75.76% reduction in seedling radicles (P<0.01; Table 1). A similar trend was found in the soil extracts, but their inhibitory effects were always lower than those of the fibrous roots'. The result indicated that the chemical compounds extracted from the fibrous roots had a major auto-inhibitory effect on the growth of Rehmannia radicle.

Bioassay of partitioned compounds of fibrous root extracts

The bioassay with petroleum ether, chloroform, ethyl acetate, and n-butanol partitions showed that the four extracts exhibited dose-dependent inhibition effects on the radicle growth and that the ethyl acetate extract had the greatest inhibitory effect (Table 2).

At the concentration of 50 mg L^{-1}, the inhibition rates of petroleum ether, chloroform, ethyl acetate, and n-butanol extracts were 8.7%, 5.8%, 64.2%, and 25.8%, respectively.

Bioassay of column chromatography fractions of ethyl acetate extract

Bioassays with each of the five fractions from ethyl acetate extract showed a significant reduction in the length of the radicle in the presence of Fr3 (P<0.01; Table 3). When the concentration was greater than 20 mg L^{-1}, the reduction reached 100%. This indicates that the chemical compounds in Fr3 had a major auto-inhibitory effect on the seedling growth, and so Fr3 was considered a candidate for further identification and characterization.

Autotoxic effect of compounds identified in fraction 3

By comparing with a GC/MS user-library spectrum of pure reference compounds, GC/MS analysis for Fr3 identified a total of

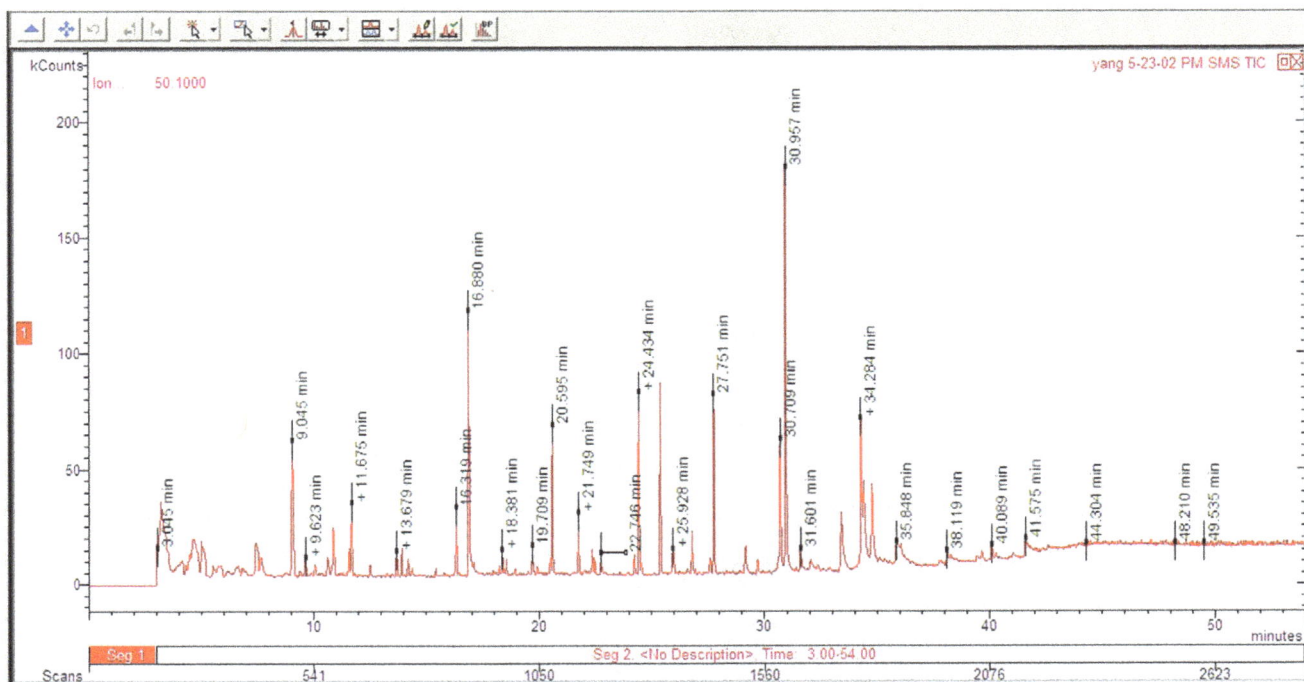

Figure 2. Total ion chromatogram of bioactive Fraction 3 of ethyl acetate extract from Rehmannia fibrous roots.

Table 4. Compounds in bioactive Fraction 3 of ethyl acetate extract from Rehmannia fibrous roots as identified by GC/MS analysis.

Rt	CAS	Scientific name	Formula
Aliphatic compounds			
22.746	5870-93-9	Butanoic acid, heptyl ester	$C_{11}H_{22}O_2$
20.595	544-63-8	Tetradecanoic acid	$C_{14}H_{28}O_2$
18.381	7132-64-1	Pentadecanoic acid, methyl ester	$C_{16}H_{32}O_2$
16.319	109-52-4	Pentanoic acid	$C_5H_{10}O_2$
31.601	143-07-7	Dodecanoic acid	$C_{12}H_{24}O_2$
33.410	55000-42-5	11-Hexadecenoic acid, methyl ester	$C_{17}H_{32}O_2$
40.089	112-95-8	Eicosane	$C_{20}H_{42}$
35.848	544-35-4	Linoleic acid ethyl ester	$C_{20}H_{36}O_2$
41.575	112-80-1	Oleic Acid	$C_{18}H_{34}O_2$
Phenolic compounds			
30.709	496-16-2	Benzofuran, 2,3-dihydro-	C_8H_8O
30.957	1135-24-6	Ferulic acid	$C_{10}H_{10}O_4$
19.709	6781-42-6	1,1'-(1,3-Phenylene)diethanone	$C_{10}H_{10}O_2$
25.928	876-02-8	4-Hydroxy-3-methylacetophenone	$C_9H_{10}O_2$
25.673	121-33-5	Vanillin	$C_8H_8O_3$
24.434	69-72-7	Salicylic acid	$C_7H_6O_3$
27.751	149-91-7	Gallic acid	$C_7H_6O_5$
34.284	99-50-3	Protocatechuic acid	$C_7H_6O_4$
35.132	99-96-7	4-Hydroxybenzoic acid	$C_7H_6O_3$
21.673	65-85-0	Benzoic acid	$C_7H_6O_2$
Terpene			
11.023	933-40-4	Cyclohexane, 1,1-dimethoxy-	$C_8H_{16}O_2$
11.675	109119-91-7	Aromadendrene	$C_{15}H_{24}$
13.679	135760-25-7	Ascaridole epoxide	$C_{10}H_{16}O_3$
15.567	77-53-2	Cedrol	$C_{15}H_{26}O$
Steroids			
14.098	546-97-4	Columbin	$C_{20}H_{22}O_6$
24.312	61834-65-9	Allopregnane-3,7,11,20-tetra-one	$C_{21}H_{28}O_4$
26.113	17673-25-5	Phorbol	$C_{20}H_{28}O_6$
27.067	52-21-1	Prednisolone Acetate	$C_{23}H_{30}O_6$
Others			
9.045	60485-45-2	Santolina epoxide	$C_{10}H_{16}O$
12.554	110-15-6	Butanedioic acid	$C_4H_6O_4$
16.880	97-67-6	L-Hydroxybutanedioic acid	$C_4H_6O_5$
18.607	86-73-7	Fluorene	$C_{13}H_{10}$
22.547	84-66-2	Phthalic acid	$C_{12}H_{14}O_4$

32 compounds in Fr3. They were classified into 5 groups: phenolics, aliphatic compounds, terpenoids, steroids, and others (Figure 2, Table 4).

Previous studies suggested that several chemicals detected in the bioactive Fr3 were possibly allelopathic compounds, which were reported as allelochemicals in other crops [4,8,19]. Based on previous reports and the availability of the chemicals in Fr3, nine individual compounds were selected to test their potential autotoxic effects on Rehmannia radicle growth (Table 5). Among the 9 chemicals examined, seven phenolic compounds showed dose-respondent inhibitory effects on the seedling growth. Specifically, the radicle of the seedlings reduced significantly by 45–76% when they were treated with the phenolic compounds (except salicylic acid) at a concentration of 50 mg·L^{-1} (P<0.01). Ferulic acid and vanillin acid were the most potent compounds in the test, and therefore, they were considered as the potential allelochemicals. All of the compounds inhibited the seedling growth even at a low concentration (e.g., 2 mg·L^{-1}, P<0.01), and IRs increased with increasing concentrations. Similarly, the compounds in the aliphatic acid family (e.g., tetradecanoic acid and oleic acid) exhibited dose-respondent inhibitory effects on the seedling growth. The extent of inhibition was around 25% for each compound (Table 5). The results demonstrate clearly that the compounds extracted from the roots of Rehmannia are potent inhibitors on seedling growth. In addition, both the phenolic acid and aliphatic acid compounds might contribute to the autotoxicity of Rehmannia.

Table 5. Effect of autotoxic chemicals on radicle growth of Rehmannia.

Autotoxic chemicals	Treated with 2 mg·L⁻¹		Treated with 2 mg·L⁻¹		Treated with 2 mg·L⁻¹	
	Radicle Length (mm)	IR (%)	Radicle Length (mm)	IR (%)	Radicle Length (mm)	IR (%)
Ferulic acid	5.15±0.12e	−46.74	3.79±0.16e	−60.89	2.37±0.17e	−75.50
Vanillic acid	5.32±0.14e	−44.98	4.33±0.23e	−55.31	2.62±0.09e	−72.91
4-Hydroxybenzoic acid	5.25±0.12e	−45.71	3.84±0.09e	−60.37	2.67±0.16e	−72.39
Protocatechuic acid	6.56±0.17d	−32.16	4.92±0.23e	−49.22	3.87±0.13d	−59.98
Benzoic acid	6.34±0.21d	−34.44	5.26±0.11d	−45.72	4.13±0.17d	−57.29
Gallic acid	7.56±0.11c	−21.82	7.11±0.16c	−26.63	5.32±0.20c	−44.98
Salicylic acid	8.38±0.15b	13.34	8.13±0.13b	−16.10	7.33±0.14b	−24.36
Oleic acid	7.65±0.14c	−20.89	7.43±0.12c	−23.32	6.89±0.18b	−28.75
Tetradecanoic acid	8.53±0.22b	−11.79	7.42±0.09c	−23.43	7.01±0.20b	−27.51
Sterilized water (control)	9.69±0.22a	-	9.69±0.22a	-	9.69±0.22a	-

Identification of autotoxic compounds in soil samples

In the fields where Rehmannia were monocultivated for 1 year and 2 years, seven phenolic compounds identified as the potential autotoxins were found at different concentrations (Table 6, Figure 3), 4-hydroxybenzoic acid being the most abundant. In the control and 1-year monocultured soil samples, only 6 phenolic compounds were detected. Thus, the findings presented here indicate that the compounds identified in Rehmannia roots can be found in relative abundance in soils previously cultivated with Rehmannia, whereas they are absent or present at much lower concentrations in non-cultivated soils. Furthermore, the higher inhibitory chemical concentration in the consecutively monocultured soil than in the newly planted soil was due to the accumulation of autotoxic compounds in the rhizosphere soil.

Discussion

Our results demonstrate that the compounds isolated from ethyl acetate-soluble extracts of Rehmannia fibrous roots had the most auto-inhibitory effects on the seedling growth. Specifically, 32 chemicals were identified by GC/MS that included 9 aliphatic, 10 phenolic, 4 terpene, 4 steroids and 5 other compounds. Among them, the 7 phenolic compounds and 2 aliphatic acids selected for testing the inhibition effects showed a significant suppressive function on the seedling

growth. The inhibitory effects were somewhat related to the concentration of the autotoxins. More importantly, all 7 bioactive phenolic compounds were detected in the top soil of the Rehmannia fields. It appeared that our study provided the first direct evidence that the autotoxic chemicals detected in the soils of different-year consecutively monocultured Rehmannia fields could be traced back to the roots of Rehmannia. During soil sample collection, we noticed that a large amount of fibrous root waste was left in the soil after harvest. It is likely that the autotoxic compounds found in soils were derived partly from the root exudates or the degraded plant tissues. Once released into the soil and allowed to accumulate, these compounds might play a major role in the autotoxic effects on the seedling growth.

In this study, we found that the inhibitory effect of each single compound was not as potent as the bioactive Fr3 (which contained all compounds). It might result from the additive or synergistic effect of the mixture compounds extracted from Rehmannia fibrous roots and its rhizosphere soil. The similar results were reported in the case of other plants [20,21].

The consecutive monoculture problems in the case study were also defined as "replanting disease" or "sick soil syndrome", and it is a very common phenomenon in many fruit trees, such as apples, pears, and plums. A wide variety of tree pathogens, including bacteria, fungi, nematodes, and viruses, have been linked to the "replanting disease" in fruit trees. These pathogens may not be

Table 6. Concentration of autotoxic compounds soil samples.

Autotoxic chemicals	Control Soil		one-year cultivated soil		two-year moncultured soil	
	Rt (min)	Conc. (mg·kg⁻¹)	Rt (min)	Conc. (mg·kg⁻¹)	Rt (min)	Conc. (mg·kg⁻¹)
Gallic acid	-	-	-	-	9.316	2.586
Protocatechuic acid	12.132	0.171	12.325	0.804	12.239	4.825
4-Hydroxybenzoic acid	18.412	3.653	18.447	11.757	18.236	12.209
Vanillic acid	22.211	0.622	20.534	1.019	20.426	5.279
Salicylic acid	24.011	0.628	24.194	7.036	23.897	8.829
Ferulic acid	28.226	2.946	28.412	3.624	28.131	8.641
Benzoic acid	42.865	1.880	42.813	3.026	44.001	3.315

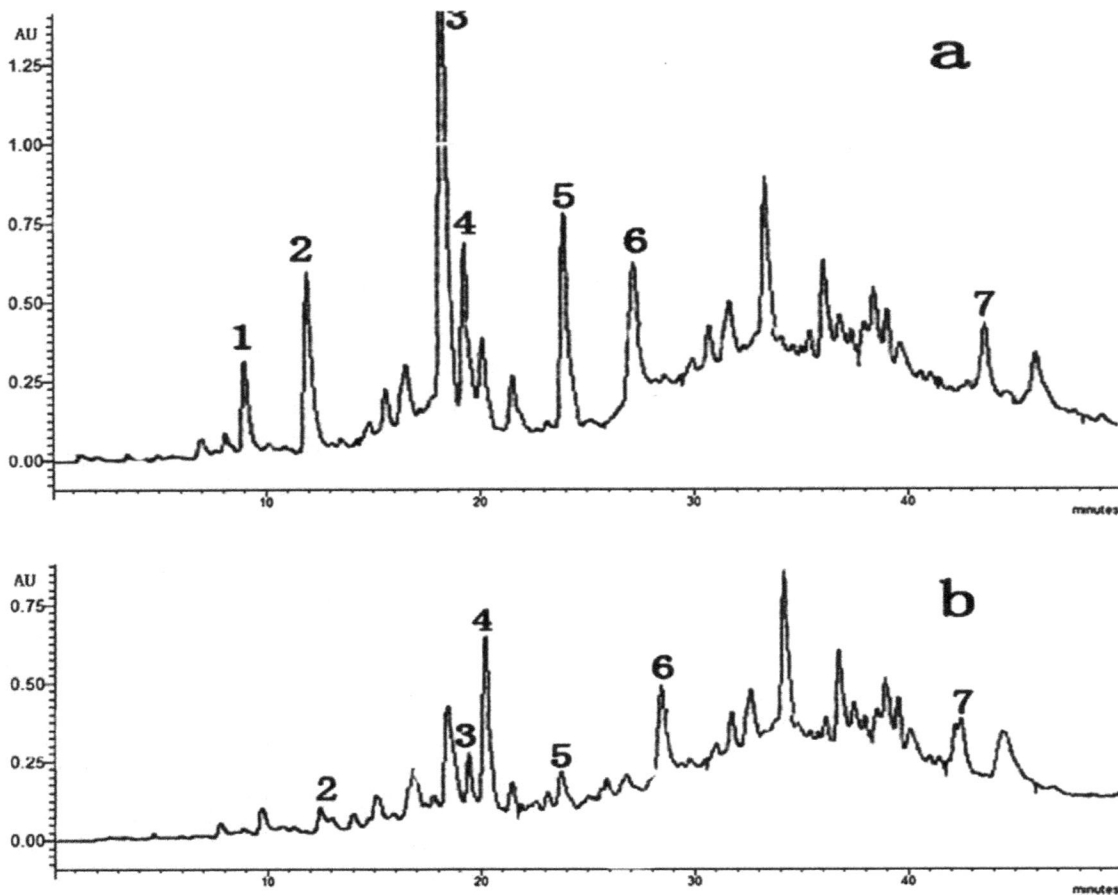

Figure 3. HPLC chromatograms of two-year consecutively monocultured soil (a), one-year cultivated soil (b). Compounds detected from samples in order of appearance in eluant: gallic acid (1), protocatechuic acid (2), 4-hydroxybenzoic acid (3), vanillic acid (4), salicylic acid (5), ferulic acid (6), and benzoic acid (7).

harmful to the mature trees, yet they retard the growth of young trees in the same field [22]. It has been reported that the presence of fungal pathogens in soils contributes to the "replanting disease" of Rehmannia [23,24]. However, this study provides evidence that the autotoxicity is another major cause of the disease [4]. Identification of the autotoxic compounds in this study might be helpful to further understand the problems associated with consecutive monoculture of Rehmannia, and it was also conducive to make the solution to effectively control the "replanting disease" for Rehmannia in consecutively cropping sequence.

Supporting Information

Figure S1   The field study picture.

Author Contributions

Conceived and designed the experiments: Z-FL W-XL. Performed the experiments: Z-FL Y-QY D-FX L-FZ Z-GZ. Analyzed the data: Z-FL Y-QY D-FX L-FZ Z-GZ W-XL. Contributed reagents/materials/analysis tools: W-XL. Wrote the paper: Z-FL.

References

1. Cui YY, Hahn EJ, Kozai T, Paek KY (2000) Number of air exchanges, sucrose concentration, photosynthetic photon flux, and differences in photoperiod and dark period temperatures affect growth of Rehmannia glutinosa plantlets in vitro. Plant Cell Tiss Org 62: 219–226.
2. Zhao HJ, Tan JF, Qi CM (2007) Photosynthesis of *Rehmannia glutinosa* subjected to drought stress is enhanced by choline chloride through alleviating lipid peroxidation and increasing proline accumulation. Plant Growth Regul 51: 255–262.
3. Zhang Z, Lei CY, Zhang LF, Yang XX, Chen R, et al. (2008) The complete nucleotide sequence of a novel *Tobamovirus*, Rehmannia mosaic virus. Arch Virol 153: 595–599.
4. Zhang ZY, Wang YP, Shao D, Yang JS, Liu DS (2005) Autotoxicity of Panax quinquefolium L. Allelopathy J 15: 67–74.
5. Hao ZP, Wang Q, Christie P, Li XL (2007) Allelopathic potential of watermelon tissues and root exudates. Sci Hortic-Amsterdam 112: 315–320.

6. Chou CH, Lin HJ (1976) Autointoxication mechanisms of *Oryza sativa*. J Chem Ecol 2: 353–367.
7. Wu HW, Pratley J, Lemerle D, An M, Liu DL (2007) Autotoxicity of wheat (*Triticum aestivum* L.) as determined by laboratory bioassays. Plant Soil 1296: 85–93.
8. Yu JQ, Shou SY, Qian YR, Zhu ZJ, Hu WH (2000) Autotoxic potential of cucurbit crops. Plant Soil 223: 147–151.
9. Chon SU, Boo HO (2005) Difference in allelopathic potential as influenced by root periderm colour of sweet potato (*Ipomoea batatas*). J Agron Crop Sci 191: 75–80.
10. Manici LM, Ciavatta C, Kelderer M, Erschbaumer G (2003) Replant problems in South Tyrol: role of fungal pathogens and microbial population in conventional and organic apple orchards. Plant Soil 256: 315–324.
11. Wilson S, Andrews P, Nair TS (2004) Non-fumigant management of apple replant disease. Sci Hortic-amsterdam 102: 221–231.

12. Chon SU, Nelson CJ, Coutts JH (2004) Osmotic and Autotoxic Effects of Leaf Extracts on Germination and Seedling Growth of Alfalfa. Agron J 96: 1673–1679.
13. Canals RM, Emeterio LS, Peralta J (2005) Autotoxicity in *Lolium rigidum*: analyzing the role of chemically mediated interactions in annual plant populations. J Theor Biol 235: 402–407.
14. Bogatek R, Gniazdowska A, Zakrzewska W, Oracz K, Gawronski SW (2006) Allelopathic effects of sunflower extracts on mustard seed germination and seedling growth. Biol Plantarum 50: 156–158.
15. Gao WW, Zhao YJ, Wang YP, Chen SL (2006) A review of research on sustainable use of medicinal plants cropland in China (in Chinese). China Journal of Chinese Materia Medica 31: 1665–1668.
16. Du JF, Yin WJ, Zhang ZY, Hou J, Huang J, et al. (2009) Autotoxicity and Phenolic Acids Content in Soils with Different Planting Interval Years of Rehmannia glutinosa (in Chinese). Chinese Journal of Ecology 48: 445–450.
17. Liu HY, Wang F, Wang YP, Lu CT (2006) The Causes and Control of Continuous Cropping Barrier in Dihuang (in Chinese). Journal of North China Agriculture 21: 131–132.
18. Lin WX, Kim KU, Shin DH (2000) Allelopathic potential in rice (*Oryza sativa L.*) and its modes of action on Barnyardgrass (Echinochloa *crusgalli L.*). Allelopathy J 7: 215–224.
19. Kalinova J, Vrchotova N, Triska J (2007) Exudation of allelopathic substances in buckwheat (Fagopyrum esculentum Moench). J Agr Food Chem 55: 6453–6459.
20. Einhellig FA (1996) Interactions involving allelopathy in cropping systems. Agron J 88: 886–893.
21. Kong C, Hu F, Xu T, Lu Y (1999) Allelopathic potential and chemical constituents of volatile oil from ageratum conyzoides. J Chem Ecol 25: 2347–2356.
22. Weller DM, Raaijmakers JM, Gardener BM, Thomashow LS (2002) Microbial populations responsible for specific soil suppressiveness to plant pathogens. Annu Rev Phytopathol 40: 309–348.
23. Xie HE, Wang JJ, Xie XH, Li JH, Chen L, et al. (2007) Identification and Morbidityu Test of Pathogenic Fungi in the Soil Samples Planted *Rehmnnia gintinosa*. Journal of Shanxi Agriculture Science 35: 59–63.
24. Li J, Wang M, Lin ZP, Yuan QJ, Yu RM, et al. (2010) Isolation and identification of endophytic fungi from different swollenroot of *Rehmannia glutinosa*. China Journal of Chinese Materia Medica 35: 1679–1683.

4

Groundwater Nitrogen Pollution and Assessment of Its Health Risks: A Case Study of a Typical Village in Rural-Urban Continuum, China

Yang Gao[1,3]*, Guirui Yu[1]*, Chunyan Luo[2], Pei Zhou[3]

1 Key Laboratory of Ecosystem Network Observation and Modeling, Institute of Geographic Sciences and Natural Resources Research, Chinese Academy of Sciences, Beijing, China, 2 Institute of Agricultural Resources and Regional Planning, CAAS, Beijing, China, 3 School of Agriculture and Biology, Shanghai Jiaotong University, Shanghai, China

Abstract

Protecting groundwater from nitrogen contamination is an important public-health concern and a major national environmental issue in China. In this study, we monitored water quality in 29 wells from 2009 to 2010 in a village in Shanghai city, whick belong to typical rural-urban continuum in China. The total N and NO_3-N exhibited seasonal changes, and there were large fluctuations in NH_4-N in residential areas, but without significant seasonal patterns. NO_2-N in the water was not stable, but was present at high levels. Total N and NO_3-N were significantly lower in residential areas than in agricultural areas. The groundwater quality in most wells belonged to Class III and IV in the Chinese water standard, which defines water that is unsuitable for human consumption. Our health risk assessments showed that NO_3-N posed the greatest carcinogenic risk, with risk values ranging from 19×10^{-6} to 80×10^{-6}, which accounted for more than 90% of the total risk in the study area.

Editor: Alex J. Cannon, Pacific Climate Impacts Consortium, Canada

Funding: This work was financially supported by National Key Basic Research Program (2010CB833504), and National Natural Science Foundation of China (40601097 and 30590381). The funders had no role in study design, data collection and analysis, decision to publish, or preparation of the manuscript.

Competing Interests: The authors have declared that no competing interests exist.

* E-mail: gaoyang0898@163.com (YG); yugr@igsnrr.ac.cn (GY)

Introduction

Groundwater is the major water supply for drinking and for the domestic, industrial, and agricultural sectors in the Shanghai region of China. One serious problem that affects the quality of the region's groundwater is leaching of nutrients from the soil, which is especially evident in areas dominated by agriculture [1–2]. Nitrogen percolates easily into the groundwater through the soil along with rainwater recharge or irrigation water. As a result, the shallow aquifers are more likely than deeper ones to initially suffer from contamination problems [3–4]. The application of large amounts of nitrogen fertilizers in regions of intensive agriculture contributes to excessive nitrogen accumulation in soils and excessive leaching into groundwater bodies [5–7]. Extensive irrigation and use of nitrogen (N) fertilizers together result in low N-use efficiency and high N loss [8]. Several studies have also reported increasing incidence of nitrogen pollution and dramatic increases in the nitrogen concentration in the groundwater of regions where intensive farming is practiced [9–11].

Because contaminated groundwater resources are often located in the vicinity of wells for drinking water, it is essential to determine how management practices in the area surrounding these wells will affect groundwater nitrogen concentrations, and particularly nitrate nitrogen (NO_3-N). Nitrate is formed from fertilizers, decaying plants, manure and other organic residues. It is found in the air, soil, water and food (particularly in vegetables) and is produced naturally within the human body. In many cases,

groundwater nitrate concentrations are currently approaching or exceeding the recommended 11.3 mg NO_3-N L^{-1} drinking water standard (e.g., [12]). Excess nitrates (levels >50 mg L^{-1}; [13]) in the drinking water cause health risks such as conversion of hemoglobin to methemoglobin, which depletes oxygen levels in the blood. Forman et al. [14] reported additional consequences among people who consumed drinking water containing high levels of nitrates: enlargement of the thyroid gland, increased incidence of 15 types of cancer and two kinds of birth defects, and even hypertension. In addition, increasing rates of stomach cancer caused by increasing nitrate intake have been reported [15].

In Shanghai, nitrogen pollution has become an increasingly serious problem. Villages in the Shanghai city are the main areas for developing urban agriculture, which can provide the main source of vegetables and fruits for many residents. Due to extensive irrigation and fertilizer use, non-point source pollution is the dominant form, and the non-point source nitrogen loading has substantially affected groundwater nitrogen concentrations [16]. Poinke and Urban [17] showed that the average nitrogen concentration in rural groundwater was five to seven times higher than that in adjacent forest-covered areas. Where groundwater is the main source of drinking, domestic, and agricultural water, potentially significant health risks are associated with the consumption of nitrate-rich groundwater. For this reason, it is important to study the nitrogen pollution problem in rural-urban continuum near Shanghai to determine the impact on food safety and health of the residents. The aim of the present study was therefore to investigate seasonal changes in levels of nitrate and other forms of nitrogen, and based on this data, to assess the health risk for a typical village in Shanghai, thereby providing a scientific basis for controlling nitrogen pollution and protecting groundwater safety.

Results

Changes in different types of nitrogen in the groundwater of agricultural areas

Figure 1 summarizes the results of the groundwater monitoring for the four types of nitrogen for wells in agricultural areas. The total N concentration was higher from June to August than during other months. The total N concentration exceeded 20 mg L^{-1} from June to August (Fig. 1a). From December to February, the total N concentration in groundwater reached its lowest value. Because NO_3-N accounted for 60 to 80% of total N, the seasonal changes in NO_3-N were similar to those for total N (Fig. 1b). According to the classification in Table 1, the groundwater quality for most wells from June to August was Class IV, with values ranging between 20 and 30 mg L^{-1}, although some wells were rated Class V, with NO_3-N exceeding 30 mg L^{-1}; in the other months, the groundwater quality was rated Class III or worse.

The degree of NH_4-N pollution was an order of magnitude lower than the NO_3-N pollution. The groundwater quality based on NH_4-N level was rated as Class III in most months, except from June to July, when the quality degraded to Class IV (Fig. 1c). From September to May of the following year, the NH_4-N concentration was relatively stable, decreasing to <0.2 mg L^{-1}. The NO_2-N concentration in groundwater was being in minimum from December to February in agricultural areas, but the change of NO_2-N was mostly rated Class IV (Fig. 1d).

Table 1. Classification standard for groundwater quality in China based on nitrogen levels. [18]

	Groundwater quality class				
	I	II	III	IV	V
Concentration (mg L^{-1})					
NH_4-N	≤0.02	≤0.02	≤0.20	≤0.50	>0.50
NO_2-N	≤0.001	≤0.010	≤0.020	≤0.100	>0.100
NO_3-N	≤2	≤5	≤20	≤30	>30

Note: if NO_3-N is class IV, it means that the concentration of NO_3-N is between 20 to 30 mg L^{-1}.

Changes in different types of nitrogen in the groundwater of residential areas

Total N concentrations in residential areas exhibited more gradual seasonal changes than in agricultural areas. In both the rainy season and the dry season, the total N concentration was significantly lower in residential areas than in agricultural areas (Fig. 2a). Although the NO_3-N levels were lower in residential than agricultural areas, none of the wells met the criteria for Class I water and the wells only met the Class II standard in December (Fig. 2b). The residential NO_3-N concentrations ranged from 5 to 20 mg L^{-1}, and were therefore graded Class III. They showed a similar pattern of change to that in the agricultural areas.

The fluctuation in NH_4-N concentrations in residential areas was large, but there was no significant seasonal change in NH_4-N (Fig. 2c). The groundwater quality based on NH_4-N levels ranged from Class IV to Class III, which was similar to the range in

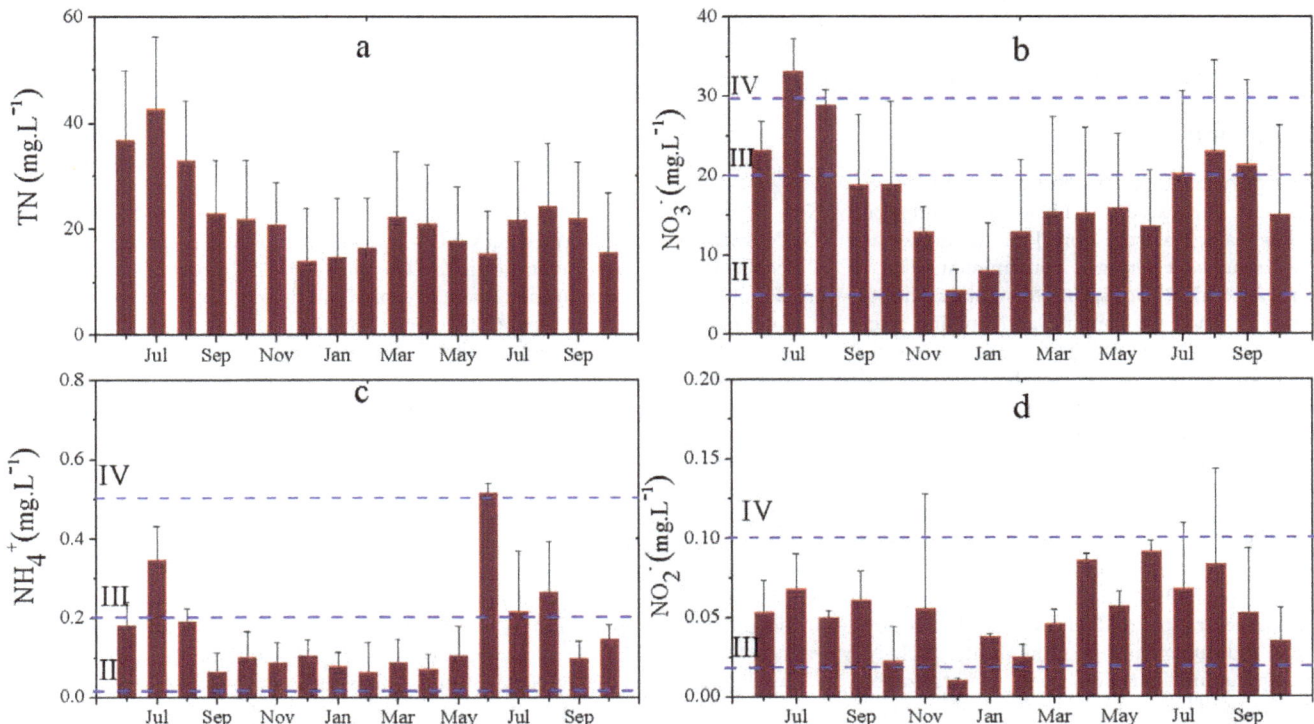

Figure 1. Seasonal changes in different types of nitrogen in the groundwater of agricultural areas of Xinchang village. Water quality grades are defined in Table 1. (a) total N; (b) nitrate nitrogen; (c) ammonia nitrogen; (d) nitrite nitrogen.

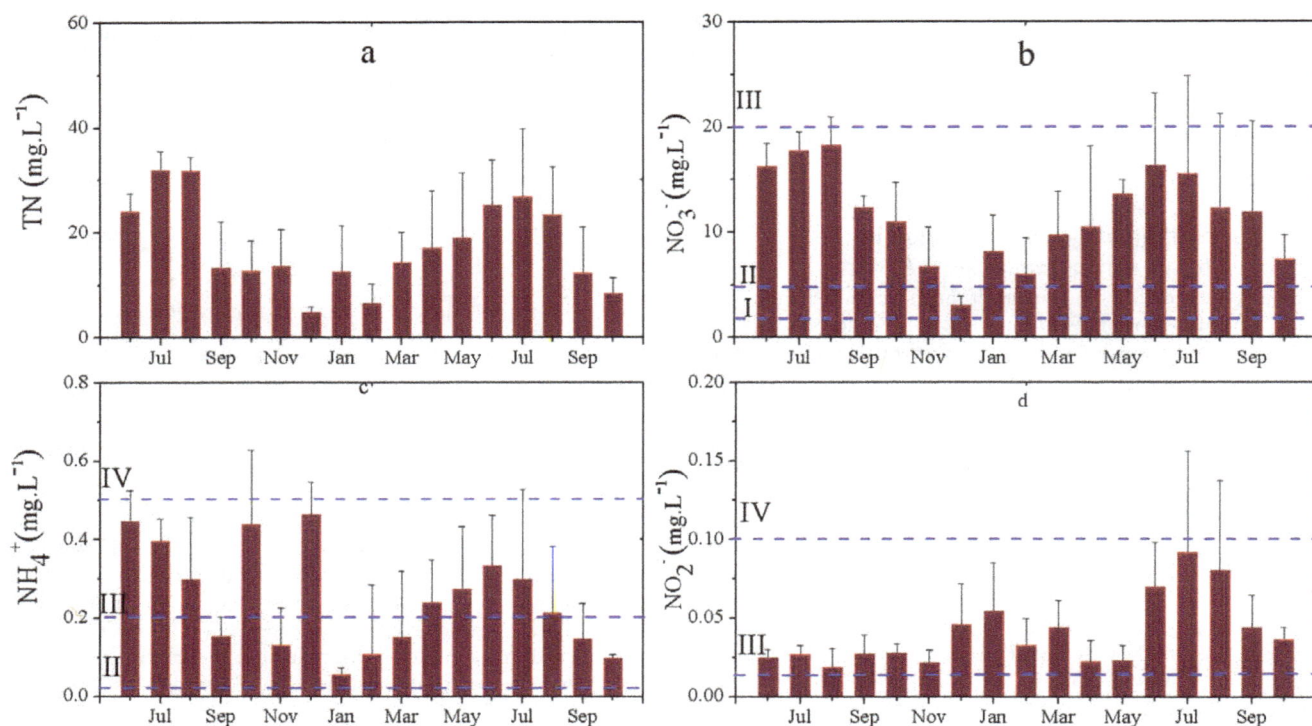

Figure 2. Seasonal changes in different types of nitrogen in the groundwater of residential areas of Xinchang village. Water quality grades are defined in Table 1. (a) total N; (b) nitrate nitrogen; (c) ammonia nitrogen; (d) nitrite nitrogen.

agricultural areas. The NO_2-N concentrations were lower than those in agricultural areas, but the groundwater quality based on this pollutant was still graded as Class III or Class IV throughout the year, with a large increase in June to August 2010 but no substantive differences during the rest of the study period (Fig. 2d).

Discussion

Nitrogen pollution in the groundwater

In the study area, the highest concentration of NO_3-N in the groundwater occurred from June to August, and NO_3-N was the most significant nitrogen contaminant. The NO_3-N concentrations were high from spring to summer and low from autumn to winter. During the jointing and booting stages of *Prunus persica* development, in March and April, most soil NO_3-N would be taken up by the trees and by other vegetables. As the rainy season began after May in the Shanghai region, the total N and NO_3-N levels in groundwater rapidly increased. From June to August, the NO_3-N concentration in the groundwater of Xinchang was close to the limit prescribed by the World Health Organization [13]. Because NO_3-N in solution is not adsorbed by soils, but NO_3-N can easily be absorbed by some tropical soils and leach into the subsurface soil and groundwater [1,5,8].The changes in NH_4-N levels were similar in agricultural and residential areas. However, the concentration of NH_4-N was slightly lower in agricultural areas, indicating that the groundwater NH_4-N was affected both by agricultural practices such as fertilization and by human habitation. The peak values of NH_4-N content in the study area appeared from May to September, during the period of tree and vegetable growth and fertilization. Other inputs may come from agricultural production and domestic wastewater. Because nitrogenous fertilizers are applied to the soil, some of the NH_4-N, which

is a reactant for denitrification, can be transformed into NO_3-N through nitrification, and some is lost as a result of denitrification to produce volatile nitrogen gas [11].

Although the concentrations of NO_2-N were considerably lower than those of NO_3-N in the groundwater, the impact of pollution by NO_2-N was worse according to the Chinese groundwater quality criteria [18]; concentrations of NO_2-N in the groundwater throughout the study area exceeded the Class III water standard, whereas NO_3-N levels occasionally approached the Class II standard. NO_2-N is not stable in water or soil, and can easily be transformed into NO_3-N or into nitrogen gas through oxidation and denitrification. Therefore, the fluctuations of NO_2-N concentrations were irregular and did not appear to be associated with seasonal changes as a result of impact factors such as changes in fertilization, rainfall, and temperature.

Effects of rainfall and land use on nitrogen pollution

We found that land use patterns (here, residential vs. agricultural use) significantly affected NO_3-N concentrations in the groundwater. Enhanced agricultural activity is often accompanied by increased incorporation of organic matter into the soil. Nitrogen compounds in the fertilizer and organic matter are transported into the groundwater by percolating water from rainfall or from irrigation [3]. Hence, the nitrogen concentrations are typically high in agricultural areas [19–21]. Another reason for this phenomenon may be that in agricultural areas, the aquifer is typically shallow, and because it is relatively close to the surface, it receives direct inputs of NO_3-rich leachate from the agricultural soils. In residential areas, the nitrogen pollution was also serious, with levels close to those in agricultural areas. This can be explained by the high nitrogen content in groundwater around livestock and feedlot areas as well as near residences with septic tanks. Komar and Anderson [22] investigated the different nitrogen sources in a rural environment using nitrogen isotopes and obtained similar results to those in our study. Another reason for our observed results may be that the aquifers in the agricultural and residential areas are close to each other, so that leaching may transport pollutants between them; as a result, the magnitude of the difference in nitrogen contents in the groundwater would decrease.

The nitrogen concentrations in groundwater are affected by both rainfall and irrigation intensity [3], so we calculate the relationship between rainfall and the nitrogen concentration in groundwater (Fig. 3). The nitrogen concentrations in groundwater differed greatly between the rainy and dry seasons. The total N and NO_3-N in the groundwater were significantly correlated with rainfall in both agricultural and residential areas, but the correlations between rainfall and NH_4-N and NO_2-N concentrations was much weaker but still significant. This can be explained by the fact that the abundant rainfall in the study area is the most important impact factor responsible for nitrogen transport through subsurface runoff into the groundwater, and by the fact that NO_3-N accounted for 60 to 80% of total N. Soil nitrogen moves easily in water, especially during the first flush, when the runoff volume is high; internal and lateral solute movement in soils carried away nitrates even more intensively than surface runoff [23]. Zhu and Wen [24] showed that NH_4-N is strongly absorbed by soil particles and is more resistant to being detached or dissolved and transported by runoff waters; this is because the ammonium ion has a positive charge and can therefore be adsorbed to cation-exchange sites on soil particles. NH_4-N is easily oxidized or lost to denitrification, and NO_2-N is not stable in water or soil. The other

reason may be that soil pH in this area is 8.2, which easily affect NH_4-N and NO_2-N transformation via equilibrium. Therefore, NH_4-N and NO_2-N did not show a strong correlation with rainfall. The correlation between different types of nitrogen and rainfall was higher in agricultural areas than in residential areas. This is likely because in agricultural areas, groundwater nitrogen pollution was strongly influenced by agricultural activities such as irrigation and fertilization; human activities in residential area have less seasonal correlation than do activities in agricultural areas.

Health risk assessment

Based on data for Shanghai from 2009 to 2010, the main parameters used in our health risk assessment had the following values: $IR = 2$ L d^{-1}, $ED = 30$ years, $EF = 365$ d year^{-1}, $BW = 70$ kg, $AT = 70$ years, $Asd = 16\ 600$ cm^2, $FE = 0.5$ times d^{-1}, $f = 1$, $k = 1$ cm h^{-1}, $t = 1$ h, and $TE = 0.4$ h. The Rfd values for NO_3-N, NO_2-N, and NH_4-N were 34, 1.6, and 0.1, respectively [25]. Potential noncarcinogenic risks for exposure to contaminants of potential concern were evaluated by comparison of the estimated contaminant intakes from each exposure route (oral, dermal, inhalation) with the RfD. The HQ assumes that there is a level of exposure (i.e., RfD) below which it is unlikely for even sensitive populations to experience adverse health effects. There may be a concern arising for the potential noncarcinogenic effects if the HQ exceeds 1×10^{-6} (unity).

Figures 4 and 5 present the noncarcinogenic risk values for dermal and oral exposures to different type of nitrogen, respectively. Drinking and contact were assumed to be the main exposure routes of humans to nitrogen pollution in our risk

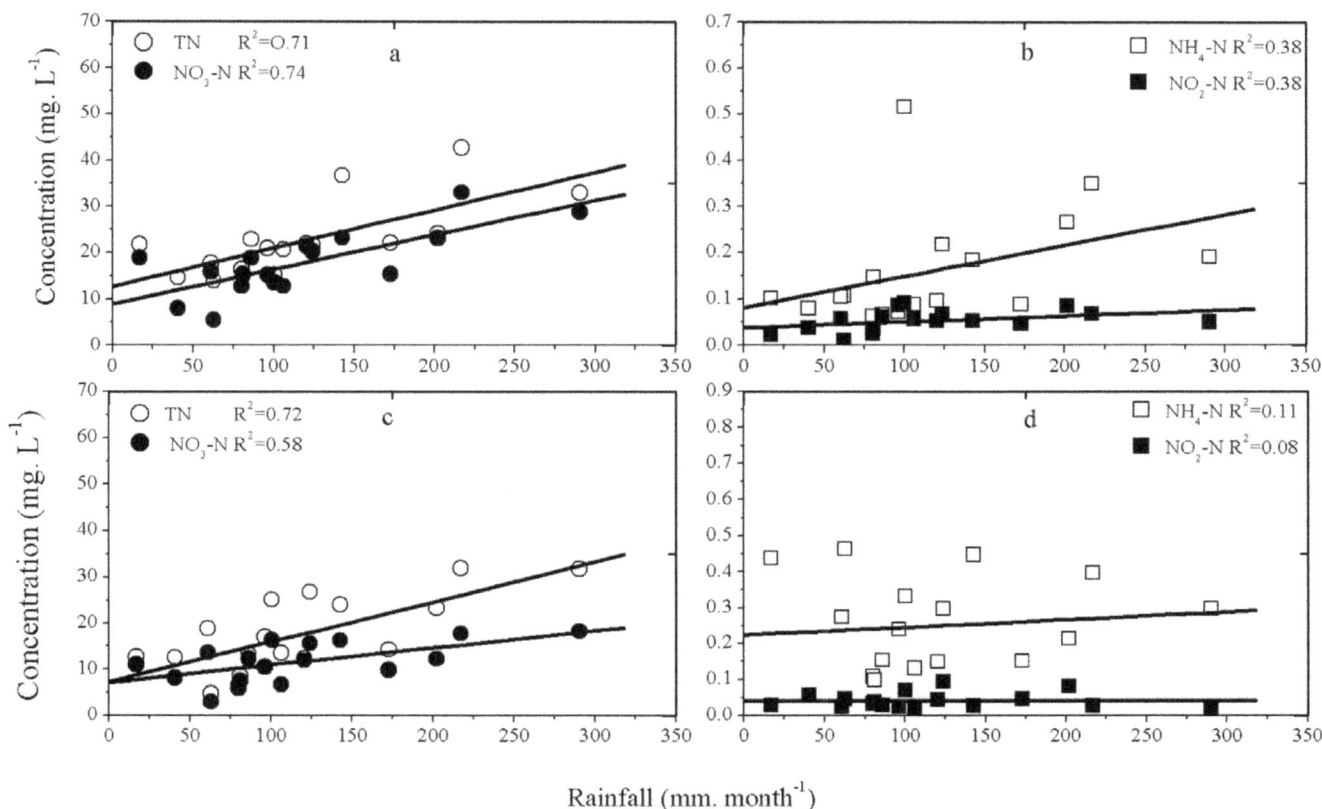

Figure 3. Relationships between the four types of nitrogen and rainfall. (a, b) agricultural areas; (c, d) residential areas.

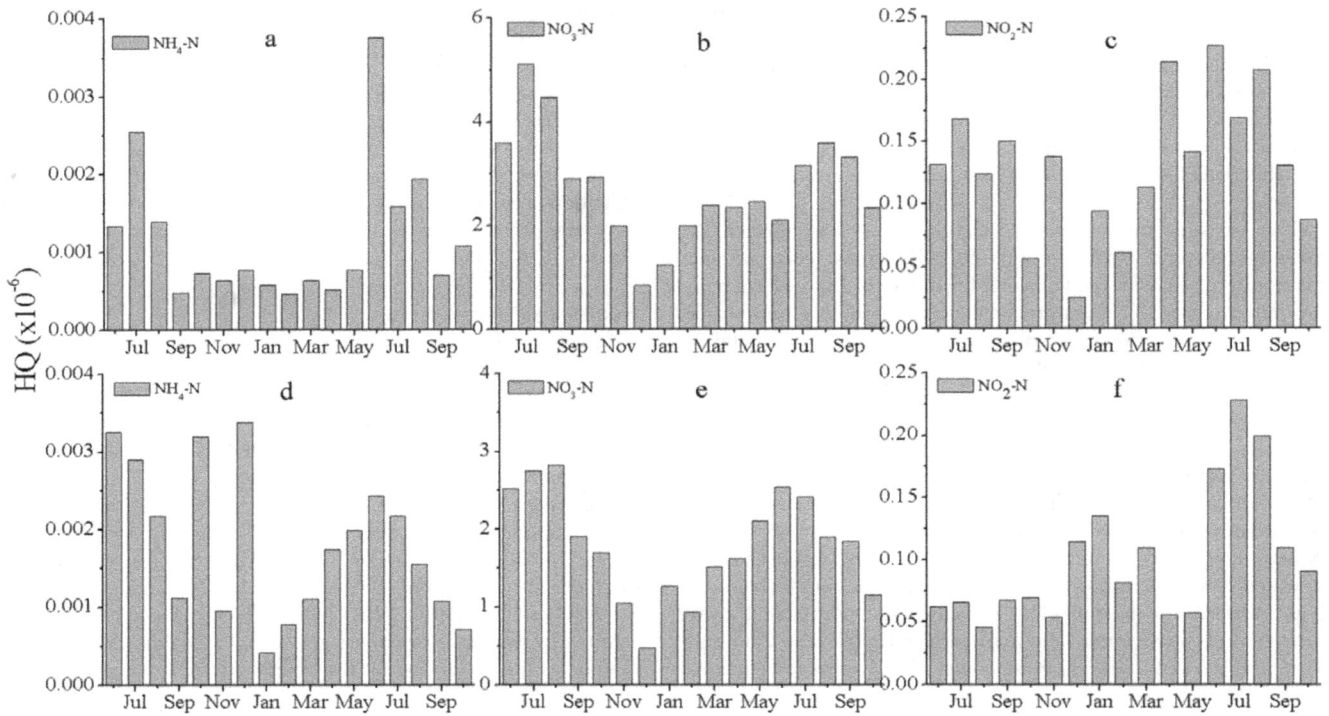

Figure 4. Noncarcinogenic dermal risk values for different types of nitrogen in the groundwater. (a, b, c) agricultural areas; (d, e, f) residential areas.

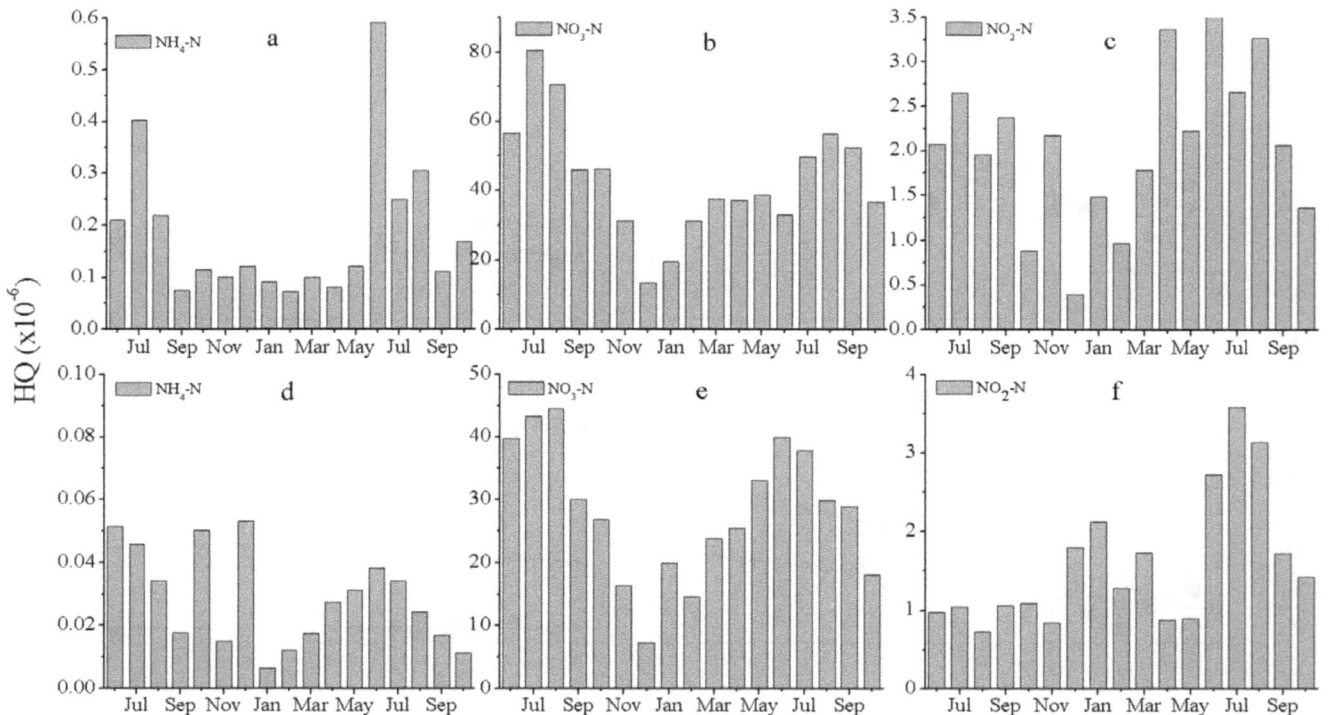

Figure 5. Noncarcinogenic oral risk values for different types of nitrogen in the groundwater. (a, b, c) agricultural areas; (d, e, f) residential areas.

assessment; we did not include inhalation as a source of exposure, which may mean that our risk estimates slightly underestimate the actual risk. The relative research showed that the levels of noncarcinogenic oral risk, toxic risk (HQ), ranged from 0.02 to 0.12×10^{-6} [26]. The noncarcinogenic dermal risks due to NO_3-N and NO_2-N showed seasonal changes, ranging from 0.8×10^{-6} to 3.5×10^{-6} (3.0 to 23.1 mg L^{-1}) and from 0.05×10^{-6} to 0.22×10^{-6} (0.01 to 0.35 mg L^{-1}), respectively. The HQ of the four types of nitrogen decreased in the following order: NO_3-N > NO_2-N > NH_4-N. NH_4-N represented the lowest noncarcinogenic dermal risk (Fig. 4). Noncarcinogenic dermal risk values were lower in residential areas than in agricultural areas.

The noncarcinogenic oral risk was two orders of magnitude higher than the noncarcinogenic dermal risk (Fig. 5). The levels of noncarcinogenic oral risk (HQ) ranged from 22×10^{-6} to 85×10^{-6}. NO_3-N posed the greatest risk, with HQ ranging from 19×10^{-6} to 80×10^{-6} (3.0 to 23.1 mg L^{-1}); this accounted for more than 90% of the total risk in the study area. Therefore, NO_3-N poses the greatest risk to human health. Daily intake of or contact with water by the local residents poses a potential health threat due to the cumulative impacts of long-term NO_3-N exposure. The difference between the dermal and oral values indicates that ingestion of water is a more critical exposure route for NO_3-N and NO_2-N. The spatial variation in oral risk was similar to the spatial patterns of the dermal risk. The noncarcinogenic oral risk values were lower in residential areas than in agricultural areas. Local farmers irrigated their crops using groundwater, which would enhance the noncarcinogenic risk for residents in Shanghai. Government statistics reported one case of cancer for every 100 woman living in Shanghai, which was the highest cancer incidence in any Chinese city, and the number of cases of cancer was double time in 2010 than that in 1960s [27].

Perspective

In China, there are many villages like Xinchang that are located in or near a big city, and which provide the main water resource and a supply of food and vegetables for city residents. Groundwater is the major source of water for drinking and for the domestic, industrial, and agricultural sectors in the Shanghai region. Therefore, protecting groundwater in this region has important implications for both food safety and human health. Because crops and animals take up nitrates from the soil and water, it will be important to quantify the nitrate contents of these foods and the quantities that are consumed in future research to determine how much this form of exposure increases the health risk to residents of Shanghai. In this study, our monitoring of the level of groundwater nitrogen pollution and our health risk assessment based on this data for Xinchang revealed that nitrogen pollution was a serious problem. Many wells exceeded the groundwater quality standard for human consumption for all forms of nitrogen, and particularly for nitrate, and higher levels of nitrogen contamination were significantly correlated with agricultural activity, human activity, and rainfall, especially in agricultural areas. The factors responsible for nitrogen pollution would be more complex in residential areas than in agricultural areas because of the greater diversity of activities. NO_3-N was the main form of nitrogen pollution of the groundwater and poses the greatest risk to the health of local residents. Long-term drinking of groundwater and irrigation using groundwater therefore pose a significant health risk for Shanghai's residents. Therefore, it is urgent to devise policy guidelines for efficient management of both the surface water and groundwater resources in this region to enhance groundwater recharge and minimize the pollution levels in both types of water to permit their safe use.

Materials and Methods

Study area

The study area was Xinchang village in the Nanhui District of Shanghai (31°03′N, 121°39′E), which is located in a typical alluvial plain of the region. The village covers an area of about 3564 ha and has an altitude ranging from 2 to 3 m asl. The farming, livestock, and agriculture in this area are well developed. The main type of land use in Xinchang is planting of peach trees (*Prunus persica*) and vegetables, which account for more than 50% of the area. The residential population is around 1000 people (Fig. 6). The village's water system belongs to the Huangpu River watershed, where there are number of crossed rivers with abundant fresh water resources. The main streams of Huangpu River include Huixin, Dazhi and Fengxin River. Groundwater storage condition in this area depends on the pore water of the unconsolidated rock, so the groundwater complement source in Xinchang village is abundant. The climate is a subtropical marine monsoon climate, with average annual rainfall of 1175 mm, an average annual temperature of 16.7°C, and 1932 h of annual sunshine. The rainfall variation is large, and 70% of the rain falls from June to August.

The soil type is yellow clay in FAO Soil Classification [28]. The soil properties are a pH of 8.2±0.2, a bulk density of 1.2±0.2 g cm^{-3}, a capillary porosity of 30.5±2.7%, a non-capillary porosity of 10.8±2.0%, a water content at field capacity of 13.6±1.0%, a total P of 0.8±0.2 g kg^{-1}, an available P of 44.7±5.9 mg kg^{-1}, an organic matter content of 20.4±0.7 g kg^{-1}, and a total N of 106.6±4.4 mg kg^{-1} (n = 29). The local combined annual application of inorganic fertilizer equals 0.018 kg N m^{-2} plus 0.011 kg P_2O_5 m^{-2}.

Water samples

There are many wells in Xinchang, and the depth to groundwater level of this area generally ranges from 1.2 to 1.5 m. We set groundwater sample point according to the land use of the village and the principle of uniform distributed points through GPS positioning, and then record the latitude, longitude and water level information for different well. We sampled water from 17 wells (red dot) surrounded by agricultural areas and 12 wells (blue dot) in residential areas (Fig. 1). The water from wells with red dot is mainly used in daily need for local resident in agricultural areas and crop irrigation, whereas the water from well with blue dot mainly meet daily need from resident in residential areas. The domestic wastewater in residential area contains large amounts of organic nitrogen, wherein part of the domestic wastewater can directly leak into groundwater from sewer, and then cause groundwater nitrogen pollution [29]. Water samples were collected from June 2009 to October 2010 at 1-month intervals period. Groundwater samples (each 500 mL in size) were collected from pumps connected to the wells. Rainfall data were obtained from the local meteorological station in Nanhui District. All data is reported as means ± S.D.

Analytical method

All water samples were passed through glass-fiber disks with a 0.70-mm pore size before analysis. To calculate total N, the water was digested in concentrated sulfuric acid using a $CuSO_4/Na_2SO_4$ mixture as a catalyst, followed by distillation of the resulting NH_4^+ into dilute boric acid and titration against a standardized 0.0025 M H_2SO_4 solution, as described by Rayment and Higginson [30]. NO_3-N was determined using an ultraviolet spectrometer [31], NO_2-N was determined by means of diazo-coupling colorimetry [11], and NH_4^+-N was determined colorimetrically using the indophenol blue method [30].

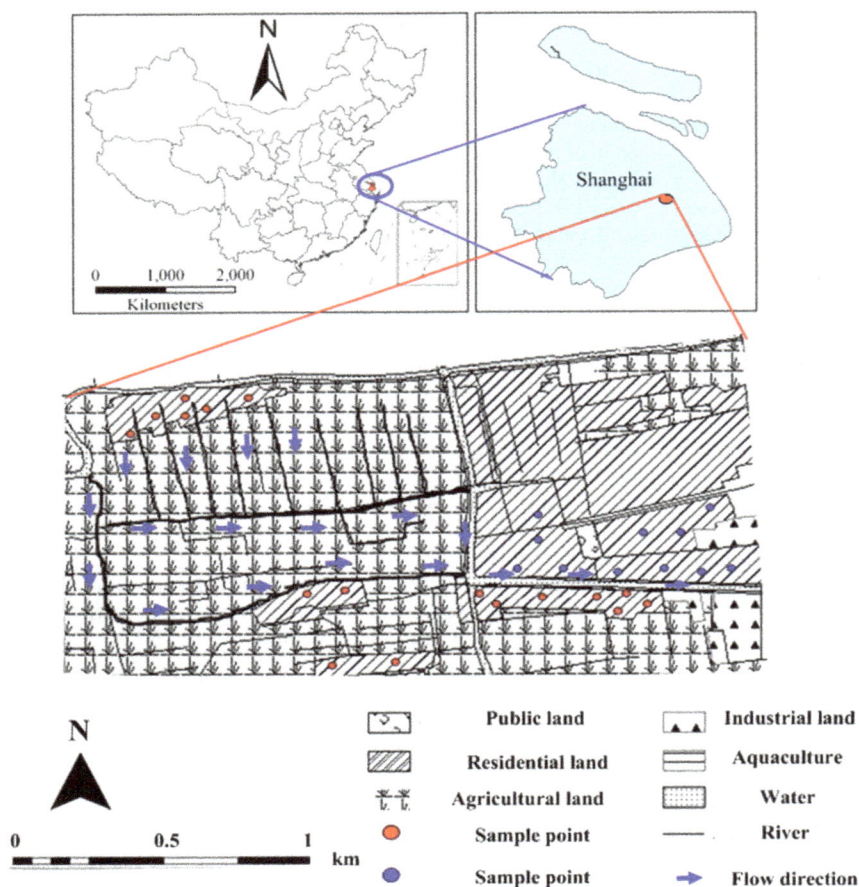

Figure 6. Location of the study area and sample points. 17 wells with red dot are in agricultural areas and 12 wells with blue dot are in residential areas.

Groundwater quality and health risk assessment

We selected NO_3-N, NO_2-N, and NH_4-N as the assessment index for groundwater nitrogen pollution. The national groundwater quality standard for nitrogen pollutants is presented in Table 1 [18]. Risk assessment is defined as the processes of estimating the probability of occurrence of an event and the probable magnitude of adverse health effects over a specified time period [32]. Human health risk assessment consists of four stages: (1) hazard identification, (2) toxicity (dose–response) assessment, (3) exposure assessment, and (4) risk characterization.

The estimated uptake of a potential toxin by the human body through contact with a contaminant is estimated using the chronic daily intake (*CDI*). The *CDI* value indicates the quantity of chemical substance ingested, inhaled, or absorbed through the skin per kilogram of body weight per day (mg kg^{-1} day^{-1}). The formulas for calculating intake are as follows:

Ingestion:

$$CDI_i = (C \times IR \times ED \times EF)/(BW \times AT) \quad (1)$$

Dermal contact:

$$CDI_i = (C \times Asd \times EF \times FE \times ED)/(BW \times AT \times f) \quad (2)$$

$$I = (2 \times 10^{-3} \times k \times C \times 6 \times t \times TE)/\pi \quad (3)$$

where i represents a specific pollutant, C is that pollutant's concentration in water (mg L^{-1}), *IR* is the drinking rate (L d^{-1}), *ED* is the exposure duration (years), *EF* is the exposure frequency (d year^{-1}), *BW* is the average body weight (kg), *AT* is the average lifespan (years), *Asd* is the human body's surface area (cm^2), *FE* is the bathing frequency (number of times d^{-1}), f is the intestinal absorption rate (unitless, $= 1$), I is pollutant adsorption by the skin when bathing (mg cm^{-2} time^{-1}), k is the adsorption parameter for the skin (cm h^{-1}), t is the lag time (h), and *TE* is the bathing time (h).

Noncarcinogenic risks

We separately characterized the risk for carcinogenic and noncarcinogenic effects, and have discussed the factors that may result in either overestimation or underestimation of the risks for the residents of Xinchang. Potential noncarcinogenic risks for exposure to contaminants were evaluated by comparison of the estimated contaminant intakes from each exposure route (oral and dermal) with the reference dose (RfD, (mg kg^{-1} day^{-1}) to produce the hazard quotient (HQ, unitless), which is defined as follows [25]:

$$HQ = CDI \times 10^{-6} / Rfd \tag{4}$$

where HQ is hazard quotient (unitless); RfD is reference dose (mg. kg^{-1} day^{-1}).

Carcinogenic risks

Carcinogenic risks were estimated as the incremental probability of an individual developing cancer over a lifetime as a result of exposure to a potential carcinogen. To do so, we used the following linear low-dose carcinogenic risk equation for each exposure route [25]:

$$CA = CDI \times slope\ factor \tag{5}$$

where CA is the carcinogenic risk and "*slope factor*" is mg kg^{-1}

day^{-1}. Slope factor can be obtained from Risk Assessment Information System [33]. If a site has multiple carcinogenic contaminants, cancer risks for each carcinogen and each exposure route can be added (based on the assumption of additivity of effects) and compared with the accepted risk.

Acknowledgments

We thank Dr.Yafeng Wang in Research Center for Eco-Environmental Sciences, Dr. Liang Mao in Shanghai Jiaotong University, and Geoffrey Hart (Montréal, Canada) for his help in writing this paper. The authors would also like to thank the anonymous reviewers for their helpful remarks.

Author Contributions

Conceived and designed the experiments: YG. Performed the experiments: CL. Analyzed the data: YG. Contributed reagents/materials/analysis tools: PZ GY. Wrote the paper: YG GY.

References

1. Zhu B, Wang T, You X, Gao MR (2008) Nutrient release from weathering of purplish rocks in the Sichuan Basin, China. Pedosphere 18(2): 257–264.
2. Gao Y, Zhu B, Zhou P, Tang JL, Wang T, et al. (2009) Effects of vegetation cover on phosphorus loss from a hillslope cropland of purple soil under simulated rainfall: a case study in China. Nutrient Cycling in Agroecosystems 85: 263–273.
3. Naik PK, Tambe JA, Dehury BN, Tiwari AN (2008) Impact of urbanization on the groundwater regime in a fast growing city in central India. Environmental Monitoring and Assessment 146: 339–373.
4. Gao Y, Zhu B, Wang T, Wang YF (2012) Seasonal change of non-point source pollution-induced bioavailable phosphorus loss: a case study of Southwestern China. Journal of Hydrology 420–421: 373–379.
5. Zhu B, Wang T, Kuang FH, Luo ZX, Tang JL, et al. (2009) Measurements of nitrate leaching from a hillslope cropland in the central Sichuan Basin, China. Soil Science Society of America Journal 73(4): 1419–1426.
6. Akhavan S, Abedi-Koupai J, Mousavi SF, Afyuni M, Eslamian SS, et al. (2010) Application of SWAT model to investigate nitrate leaching in Hamadan–Bahar Watershed, Iran. Agriculture Ecosystems and Environment 139: 675–688.
7. Gao Y, Zhu B, Wang T, Tang JL, Zhou P, et al. (2010) Bioavailable phosphorus transport from a hillslope cropland of purple soil under natural and simulated rainfall. Environmental Monitoring and Assessment 171: 539–550.
8. Chen SF, Wu WL, Hu KL, Li W (2010) The effects of land use change and irrigation water resource on nitrate contamination in shallow groundwater at county scale. Ecological Complexity 7: 131–138.
9. Adhikary PP, Chandrasekharan H, Chakraborty D, Kamble K (2010) Assessment of groundwater pollution in West Delhi, India using geostatistical approach. Environmental Monitoring and Assessment 167: 599–615.
10. Hu KL, Huang YF, Li H, Li BG, Chen DL, et al. (2005) Spatial variability of shallow groundwater level, electrical conductivity and nitrate concentration, and risk assessment of nitrate contamination in North China Plain. Environment International 31: 896–903.
11. Chen XM, Wo F, Chen C, Fang K (2010) Seasonal changes in the concentrations of nitrogen and phosphorus in farmland drainage and groundwater of the Taihu Lake region of China. Environmental Monitoring and Assessment 169: 159–168.
12. Jackson BM, Browne CA, Butler AP, Peach D, Wade AJ, et al. (2008) Nitrate transport in chalk catchments: monitoring, modelling and policy implications. Environmental Science and Policy 11: 125–135.
13. WHO (2008) Guidelines for drinking-water quality- Third Edition Incorporating the first and second addenda. Volume 1. Recommendations. World Health Organization, Geneva website. Available: http://www.who.int/water_sanitation_health/dwq/GDWPRecomdrev1and2.pdf. Accessed 10 Sept 2010.
14. Forman D, Al-Dabbagh S, Doll R (1985) Nitrates, nitrites and gastric cancer in Great Britain. Nature 313: 620–625.
15. Payne MR (1993) Farm waste and nitrate pollution. In: Jones JG, ed. Agriculture and the environment 63–73. Horwood, New York.
16. Huang HB, Gao Y, Cao JJ, Huang HY, Zhang X, et al. (2010) Nonpoint source pollution of nitrogen in groundwater and risk assessment in urban agricultural region of Shanghai. Journal of Soil and Water Conservation 24(3): 56–70. (In Chinese).
17. Poinke HB, Urban JB (1985) Effect of agricultural land-use on groundwater quality in a small Pennsylvania watershed. Ground Water 23: 68–80.
18. State Environmental Protection Administration of China (1994) Chinese groundwater quality criteria of classification, (GB/T14848-93), (pp. 23–32). Beijing: China Environmental Science Press (in Chinese).
19. Vidal M, Melgar J, Opez AL, Santoalla MC (2000) Spatial and temporal hydrochemical changes in groundwater under the contaminating effects of fertilizers and wastewater. Journal of Environmental Management 60: 215–225.
20. Cepuder P, Shukla MK (2002) Groundwater nitrate in Austria: A case study in Tullnerfeld. Nutrient Cycling in Agroecosystems 64: 301–315.
21. Reddy AGS, Kumar KN, Rao DS, Rao SS (2009) Assessment of nitrate contamination due to groundwater pollution in north eastern part of Anantapur District, A.P. India. Environmental Monitoring and Assessment 148: 463–476.
22. Komar SC, Anderson HW (1993) Nitrogen isotopes as indicators of nitrate source in Minnesota sand plain aquifers. Groundwater 31(2): 250–270.
23. Yang JL, Zhang GL, Shi XZ, Wang HJ, Cao ZH, et al. (2009) Dynamic changes of nitrogen and phosphorus losses in ephemeral runoff processes by typical storm events in Sichuan Basin, Southwest China. Soil & Tillage Research 105: 292–299.
24. Zhu ZL, Wen QX (1992) Nitrogen of Chinese Soil. Jiangsu Science and Technology Press, Nanjing.
25. USEPA (1989) Risk assessment guidance for superfund, Vol. I, human health evaluation manual. Part A: (interim final), EPA/540/1–89/002. Washington, DC: Office of Emergency and Remedial Response, U.S. Environmental Protection Agency.
26. Li YL, Liu JL, Cao ZG, Lin C, Yang ZF (2010) Spatial distribution and health risk of heavy metals and polycyclic aromatic hydrocarbons (PAHs) in the water of the Luanhe River Basin, China. Environmental Monitoring and Assessment 163: 1–13.
27. Shanghai Municipal Center for Disease Control & Prevention (2010) website. Available: http://www.scdc.sh.cn/. Accessed 10 Sept 2010. (in Chinese).
28. Gong ZT Chinese Soil Taxonomy, Science Press, Beijing. (In Chinese).
29. Cao JJ, Gao Y, Huang HB, Huang HY, Mao L, Zhang X, et al. (2010) Output characteristics of non-point nitrogen from a typical village region in Yangtze Delta under an individual rainfall event. Environmental Science 31(11): 2587–2593. (In Chinese).
30. Rayment GE, Higginson FR (1992) Australian laboratory handbook of soil and water chemical methods. Inkata Press, Sydney.
31. Committee of Analytical Method of Water and Wastewater (1989) Analytical Method of Water and Wastewater. China Environmental Science Press, Beijing. (In Chinese).
32. Kolluru RV, Bartell SM, Pitblado RM, Stricoff RS (1996) Risk assessment and management handbook. McGraw-Hill, New York.
33. USEPA Risk Assessment Information System [EB/OL] website. Available: http://rais. ornl. gov/cgi-bin /tox / TOX- select ? select = nrad. Accessed 10 Sept 2010.

5

Effects of Tillage and Nitrogen Fertilizers on CH$_4$ and CO$_2$ Emissions and Soil Organic Carbon in Paddy Fields of Central China

Li Cheng-Fang[1]◑, Zhou Dan-Na[2]◑, Kou Zhi-Kui[1], Zhang Zhi-Sheng[1], Wang Jin-Ping[1], Cai Ming-Li[1], Cao Cou-Gui[1]*

1 College of Plant Science and Technology, Huazhong Agricultural University, Wuhan, Hubei, China, **2** Institute of Animal Husbandry and Veterinary Science, Hubei Academy of Aguicultural Sciences, Wuhan, Hubei, China

Abstract

Quantifying carbon (C) sequestration in paddy soils is necessary to help better understand the effect of agricultural practices on the C cycle. The objective of the present study was to assess the effects of tillage practices [conventional tillage (CT) and no-tillage (NT)] and the application of nitrogen (N) fertilizer (0 and 210 kg N ha^{-1}) on fluxes of CH$_4$ and CO$_2$, and soil organic C (SOC) sequestration during the 2009 and 2010 rice growing seasons in central China. Application of N fertilizer significantly increased CH$_4$ emissions by 13%–66% and SOC by 21%–94% irrespective of soil sampling depths, but had no effect on CO$_2$ emissions in either year. Tillage significantly affected CH$_4$ and CO$_2$ emissions, where NT significantly decreased CH$_4$ emissions by 10%–36% but increased CO$_2$ emissions by 22%–40% in both years. The effects of tillage on the SOC varied with the depth of soil sampling. NT significantly increased the SOC by 7%–48% in the 0–5 cm layer compared with CT. However, there was no significant difference in the SOC between NT and CT across the entire 0–20 cm layer. Hence, our results suggest that the potential of SOC sequestration in NT paddy fields may be overestimated in central China if only surface soil samples are considered.

Editor: Kurt O. Reinhart, USDA-ARS, United States of America

Funding: The study was supported by the National Technology Project for High Food Yield of China (2011BAD16B02), National Natural Science Foundation of China (31100319), Fundamental Research Funds for the Central Universities (2010QC032) and the Foundation of Hubei Key Laboratory of Animal Embryo Engineering and Molecular Breeding (2011ZD152). The funders had no role in study design, data collection and analysis, decision to publish, or preparation of the manuscript.

Competing Interests: The authors have declared that no competing interests exist.

* E-mail: ccgui@mail.hzau.edu.cn

◑ These authors contributed equally to this work.

Introduction

Global surface temperatures have increased by 0.88°C since the late nineteenth century [1]. The observed climate changes are caused by the emission of greenhouse gases (GHGs) mainly through anthropogenic activities. Methane and CO$_2$ are the most important GHGs, respectively contributing 15% and 60% to the anthropogenic GHG effect [2]. Rice paddies are an important source of atmospheric CH$_4$. The amount of CH$_4$ emitted from wetland paddy fields accounts for 10% to 20% of the total CH$_4$ emissions (i.e. 50 Tg yr^{-1} to 100 Tg yr^{-1}) [2]. The rice production of China exceeds that of any other country, accounting for 30% of the world total [3]. Agricultural activity affects CH$_4$ and CO$_2$ emissions, contributing 39% of the excess CH$_4$ and 1% of the excess CO$_2$ to global emissions [4]. Hence, CH$_4$ emissions from paddy fields under different agricultural management practices in China are relevant to the discussion of the global C cycle and climate changes.

The entire process of CH$_4$ emission from rice fields, including production, oxidation, and transport into the atmosphere is influenced by agricultural management practices, such as tillage and N fertilizer use [5–7]. Tillage affects a range of biological,

chemical, and physical properties, thereby affecting the release of CH$_4$ [8]. No-tillage (NT) has been reported to reduce CH$_4$ emissions from paddy soils because rice straw is placed on the soil surface under NT and the soil conditions are more oxidative than those of conventional tillage (CT) [7,9]. CH$_4$ emissions from paddy fields are reportedly affected by the form and amount of N fertilizer applied [10]. Overall, the effects of N fertilizer application on CH$_4$ fluxes from paddy fields are mostly unclear. Therefore, more research on the effects of N addition on CH$_4$ emissions is needed.

Tillage practices can affect soil biochemical and physical properties, consequently influencing the release of CO$_2$ [8]. However, there is no consensus on the differences in the soil CO$_2$ emissions between NT- and CT-treated paddy fields. Some authors have reported similar soil CO$_2$ fluxes from NT- and CT-treated paddy fields [7]. However, Liang et al. [9] reported higher soil CO$_2$ emissions from CT-treated paddy fields than from the NT paddy fields. Nitrogen supplied by commercial fertilizers can be expected to affect soil CO$_2$ flux by increasing the C input from enhanced plant productivity and crop residues returned to the soil [11]. However, studies on the effects of N fertilizer on soil CO$_2$ emissions reveal diverse results [12]. Within the past few years, Iqbal et al. [13] and Xiao et al. [14] observed increased CO_2 emissions from paddy soils because of a positive effect of N fertilization on plant biomass. However, Burton et al. [15] and DeForest et al. [16] found that the use of N reduced extracellular enzymatic activities and fungal populations, resulting in decreased soil CO_2 flux. The effect of N fertilization on variation in CO_2 emission under anaerobic conditions in paddy soils remains unknown.

Land management practices are increasingly thought to affect soil carbon levels and may partially ameliorate CO_2 emissions and climate change [17,18]. Studies have indicated that NT can increase C sequestration in paddy soils compared with CT [19–21]. In 2007, Tang et al. [20] indicated that the NT could sequester 112.3 kg C ha^{-1} yr^{-1} in the top 20 cm of purple paddy soil in the Beipei district of Chongqing City, China. In a 12-year study, Gao et al. [21] reported that NT could sequester 26.68 kg C ha^{-1} yr^{-1} in gray fluvoaguic paddy soils to a depth of 30 cm in Zhangjiagang City, Jiangsu Province, China. However, Six et al. [22] and Su [23] indicated that the effects of NT on SOC sequestration depend on the soil type. In a 5-year study, He et al. [24] indicated that NT did not increase the SOC sequestration of paddy fields in the 20 cm layer of sandy silty loam in Ningxiang country, Hunan Province. However, Angers and Eriksen-Hamel [25] reviewed the related literature and concluded that soil variables do not affect the tillage effects on soil C sequestration. Hence, further research is needed to clarify the effects of soil type on C sequestration in NT-treated soils.

No-tillage may influence SOC accumulation when soil surface layers are considered, but the effect may not be detected more deeply [22]. The influence of NT on SOC sequestration is still unclear. Hence, Baker et al. [26] analyzed sampling strategies on the potential of SOC sequestration under conservation tillage and indicated that SOC sequestration under this tillage varied with soil depth. Thus, shallow sampling may not be sufficient to assess the differences in SOC sequestration between NT- and CT-treated soils, and further research on the effects of deeper soil sampling on SOC sequestration in NT-treated soils should be performed.

Application of N fertilizer may play a significant role in the soil C sequestration [17]. Application of N fertilizer affects the soil C stock in two ways. These compounds can increase the crop biomass and influence the microbial decomposition of crop residues by affecting the N availability [27]. However, a meta-analysis of 111 studies covering 12 soil types of divergent ecosystems indicated that the effects of N fertilizer application on soil C content vary with the soil type although N fertilizer application consistently increases the crop biomass [28]. For example, Tong et al. [29] found in a 17-year study published in 2009 that the use of chemical N fertilizers did not increase the SOC content in a hydromorphic paddy soil in Hunan Province compared with no fertilizer use. By contrast, Shang et al. [30] found in the same province that increased N fertilization increased the SOC sequestration in paddy soils derived from quaternary red clay.

Central China is one of the major rice-producing regions in the country, comprising 28% of the total area cultivated with rice in China [31]. Recently, NT practices have become increasingly popular in this region. However, to our knowledge, relatively few studies have been performed on the effects of tillage and N fertilizer on CH_4 and CO_2 emissions as well as on SOC sequestration in the paddy fields in this region. We hypothesized that tillage practices and N fertilizer use affect CH_4 and CO_2 emissions as well as soil C sequestration in hydromorphic paddy fields in this region. We specifically tested the effects of tillage practices and N fertilizer use on SOC in soils from 0 cm to 5 cm,

as well as from 0 cm to 20 cm, during the 2009 and 2010 rice growing seasons. This paper also aimed to evaluate the effects of tillage and N fertilizer on CH_4 and CO_2 emissions during the rice growing seasons.

Results

Temperature

The air temperature in the experimental site is shown in Table 1. The mean monthly air temperature ranged from 21.4°C to 28.9°C and from 19.7°C to 29.8°C during the 2009 and 2010 rice growing season, respectively. The mean monthly air temperature during rice growing seasons in 2009 was slightly lower than that in 2010. The mean air temperature from June to September, except for August, was significantly higher ($P<0.05$) in 2010 than in 2009.

CH_4 and CO_2 Emissions

The pattern of seasonal CH_4 emission fluxes was similar across NT and CT treatments during the 2009 and 2010 rice growing seasons (Fig. 1). In both years, the CH_4 emission fluxes in the four treatment groups were all initially low, increased gradually, and then peaked in mid-July (about 4–5 weeks after sowing). Thereafter, the CH_4 emission fluxes declined gradually and remained relatively low until harvesting when the CH_4 emission fluxes were lowest.

Application of N fertilizer significantly increased CH_4 emissions by 13%–66% in 2009 and 2010 ($P<0.05$) (Table 2). Tillage significantly affected CH_4 emissions, where NT significantly decreased CH_4 emissions by 10%–36% compared with CT ($P<0.05$). No significant effect of tillage×fertilizer on the cumulative CH_4 emissions was observed in 2009 or 2010. The cumulative CH_4 emissions in 2010 were 1.39–2.45 times those recorded in 2009.

Tillage treatments exhibited clear seasonal variations in soil CO_2 fluxes in the 2009 and 2010 rice growing seasons (Fig. 2). The soil CO_2 fluxes remained relatively low for the first two weeks after tillage, increased rapidly, stayed relatively high until about the middle 10 days of July, and then decreased to relatively low levels. Just one day after tillage (June 9, 2009 and June 13, 2010), the soil CO_2 fluxes from CT were 1.40–4.60 times higher than those from NT ($P<0.05$).

The cumulative CO_2 emissions from NT were 1.30–1.33 times those of CT ($P<0.05$) (Table 2). The application of N fertilizer had no significant effect on cumulative CO_2 emissions. We observed a significant effect of tillage×fertilizer on CO_2 emissions in 2009 ($P<0.05$) but not in 2010. In addition, cumulative CO_2 emissions in 2010 were 2.44–2.93 times those in 2009.

Table 1. Mean monthly air temperature during rice growing season in the experimental site/°C.

Time	2009	2010
June	26.1 b	27.1 a
July	28.9 a	29.1 a
August	28.0 b	28.8 a
September	24.7 b	25.6 a
October	21.4 a	19.7 a
Mean air temperature during the rice growing season	26.7 a	27.4 a

Different letters in a line mean significant differences at the 5% level.

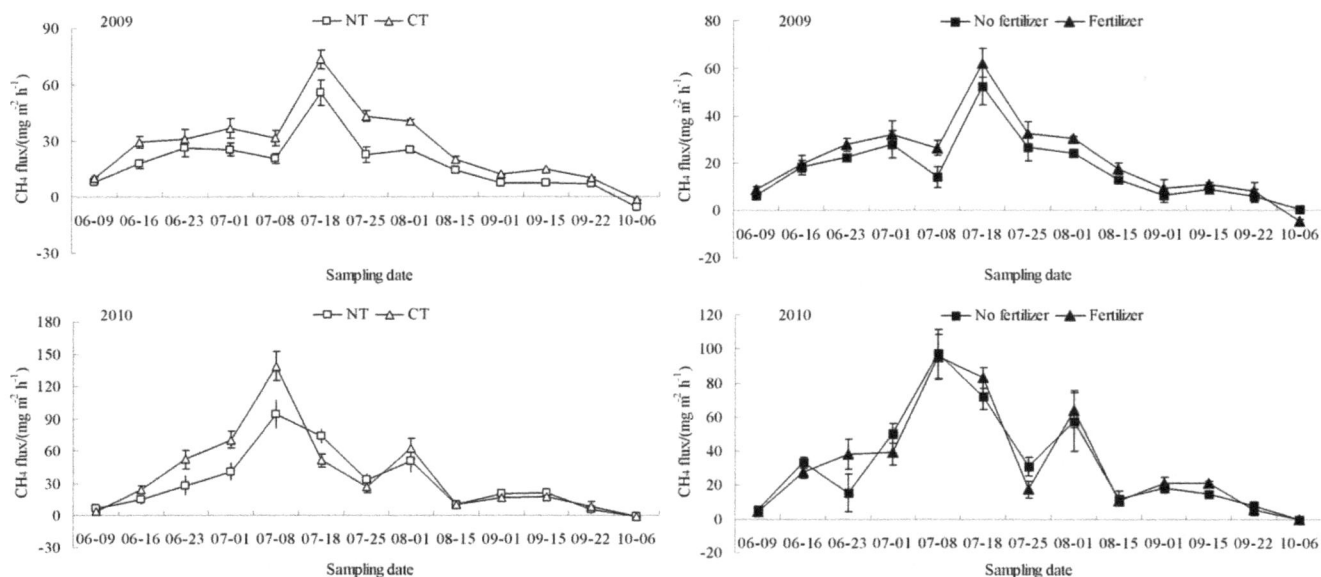

Figure 1. Changes in CH₄ emission fluxes from paddy fields under different management practices during the 2009 and 2010 rice growing seasons. The vertical bars are standard deviations of the mean, n = 3.

Soil Organic C and Bulk Density

As shown in Table 3, neither tillage nor N fertilizer application had any significant effect on bulk density before tilling fields or at harvesting in either year irrespective of the soil sampling depth. The SOC contents were significantly higher at 0–5 cm depth than at 0–20 cm depth under NT. In both years, application of N fertilizer significantly increased the SOC content by 4%–9% at harvesting and the SOC at the end of the growing seasons in 2009 and 2010 (21–94%) irrespective of soil sampling depths. Though NT had slightly higher SOC at the end of the growing seasons at 0–20 cm depth than CT in 2009 and 2010, we observed no significant effect of tillage or tillage×fertilizer in either year. However, across both years, tillage affected the SOC at 0–5 cm depth at harvesting, where NT significantly increased SOC contents by 12%–15% and SOC sequestration by 102%–270% than CT.

Based on the SOC content and bulk density at harvesting in the plow layer (0–20 cm; Table 3), we estimated SOC at harvesting in the plow layer to be 27.0–29.5 t C ha^{-1} in 2009 and 2010. Correspondingly, annual SOC accumulation rate in the plow layer

Table 2. Cumulative CH₄ and CO₂ emissions (g m^{-2}) from different tillage treatments in the 2009 and 2010 rice growing seasons, n = 3.

Tillage	N fertilizer	Cumulative CH₄ emissions		Cumulative CO₂ emissions	
		2009	2010	2009	2010
NT	No fertilizer	2.74 (0.57)	6.72 (0.91)	125.7 (10.6)	326.1 (15.6)
	Fertilizer	4.54 (0.44)	7.56 (1.02)	140.1 (6.6)	386.8 (10.5)
CT	No fertilizer	4.28 (0.27)	7.49 (0.33)	103.4 (7.2)	252.8 (12.2)
	Fertilizer	6.76 (0.40)	9.40 (0.60)	100.3 (4.3)	293.8 (14.1)
Analysis of variance					
T		*	*	**	*
F		*	*	NS	NS
T×F		NS	NS	*	NS

T, tillage;
F, application of N fertilizer;
*, significant at the 0.05 probability level;
**, significant at the 0.01 probability level;
NS, not significant;
The values in brackets are standard deviations of the mean.

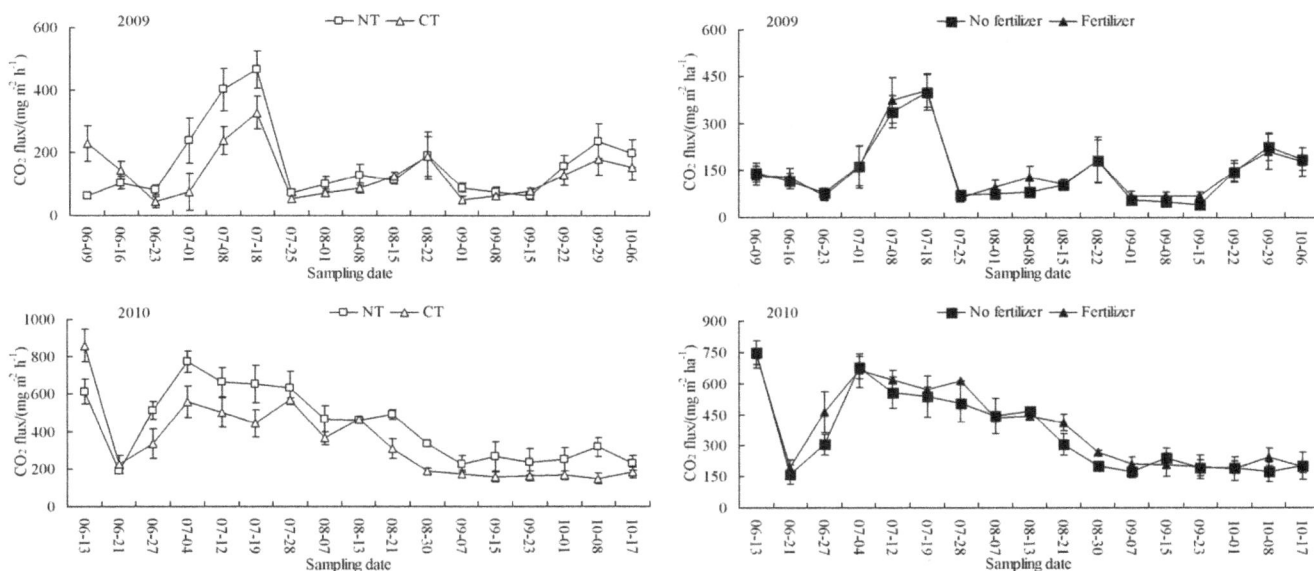

Figure 2. Changes in CO_2 emission fluxes from paddy fields under different management practices during the 2009 and 2010 rice growing seasons. The vertical bars are standard deviations of the mean, n = 3.

was estimated to be 0.06–0.14 t C ha^{-1} yr^{-1} for no fertilizer treatments and 0.25–0.47 t C ha^{-1} yr^{-1} for fertilizer treatments, with an average of 0.23 t C ha^{-1} yr^{-1} over the period 2009–2010.

Discussion

CH$_4$ Emission

Application of N fertilizer in the present study increased CH_4 emissions from paddy fields because of the promotion of rice growth, providing additional C sources and emission pathways [32]. Lindau and Bollich [33], in a study on a Louisiana rice field, which also had a humid subtropical climate, reported similar results from silt loam soil. However, Wassmann et al. [34] and Lu et al. [35] indicated no significant effect of N fertilizer application on CH_4 emissions from paddy fields in Zhejiang Province, China. Schütz et al. [36] found that the application of urea significantly decreased CH_4 emissions from paddy fields in Italy. Results varied among studies because of the differences in soil texture or climate. These findings show that further study is needed to understand the functioning of these complex and dynamic systems.

No-tillage significantly decreased CH_4 emissions relative to CT in the present study. This is in accordance with the findings reported by Harada et al. [7] and Liang et al. [9]. The decrease in CH_4 emissions under NT may be attributed to the differences regarding the size and activity of the methanotrophic community between tillage treatments [37]. Tillage also affects gaseous diffusivity and the rate of supply of atmospheric CH_4 [38]. By contrast, NT improves macroporosity and maintains its continuity [39]. The improvement probably allows greater air diffusion, increasing CH_4 uptake and decreasing CH_4 emissions.

CO$_2$ Emissions

Application of N fertilizer increases plant biomass production, stimulating soil biological activity, and consequently, CO_2 emission [40]. Wilson and Al-Kaisi [41], as well as Iqbal et al. [13], observed increased CO_2 emissions caused by N fertilizer

application. By contrast, Burton et al. [15] and DeForest et al. [16] indicated that reduced extracellular enzyme activities and fungal populations resulting from N fertilizer application resulted in decreased soil CO_2 emissions. We observed no significant effect of N fertilizer application on cumulative CO_2 emissions (Table 2), consistent with the results reported by Almaraz et al. [42]. This finding may be due to the fact that CO_2 is reduced to CH_4 under anaerobic conditions, thus leading to significant differences in CH_4 emissions rather than in CO_2 emissions between fertilized and unfertilized treatment areas (see Table 2).

We observed greater CO_2 emissions from NT than from CT during the 2009 and 2010 rice growing seasons (Table 2). Similar results were obtained by Liu et al. [43] and Oorts et al. [8]. The differences between the soil CO_2 emissions under the tillage treatments may have been caused by variation in soil C mineralization. Our own previously published work and those of other researchers indicated greater soil C mineralization under NT [7,8,44]. Increased SOC (Table 3) and higher microbial activity on the soil surface under NT [39] also resulted in greater soil CO_2 emissions for NT than CT. However, CT is generally reported to increase CO_2 emissions by exposing organic matter to more oxidizing conditions of the topsoil and accelerating the decomposition of aggregate-associated soil organic matter [38,45]. The increased levels of surface crop residues in NT probably serve as a barrier for CO_2 emissions from soil, decreasing the decomposition of crop residues because of reduced soil temperature and minimum soil-residue contact [46]. The inconsistent tillage effects on soil CO_2 fluxes suggest that tillage is not the only factor affecting CO_2 flux and that other factors are also involved. As suggested by Mosier et al. [47], CO_2 emissions caused by NT may be similar or slightly lower than those caused by CT if entire growing and fallow seasons are considered.

We observed a significant effect of tillage×N fertilizer on cumulative CO_2 emissions in 2009, in accordance with the results reported by Roberson et al. [48]. The cumulative CO_2 emissions during the rice growing seasons in the present study were 1003–1401 kg C ha^{-1} in 2009 and 2528–3868 kg C ha^{-1} in 2010. These values were greater than 363–371 and 506–926 kg C ha^{-1}

Table 3. SOC contents (g kg^{-1}) and bulk density (g cm^{-3}) before tillage and at harvesting, and SOC sequestration (kg C ha^{-1}) based on soil sampling depths from different tillage treatments in the 2009 and 2010 rice growing seasons, n = 3.

2009

Tillage	N fertilizer	0–20 cm					0–5 cm				
		Bulk density before tillage	Bulk density at harvesting	SOC contents before tillage	SOC contents at harvesting	SOC sequestration	Bulk density before tillage	Bulk density at harvesting	SOC contents before tillage	SOC contents at harvesting	SOC sequestration
NT	No fertilizer	1.19 (0.02)	1.19 (0.01)	18.40 (0.78)	20.16 b (1.12)	2318 (129)	1.20 (0.02)	1.22 (0.03)	18.92 (1.06)	22.55 a (1.22)	1389 (115)
	Fertilizer	1.19 (0.05)	1.17 (0.03)	18.89 (0.91)	21.77 b (1.00)	3439 (271)	1.21 (0.04)	1.23 (0.02)	19.47 (1.23)	23.90 a (4.15)	1685 (162)
CT	No fertilizer	1.17 (0.03)	1.18 (0.04)	18.10 (0.85)	19.70 a (0.89)	2187 (148)	1.18 (0.03)	1.20 (0.06)	18.32 (1.73)	19.68 a (2.11)	559 (90)
	Fertilizer	1.18 (0.04)	1.20 (0.05)	18.56 (1.02)	20.59 a (1.14)	3146 (347)	1.19 (0.05)	1.19 (0.06)	18.85 (1.15)	20.81 a (2.94)	835 (117)
Analysis of variance											
T		–	NS	–	NS	NS	–	NS	–	**	**
F		–	NS	–	*	*	–	NS	–	*	**
T×F		–	NS	–	NS	NS	–	NS	–	NS	NS

2010

Tillage	N fertilizer	0–20 cm					0–5 cm				
		Bulk density before tillage	Bulk density at harvesting	SOC contents before tillage	SOC contents at harvesting	SOC sequestration	Bulk density before tillage	Bulk density at harvesting	SOC contents before tillage	SOC contents at harvesting	SOC sequestration
NT	No fertilizer	1.19 (0.04)	1.25 (0.07)	18.76 (0.88)	19.18 b (1.21)	2102 (123)	1.19 (0.03)	1.25 (0.06)	19.96 (1.25)	21.92 a (1.42)	1032 (66)
	Fertilizer	1.18 (0.04)	1.25 (0.04)	18.85 (1.06)	19.81 b (1.44)	3630 (310)	1.21 (0.04)	1.23 (0.08)	19.93 (1.25)	23.87 a (1.64)	1492 (122)
CT	No fertilizer	1.18 (0.06)	1.24 (0.06)	18.63 (0.76)	18.81 a (1.35)	1949 (251)	1.19 (0.04)	1.21 (0.07)	19.05 (0.97)	19.53 a (1.54)	279 (57)
	Fertilizer	1.17 (0.03)	1.27 (0.04)	18.80 (1.11)	19.33 b (1.51)	2877 (346)	1.20 (0.05)	1.21 (0.07)	19.70 (1.33)	21.15 a (1.71)	542 (115)
Analysis of variance											
T		–	NS	–	NS	NS	–	NS	–	*	**
F		–	NS	–	*	**	–	NS	–	**	**
T×F		–	NS	–	NS	NS	–	NS	–	NS	NS

T, tillage; F, application of N fertilizer;
*, significant at the 0.05 probability level;
**, significant at the 0.01 probability level; NS, not significant; SOC, soil organic C.
Different letters in a year at different depths mean significant differences at the 5% level.
The values in brackets are standard deviations of the mean.

of cumulative CO_2 emissions from different rice tillage systems at an Ogata farm (Japan) and the Hailun Experimental Station of Ecology (Heilongjiang Province, China), respectively [7,9]. The differences in the emissions are possibly related to the dissimilar climates. The experimental field (humid mid-subtropical monsoon climate) in the present study is located at a lower latitude than those of the aforementioned studies.

We observed only one peak of CH_4 or CO_2 emission at the complete tillering stage, in contrast to the two or three emission peaks observed by other researchers [7,31,32]. The discrepancies are likely related to the different rice cropping systems (e.g. single, early, or late rice cropping), field pre-cropping management (e.g. rape and wheat), soil properties, weather conditions, and the use of N fertilizer [31]. The peak of CH_4 emission in the present study may be attributed to (1) the higher availability of substrates through root exudation or decayed plant residues for methanogenic bacteria in the rice rhizosphere [49,50] and (2) vigorous respiration by rice plants during this stage [51]. These processes promote CH_4 emission because most of the CH_4 is emitted through plants [52]. The peak of soil CO_2 emission might be attributed to the increased availability of substrates from root exudation or microbial decomposition of left-over plant residues at the active vegetative growth stage.

Higher cumulative CH_4 and CO_2 emissions (Table 2) were observed during the rice growing season in 2010 than in 2009. Similar interannual differences between CH_4 and CO_2 emissions have been found by other researchers [31,53]. Although these interannual differences in emissions are difficult to explain, discrepancies in climatic conditions and pre-crop residue management are probably involved. Residues of rapeseed were burnt before the experiment was started in 2009, which may be an important reason for the significantly lower emissions observed in this year. Higher mean air temperatures from June to August in 2010 than 2009 may be another important factor that led to higher cumulative CH_4 and CO_2 emissions.

Soil Organic C

In the present study, N fertilizer application had a positive effect on SOC (Table 3). This is attributed to more rice biomass and in turn more residue input to soil under the N fertilized treatments [19]. Others [19,30,54,55] also reported similar results. However, there were other reports indicating that application of chemical N fertilizers caused no significant or even negative effects on SOC [56–59]. The inconsistent results might depend on differences in the climatic and soil conditions, crop residue management, tillage regime, and experimental duration [60].

Here topsoil SOC (27.0–29.5 t C ha^{-1}) was comparable to the results of Pan et al. (27.9–30.9 t C ha^{-1}) [61], but lower than previous estimates of SOC of double-rice paddy soils reported by Shang et al. (36.4–48.2 t C ha^{-1}) [30] and Wang et al. (32.7–41.9 t C ha^{-1}) [62]. The SOC accumulation rate averaged 0.23 t C ha^{-1} yr^{-1} over the period 2009–2010 in the present study, generally lower than previous estimates in some double-rice paddy soils under short- or long-term chemical N fertilizer application [30,61,62]. However, it falls within the SOC sequestration rate range of 0.13–2.20 t C ha^{-1} yr^{-1} estimated by Pan et al. [63]. Lower levels of SOC in the present study could be attributed to differences in crop rotation systems. The decomposition rate of SOC in the single rice paddy-upland rotation system was higher than double rice-cropping paddy soils primarily dominated by surface waterlogging [30].

NT significantly increased SOC contents relative to CT at 0–5 cm depth but not at 0–20 cm depths (Table 3). A possible reason could be the return of moderately higher residues and root

biomass to the soil surface, instead of migrating deeper into the soil under NT. CT incorporates residues into a greater soil volume [64,65], resulting in relatively high SOC contents at deeper depths than NT [44]. Consequently, the lower SOC content at deeper depths under NT may weaken the tillage effects on SOC contents in the 20 cm layer. Similar observations were reported by other researchers [66,67].

The present results indicate that tillage has different effects on SOC sequestration based on the soil sampling depth (Table 3). NT significantly increased SOC compared with CT only at 0–5 cm but not at 0–20 cm. This result is likely caused by the residue accumulation on the soil surface. Similar results were observed by Wright et al. [65] and Wright and Hons [68]. Our results were in contrast to the results reported by other researchers [69–71]. Nyamadzawo et al. [69] found that NT had more SOC than CT at 0–20 cm depth. Deen and Kataki [70] reported that, compared to CT, NT increased SOC storage only for the surface layer (0–5 cm) but had significantly lower SOC for the entire soil profile (0–40 cm). However, Christopher et al. [71] found that NT had similar amounts of SOC to CT across the entire soil profile (0–60 cm). We can speculate that the potential of SOC sequestration under NT paddy fields in the present study may be overestimated at deeper soil depths (>20 cm). Further research is needed to understand the sequestration of SOC under CT and NT systems based on different soil sampling depths.

Materials and Methods

Site Description

The experimental site is situated at an experimental farm in Zhonggui Country, Dafashi Town, Wuxue City, Hubei Province, China (29°55′ N, 115°30′ E). This region has a humid mid-subtropical monsoon climate, an average annual temperature of 16.8°C, and a mean annual precipitation of 1360.6 mm. Rainfall mostly occurred between April and August in the past 5 years. The paddy field soil is a hydromorphic paddy soil, which is silty clay loam (3% sand, 50% silt, and 47% clay) derived from quaternary yellow sediment. The main soil properties (0–20 cm depth) of the site are as follows: pH (extracted by H_2O; soil: water = 1:2.5), 6.58; organic C, 18.29 g kg^{-1}; total N, 1.05 g kg^{-1}; NO_3^--N, 4.37 mg kg^{-1}; NH_4^+-N, 2.43 mg kg^{-1}; total P, 0.70 g kg^{-1}; Bray-P, 3.65 mg kg^{-1}; and available K (extracted by CH_3COONH_4), 111 mg kg^{-1}.

The rice variety planted was *Liangyoupeijiu* (*Oryza sativa* L.), a mid-season rice variety. The experimental site was cultivated with a rape (*Brassica napus*)–rice (*Oryza sativa* L.) rotation. Rice was directly seeded from May to October each year and rape was planted from October to May the following year for the past 30 years.

Experimental Design

Implementation of NT was initiated in 2006. Treatments were established following a split-plot design of a randomized complete block with standard tillage practices in the main plot and N fertilizers in the sub-plots. Each treatment had three replications. Each plot was isolated with a plastic film driven to a depth of 40 cm along the inner edge of the field ridge (30 cm at the base and 30 cm in height) in order to prevent lateral water movement caused by either leakage or permeable lateral flow. Each plot had an area of 45 m^2 and an inlet for irrigation as well as an outlet for drainage. Two water meters were installed at the inlet and outlet to record water flow.

The weeds were controlled by spraying 36% glyphosate at 3 L ha^{-1} on June 4, 2009 and June 10, 2010. The field was then flooded on June 5, 2009 and June 12, 2010, respectively. Thereafter, the CT treatments were cultivated to 8–10 cm depth by hoeing, and were subsequently mouldboard ploughed twice to 20 cm depth before sowing. There was no tillage in the NT-treated subplots. Before sowing, rice seeds were soaked in water for 12 h and mixed with Dry-Raised Nurse (provided by Yangzhou Lvyuan Biochemial Co., LTD), a biological seed coat agent that can promote rice seed germination at a ratio of 1:3 ratio. Rice seeds were sown manually at a rate of 22.5 kg ha^{-1} on June 8, 2009 and June 12, 2010. The crops were then harvested on October 8, 2009 and October 17, 2010, respectively. Commercial inorganic N–phosphorus (P)–potassium (K) fertilizer (15% N, 15% P_2O_5, 15% K_2O), urea (46% N), single superphosphate (16% P_2O_5) and potassium chloride (60% K_2O) were used to furnish 210 kg N ha^{-1}, 135 kg P_2O_5 ha^{-1} and 240 kg K_2O ha^{-1} during the rice growing season. Nitrogen fertilizers were broadcast at a rate of 84 kg N ha^{-1} as basal fertilizers immediately after sowing. The P and K fertilizers were only used as basal fertilizers immediately after seeding. The remaining N fertilizers were split into three doses of 42 kg N ha^{-1} on June 24, July 19 and August 12, 2009, as well as June 25, July 21, and August 14, 2010. The irrigation and application of pesticide were the same in all experimental treatments. According to local conventional irriga-tion-drainage practices, the plots were irrigated immediately upon the germination of rice seeds. Thereafter, the plots were reirrigated to a depth of 10 cm whenever that the water depth decreased to 1 cm to 2 cm above the soil surface during the growing season. The fields were not flooded for the entire 2 weeks before the rice was harvested.

Methane Emission

Closed steel cylinders with diameters of 58 cm and height of 110 cm were used to quantify the CH_4 fluxes from all plots during the rice growing seasons [72]. CH_4 gas samples were collected from June 9 to October 8, 2009 and from June 12 to October 17, 2010. Two permanent rings were placed below water level to create a seal in each treatment plot and chambers were temporarily placed on these rings to measure the gas fluxes. Fans installed on the tops of the chambers were run for 1 min to mix the air within the chamber before each gas sample was taken. Then the gases in the chamber were drawn off with a syringe and immediately transferred into a 20 ml vacuum glass container. Three gas samples from the chamber headspace were collected at 8 min intervals using 25 ml plastic syringes during a half-hour period. Measurements of CH_4 fluxes were conducted twice a day in the morning (9:00 to 11:00) and afternoon (15:00 to17:00). The morning and afternoon measurements from each plot were then averaged and considered as representative of that plot. The gas samples were collected 1 day after each N fertilizer application, and weekly.

We measured CH_4 concentrations with gas chromatograph meter (Shimadzu GC-14B), fitted with a 6′ to 1/8′ stainless steel column (Porapack N, length×inner diameter: 3 m×2 mm) and a flame ionization detector as previously presented [73]. For determination of CH_4, N_2 (flow rate: 330 ml min^{-1}), H_2 (flow rate: 30 ml min^{-1}) and zero air (flow rate: 400 ml min^{-1}) were used as the carrier, fuel, and supporting gas, respectively. The temperatures of the column, injector, and detector were set at 55, 100, and 200°C, respectively. The changes in CH_4 concentrations remained linear throughout the sampling period. The gas emission flux was calculated from the difference in the gas concentration according to the equation given by Zheng et al. [74]:

$$F = \rho \times h \times dC/dt \times 273 \div (273 + T)$$

where F is the gas emission flux (mg m^{-2} h^{-1}), ρ is the gas density at the standard state, h is the height of the chamber above the soil (m), C is the gas mixing ratio concentration (mg m^{-3}), and T is the mean air temperature inside the chamber during sampling.

Carbon Dioxide Emission

The soil CO_2 flux was measured using the soil respiration method described by Parkinson [75]. In this method, a cylinder static chamber of 20 cm in diameter and 30 cm in height was placed on the soil and the rate of increase in CO_2 concentration within the chamber was monitored using a LI–6400 portable photosynthesis analyzer (Li–Cor Inc., Lincoln, NE). We measured soil fluxes from 2 h measurements between 9:00 and 11:00 (a representative time of daily averages in this region described by Lou et al. [76]). The soil CO_2 flux in the present study was measured in this way. The soil CO_2 fluxes were measured 17 times at weekly intervals from June 9 to October 6, 2009, and 17 times at a 7–10 day interval from June 13 to October 17, 2010.

Each soil CO_2 flux was determined every 1 min for 20 min. Three measurements were performed for each plot on each sampling day, and soil CO_2 flux was the average of three individual measurements. Meteorological data were collected from the weather station in Wuxue City, 1 km from the experimental site.

Changes in the concentration with the sampling time were used to calculate the soil CO_2 flux rate. The flux rate was calculated by simple linear regression when the concentration of gas inside the chamber varied linearly over time. Otherwise, the rate flux was calculated by nonlinear regression [77]. For the nonlinear regression, a model based on Fick's law was fitted to the chamber data:

$$C(t) = C_{max} - (C_{max} - C_0) \times \exp^{(-k \times t)}$$

where the regression parameter C_0 is the air concentration at time $t = 0$; C_{max} is the maximum concentration that can be reached in the chamber, and k is a rate constant. The values of C_{max}, C_0 and k were estimated iteratively using the observed concentration versus time data. Methane and CO_2 fluxes were both expressed as mg m^{-2} h^{-1}.

The cumulative CH_4 and CO_2 emissions were calculated for each plot by linearly interpolating the gas emissions between sampling dates under the assumption that the measured fluxes represented the average daily fluxes. The cumulative emissions were calculated according to the following equation:

$$CE = \sum \left\{ (F_i + F_{i+1}) \div 2 \times 10^{-3} \times t \times 24 \right\}$$

where CE is the cumulative emissions (g m^{-2}), F_i and F_{i+1} are the measured fluxes of two consecutive sampling days (mg m^{-2} h^{-1}), and t is the number of days between two consecutive sampling days (d).

Sampling and Analytical Methods

Paddy soil samples (0–5 or 0–20 cm depth) were collected using a soil sampler with a diameter of 5 cm at eight random positions in each plot 1 day before the field was tilled and immediately after rice was harvested. The SOC were determined by dichromate oxidation and titration with ferrous ammonium sulfate [78]. The soil bulk density was determined by the method as described by Bao [78]. Soil bulk density samples for 0–5 or 0–20 cm soil layers were collected from each plot using metallic cores of 5.3 cm in diameter and 5.0 cm tall or 5.3 cm in diameter and 20 cm tall. Three soil cores were collected from each plot at 0–5 cm depth. The soil bulk density was computed as the weight to volume ratio of oven-dried ($105°C$) soil. Each measurement was replicated thrice. The SOC density (kg C ha^{-1}) at the soil depth was evaluated by the methods described by Lu et al. [19]. The SOC density was calculated as follows:

$$SOCD = SOCC \times BD \times H$$

where *SOCD* and *SOCC* are the SOC density (kg C ha^{-1}) and SOC concentration (g kg^{-1}), respectively; *BD* is the soil bulk density, and *H* is the soil sampling depth in the paddy field.

Statistical Analysis of Data

The SPSS 16.0 analytical software package was used for all statistical analyses. All data (mean$\pm$SE, $n=3$) were checked for normal distribution. Statistical analysis was performed by two–way ANOVA to analyze the effects of N fertilizer and tillage on the CH$_4$ and CO$_2$ flux, as well as other C indices, using the SPSS general linear model procedure. The least significant difference (LSD) was calculated only when the ANOVA F-test was found to be significant at the $P<0.05$ probability level.

Author Contributions

Conceived and designed the experiments: LCF CCG. Performed the experiments: ZDN KZK ZZS. Analyzed the data: LCF ZDN KZK CCG. Contributed reagents/materials/analysis tools: ZDN WJP CML. Wrote the paper: LCF ZDN CCG.

References

1. IPCC (2007) Climate change 2007. impacts Climatechange, ed. adaptation and vulnerability. Working Group II. Geneva, Switzerland: IPCC.
2. Reiner W, Milkha SA (2000) The role of rice plants in regulating mechanisms of methane missions. Biol Fertil Soils 31: 20–29.
3. IRRI, International Rice Research Institute (2004) Rice Stat. Database, Los Banõs, Philippines. Available: http://www.irri.org/science/ricestat/index.asp. Accessed 2011 Sep 13.
4. OECD (2000) Environmental indicators for agriculture methods and results. Executive Summary, Paris.
5. Chu H, Hosen Y, Yagi K (2007) NO, N$_2$O, CH$_4$ and CO$_2$ fluxes in winter barley field of Japanese Andisol as affected by N fertilizer management. Soil Biol Biochem 39: 330–339.
6. Guo J, Zhou C (2007) Greenhouse gas emissions and mitigation measures in Chinese agroecosystems. Agric Forest Meteorol 142: 270–277.
7. Harada H, Kobayashi H, Shindo H (2007) Reduction in greenhouse gas emissions by no-tilling rice cultivation in Hachirogata polder, northern Japan: life-cycle inventory analysis. Soil Sci Plant Nutr 53: 668–677.
8. Oorts K, Merckx R, Gréhan E, Labreuche J, Nicolardot B (2007) Determinants of annual fluxes of CO$_2$ and N$_2$O in long–term no–tillage and conventional tillage systems in northern France. Soil Till Res 95: 133–148.
9. Liang W, Shi Y, Zhang H, Yue J, Huang GH (2007) Greenhouse gas emissions from northeast China rice fields in fallow season. Pedosphere 17(5): 630–638.
10. Minami K (1995) The effect of nitrogen fertilizer use and other practices on methane emission from flooded rice. Fertil Res 40: 71–84.
11. Paustian K, Collins HP, Paul EA (1997) Management controls on soil carbon. In: Paul EA, Paustian K, Elliot ET, Cole CV (Eds.) *Soil Organic Matter in Temperate Agroecosystems - Long-term Experiments in North America*, CRC Press, Boca Raton, FL, 15–49.
12. Lee DK, Doolittle JJ, Owens VN (2007) Soil carbon dioxide fluxes in established switch grass land managed for biomass production. Soil Biol Biochem 39: 178–186.
13. Iqbal J, Hu RG, Lin S, Hatano R, Feng ML, et al. (2009) CO$_2$ emission in a subtropical red paddy soil (Ultisol) as affected by straw and N fertilizer applications: a case study in Southern China. Agric Ecosyst Environ 131: 292–302.
14. Xiao Y, Xie G, Lu G, Ding X, Lu Y (2005) The value of gas exchange as a service by rice paddies in suburban Shanghai, PR China. Agric Ecosyst Environ 109: 273–283.
15. Burton AJ, Pregitzer KS, Crawford JN, Zogg GP, Zak DR (2004) Simulated chronic NO$_3$-deposition reduces soil respiration in Northern hardwood forests. Global Change Biol 10: 1080–1091.
16. DeForest JL, Zak DR, Pregitzer KS, Burton AJ (2004) Atmospheric nitrate deposition, microbial community composition, and enzyme activity in Northern hardwood forests. Soil Sci Soc Am J 68: 132–138.
17. Lal R (2004) Soil carbon sequestration impacts on global climate change and food security. Science 304: 1623–1627.
18. DeLuca TH, Zabinski CA (2011) Prairie ecosystems and the carbon problem. Front Ecol Environ 9: 407–413.
19. Lu F, Wang XK, Han B, Ouyang ZY, Duan XN, et al. (2009) Soil carbon sequestrations by nitrogen fertilizer application, straw return and no-tillage in China's cropland. Global Change Biol 15: 281–305.
20. Tang XH, Shao JA, Gao M, Wei CF, Xie DT, et al. (2007) Effects of conservational tillage on aggregate composition and organic carbon storage in purple paddy soil. Chin J Appl Ecol 18: 1027–1032. (in Chinese).
21. Gao YJ, Zhu PL, Huang DM, Wang ZM (2000) Long-term impact of different soil management on organic matter and total nitrogen in rice-based cropping system. Soil Environ Sci 9: 27–30. (in Chinese).
22. Six J, Feller C, Denef K, Ogle SM, Moraes Sa JC, et al. (2002) Soil organic matter, biota and aggregation in temperate and tropical soils – effects of no-tillage. Agronomie 22: 755–775.
23. Su YZ (2007) Soil carbon and nitrogen sequestration following the conversion of cropland to alfalfa forage land in northwest China. Soil Till Res 92: 181–189.
24. He YY, Zhang HL, Sun GF, Tang WG, Li Y, et al. (2010) Effect of different tillage on soil organic carbon and the organic carbon storage in two-crop paddy field. J Agro-Environ Sci 29(1): 200–204. (in Chinese).
25. Angers DA, Eriksen-Hamel NS (2008) Full-inversion tillage and organic carbon distribution in soil profiles: a meta-analysis. Soil Sci Soc Am J 72: 1370–1374.
26. Baker JM, Ochsner TE, Venterea RT, Griffis TJ (2007) Tillage and soil carbon sequestration – what do we really know? Agric Ecosyst Environ 118: 1–5.
27. Green CJ, Blackmer AM, Horton R (1995) Nitrogen effects on conservation of carbon during corn residue decomposition in soil. Soil Sci Soc Am J 59: 453–459.
28. Alvarez R (2005) A review of nitrogen fertilizer and conservation tillage effects on soil organic carbon storage. Soil Use Manage 21: 38–52.
29. Tong CL, Xiao HA, Tang GY, Wang HQ, Huang TP, et al. (2009) Long-term fertilizer effects on organic carbon and total nitrogen and coupling relationships of C and N in paddy soils in subtropical China. Soil Till Res 106: 8–14.
30. Shang QY, Yang XX, Gao CM, Gao CM, Wu PP, et al. (2011) Net annual global warming potential and greenhouse gas intensity in Chinese double rice-cropping systems: a 3-year field measurement in long-term fertilizer experiments. Global Change Biol 17: 2196–2210.
31. Wang MX, Li J (2002) CH$_4$ emission and oxidation in Chinese rice paddies. Nutr Cy Agroecosyst 64: 43–55.
32. Neue HU, Roger PA (2000) Rice agriculture: factors controlling emissions. In: Khalil MAK (ed.), Atmospheric Methane. Its Role in the Global Environment, 134–169.
33. Lindau CW, Bollich PK (1993) Methane emissions from Louisiana first and Ratoon crop rice. Soil Sci 156: 42–48.
34. Wassmann R, Schüetz H, Papen H, Rennenberg H, Seiler W, et al. (1993) Quantification of methane emissions from Chinese rice fields (Zhejiang Province) influenced by fertilizer treatment. Biogeochemistry 20: 83–101.
35. Lu WF, Chen W, Duan WM, Lu Y, Lantin RS, et al. (2000) Methane emission and mitigation options in irrigated rice fields in southeast China. Nutr Cy Agroecosyst 58: 65–73.
36. Schütz H, Holzapfel-Pschorn A, Conrad R, Rennenberg H, Seiler W (1989) A 3-year continuous record on the influence of daytime, season and fertilizer treatment on methane emission rate from an Italian rice paddy. J Geophys Res 94: 16406–16416.
37. Ussiri DAN, Lal R, Jarecki MK (2009) Nitrous oxide and methane emissions from long–term tillage under a continuous corn cropping system in Ohio. Soil Till Res 104: 247–255.
38. Hütsch BW (1998) Tillage and land use effects on methane oxidation rates and their vertical profiles in soil. Biol Fertil Soils 27: 284–292.
39. Ball BC, Scott A, Parker JP (1999) Field N$_2$O, CO$_2$ and CH$_4$ fluxes in relation to tillage, compaction and soil quality in Scotland. Soil Till Res 53: 29–39.
40. Dick RP (1992) A review: long term effects of agricultural systems on soil biochemical and microbial parameters. Agric Ecosyst Environ 40: 25–36.
41. Wilson HM, Al-Kaisi MM (2008) Crop rotation and nitrogen fertilization effect on soil CO$_2$ emissions in central Iowa. Appl Soil Ecol 39: 264–270.
42. Almaraz JJ, Zhou XM, Mabood F, Madramootoo C, Rochette P, et al. (2009) Greenhouse gas fluxes associated with soybean production under two tillage systems in southwestern Quebec. Soil Till Res 104: 134–139.
43. Liu XJ, Mosier AR, Halvorson AD, Zhang FS (2006) The impact of nitrogen placement and tillage on NO, N$_2$O, CH$_4$ and CO$_2$ fluxes from a clay loam soil. Plant Soil 280: 177–188.

44. Li CF, Kou ZK, Yang JH, Cai ML, Wang JP, et al. (2010) Soil CO_2 fluxes from direct seeding rice fields under two tillage practices in central China. Atmos Environ 44: 2696–2704.

45. Reicosky DC, Dugas WA, Torbert HA (1997) Tillage–induced carbon dioxide loss from different cropping systems. Soil Till Res 41: 105–118.

46. Omonode RA, Vyn TJ, Smith DR, Hegymegi P, Gál A (2007) Soil carbon dioxide and methane fluxes from long–term tillage systems in continuous corn and corn–soybean rotations. Soil Till Res 95: 182–195.

47. Mosier AR, Halvorson AD, Peterson GA, Robertson GP, Sherrod L (2005) Measurement of net global warming potential in three agroecosystems. Nutr Cy Agroecosyst 72: 67–76.

48. Roberson T, Reddy KC, Reddy SS, Nyakatawa EZ, Raper RL, et al. (2008) Carbon dioxide efflux from soil with poultry litter applications in conventional and conservation tillage systems in northern Alabama. J Environ Qual 37: 535–541.

49. Xu H, Cai ZC, Jia ZJ, Tsuruta H (2000) Effect of land management in winter crop season on CH_4 emission during the following flooded and rice growing period. Nutr Cy Agroecosyst 58: 12–18.

50. Mitra S, Aulakh MS, Wassmann R, Olk DC (2005) Triggering of methane production in rice soils by root exudates: effects of soil properties and crop management. Soil Sci Soc Am J 69: 563–570.

51. Zhan M, Cao CG, Wang JP, Jiang Y, Cai ML, et al. (2011) Dynamics of methane emission, active soil organic carbon and their relationships in wetland integrated rice-duck systems in Southern China. Nutr Cy Agroecosyst 89: 1–13.

52. Butterbach-Bahl K, Papen H, Rennenberg H (1997) Impact of gas transport through rice cultivars on methane emission from rice paddy fields. Plant Cell Environ 20: 175–1183.

53. Drury CF, Reynolds WD, Tan CS, Welacky TW, Calder W, et al. (2006) Emissions of nitrous oxide and carbon dioxide: influence of tillage type and nitrogen placement depth. Soil Sci Soc Am J 70: 570–581.

54. Nayak P, Patel D, Ramakrishnan B, Mishra AK, Samantaray RN (2009) Long-term application effects of chemical fertilizer and compost on soil carbon under intensive rice–rice cultivation. Nutr Cy Agroecosyst 83: 259–269.

55. Wang CJ, Pan GX, Tian YG, Li LQ, Zhang XH, et al. (2010) Changes in cropland topsoil organic carbon with different fertilizations under long-term agro-ecosystem experiments across mainland China. Sci China Life Sci, 53: 858–867.

56. López-Bellido RJ, Fontán JM, López-Bellido FJ, López-Bellido L (2010) Carbon sequestration by tillage, rotation, and nitrogen fertilization in a Mediterranean Vertisol. Agron J 102: 310–318.

57. Li JT, Zhang B (2007) Paddy soil stability and mechanical properties as affected by long-term application of chemical fertilizer and animal manure in subtropical China. Pedosphere 17(5): 568–579.

58. Halvorson AD, Wienhold BJ, Black AL (2002) Tillage, nitrogen, and cropping system effects on soil carbon sequestration. Soil Sci Soc Am J 66: 906–912.

59. Khan SA, Mulvaney RL, Ellsworth TR, Boast CW (2007) The myth of nitrogen fertilization for soil carbon sequestration. J Environ Qual 36: 1821–1832.

60. Lou YL, Xu MG, Wang W, Sun XL, Zhao K (2011) Return rate of straw residue affects soil organic C sequestration by chemical fertilization. Soil Till Res 113: 70–73.

61. Pan GX, Zhou P, Li ZP, Smith P, Li LQ, et al. (2009) Combined inorganic/organic fertilization enhances N efficiency and increases rice productivity through organic carbon accumulation in a rice paddy from the Tai Lake region, China. Agr Ecosyst Environ 131: 274–280.

62. Wang SX, Liang XQ, Luo QX, Fan F, Chen YX, et al. Fertilization increases paddy soil organic carbon density. J Zhejiang Uni Sci B Doi:10.1631/jzus.B1100145, in press.

63. Pan G, Li L, Wu L, Zhang X (2003) Storage and sequestration potential of topsoil organic carbon in China's paddy soils. Global Change Biol 10: 79–92.

64. Six J, Elliot ET, Paustian K (1999) Aggregate and soil organic matter dynamics under conventional and no-tillage systems. Soil Sci Soc Am J 63: 1350–1358.

65. Wright AL, Hons FM, Lemon RG, McFarland ML, Nichols RL (2008) Microbial activity and soil C sequestration for reduced and conventional tillage cotton. Appl Soil Ecol 38: 168–173.

66. Causarano HJ, Franzluebbers AJ, Reeves DW, Shaw JN (2006) Soil organic carbon sequestration in cotton production systems of the southeastern United States: a review. J Environ Qual 35: 1374–1383.

67. Dolan MS, Clapp CE, Allmaras RR, Baker JM, Molina JAE (2006) Soil organic carbon and nitrogen in a Minnesota soil as related to tillage, residue and nitrogen management. Soil Till Res 89: 221–231.

68. Wright AL, Hons FM (2005) Carbon and nitrogen sequestration and soil aggregation under sorghum cropping sequences. Biol Fertil Soils 41: 95–100.

69. Nyamadzawo G, Chikowo R, Nyamugafata P, Nyamangara J, Giller KE (2008) Soil organic carbon dynamics of improved fallow-maize rotation systems under conventional and no-tillage in Central Zimbabwe. Nutr Cy Agroecosyst 81: 85–93.

70. Deen K, Kataki PK (2003) Carbon sequestration in a long-term conventional versus conservation tillage experiment. Soil Till Res 74: 143–150.

71. Christopher S, Lal R, Mishra U (2009) Long-term no-till effects on carbon sequestration in the Midwestern U.S. Soil Sci Soc Am J 73: 207–216.

72. Crill PM, Bartlett KB, Harriss RC, Gorham E, Verry ES, et al. (1988) Methane flux from Minnesota peatlands. Global Biogeochem Cy 2: 371–384.

73. Li CF, Cao CG, Wang JP, Zhan M, Yuan WL, et al. (2009) Nitrous oxide emissions from wetland rice–duck cultivation systems in southern China. Arch Environ Contam Toxicol 56: 21–29.

74. Zheng XH, Wang MX, Wang YS, Shen RX, Li J (1998) Comparison of manual and automatic methods for measurement of methane emission from rice paddy fields. Adv Atmos Sci 15(4): 569–579.

75. Parkinson KJ (1981) An improved method for measuring soil respiration in the field. J Appl Ecol 18: 221–228.

76. Lou YS, Li ZP, Zhang TL (2003) Carbon dioxide flux in a subtropical agricultural soil of China. Water Air Soil Pollut 149: 281–293.

77. Kroon PS, Hensen A, van den Bulk WCM, Jongejan PAC, Vermeulen AT (2008) The importance of reducing the systematic error due to non-linearity in N_2O flux measurements by static chambers. Nutr Cy Agroecosyst 82: 175–186.

78. Bao SD (2000) Analytical Method for Soil and Agricultural Chemistry. Beijing: Chinese Agriculture Press, 42–56. (in Chinese).

6

Persistence of *Escherichia coli* O157:H7 and Its Mutants in Soils

Jincai Ma[1,2], A. Mark Ibekwe[1]*, Xuan Yi[3], Haizhen Wang[1,2,4,5], Akihiro Yamazaki[3], David E. Crowley[2], Ching-Hong Yang[3]

1 United States Salinity Laboratory, Agriculture Research Service, United States Department of Agriculture, Riverside, California, United States of America, 2 Department of Environmental Sciences, University of California Riverside, Riverside, California, United States of America, 3 Department of Biological Sciences, University of Wisconsin, Milwaukee, Wisconsin, United States of America, 4 Institute of Soil and Water Resources and Environmental Science, Zhejiang University, Hangzhou, China, 5 Zhejiang Provincial Key Laboratory of Subtropical Soil and Plant Nutrition, Zhejiang University, Hangzhou, China

Abstract

The persistence of Shiga toxin-producing *E. coli* O157:H7 in the environment poses a serious threat to public health. However, the role of Shiga toxins and other virulence factors in the survival of *E. coli* O157:H7 is poorly defined. The aim of this study was to determine if the virulence factors, stx_1, stx_2, stx_{1-2}, and *eae* in *E. coli* O157:H7 EDL933 play any significant role in the growth of this pathogen in rich media and in soils. Isogenic deletion mutants that were missing one of four virulence factors, stx_1, stx_2, stx_{1-2}, and *eae* in *E. coli* O157:H7 EDL933 were constructed, and their growth in rich media and survival in soils with distinct texture and chemistry were characterized. The survival data were successfully analyzed using Double Weibull model, and the modeling parameters of the mutant strains were not significantly different from those of the wild type. The calculated T_d (time needed to reach the detection limit, 100 CFU/g soil) for loamy sand, sandy loam, and silty clay was 32, 80, and 110 days, respectively. It was also found that T_d was positively correlated with soil structure (*e.g.* clay content), and soil chemistry (*e.g.* total nitrogen, total carbon, and water extractable organic carbon). The results of this study showed that the possession of Shiga toxins and intimin in *E. coli* O157:H7 might not play any important role in its survival in soils. The double deletion mutant of *E. coli* O157:H7 ($stx_1^- stx_2^-$) may be a good substitute to use for the investigation of transport, fate, and survival of *E. coli* O157:H7 in the environment where the use of pathogenic strains are prohibited by law since the mutants showed the same characteristics in both culture media and environmental samples.

Editor: Olivier Neyrolles, Institut de Pharmacologie et de Biologie Structurale, France

Funding: This research was supported by the National Research Initiative/National Institute of Food and Agriculture Agreement No. 2008-35201-18709 and the 206 Manure and Byproduct Utilization Project of the United States Department of Agriculture-Agriculture Research Service. The funders had no role in study design, data collection and analysis, decision to publish, or preparation of the manuscript.

Competing Interests: The authors have declared that no competing interests exist.

* E-mail: Mark.Ibekwe@ars.usda.gov

Introduction

Escherichia coli O157: H7 was initially identified as an important human pathogen in 1982 during an investigation into a food-borne disease outbreak in the United States [1]. Since then, an increasing number of *E. coli* O157:H7 outbreaks have been reported in the United States. It is estimated that in the United States *E. coli* O157: H7 alone is responsible for a total of 73,480 cases of disease per year, among which, there are more than 1,800 cases of hospitalizations and 52 deaths. Evidence has shown that *E. coli* O157:H7 is one of the most commonly isolated bacterial pathogens from meat and fresh produce after *Campylobacter*, *Salmonella*, and *Shigella* spp [2]. In addition to the USA, many large outbreaks of *E. coli* O157:H7 infections have also been reported in many countries making *E. coli* O157:H7 an increasing public health concern worldwide. The infectious threshold of *E. coli* O157:H7 is very low, and ingestion of 10 cells may be enough to cause severe gastrointestinal illness [3]. The typical clinical symptoms of *E. coli* O157:H7 infections are watery diarrhea and hemorrhagic colitis [1], which can progressively develop into life-threatening hemolytic uremic syndrome (HUS) [4,5].

Outbreaks of *E. coli* O157: H7 infections are always traced back to consumption of food that has been directly or indirectly contaminated by manure/water containing *E. coli* O157:H7. Animals including deer, horses, dogs, and birds [6,7,8,9] are known to be *E. coli* O157:H7 carriers. However, cattle are thought to be the main carrier of *E. coli* O157:H7 [10,11]. *E. coli* O157:H7 in the environment may originate from farms where manure amendments are used as fertilizer. The pathogen could be mobilized through irrigation water, providing an opportunity for the pathogen to spread out into its secondary reservoir, typically water and soil. The persistence and regrowth in these habitats may increase the potential for the pathogen to enter into the food chain and thereby constitute a public health risk. There have been some cases of infection from direct contact with *E. coli* O157:H7 contaminated soil, and more cases of food poisoning caused by or consumption of vegetables grown in soils contaminated by *E. coli* O157:H7 [12,13].

The survival of *E. coli* O157:H7 in water [14,15,16,17,18,19], manure and manure slurry [20,21,22], manure-amended soil [23,24,25], and sediment [26,27], is well documented with sporadic reports in natural soils [28,29]. More direct results could be obtained by applying pathogenic strain in the survival experiments [30], however, most of the studies used nonpathogenic *E. coli* O157:H7 strains [22,24,31,32] due to environmental safety and regulations. This raises the question on how well those indirect results can represent results using pathogenic strains for comparison. Therefore, additional evidence is needed to clearly understand the role of *stx* genes and other virulence factors in the survival of pathogenic *E. coli* O157:H7 in the environment. Previous work [20] showed that there was a similar survival pattern between a Shiga toxin negative *E. coli* O157:H7 strain and a Shiga toxin positive *E. coli* O157:H7 strain. However, these strains were not isogenic, and the minor differences in survival might be attributed to other factors, such as the differences in their genomic DNA. Indeed, the variability in growth and survival of *E. coli* in soils has been shown to be strain-dependent [28].

In the current study, we chose *E. coli* O157:H7 EDL933 as the model pathogenic *E. coli* since its genome has been fully sequenced and annotated [42]. *E. coli* O157:H7 EDL933 and its isogenic mutant derivatives that are missing one of the following virulence factors, stx_1, stx_2, stx_{1-2}, and *eae*, were constructed, their growth in rich medium and survival in soils compared to that of the wild type parental strain (Fig. 1). We hypothesized that since all of the strains are isogenic the results will provide insights into the role of *stx* and *eae* genes in the survival of *E. coli* O157:H7 in soils. Additionally, the survival of the *E. coli* O157:H7 EDL933 in soils will correlate with the survival of pathogenic *E. coli* O157:H7 strains in the environment.

Materials and Methods

Bacterial strains, construction and growth of mutants

The bacteria and plasmids used in this study are listed in Table 1. In order to facilitate the enumeration of *E. coli* O157:H7 EDL933 on selective media, the *E. coli* O157:H7 wild type was tagged with nalidixic acid in addition to rifampicin resistance, and its growth curve in LB (Luria-Bertani) broth was found to be identical to that of the non-tagged wild-type strain.

Mutants lacking Stx_1, Stx_2, and Eae were generated by allelic exchange protocol [33]. The flanking regions were amplified by PCR with specific primers (Table 1), among which primers B and C (e.g. stx1_B, stx1_C) have the linkers at the 5' end that are complimentary to primers P1 and P2 [14], respectively, for crossover PCR. The kanamycin (Km) cassette was amplified from pKD4 (GenBank accession #, AY048743.1) and the chloramphenicol (Cm) cassette was amplified from pKD3 (GenBank accession #, AY048742.1) using the universal primer set consisting of forward primer P1 and reverse primer P2. Three-way crossover PCR was performed using the flanking regions and Km or Cm cassette as templates, and primers A and D (e.g. stx1_A, stx1_D) were used in this process. The PCR product was then cloned into pWM91 digested with *Xcm*I (T-vector). The resulting plasmid was transformed into *E. coli* S17-1 λ *pir*, and then introduced into EDL933 by transconjugation. Recombinants resulting from double crossover events were obtained by *sacB* and sucrose positive selection. All the mutant strains and the wild type strain were separately stored under −80°C on cryoprotective beads in MicroBank microbial storage tubes (Pro-Lab Diagnostics, Ontario, Canada).

The *stx* and *eae* mutants, together with the wild type strain were inoculated into 100 ml of LB broth, and grew under 37°C with a rotation rate of 250 rpm. The optical density at 610 nm ($OD_{610\ nm}$) was monitored using a VIS-UV spectrophotometer (Pharmacia Biotech Inc. NJ). The $OD_{610\ nm}$ was plotted against incubation time, and the apparent growth rate (k, h^{-1}) was calculated using the following equation,

$$k = (OD_2 - OD_1)/(t_2 - t_1)$$

where OD_1, OD_2 are the optical density measured at time t_1 and time t_2, respectively, k is the apparent growth rate (h^{-1}).

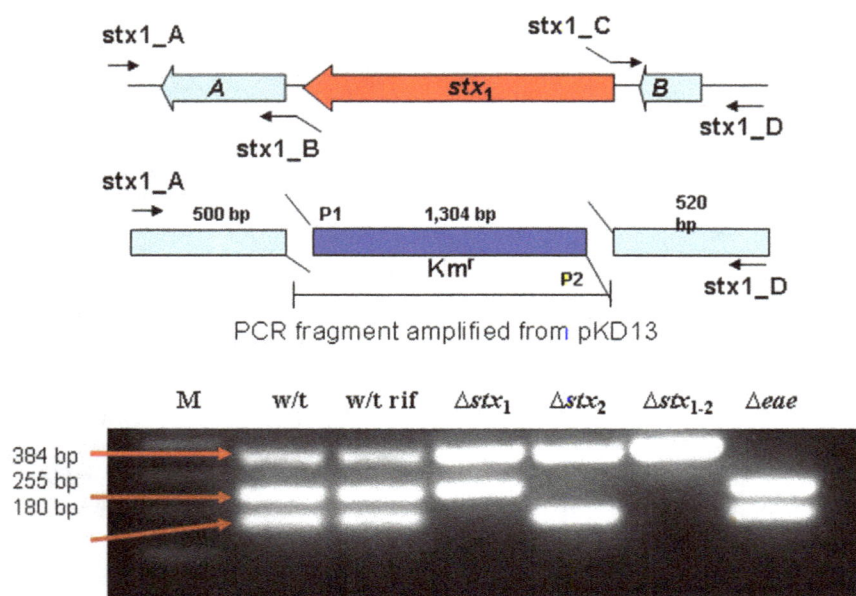

Figure 1. Construction of *stx* and *eae* mutants (top) and multiplex PCR confirmation of the mutant constructed (bottom). M represents 100 bp λDNA ladder.

Table 1. Bacterial strains and plasmids.

Strain or plasmid	Relevant characteristics	Source or reference
strains		
E. coli DH5α	General laboratory strain	Gibco-BR
E. coli S17-1	General laboratory strain	Simon *et al.* 1983
E. coli EDL933	wild type	ATCC 43895
E. coli EDL933	rifampicin tagged, Rif^r	This study
E. coli EDL933	*stx*1(del), , Km, Km^r	This study
E. coli EDL933	*stx*2(del), , Cm, Cm^r	This study
E. coli EDL933	*stx*1–2(del), , KmCm, Km^{r}Cm^r	This study
E. coli EDL933	*eae*(del), , Km, Km^r	This study
plasmids		
pWM91	Suicide vector, Ap^r	Metcalf *et al.*, 1996
pKD3	plasmid carrying Cm resistance cassette, Cm^r	Datsenko and Wanner, 2000
pKD4	plasmid carrying Km resistance cassette, Km^r	Datsenko and Wanner, 2000

Rif^r, rifampicin resistance; Km^r, kanamycin resistance; Cm^r, chloramphenicol; Ap^r, ampicillin resistance.

Multiplex PCR confirmation of mutants

Multiplex PCR was performed on the mutants and the wild type to confirm the deletions of stx_1, stx_2, stx_{1-2}, and *eae* genes in their genomes. PCR was performed using Ready-to-Go PCR beads with the three primer pairs (Table 2) targeting stx_1, stx_2, and *eae* gene [34]. Thermocycler protocol included an initial denaturation at 95°C for 10 min, followed by 35 cycles of denaturation at 94°C for 30 s, annealing at 55°C for 30 s, and extension at 72°C for 40 s, and a final extension at 72°C for 5 min. The PCR product was resolved by electrophoresis on a 1.0% agarose gel. The gel was then stained with ethidium bromide, visualized and photographed using a gel imaging system (Bio-Rad Lab., Irvine, CA). The PCR products with the correct sizes were cloned into TOPO TA cloning kit (Invitrogen, Carlsbad, CA) according to manufacture's protocol, and the resulting plasmids were sequenced. DNA sequence analysis was performed using DNAStar software (Lasergene, Madison, WI). Database searches were conducted with identified open reading frames (ORFs) by using the BLAST algorithm (http://blast.ncbi.nlm.nih.gov) to confirm the deletion of the corresponding gene(s).

Collection, characterization, and inoculation of soils samples

Dello loamy sand, Arlington sandy loam, and Willow silty clay were collected from Santa Ana River bed, fallow field at the University of California-Riverside, and Mystic Lake dry bed, California, respectively (Table 3). Arlington sandy loam is a typical agricultural soil found in Riverside, CA, while the other two soils are typical soil types used for cattle production in eastern and western Riverside County, USA. Permit was obtained from the University of California Riverside to collect the Arlington sandy loam. The soil from the Mystic Lake dry bed has high clay content (71%), and the soil from Santa Ana River bed has high sand content (99%). The texture and chemistry of the three soils are listed in Table 3. Soil samples were collected, sieved (2 mm), put into plastic bags, and stored at 4°C in dark. Soil properties characterized included, clay, silt, and sand content, water content, water holding capacity (WHC), soil organic carbon (OC), and total nitrogen (T-N) [35]. Soil microbial biomass carbon (MBC) was extracted by the chloroform-fumigation-extraction method [36],

and water extractable organic carbon (WEOC) was measured by a total organic carbon analyzer (TOC-500, Shimadzu Corp., Kyoto, Japan) according to the method by Liang et al. [37]. The assimilable organic carbon (AOC) fraction in WEOC was determined using a luminous bacterium strain, *Vibrio harveyi* (Ma et al., unpublished).

One cryoprotective bead from MicroBank microbial storage tube containing *E. coli* O157:H7 was aseptically transferred to a 15 ml tube containing 5.0 ml LB broth and incubated at 37°C for 18 h. From the overnight culture, a 1.0 ml aliquot was transferred into a 250 ml flask containing 100 ml LB broth, and incubated at 37°C for 18 h to achieve early stationary phase. Stationary phase cells were used because in the natural environment, the majority of bacteria exist in this condition [38]. The cells were harvested by centrifugation at 3500 g (Beckman, Brea, CA), washed three time using phosphate buffer (10 mM, pH 7.2), and finally resuspended in sterile deionized water. The wash step was essential to remove the nutrient, typically organic carbon from the LB broth, since *E. coli* O157 is able to grow at low carbon concentrations in freshwater [39].

Cell from stock cultures were streaked on LB agar (without antibiotics), and incubated 37°C overnight. Single colonies were picked and restreaked onto LB agar with appropriate antibiotics. Single colonies were streaked onto SMAC (sorbitol MacConkey) agar supplemented with BCIG (5-bromo-4-chloro-3-indoxyl-ß-D-glucuronide) (Lab M, Lancashire, UK). The isolated colonies were inoculated into 100 ml LB broth with appropriate antibiotics (Table 1), and incubated at 37°C for about 16 h. The overnight culture were harvested by centrifugation at 4°C, washed three times with phosphate buffer (10 mM, pH 7.2), resuspended in sterile deionized water, and inoculated into soil samples. Cell concentrations in soils were about 0.5×10^7 CFU per gram soil (g/dw) according to Franz et al. [22]. Briefly, the cell suspension was thoroughly mixed with soil in a plastic bag and 500 gram of the inoculated soil was transferred to a top perforated plastic bag for air exchange. The same amount of non-inoculated soil was put into another plastic bag, which was used as uninoculated control, with deionized water added instead of cell suspension. The experiment use triplicate bags of soils. The plastic bags were weighed and incubated at 10°C in darkness. Moisture content of

Table 2. Primers for mutants' construction and multiplex PCR.

Primers ID	Nucleotide sequence (5′ end to 3′ end)	Predicted product size (bp)	Source or reference
stx1_A	GGGTCCGGACGGTCATATGT	827	This study
stx1_B	gaagcagctccagcctacacTCAGTGAAAATAGCAGGCGC		
stx1_C	ctaaggaggatattcatatGACCCCCTGAAGGACGGCGTTTT	814	This study
stx1_D	CACCCATTGCCGCCGGATTT		
stx2_A	CATGCTGATGATGCTGGGAGTG	781	This study
stx2_B	gaagcagctccagcctacacGCGCGTTGTACTGGATTCGA		
stx2_C	ctaaggaggatattcatatGAACCTGATTCGTGGTATGTGGG	801	This study
stx2_D	TGGATCAGGGCTGTCGAATG		
eae_A	GCAATAACCAAATCATATCCGC	852	This study
eae_B	gaagcagctccagcctacacAACCACCCCGGCTAAAATATGT		
eae_C	ctaaggaggatattcatatGCTCGAGTTTTTCAGGGGTAGCA	799	This study
eae_D	TCCAGCATAGGGACCGTGCA		
P1	GTGTAGGCTGGAGCTGCTTC	1463 or 1014	Datsenko and Wanner, 2000
P2	CATATGAATATCCTCCTTAGTTCC		
stx1_F	ATAAATCGCCATTCGTTGACTAC	180	Paton and Paton, 1998
stx1_R	AGAACGCCCACTGAGATCATC		
stx2_F	GGCACTGTACTGAAACTGCTCC	255	Paton and Paton, 1998
stx2_R	TCGCCAGTTATCTGACATTCTG		
eae_F	GACCCGGCACAAGCATAAGC	384	Paton and Paton, 1998
eae_R	CCACCTGCAGCAACAAGAGG		

the soil sample were adjusted to 60% water holding capacity (WHC), and water concentration was maintained during the course of experiment by adding additional deionized water weekly to obtain the original weight. Antibiotics were added into the agar media at the following concentrations, kanamycin (Km), 50 μg/ml; chloramphenicol (Cm), 25 μg/ml; rifampicin (Rif), 100 μg/ml; and nalidixic acid (Nal), 25 μg/ml.

Sampling and enumeration

The inoculated soils were sampled periodically to determine the survival of the wild-type and mutant strains over time. At each point, two samples (1.0 g) of each triplicate bag was removed from the middle of the soil sample and put into pre-weighed dilution tubes. The tubes containing soil samples were weighted to calculate the exact size of soil sample. A 5.0 ml of 0.1% peptone buffer (Lab M, Lancashire, UK) was added to the test tube containing the soil sample, and the soil was thoroughly mixed with the buffer by inverting the tube several times and then vortexed for 2×20 s. The resulting soil paste (cell suspension) was then

subjected to 10-fold serial dilutions. Fifty μl of the two highest dilutions were plated in duplicate on SMAC/BCIG agar with appropriate antibiotics for enumeration. The inoculated SMAC agar plates were incubated at 37°C for 16 h, and the results expressed as log colony forming units per gram dry weight (CFU g/dw). The detection limit of the plating method was approximately 100 CFU g/dw. Our preliminary experiments showed that the average cell recovery rate of the method was from 90 to 110% of the theoretical value.

Survival data

Survival of *E. coli* O157:H7 was modeled by fitting the experimental data to the double Weibull survival model proposed by Coroller et al. [40] using GInaFiT version 1.5 developed by Dr. Annemie Geeraerd at Katholieke Universiteit, Leuven, Belgium [41]. The double Weibull survival model was constructed based on the hypothesis that the population is composed of two subpopulations differing in their capability on resistance to stress, and deactivation kinetics of both subpopulations follows a Weibull

Table 3. Soil texture and chemistry.

Soil type	Sand (%)	Silt (%)	Clay (%)	Bulk density (g/cm)	WHC (%)	pH	T-N (g/kg)	OC (g/kg)	WEOC (mg/kg)	MBC ((mg/kg)	AOC ((mg/kg)
Dello loamy sand	99.1	0.2	0.7	1.67	17	7.1	0.07	0.58	10	11	0.20
Arlington sandy loam	70.9	20.8	8.3	1.54	21	7.2	0.61	5.40	44	56	0.90
Willow silty clay	3.7	49.1	47.2	1.51	63	7.2	1.61	20.4	242	278	4.94

WHC, water holding capacity; T-N, total nitrogen; OC, organic carbon; MBC, microbial biomass carbon; AOC, assimilable organic carbon.

distribution. The size of the surviving population can be calculated using equation 1,

$$N_t = \frac{N_0}{1 + 10^\alpha} \left[10^{-\left(\frac{t}{\delta_1}\right)^p + \alpha} + 10^{-\left(\frac{t}{\delta_2}\right)^p} \right] \qquad (1)$$

$$\alpha = \log_{10}\left(\frac{f}{1-f}\right) \qquad (2)$$

Where N is the number of survivors, N_0 is the inoculum size; t is the time; p is the shape parameter, when $p>1$ a convex curve is observed; when $p<1$ a concave curve is observed, when $p=1$ a linear curve is observed. The scale parameter, δ, represents the time needed for first decimal reduction; f, varying from 0 to 1, is the fraction of subpopulation 1 in the population. Another parameter, α, varying from negative infinity to positive infinity, is obtained by logit transformation of f as shown in equation 2. The strong correlation between the scale (δ) and the shape (p) parameters makes the double Weibull model to fit most of the shapes of deactivation curves. Previous study proved that the double Weibull model can successfully describe a biphasic shape with nonlinear decrease, which can not be described by other survival models [40]. Additionally, when $\delta_1 = \delta_2$, the double Weibull model can be simplified into a single Weibull model, and the survival curve can be described by only three parameters. A very important and useful parameter, Td (time needed to reach detection limit, 100 CFU g/dw) can also be calculated when using GInaFiT to fit the experimental survival data.

Statistical analysis

Analysis of variance (ANOVA) was performed to investigate the differences in growth in rich medium, and the survival in soils using SPSS 16.0 software package (Chicago, IL).

Results

Mutant construction and confirmation

The genome of *E. coli* O157:H7 EDL933 has been fully sequenced and annotated [42], which makes it possible to knock out the genes of interest. The multiplex PCR assay (Fig. 1) clearly showed that the wild type strains displayed three bands representing the amplicons from *eae*, stx_2, and stx_1 genes, from top to bottom, with predicted sizes of 384, 255, and 180 bp, respectively. For the mutant derivatives, there was one band missing for Δstx_1, Δstx_2, and Δeae, and two bands missing for the double mutant construct Δstx_{1-2}, compare to the wild type strain. Figure 1 clearly showed that the virulence factors were successfully deleted as evidenced by the missing of the corresponding bands on the agarose gel.

Growth in rich medium and survival in soils

The growths of the mutant strains in LB broth were compared with that of the wild type (Fig. 2). The results showed that the *E. coli* O157:H7 EDL933 mutant derivatives growth was not significantly different from that of the wild type. Overall, there was about 1.5 h lag time followed by an exponential phase (5 h from incubation), then stationary phase (8 h post inoculation), and no decay phase was observed until 25 h of growth in LB broth under 37°C. The calculated apparent growth rates (r) of wild type, Δstx_1, Δeae, Δstx_2, and Δstx_{1-2} were as following, 0.46±0.02, 0.46±0.03, 0.46±0.03, 0.47±0.04, and 0.46±0.03.

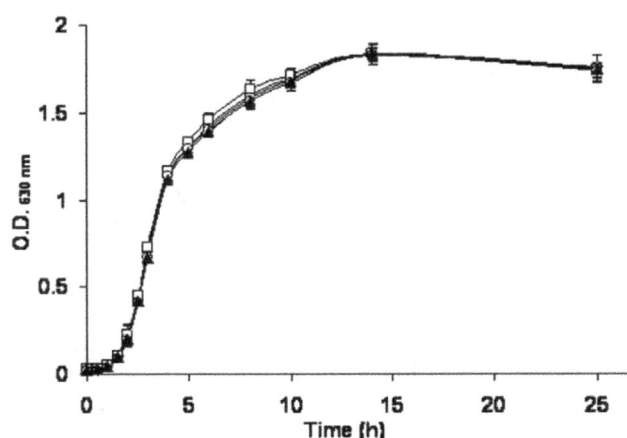

Figure 2. Growth curves of wild type strain (●) and its derivative mutants strains, Δstx_1 (◇), Δeae (▲), Δstx_2 (□), and Δstx_{1-2} (○). The data represent the average of triplicate measurements.

The wild type strain and the mutant strains were inoculated in soils to test their survival at 10°C. The results (Fig. 3) showed that within the same soil, there were no significant differences in deactivation profiles between the mutant and the wild type strains. It was also observed that the survival varies greatly in different soils. The cells survived shortest (32 day) in loamy sand with less nutrients (Fig. 3A), longest survival (113 day) was found in silty clay soil where there are more finer particles and more nutrient (e.g., organic carbon, nitrogen) (Fig. 3C), while the survival length was intermediate (82 day) in sandy loam soil. In loamy sand (Fig. 3A), there was a sharp decline of cell population within the first two weeks post inoculation, followed by a steady decrease until cell concentration dropped below detection limit. In sandy loam (Fig. 3B), a similar trend was also observed, a quick drop during first two weeks followed by a progressive decline. While in the silty clay (Fig. 3C), cells survived longer, because the cell concentrations did not decline significantly until four weeks post inoculation. Here after, cells started a very slow decline and dropped below detection limit (100 CFU g dw^{-1}) after 113 days.

Modeling of survival data

To accurately compare the survival kinetics between the wild type and mutant strains, survival data were modeled using a double Weibull equation as shown in Fig. 4. Similar modeling parameters (α, δ, and p) from mutant strains and the wild type strain were calculated when they were inoculated into the same soil. However, more variations in these parameters were observed from different soils, especially the δ values. When these strains were characterized in loamy sand and sandy loam soils, distinct δ_1 and δ_2 were observed indicating that the two subpopulations behave differently in both soils. The subpopulation with greater δ value declines slower than the one with smaller δ value. In contrast, almost identical δ_1 and δ_2 values were calculated from the survival data in silty clay soil indicating that the two subpopulations of cells in this soil likely behave similarly, thus the survival data in silty clay might be simplified into one Weibull model that can be described by only three parameters, α, δ and p. The initial sharp decrease in cell numbers in loamy sand soil might largely be attributed to the faster decline of subpopulation with smaller δ. However, with the time, the subpopulation with greater δ dominated the cell population, leading to a slower and steadier decline of the cell concentrations. A similar trend was also

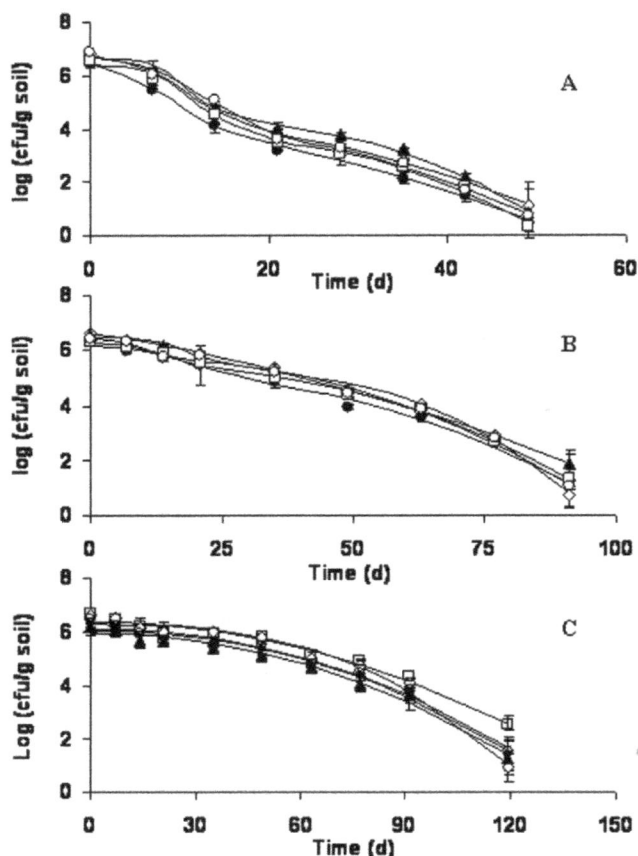

Figure 3. Survival of the wild type (●) and its mutant derivatives, Δstx₁ (◇), Δeae (▲), Δstx₂ (□), and Δstx₁₋₂ (○), in loamy sand (3A), sandy loam (3B), and silty clay (3C). The data represent the average of triplicate experiments.

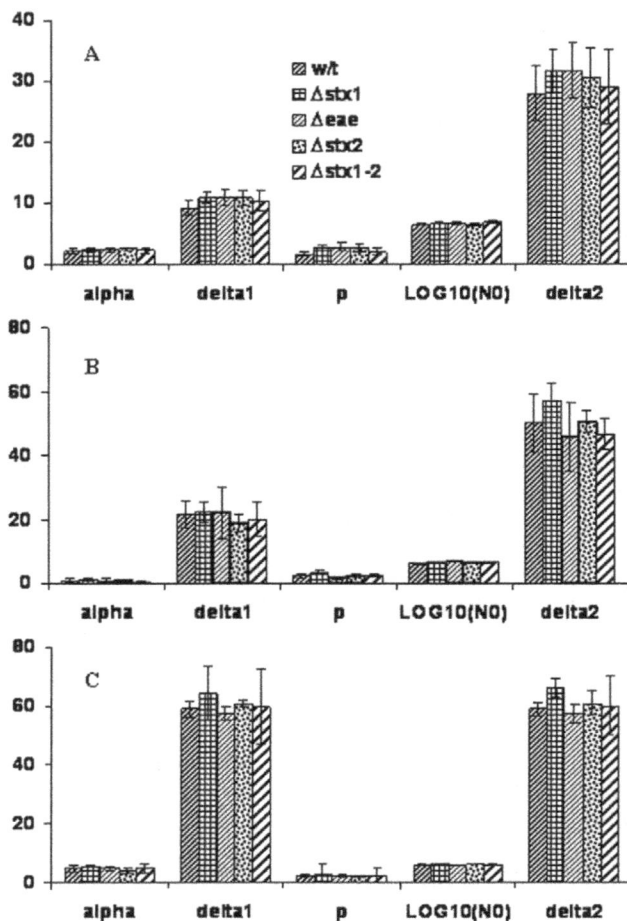

Figure 4. Double Weibull Model parameters of wild type strain and its mutant derivatives in loamy sand (4A), sandy loam (4B), and silty clay (4C).

observed in sandy loam soil. However, in silty clay soil, the cell concentrations did not change until 3 weeks post inoculation. This was followed by a steady decrease in cell concentrations until the population dropped below the detection limit (57–66 days).

The time to reach detection limit (*Td*) between the wild type and mutant strains in the same soil was not significantly different (*P* = 0.05) (Fig. 5). *Td* values in soils follow the order of, silty clay>sandy loam>loamy sand, which is consistent with the order of fine particle and nutrient levels in the soils. The effect of soil properties on the survival of *E. coli* O157:H7, and the time that it takes for the pathogen to reach detection limit was determined (Fig. 6). The results showed that with the increase in clay content, total organic carbon, total nitrogen, and water extractable organic carbon, there was a corresponding increase in *Td* values.

Discussion

The most significant finding of this work is Shiga toxins and intimin have no influence on the survival of pathogenic *E. coli* O157:H7 EDL933 in the three soils. The Shiga toxins *stx₁*, *stx₂* genes, and *eae* gene in *E. coli* O157:H7 have been intensively investigated [43,44]. Shiga toxins might induce an advantage in *E. coli* O157:H7 survival in the environment [44,45,46]. However, the role of these genes in survival of the pathogenic *E. coli* O157:H7 is still not completely understood [20]. Most of the previous survival studies used nonpathogenic *E. coli* O157:H7

strains [30,24,31,32], and the survival data based on pathogenic strains in the environment are not available due to regulations and safety concerns [27,28]. The typical nonpathogenic *E. coli* O157:H7 strain widely used in the literature include a green

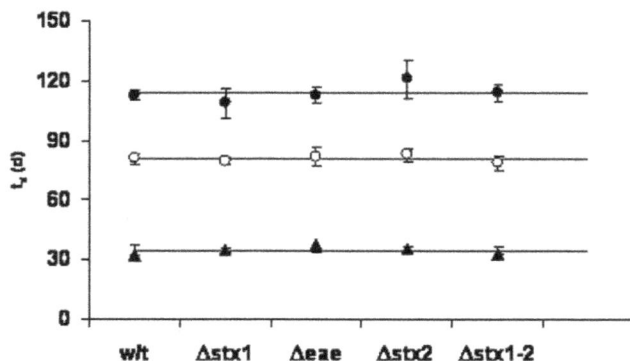

Figure 5. *Td* values calculated from the double Weibull model for wild type and its mutants derivatives in loamy sand (▲), sandy loam (○), and silty clay (●). The data represent the average of triplicate modeling of the raw survival data.

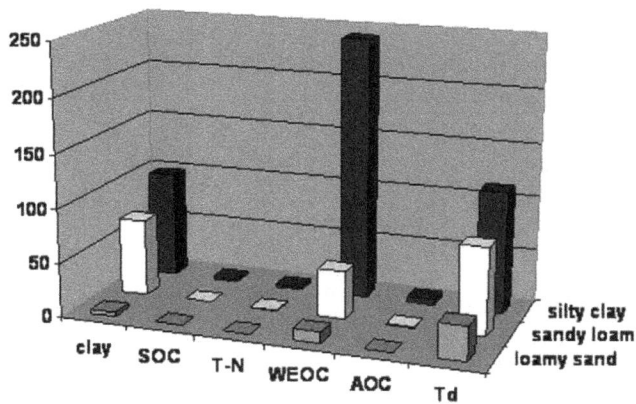

Figure 6. Effects of clay content (%), soil organic carbon (OC, %), total nitrogen (T-N, %), water extractable organic carbon (WEOC, mg/kg) _Td_. Gray, white, and black columns represent loamy sand, sandy loam, and silty clay, respectively.

fluorescence protein labeled strain (*E. coli* O157:H7 B6-914 GFP-91) constructed by Fratamico et al. [47] and a bioluminescent construct (*E. coli* O157:H7 Tn*5*, *luxCDABE*) by Ritchie et al. [31], both of which have been shown to survive in soil for months. Researchers used these nonpathogenic strains in the survivals studies on the assumption that they behave the same with the pathogenic strains. Therefore, comparative studies relating the survival of pathogenic *E. coli* O157:H7 strains to nonvirulent strains are needed to make a firm conclusion. The study by Kudva and colleagues [7] revealed that identical or very similar survival patterns were observed within a Shiga toxin positive *E. coli* O157:H7 strain (ATCC 43894) and Shiga toxin negative *E. coli* O157:H7 strain (ATCC 43888), indicating that Shiga toxins might have little or no influence on *E. coli* O157:H7 survival in manure and manure slurry. However, the strains used in their analysis were not isogenic, and factors other than Shiga toxins may have contributed to the minor survival differences observed in that study. In the present study, we have constructed a cluster of mutant derivatives from *E. coli* O157:H7 EDL933, with one of the following virulence factors deleted, stx_1, stx_2, stx_{1-2}, and *eae*. The indistinguishable growth curves between the mutants and the wild type strains in rich medium, in combination with their similar survival profiles in three different soils, offer strong evidence that the Shiga toxin genes and *eae* gene do not likely play important role in the survival of *E. coli* O157:H7 in soils.

In the current study, the survival data were successfully modeled by the double Weibull model. Different models were fitted into the survival data, but the best fit was obtained by applying double Weibull model. Since double Weibull model was based on the assumption that there are two subpopulations, and they differs in level of resistance to stress, and the survival of both subpopulations follow a Weibull distribution. Subpopulation with smaller δ die off faster compared to the other subpopulation with greater δ. In loamy sand, and sandy loam soils, distinct δ values, *i.e.* $\delta_1 \neq \delta_2$, were obtained for the two subpopulations, indicating that the two subpopulation exhibit different resistant capability in both soils. On the other hand, almost identical δ values, *i.e.* $\delta_1 \approx \delta_2$, were observed for the two subpopulations in silty clay soil, implying that the two subpopulations show a similar survival behavior in silty clay soil.

The persistence of *E. coli* O157:H7 is highly dependent on soil types, since distinct persistence time (*Td*) of this pathogen varies

significantly in different soils in terms of soil chemistry and texture. The longest survival was observed in silty clay soil, while the shortest survival was found in loamy sand soil. The results of soil characterization revealed that the silty clay soil is most abundant in clay, organic carbon, total nitrogen, and water extractable organic carbon, while the least abundant of those fractions is found in loamy sand soil. The variation in *Td* was best explained by the clay content in soils, since *Td* was closely correlated with the clay content. This agrees with the fact that the pathogens survived longer in finer-textured (clayey) than in courser (sandy) soils under similar environmental conditions [48]. Colonization of soil particles and aggregates is thought to be critical for the inoculated bacteria to survive in soil [49]. Finer textured soils (clayey) compared to coarser textured soils (sandy) may provide protective pore spaces to improve the survival of soil bacteria [50]. Indeed, the survival of a bacterial pathogen in 23 soils types was found to be positively correlated with soil clay content, in addition to other factors [51]. Indeed, greater survival of *E. coli* in sediment rich in clay (>25%) has been observed [25].Similarly, survival of *E. coli* O157:H7 was primarily determined by the soil texture, with prolonged survival associated with more clay particles compared with sand particles [22,52,53]. In addition to soil texture, soil chemistry characteristics, such as organic carbon, total nitrogen, and water extractable organic carbon, were also found to be positively related to survival of *E. coli* O157:H7. In our study, the availability of nutrient, such as nitrogen and organic carbon in soil were found to correlate with the pathogen survival in soils. Recently, Franz et al. [22] showed that the survival of *E. coli* O157 in 36 soils can best be explained by dissolved organic carbon and the ratio of dissolved organic carbon to microbial biomass carbon. In addition to soil texture and soil chemistry, biological factors cannot be neglected when interpreting the survival data of *E. coli* O157:H7 in soils. Overall, soils that are rich in clay or organic carbon might be a good secondary medium for extended persistence of *E. coli* O157:H7. Special attention should be paid to such soils when evaluating the environmental risk associated with *E. coli* O157:H7. The studies by the above authors and a recent review [54], to the best of our knowledge, have produced the most up to date data on survival of *E. coli* O157:H7 in soil. The review showed that temperature, soil structure, and microbial communities are the most important factors affecting survival. These authors showed from their previous studies [32] that the survival of *E. coli* O157:H7 was inversely proportional to the diversity of the microbial community established through differential fumigation and regrowth activities. Niche dependency strategy has also been suggested as a mechanism for *E. coli* O157:H7 survival in the open environment [55] rather than the biphasic growth model tested in this study. This argument is based on nutrient availability as the most important physiological factor for survival of *E. coli* O157 in nutrient-limited environment. However, we did not test this phenomenon in this study, but further studies in our laboratory will be looking at this in the nearest future.

In summary, stx_1, stx_2, and *eae* genes conferred in *E. coli* O157:H7 EDL933 did not play any direct role in survival of this pathogen in soil because the isogenic mutant strains showed indistinguishable survival profiles in three soils with distinct soil chemistry. The survival results obtained based on the non-pathogenic isogenic *E. coli* O157 strains from this study might be safely extrapolated to be equivalent to data obtained from pathogenic strains since the survival data from pathogenic strains in the environment are not available due to regulations and safety concerns. However, other conditions should be considered, *e.g.*, genes other than *stx* and *eae* that might be important in *E. coli*

O157 survival in the environment. Best management practices (BMPs) and good agricultural practices (GAPs) must be followed when leafy greens are grown in soils with high clay and organic carbon contents to reduce the risk of such soils being contaminated with *E. coli* O157:H7.

Acknowledgments

Mention of trademark or propriety products in this manuscript does not constitute a guarantee or warranty of the property by the United States Department of Agriculture and does not imply its approval to the exclusion of other products that may also be suitable.

Author Contributions

Conceived and designed the experiments: JM AMI. Performed the experiments: JM HW XY AY. Analyzed the data: JM HW AMI. Contributed reagents/materials/analysis tools: AMI DEC CHY. Wrote the paper: JM AMI.

References

1. Riley L, Remis RS, Helgerson SD, McGee HB, Wells JG, et al. (1983) Hemorrhagic colitis associated with a rare *Escherichia coli* serotype. N Engl J Med 308: 981–685.
2. Mead PS, Slutsker L, Dietz V, McCaig LF, Bresee JS, et al. (1999) Food-related illness and death in the United States. Emerg Infect Dis 5: 607–625.
3. Griffin PM, Tauxe RV (1991) The epidemiology of infections caused by *Escherichia coli* O157:H7, other enterohemorrhagic *E. coli*, and the associated hemolytic uremic syndrome. Epidemiol Rev 13: 60–98.
4. Carter AO, Borczyk AA, Carlson JA, Harvey B, Hockin JC, et al. (1987) A severe outbreak of *Escherichia coli* O157:H7.associated hemorrhagic colitis in a nursing home. N Engl J Med 317: 1496–500.
5. Karmali MA, Petric M, Lim C, Fleming PC, Steele BT (1983) *Escherichia coli* cytotoxin, haemolytic–uraemic syndrome, and haemorrhagic colitis. Lancet 2: 1299–1300.
6. Chapman PA, Siddons CA, Cerdan-Malo AT, Harkin MA (1997) A 1-year study of *Escherichia coli* O157 in cattle, sheep, pigs and poultry. Epidemiol Infect 119: 245–250.
7. Kudva IT, Hatfield PG, Hovde CJ (1997) Characterisation of *Escherichia coli* O157:H7 and other shiga toxin-producing *E. coli* serotypes isolated from sheep. J Clin Microbiol 35: 892–899.
8. Rice DH, Hancock DD, Besser TE (1995) Verotoxigenic *E. coli* colonisation of wild deer and range cattle. Vet Rec 137: 524.
9. Wallace JS, Cheasty T, Jones K (1997) Isolation of verocytotoxin producing *Escherichia coli* O157 from wild birds. J Appl Microbiol 82: 399–404.
10. Armstrong GL, Hollingsworth J, Morris Jr. JG (1996) Emerging foodborne pathogens, *Escherichia coli* O157:H7 as a model of entry of a new pathogen into the food supply of the developed world. Epidemiol Rev 18: 29–51.
11. Chase-Topping M, Gally D, Low C, Mathews L, Woolhouse M (2008) Super-shedding and the link between human infection and livestock carriage of *Escherichia coli* O157. Nature Rev Microbiol 6: 904–912.
12. Cieslak PR, Barrett TJ, Griffin PM, Gensheimer KF, Beckett G, et al. (1993) *Escherichia coli* O157:H7 infection from a manured garden. Lancet 342: 367.
13. Morgan GM, Newman C, Palmer SR, Allen JB, Shepherd W, et al. (1988) First recognized community outbreak of haemorrhagic colitis due to verotoxin-producing *Escherichia coli* O157 in the UK. Epidemiol Infect 101: 83–91.
14. Datsenko KA, Wanner BL (2000) One-step inactivation of chromosomal genes in *Escherichia coli* K-12 using PCR products. Proc Natl Acad Sci U S A 97: 6640–6645.
15. Artz RRE, Killham K (2002) Surivival of *Escherichia coli* O157:H7 in private drinking water wells influences of protozoan grazing and elevated copper concentration. FEMS Microbiol Lett 216: 117–122.
16. Avery LM, Williams AP, Killham K, Jones DL (2007) Survival of *Escherichia coli* O157:H7 in waters from lakes, rivers, puddles and animal-drinking troughs. Sci Total Environ 389: 378–385.
17. McGee P, Bolton DJ, Sheridan JJ, Earley B, Kelly G, et al. (2002) Survival of *Escherichia coli* O157:H7 in farm water, its role as a vector in the transmission of the organism within herds. J Appl Microbiol 93: 706–13.
18. Ravva SV, Sarreal CZ, Duffy B, Stanker LH (2006) Survival of *Escherichia coli* O157:H7 in wastewater from dairy lagoons. J Appl Microbiol 101: 891–902.
19. Watterworth L, Rosa B, Schraft H, Topp E, Leung KT (2006) Survival of various ERIC-genotypes of Shiga toxin-producing *Escherichia coli* in well water. Water, Air, Soil Pollut 177: 367–382.
20. Bolton DJ, Byrne CM, Sheridan JJ, McDowell DA, Blair IS (1999) The survival characteristics of a non-pathogenic strain of *Escherichia coli* O157:H7. J Appl Microbiol 86: 407–411.
21. Kudva IT, Blanch K, Hovde CJ (1998) Analysis of *Escherichia coli* O157:H7 survival in ovine or bovine manure and manure slurry. Appl Environ Microbiol 64: 3166–3174.
22. Williams AP, McGregor KA, Killham K, Jones DL (2008) Persistence and metabolic activity of *Escherichia coli* O157:H7 in farm animal faeces. FEMS Microbiol Lett 287: 168–173.
23. Franz E, Semenov AV, Termorshuizen AJ, de Vos OJ, Bokhorst JG, et al. (2008) Manure-amended soil characteristics affecting the survival of *E. coli* O157:H7 in 36 Dutch soils. Environ Microbiol 10: 313–327.
24. Jiang X, Morgan J, Doyle MP (2002) Fate of *Escherichia coli* O157:H7 in Manure-Amended Soil. Appl Environ Microbiol 68: 2605–2609.
25. Semenov A, Franz E, van Overbeek L, Termorshulizen AJ, et al. (2008) Estimating the stability of *Escherichia coli* O157:H7 survival in manure-amended with different management histories. Environ Microbiol 10: 1450–1459.
26. Burton Jr. GA, Gunnison D, Lanza GR (1987) Survival of pathogenic bacteria in various freshwater sediments. Appl Environ Microbiol 53: 633–638.
27. LaLiberte P, Grimess DJ (1982) Survival of *Escherichia coli* in lake bottom sediment. Appl Environ Microbiol 43: 623–628.
28. Campbell GR, Prosser J, Glover A, Killham K (2001) Detection of *Escherichia coli* O157:H7 in soil and water using multiplex PCR. J Appl Microbiol 91: 1004–1010.
29. Topp E, Welsh M, Tien Y-C, Dang A, Lazarovits G, et al. (2003) Strain-dependent variability in growth and survival of *Escherichia coli* in agricultural soil. FEMS Microbiol Ecol 44: 303–308.
30. Vidovic S, Block HC, Korber DR (2007) Effect of soil composition, temperature, indigenous microflora, and environmental conditions on the survival of *Escherichia coli* O157:H7. Can J Microbiol 53: 822–829.
31. Ritchie JM, Campbell GR, Shepherd J, Beaton Y, Jones D, et al. (2003) A Stable bioluminescent construct of *Escherichia coli* O157:H7 for hazard assessments of long-term survival in the environment. Appl Environ Microbiol 69: 3359–3367.
32. van Elsas JD, Hill P, Chroňáková A, Grekova M, Topalova Y, et al. (2007) Survival of genetically marked *Escherichia coli* O157:H7 in soil as affected by soil microbial community shifts. ISME J 1: 204–214.
33. Metcalf WW, Jiang WH, Daniels LL, Kim SK, Haldimann A, et al. (1996) Conditionally replicative and conjugative plasmids carrying *lacZ* alpha for cloning, mutagenesis, and allele replacement in bacteria. Plasmid 35: 1–13.
34. Paton AW, Paton JC (1998) Detection and characterization of Shiga toxigenic *Escherichia coli* by using multiplex PCR assays for stx_1, stx_2, eaeA, enterohemorrhagic *E. coli* hlyA, rfb_{O111}, and rfb_{O157}. J Clin Microbiol 36: 598–602.
35. Klute A (1996) Methods of soil analysis, part 1, physical and mineralogical methods (2nd edition), American Society of Agronomy, Agronomy Monographs 9(1), Madison, Wisconsin. .
36. Vance ED, Brookes PC, Jenkinson DS (1987) An extraction method for measuring soil microbial biomass C. Soil Biol Biochem 19: 703–707.
37. Liang BC, MacKenzie AF, Schnitzer M, Monreal CM, Voroney PR, et al. (1998) Management-induced change in labile soil organic matter under continuous corn in eastern Canadian soils. Biol Fertil Soils 26: 88–94.
38. Kolter R, Siegele DA, Tormo A (1993) The stationary phase of the bacterial cell. Annu Rev Microbiol 47: 855–874.
39. Vital M, Hammes F, Egli T (2008) *Escherichia coli* O157 can grow in natural freshwater at low carbon concentrations. Environ Microbiol 10: 2387–2396.
40. Coroller L, Leguerinel I, Mettler E, Savy N, Mafart P (2006) General model, based on two mixed Weibull distributions of bacterial resistance, for describing various shapes of inactivation curves. Appl Environ Microbiol 72: 6439–6502.
41. Geeraerd AH, Valdramidis VP, van Impe JF (2005) GInaFiT, a freeware tool to assess non-log-linear microbial survivor curves. Intl J Food Microbiol 102: 95–105.
42. Perna N, Plunkett III G, Burland V, Mau B, Glasner JD, et al. (2001) Genome sequence of enterohaemorrhagic *Escherichia coli* O157:H7. Nature 409: 529–533.
43. Besser RE, Griffin PM, Slustsker L (1999) *Escherichia coli* O157:H7 gastroenteritis and the hemolytic uremic syndrome: an emerging infection disease. Annu Rev Med 50: 355–367.
44. Kaper JB, Nataro JP, Mobley HL (2004) Pathogenic *Escherichia coli*. Nat Rev Microbiol 2: 123–40.
45. Lainhart W, Stolfa G, Koudelka GB (2009) Shiga Toxin as a bacterial defense against a eukaryotic predator, *Tetrahymena thermophila*. J Bacteriol 191: 5116–5122.
46. O'Loughlin E, Robins-Browne RM (2001) Effect of Shiga toxins on eukaryotic cells. Microbes Infect 3: 493–507.
47. Fratamico PM, Deng MY, Strobaugh TP, Palumbo SA (1997) Construction and characterization of *Escherichia coli* O157:H7 strains expressing firefly luciferase and Green fluorescent protein and their use in survival Studies. J Food Prot 60: 1167–1173.
48. van Elsas JD, Dijkstra AF, Govaert JM, van Veen JA (1986) Survival of *Pseudomonas fluorescens* and *Bacillus subtilis* introduced into two soils of different texture in field microplots. FEMS Microbiol Ecol 38: 151–160.
49. Hattori T, Hattori R (1976) The physical environment in soil microbiology, an attempt to extend principles of microbiology to soil microorganisms. CRC Crit Rev Microbiol 4: 423–461.

50. van Veen JA, van Overbeek LS, van Elsas JD (1997) Fate and activity of microorganisms introduced into soil. Microbiol Mol Biol Rev 61: 121–135.

51. Bashan Y (1986) Enhancement of wheat root colonization and plant development by *Azospirillum brasilense* Cd following temporary depression of rhizosphere microflora. Appl Environ Microbiol 51: 1067–1071.

52. Mubiru DN, Coyne MS, Grove JH (2000) Mortality of *Escherichia coli* O157:H7 in two soils with different physical and chemical properties. J Environ Qual 29: 1821–1825.

53. Nicholson FA, Groves SJ, Chambers BJ (2005) Pathogen survival during livestock manure storage and following land application. Bioresource Technol 96: 135–143.

54. van Elsas JD, Semenov AV, Costa R, Trevors JT (2011) Survival of *Escherichia coli* in the environment: fundamental and public health aspects. ISME J 5: 173–183.

55. Franz E, van Bruggen AHC (2008) Ecology of *E. coli* O157:H7 and Salmonella enterica in the primary vegetable production chain. Crit Rev Microbiol 34: 143–161.

7

Effect of Stocking Rate on Soil-Atmosphere CH$_4$ Flux during Spring Freeze-Thaw Cycles in a Northern Desert Steppe, China

Cheng-Jie Wang[1][*][◑], Shi-Ming Tang[1][◑], Andreas Wilkes[2], Yuan-Yuan Jiang[1], Guo-Dong Han[1], Ding Huang[3][◑]

1 College of Ecology and Environmental Science, Inner Mongolia Agricultural University, Huhhot, China, 2 World Agroforestry Centre, 12 Zhongguancun, Beijing, China, 3 Institute of Grassland Science, China Agricultural University, Beijing, China

Abstract

Background: Methane (CH$_4$) uptake by steppe soils is affected by a range of specific factors and is a complex process. Increased stocking rate promotes steppe degradation, with unclear consequences for gas exchanges. To assess the effects of grazing management on CH$_4$ uptake in desert steppes, we investigated soil-atmosphere CH$_4$ exchange during the winter-spring transition period.

Methodology/Main Finding: The experiment was conducted at twelve grazing plots denoting four treatments defined along a grazing gradient with three replications: non-grazing (0 sheep/ha, NG), light grazing (0.75 sheep/ha, LG), moderate grazing (1.50 sheep/ha, MG) and heavy grazing (2.25 sheep/ha, HG). Using an automatic cavity ring-down spectrophotometer, we measured CH$_4$ fluxes from March 1 to April 29 in 2010 and March 2 to April 27 in 2011. According to the status of soil freeze-thaw cycles (positive and negative soil temperatures occurred in alternation), the experiment was divided into periods I and II. Results indicate that mean CH$_4$ uptake in period I (7.51 μg CH$_4$–C m^{-2} h^{-1}) was significantly lower than uptake in period II (83.07 μg CH$_4$–C m^{-2} h^{-1}). Averaged over 2 years, CH$_4$ fluxes during the freeze-thaw period were −84.76 μg CH$_4$–C m^{-2} h^{-1} (NG), −88.76 μg CH$_4$–C m^{-2} h^{-1} (LG), −64.77 μg CH$_4$–C m^{-2} h^{-1} (MG) and −28.80 μg CH$_4$–C m^{-2} h^{-1} (HG).

Conclusions/Significance: CH$_4$ uptake activity is affected by freeze-thaw cycles and stocking rates. CH$_4$ uptake is correlated with the moisture content and temperature of soil. MG and HG decreases CH$_4$ uptake while LG exerts a considerable positive impact on CH$_4$ uptake during spring freeze-thaw cycles in the northern desert steppe in China.

Editor: Gil Bohrer, Ohio State University, United States of America

Funding: This research was funded by the Chinese National Natural Science Foundation (31160109, 30960072, 31070413), Inner Mongolia Palmary Youth Project (2010JQ04), and National Commonweal Project (200903060, 201003019), Inner Mongolia Agricultural University Innovation Group (NDTD2010-5) and Management and Policy Support to Combat Land Degradation Project: PRC-GEF Partnership on Land Degradation in Dryland Ecosystems. The funders had no role in study design, data collection and analysis, decision to publish, or preparation of the manuscript.

Competing Interests: The authors have declared that no competing interests exist.

* E-mail: cjwang3@sohu.com

◑ These authors contributed equally to this work.

Introduction

Methane (CH$_4$) is the second most important long-living greenhouse gas (GHG) after carbon dioxide (CO$_2$). It is widely recognized that an increases in atmospheric CH$_4$ concentration may cause global warming. The Intergovernmental Panel on Climate Change (IPCC) reported that global warming potential-weighted emissions of GHG increased by approximately 70% from 1970 to 2004, including emissions of CH$_4$ which have risen by about 40%. The global warming potential of CH$_4$ over a 100-year timeframe is estimated at 25 times that of CO$_2$ [1].

Chinese grasslands cover 41.7% of China's land area, and are distributed mainly in Inner Mongolia, Xinjiang, Gansu, and the Qinghai-Tibet plateau [2]. Approximately 1.42 million ha of grassland in Inner Mongolia is classified as desert steppe, accounting for 18% of the total steppe area in China [3]. In recent decades, long-term high intensity grazing has led to severe degradation and desertification in this natural grassland ecosystem, with notable impacts on agricultural activities. Steppe soils are known to function as a significant sink for atmospheric CH$_4$ [4]. It has been reported that in Inner Mongolian grasslands grazing changes soil moisture holding capacity, which in turn affects GHG emissions [4,5,6]. However, most previous studies reported measurements carried out on typical grassland during the growing season. Few studies have addressed CH$_4$ uptake during the freeze-thaw period [7]. Mosier et al. (1996) reported for North American prairie systems that winter fluxes may contribute 30–40% of the annual N$_2$O and CH$_4$ fluxes [8]. Therefore, current estimates for annual exchange rates of CH$_4$ between steppe soils and the atmosphere are still uncertain. To combat this uncertainty, CH_4 exchange should be observed during the winter and winter-spring transition period in Inner Mongolian steppes.

Soil freezing and thawing events affect the soil physical structure and solute distribution as well as nutrient availability. This causes secondary effects on microorganism activity [7]. Nutrient and moisture conditions in frozen soil not only allow physiological activity of microorganisms even under strong frost conditions [9], and can also greatly promote their activity, especially during the spring thaw period [7]. The desert steppe is vulnerable to the impacts of livestock because of its relatively short and sparse ground cover and sandy soil. Wolf et al. (2010) reported that increased stocking rates in a continental steppe led to reductions in natural N_2O release during the spring thawing period [6], but the impacts of grazing on CH_4 exchange remain unknown, especially in the Inner Mongolian desert steppe. Our aim here was to investigate the characteristics of soil-atmosphere CH_4 exchange across a gradient of stocking rates during the winter-spring transition, and to thus answer whether stocking rates influence CH_4 fluxes in the desert steppe ecosystem in Inner Mongolia. We hypothesize: 1) grazing management affects CH_4 uptake capacity of steppe soils during the spring thawing period, and 2) the moisture content and temperature of soil are the principal factors controlling CH_4 uptake.

Materials and Methods

Study site description

The study was conducted at an experimental site at the Inner Mongolia Academy of Agriculture and Animal Husbandry Research Station ($41°47'17''$N, $111°53'46''$E). The site has an elevation of 1450 m and is in a temperate continental climate, characterized by a short growing season and long cold winter with a frost-free period of 175 days. January is the coldest month with an average temperature of $-15.1°C$ while July is the warmest month with an average temperature of $19.6°C$. Average annual precipitation is approximately 280 mm, of which nearly 75% falls during June through September. The grassland is dominated by *Stipa breviflora* Griseb., *Artemisia frigida* Willd., *Cleistogenes songorica* (Roshev.) Ohwi, and accompanied by *Convolvulus ammannii* Desr., *Heteropappus altaicus* (Willd.) Novopokr., *Neopallasia petinata* (Pall.) Poljak., *Bassia prostrate* (L.) A.J. Scott, *Caragana stenophylla* Pojark., *Leymus chinensis* (Trin.) Tzvelev. The dominant soil types are Kastanozem (FAO soil classification) or Brown Chernozem (Canadian Soil Classification) with a loamy sand texture [10].

Measurements of CH4 and other factors

CH_4 fluxes were measured in twelve grazing plots during March 1–April 29 in 2010 and March 2–April 27 in 2011. The areas are also used for a grazing experiment started in 2002 with four stocking rates (non-grazing, 0 sheep/ha, NG; light grazing, 0.75 sheep/ha, LG; moderate grazing, 1.50 sheep/ha, MG; and heavy grazing, 2.25 sheep/ha, HG) with three replications each. The areas have been grazed from June to October only. The stocking rate was calculated on the basis of species composition and ground cover. The main characteristics of the grazing areas are shown in Table 1. CH_4 fluxes in each plot with three fixed points were measured at the same time (i.e. 10:30 to 12:30) at an interval of one day, using an automatic cavity ring-down spectrophotometer (Picarro G1301, Santa Clara, CA, USA). The principle of the measurement is wavelength scanning optical cavity ring-down spectroscopy (WS-CRDS) technology. CH_4 fluxes were calculated according to the following equation:

$$F = \rho \cdot \frac{V}{A} \cdot \frac{\Delta C}{\Delta T}$$

where F is the flux ($mg\ m^{-2}\ h^{-1}$) of gas; ρ is the density of gas; $\Delta C/\Delta T$ is the slope of the linear regression for gas concentration gradient through time, negative values indicating CH_4 uptake into soil from atmosphere; V and A are volume (m^3) and the hood base area (m^2), respectively.

Soil temperature (5 cm depth) and moisture (0–5 cm) were measured by thermocouples and a hand-held reader (HH-25TC, OMEGA Engineering Inc., Stamford, CT) and a portable TDR probe (HH2, Delta-T Devices, Cambridge, UK) at the same time as the measurement of CH_4 flux.

Partition of the measurement period

Soil temperature remained below zero and was characterized by severe frost from March 1 to March 12 in 2010 and from March 2 to March 10 in 2011. Therefore, soil moisture could not be measured due to the frozen topsoil. This period is referred to as period I. Plants started germination soon after positive and negative soil temperatures occurred in alternation, and the soil became unfrozen down to a depth of 5 cm (Figure 1A, B), as freeze-thaw events occurred. This period is referred to as period II.

Statistical analysis

CH_4 fluxes were analyzed using MIXED procedure of the Statistical Package for Social Science (SPSS 13.0 for Windows, 2003) [11], to test experimental different in CH_4 fluxes. Replicate flux measurements were averaged by the sampling point for each grazing plot. Stocking rate, year, period and all possible interactions were treated as fixed effects, with the grazing plot of each stocking rate as a random effect, and sampling date as the repeated measure with the grazing plot used as the subject. The best fit covariance structure was determined to be compound symmetry. The data was examined for homogeneity of variances (Levene statistic test) and for normal distribution (Kolmogorov-Smirnov test) before analysis. All data was adjusted using log transformation (+1) after test. This transformation is usually applied when observed values were zero. The model testing

Table 1. Main characteristics of studied grazing areas.

Items	NG	LG	MG	HG
Plots sizes (ha)	4.0	4.0	4.0	4.0
Stocking rate (sheep ha^{-1})	0.00	0.75	1.50	2.25
Vegetation characteristics (growing season)				
Aboveground biomass (g m^{-2}, DM)	57.84	45.13	27.24	13.14
Plant cover (%)	21	22	19	18
Soil characteristics				
Bulk density, 0–10 cm (g cm^{-3})	1.23	1.29	1.31	1.38
C to N ratio, 0–10 cm	7.75	7.82	7.75	6.83
Organic C content, 0–10 cm (g kg^{-1})	13.09	12.91	12.55	11.68
Soil texture, 0–10 cm				
Sand (%)	59.2	63.8	69.3	69.1
Silt (%)	29.7	26.9	20.1	19.8
Clay (%)	11.0	9.3	10.6	11.1

NG, non-grazing; LG, light grazing; MG, moderate grazing; HG, heavy grazing.

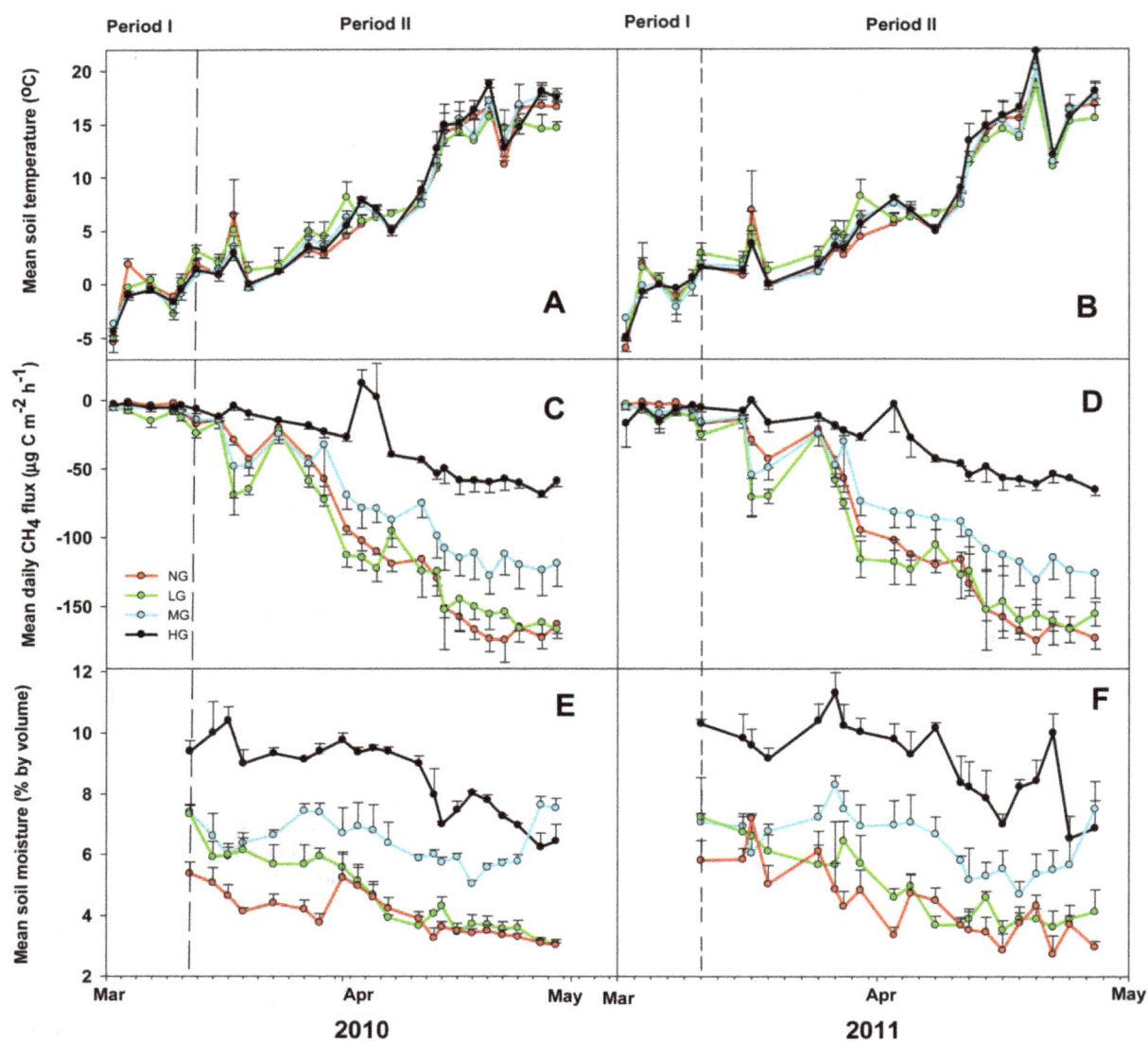

Figure 1. CH$_4$ fluxes, soil temperature and soil moisture (5 cm depth) under different stocking rates. Measurement periods were from March 1 to April 27 in 2010 and March 2 to April 27 in 2011. The left side of the dashed line in the graph indicates Period I (March 1–12 in 2010 and March 2–10, 2011), and the right part period II (March 15 to April 29 in 2010 and March 11 to April 27 in 2011) (NG, non-grazing; LG, light grazing; MG, moderate grazing; HG, heavy grazing). The treatment symbols represent mean values for three plots (9 points). A pertain to soil temperature in 2010, B soil temperature in 2011, C CH$_4$ fluxes in 2010, D CH$_4$ fluxes in 2011, E soil moisture in 2010, and F soil moisture in 2011.

interactive effects of stocking rate, period and year for CH$_4$ flux and their degrees of freedom are shown in Table 2. Paired means of significant differences in treatments were determined using Fisher's least significant difference (LSD) statistic. To test the correlations between soil temperature, moisture and CH$_4$ fluxes, Pearson's correlation analysis was performed. Linear and quadratic regression analysis was used to test the possible dependency of CH$_4$ fluxes on soil moisture and temperature. R^2 (square of Pearson correlation coefficient) value was used to decide the best fitted function (linear or quadratic). All significances mentioned in this paper are at the P = 0.05 level.

Table 2. MIXED model testing interactive effects of stocking rate, period and year for CH$_4$ flux and their degrees of freedom.

Effect	Degrees of freedom	P value
Stocking rate (S)	3	<0.0001
Period (P)	1	<0.0001
Year (Y)	1	0.646
S×P	3	<0.0001
S×Y	3	0.941
P×Y	1	0.608
S×P×Y	3	0.998

Results

Effect of freeze-thaw events on CH$_4$ exchange

Soil CH$_4$ fluxes were not affected (P = 0.646) by year in our study. Averaged over 2 years, CH$_4$ fluxes in the freeze-thaw period were -67.13 µg CH$_4$-C m^{-2} h^{-1} and -66.40 µg CH$_4$-C m^{-2} h^{-1} for periods I and II respectively (Figure 2A).

CH$_4$ uptake of soil was affected by freeze-thaw events. Mean CH$_4$ fluxes were significantly increased in period II (-83.07 µg CH$_4$-C m^{-2} h^{-1}) compared to period I (-7.51 µg CH$_4$-C m^{-2} h^{-1}) (Figure 2B). The soil surface was covered with patched snow until the start of period II. The soil surface of the NG and LG plots even had a thin layer of ice derived from thaw-refreezing that could be observed during period I. We did not measure CH$_4$ fluxes in the snow and ice sites for this experiment. Therefore, it remains unclear whether these snow patches and ice layers affect CH$_4$ fluxes.

Influence of stocking rates on CH$_4$ exchange

The measured mean CH$_4$ fluxes for NG, LG, MG and HG plots in 2010 and 2011 were shown in Table 3. CH$_4$ fluxes during the freeze-thaw period, averaged over 2 years, were illustrated in Figure 2C. The strength of soil-atmospheric CH$_4$ uptake decreased with an increase in stocking rate and the grazing areas were a CH$_4$ sink during the entire period investigated. The uptake and emission of CH$_4$ ranged from -197.93 to 42.03 µg CH$_4$-C m^{-2} h^{-1} in the measured period. The main uptake of CH$_4$ occurred at the NG and LG plots indicating a strong sink (Figure 1C, D). CH$_4$ uptake in LG plots was higher compared to uptake in the other plots during period I, and uptake of NG and LG plots was higher compared to MG and HG plots during period II (Table 3).

Relationship between CH$_4$ uptake and other factors

The measured daily mean soil temperature (5 cm) and soil moisture (0–5 cm, for period II only) are illustrated in Figure 1 (A, B, E, F). The relationship between CH$_4$ fluxes and soil temperature is best represented by a quadratic function

$[F(x) = 0.18x^2 - 9.31x - 20.39, \ R^2 = 0.60, \ P < 0.0001]$ (Figure 3 and Table 4). Soil moisture increased with an increase in stocking rate during the spring thawing period (Figure 1E, F). Using the linear regression function, we found that the soil moisture dependencies were different between the grazing areas (Figure 4 and Table 4). CH$_4$ uptakes were negatively correlated with soil moisture in this study. The slope value (b) of the functions also indicated the difference of CH$_4$ uptakes at the different grazing plots (Table 4).

Discussion

Impacts of freeze-thaw events on CH$_4$ exchange

The most important natural sink of CH$_4$ is in natural upland and forest soils, where methanotrophic activity occurs. The exchange of soil-atmosphere CH$_4$ is comprised of two parts: CH$_4$ produced by soil methanogenesis and atmospheric CH$_4$ oxidation by soil [12]. Physiological activity of microorganisms is limited by temperature, moisture, physical structure, and nutrient pools of soil during the freeze-thaw period [9,13,14]. In our study, CH$_4$ uptake was significantly influenced by freeze-thaw cycles in the desert grassland. This is inconsistent with a previous study conducted in a typical grassland area of Inner Mongolia [7]. Their results found that soil temperature was not a main constraint on CH$_4$ uptake during the freeze-thaw period. We assume that the activity of microorganisms increased with the increase of environmental temperature. This is supported by Sharma et al. (2006) whose research in Southern Germany found that freeze-thaw causes significant physical and biological changes in the soil [13]. On the other hand, CH$_4$ fluxes are also affected by different textured model soils. Low clay content in desert grassland soil is prone to CH$_4$ uptake since it promotes methanotrophic activity of soils [15]. Further studies are required to elucidate the effect of grassland type on CH$_4$ consuming microbes in soils during the freeze-thaw cycle.

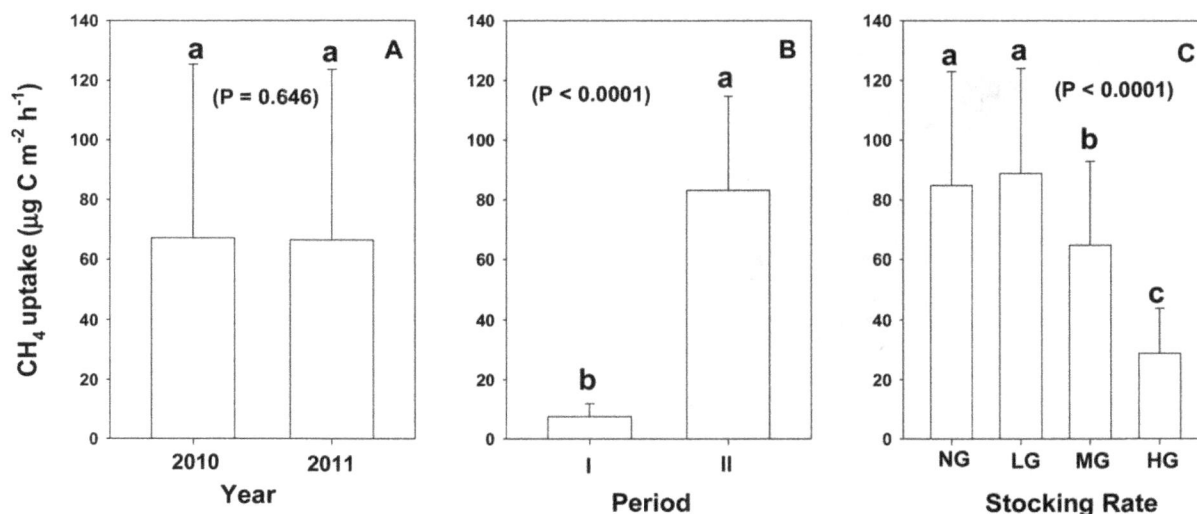

Figure 2. Comparison of CH$_4$ uptake in desert steppes during the spring thaw period. The data shown in (A) represent the mean values for two years at different sites. The data illustrated in (B) are mean values for two periods at different sites. The data shown in (C) represent the mean values for four treatment groups. Different lowercase letters indicate significant differences among groups. NG, LG, MG, and HG are non-grazing, light grazing, moderate grazing, and heavy grazing, respectively.

Table 3. Mean fluxes of CH_4 (μg CH_4-C m^{-2} h^{-1}) under different stocking rates during spring freeze-thaw period.

Period	2010				2011			
	NG	LG	MG	HG	NG	LG	MG	HG
I	−6.04	−11.95	−7.37	−4.24	−3.95	−9.86	−7.20	−9.46
II	−110.24	−112.55	−82.02	−35.38	−103.10	−107.74	−79.13	−34.43
I+II	−86.19	−89.33	−64.79	−28.19	−83.27	−88.16	−64.74	−29.44

Period I: March 1–12 2010 and March 2–10 2011; period II: March 15 to April 29 2010 and March 11 to April 27 2011.
NG, non-grazing; LG, light grazing; MG, moderate grazing; HG, heavy grazing.

Stocking rate effects on CH_4 exchange

The investigated areas acted as a sink for atmospheric CH_4 during the freeze-thaw period in this study. This is in agreement with other studies conducted in other ecosystems [4,7,8,16]. In our study, mean CH_4 uptake during the measured period was 84.76 μg CH_4-C m^{-2} h^{-1} (NG), which is significantly higher than in another study in an Inner Mongolian typical steppe [7]. Wang et al. (2005) reported CH_4 uptake of 28 ± 42 μg CH_4-C m^{-2} h^{-1} and 22 ± 19 μg CH_4-C m^{-2} h^{-1} for non-grazed and grazed *Leymus chinensis* steppe during the non-growing season [17]. However, differences in grassland type, soil property, grazing duration, grazing density, measurement frequencies and method make it difficult to compare those results with ours.

Previous studies indicated that an increase in stocking rate induces a reduction in CH_4 uptake [7,8]. Simulating the effects of grazing management with the PaSim model, Soussana et al. (2004) also suggested that a decline in the GHG sink activity of managed steppes occurs with increased stocking density [18]. We also found a significantly negative correlation between stocking rate and soil CH_4 uptake. However, our results indicate that LG induced a slight increase in CH_4 uptake by soils compared to the NG areas during the spring thaw period. We assume that LG may enhance CH_4 uptake by increasing the population size of CH_4 oxidizing bacteria during the freeze-thaw period in the desert steppe. This is supported by research from a typical steppe in Inner Mongolia [19]. Similarly, Chen et al. (2011) also reported light-to-moderate (stocking rate ≤ 1 sheep ha^{-1} yr^{-1}) grazing did not significantly change CH_4 uptake compared with non-grazed typical steppes [20]. Our results are therefore likely to be pertinent in the context of the current debate on the global magnitude of CH_4 emissions from agricultural ecosystems.

Two mechanisms are noteworthy with regard to the significant effect of HG on CH_4 fluxes. 1) soil bulk density (0–10 cm) increased along the grazing gradient in our study (Table 1), which may result in a reduction of CH_4 diffusion in soil due to reduced pore continuity under HG. This explanation was supported by Chen et al. (2011) who showed a significant linear dependence of topsoil air permeability (AP, cm s^{-1}) on stocking rates (SR, sheep ha^{-1} yr^{-1}; i.e., AP $= -1.46$SR$+4.45$) [20]; 2) although the nutrient pools (e.g. organic carbon, C to N ratio) of soils were hardly affected by HG in this study (Table 1), extraction from grazing animal had the higher potential for CH_4 emission [21], which could offset CH_4 uptake of soil in intensively grazed steppe.

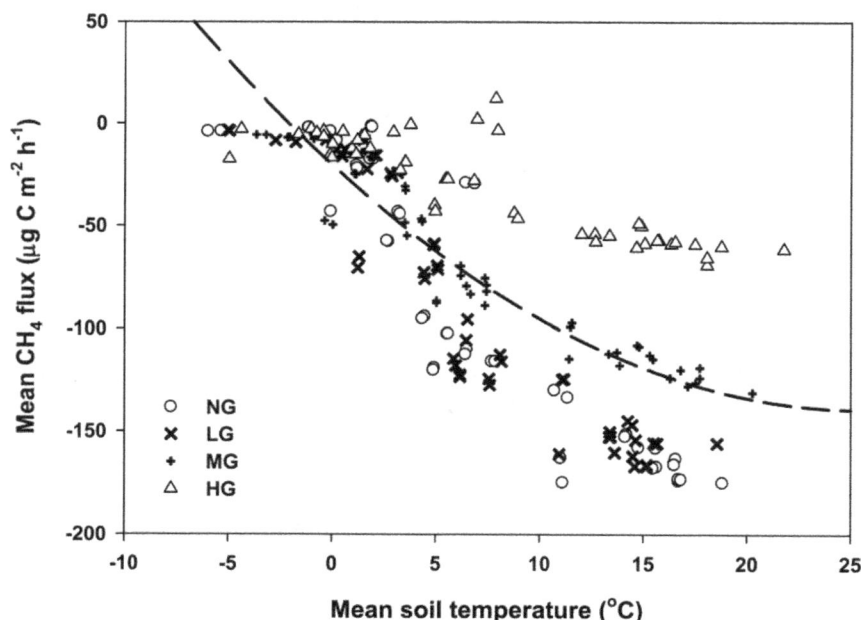

Figure 3. Dependency of CH_4 fluxes on soil temperature (5 cm depth) under different stocking rates. NG, LG, MG, and HG are non-grazing, light grazing, moderate grazing, and heavy grazing, respectively.

Table 4. Parameters of the equations between daily CH_4 fluxes ($\mu g\ C\ m^{-2}\ h^{-1}$) and daily mean soil (5 cm) temperature (°C) or daily soil (0–5 cm) moisture (% by volume) at each site.

Items	a	b	c	R^2	P
T [fitting with $F = ax^2+bx+c$]					
	0.18	−9.31	−20.39	0.60	<0.0001
M [fitting with $F = bx+c$]					
NG	-	48.55	−306.77	0.47	<0.0001
LG	-	35.11	−276.59	0.75	<0.0001
MG	-	27.56	−250.23	0.51	<0.0001
HG	-	13.68	−156.11	0.59	<0.0001

T and M represent the soil temperature and soil moisture, respectively; NG, non-grazing; LG, light grazing; MG, moderate grazing; HG, heavy grazing.

Very low CH_4 emissions were observed during the freeze-thaw period (Period I) in our study. This could be related to snowfall occurring before this experiment. The thaw of snow patches increased topsoil moisture content. A high moisture content of soil allows the development of methanogenic activity [15,22]. This could be an important issue to address in efforts to improve estimates of the CH_4 sink and source potential of desert steppe soils.

Relationships between CH_4 flux and other factors

Soil temperature and moisture content were considered as potential influencing factors on CH_4 fluxes. Although some studies have reported that there is a strong relationship between soil temperature and CH_4 uptake [4,17,20], in agreement with our findings, most studies found this relationship is limited by soil texture, bulk density, and soil moisture, placing greater impor-

tance with controls on gas diffusivity as the determining variable [7,23].

Soil moisture increased along a grazing gradient during the freeze-thaw period (Period II) (Figure 1E, F), which could be caused by an increase in soil compaction due to the high stocking rate [24], leading to a decrease in soil moisture evaporation during the freeze thaw period. This indicates that the effect of grazing events on soil moisture is different between the growing season and the thawing period. An important factor determining CH_4 uptake is gas transport resistance, which is influenced by soil wetness [7,8,17,22]. This view is supported by our findings (Period II) that CH_4 uptake decreased with an increase in soil moisture (Figure 4). Therefore, soil moisture is a major predicting variable for estimation of soil CH_4 dynamics in desert grassland ecosystems.

The possible effects of factors other than soil temperature and moisture on CH_4 uptake were not considered in this study. Consequently, we did not make observation for these other processes in our research area. However, they may produce great influences on CH_4 uptake, and should not be excluded elsewhere. The practice of grazing sheep is a complex event in Inner Mongolian desert grasslands. A range of specific factors, such as fecal and urine deposition, shifts in plant rhizosphere exudation, shifts in plant species, and changes in soil structure and aerobicity can change soil characteristics [25]. Pol-van Dasselaar et al. (1998) reported that CH_4 uptake may be promoted by reduced inorganic nitrogen content [22]. Aronson et al. (2010) also reported that smaller amounts of N tended to stimulate CH_4 uptake while larger amounts tend to inhibit uptake by soil [26]. Therefore, further studies involving the effects of nitrogen, as well as long-term and intensive measurements of soil-atmosphere CH_4 exchange are needed in desert grasslands.

Conclusion

Our findings offer substantial evidence for differences in soil-atmospheric CH_4 exchange during spring freeze-thaw cycles under different stocking rates in a desert steppe ecosystem. These

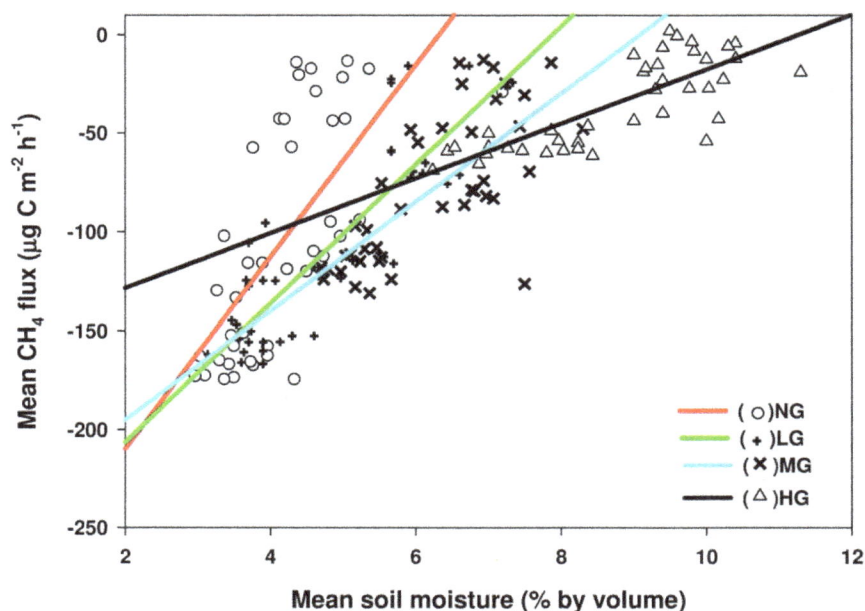

Figure 4. Dependence of CH_4 fluxes on soil moisture (0–5 cm depth) under different stocking rates. NG, LG, MG, and HG are non-grazing, light grazing, moderate grazing, and heavy grazing, respectively. Each data point of daily CH_4 flux measurement is the mean of three replicates.

results are in accordance with other findings that suggest the degree of steppe degradation should be considered for annual CH_4 budgets. CH_4 uptake by soil significantly decreases with an increase in stocking rates. LG can exert a considerable positive impact on CH_4 uptake in the desert steppe at a regional scale. Hence, optimal grazing management practices could play a key role in controlling CH_4 uptake in temperate desert steppes.

Acknowledgments

The authors gratefully acknowledge Inner Mongolia Academy of Agriculture and Animal Husbandry Research Station staff for their assistance in gas sample collection. We also thank A.C.L. Postnikoff for correction of English and grammar.

Author Contributions

Conceived and designed the experiments: SMT CJW GDH DH. Performed the experiments: SMT YYJ. Analyzed the data: SMT CJW. Contributed reagents/materials/analysis tools: CJW AW GDH. Wrote the paper: SMT CJW AW DH.

References

1. IPCC (2007) Climate Change 2007: Mitigation. Contribution of Working Group III to the Fourth Assessment Report of the Intergovernmental Panel on Climate Change. Cambridge University Press, Cambridge, United Kingdom and New York.
2. NSBC (National Statistics Bureau of China) (2002) China Statistics Yearbook 2002 (in Chinese). Chinese Statistics Press, Beijing, China. 6 p.
3. Liao G, Jia Y (1996) Grassland Resource of China. Chinese Science and Technology Publisher, Beijing, P. R. China (in Chinese).
4. Liu C, Holst J, Brüggemann N, Butterbach-Bahl K, Yao Z, et al. (2007) Winter-grazing reduces methane uptake by soils of a typical semi-arid steppe in Inner Mongolia, China. Atmos Environ 41: 5948–5958.
5. Qi Y, Dong Y, Yang X, Geng Y, Liu L, et al. (2005) Effects of grazing on carbon dioxide and methane fluxes in typical temperate grassland in Inner Mongolia, China. Resour Sci 27: 103–109 (in Chinese).
6. Wolf B, Zheng XH, Brüggemann N, Chen WW, Dannenmann M, et al. (2010) Grazing-induced reduction of natural nitrous oxide release from continental steppe. Nature 08931: 881–884.
7. Holst J, Liu C, Yao Z, Brüggemann N, Zheng X, et al. (2008) Fluxes of nitrous oxide, methane, carbon dioxide during freezing- thawing cycles in an Inner Mongolian steppe. Plant Soil 308: 105–117.
8. Mosier AR, Parton WJ, Valentine DW, Ojima DS, Schimel DS, et al. (1996) CH_4 and N_2O fluxes in the Colorado shortgrass steppe: 1. Impact of landscape and nitrogen addition. Global Biogeochem Cy 10: 387–399.
9. Rivkina EM, Friedmann EI, Mckay CP, Gilichinsky DA (2000) Metabolic activity of permafrost bacteria below the freezing point. Appl Environ Microb 8: 3230–3233.
10. Li CL, Hao XY, Zhao ML, Han GD, Willms WD (2008) Influence of historic sheep grazing on vegetation and soil properties of a Desert Steppe in Inner Mongolia. Agr Ecosyst Environ 128: 109–116.
11. SPSS (Statistical Package for Social Science) for Windows XP (2003) Release 16.0, Copyright. SPSS Inc.
12. Wager C, GriessHammer A, Drake HL (1996) Acetogenic capacities and the anaerobic turnover of carbon in Kansas prairie soil. Appl Soil Ecol 62: 494–500.
13. Sharma S, Szele Z, Schilling R, Munch JC, Schloter M (2006) Influence of freeze-thaw stress on the structure and function of microbial communities and denitrifying populations in soil. Appl Environ Microb 72: 2148–2154.
14. Kotelnikova S (2002) Microbial production and oxidation of methane in deep subsurface. Earth-Sci Rev 58: 367–395.
15. Mer JL, Roger P (2001) Production, oxidation, emission and consumption of methane by soils: A review. Eur J Soil Sci 37: 25–50.
16. Dörsch P, Palojärvi A, Mommertz S (2004) Overwinter greenhouse gas fluxes in two contrasting agricultural habitats. Nutr Cycl Agroecosys 70: 117–133.
17. Wang YS, Xue M, Zheng XH, Ji BM, Du R, et al. (2005) Effects of environmental factors on N_2O emission and CH_4 uptake by the typical grasslands in the Inner Mongolia. Chemosphere 58: 205–215.
18. Soussana JF, Loiseau P, Vuichard N, Ceschia E, Balesdent J, et al. (2004) Carbon cycling and sequestration opportunities in temperate grasslands. Soil Use Manage 20: 219–2130.
19. Zhou XQ, Wang YF, Huang XZ, Tian JQ, Hao YB (2008) Effect of grazing intensities on the activity and community structure of methane-oxidizing bacteria of grassland soil in Inner Mongolia. Nutr Cycl Agroecosys 80: 145–152.
20. Chen WW, Wolf B, Zheng XH, Yao ZS, Butterbach- Bahl K, et al. (2011) Annual methane uptake by temperate semiarid steppes as regulated by stocking rates, aboveground plant biomass and topsoil air permeability. Global Change Biol 17: 2803–2816.
21. Jiang YY, Tang SM, Wang CJ, Zhou P, Tenuta M, et al. (2012) Contribution of urine and dung patches from grazing sheep to methane and carbon dioxide fluxes in an Inner Mongolian desert grassland. Asian-Aust J Anim Sci 25: 207–212.
22. Pol-van Dasselaar A, Beusichem ML, Oenema O (1998) Effects of soil moisture content and temperature on methane uptake by grasslands on sandy soils. Plant Soil 204: 213–222.
23. Smith KA, Ball T, Conen F, Dobble KE, Massheder J, et al. (2003) Exchange of greenhouse gases between soil and atmosphere: interactions of soil physical factors and biological processes. Eur J Soil Sci 54: 779–791.
24. Pietola L, Horn R, Yli-Halla M (2005) Effects of trampling by cattle on the hydraulic and mechanical properties of soil. Soil Till Res 82: 99–108.
25. Clegg CD (2006) Impact of cattle grazing and inorganic fertilizer additions to managed grasslands on the microbial community composition of soils. Appl Soil Ecol 31: 73–82.
26. Aronson EL, Helliker BR (2010) Methane flux in non-wetland soils in response to nitrogen addition: a meta-analysis. Ecology 91: 3242–3251.

8

Comparison of Human and Soil *Candida tropicalis* Isolates with Reduced Susceptibility to Fluconazole

Yun-Liang Yang[1,2], Chih-Chao Lin[3], Te-Pin Chang[3], Tsai-Ling Lauderdale[3], Hui-Ting Chen[2,3], Ching-Fu Lee[4], Chih-Wen Hsieh[4], Pei-Chen Chen[3], Hsiu-Jung Lo[3,5]*

1 Department of Biological Science and Technology, National Chiao Tung University, Hsinchu, Taiwan, 2 Institute of Molecular Medicine and Bioengineering, National Chiao Tung University, Hsinchu, Taiwan, 3 National Institute of Infectious Diseases and Vaccinology, National Health Research Institutes, Miaoli, Taiwan, 4 Department of Applied Science, National Hsinchu University of Education, Hsinchu, Taiwan, 5 School of Dentistry, China of Medical University, Taichung, Taiwan

Abstract

Infections caused by treatment-resistant non-albicans *Candida* species, such as *C. tropicalis*, has increased, which is an emerging challenge in the management of fungal infections. Genetically related diploid sequence type (DST) strains of *C. tropicalis* exhibiting reduced susceptibility to fluconazole circulated widely in Taiwan. To identify the potential source of these wildly distributed DST strains, we investigated the possibility of the presence in soil of such *C. tropicalis* strains by pulsed field gel electrophoresis (PFGE) and DST typing methods. A total of 56 *C. tropicalis* isolates were recovered from 26 out of 477 soil samples. Among the 18 isolates with reduced susceptibility to fluconazole, 9 belonged to DST149 and 3 belonged to DST140. Both DSTs have been recovered from our previous studies on clinical isolates from the Taiwan Surveillance of Antimicrobial Resistance of Yeasts (TSARY) program. Furthermore, these isolates were more resistant to agricultural azoles. We have found genetically related *C. tropicalis* exhibiting reduced susceptibility to fluconazole from the human hosts and environmental samples. Therefore, to prevent patients from acquiring *C. tropicalis* with reduced susceptibility to azoles, prudent use of azoles in both clinical and agricultural settings is advocated.

Editor: Kirsten Nielsen, University of Minnesota, United States of America

Funding: This study was supported by 99A1-ID-PP-04-014 and 00A 1-ID-PP-04-014 from National Health Research Institutes and NSC 99-2320-B-400-006-MY3 and NSC 99-2320-B-009-001-MY3 from National Science Council in Taiwan. The funders had no role in study design, data collection and analysis, decision to publish, or preparation of the manuscript.

Competing Interests: The authors have declared that no competing interests exist.

* E-mail: hjlo@nhri.org.tw

Introduction

Due to the elevated number of risk populations, the prevalence of fungal infections has increased significantly in past two decades. *Candida* species are the most frequently isolated fungal pathogens causing morbidity and mortality in seriously immunocompromised hosts. Although *Candida albicans* is the most prevalent species in hospitalized individuals and in nosocomial infections, there has been a shift toward the more treatment-resistant non-albicans *Candida* species [1,2,3,4,5]. This has become an emerging issue in the management of fungal infection. The prevalence of these species differed significantly in various geographic areas [4,5,6,7,8,9,10]. *Candida glabrata* was the most frequently isolated species in Western countries [4,11,12], whereas *C. tropicalis* predominated in Asia [5,13,14,15]. For the treatment, azoles, echinocandins, polyenes, and 5-flucytosine are the four major classes of antifungal drugs. Due to low cost and less side effects, fluconazole has become one of the most commonly prescribed drugs.

Among the phenomena associated with azole resistance, 'trailing' describes the reduced but persistent growth that some isolates exhibit at drug concentrations above the minimum inhibitory concentrations (MICs) in broth dilution tests [16]. When the MIC of an isolate measured after 48 hours (h) incubation is approximately 4-fold higher than that at the 24 h point [17], the isolate is defined to have trailing growth. Thus, in the present study, isolates with fluconazole MICs$\geq$64 mg/L or voriconazole MICs$\geq$4 mg/L after 48 h incubation were considered to have reduced susceptibility to azole drugs.

In order to monitor the trends of species distribution and drug susceptibilities of yeast pathogens, the Taiwan Surveillance of Antimicrobial Resistance of Yeasts (TSARY) program was initiated in 1999 [18]. Subsequently, two more rounds of TSARY were conducted in 2002 and 2006. Previously, we found 23 of the 162 *C. tropicalis* isolates collected from TSARY in 1999 with fluconazole MICs$\geq$64 mg/L to be closely related despite being collected from different hospitals throughout Taiwan [19,20]. Furthermore, 5 of the 23 isolates exhibiting reduced susceptibility to fluconazole were from hospital N4 and all belonged to the same diploid sequence type (DST), DST140 (allele combination, 1, 3, 3, 17, 54, and 3 for *ICL1*, *MDR1*, *SAPT2*, *SAPT4*, *XYR1*, and *ZWF1a*, respectively). However, we did not find any evidence to suggest that these five isolates were transmitted horizontally from person to person [21]. In addition, 2 genetically closely related DST *C. tropicalis* strains, DST140 and DST98 (allele combination, 1, 3, 3, 17, 9, and 3), exhibited reduced susceptibility to fluconazole were identified in TSARY 1999 and 2006. Among them, 18 DST140 isolates were recovered from 10 different hospitals located in all 4 geographic regions in Taiwan and 7 DST98 ones were recovered from 4 different hospitals, two each in northern and southern Taiwan. There were also 3 DST149 (allele combination, 1, 44, 3, 7, 58, and 3) isolates exhibiting reduced susceptibility to fluconazole and from 3 hospitals located in northern Taiwan were identified. These results indicated that those DST strains exhibiting reduced susceptibility to fluconazole circulated widely in Taiwan from 1999 to 2006 and their presence was not a result of outbreaks in certain hospitals or geographic regions [21].

Exposure to azole compounds paves the way for the selection and enrichment of fungal isolates exhibiting reduced susceptibility to drugs, which may occur in patients receiving azole treatments. However, emergence of person to person transmission of isolates exhibiting reduced susceptibility to fluconazole during medical treatment is unlikely. Alternatively, the use of azole compounds in the environment may select organisms exhibiting reduced susceptibility to drugs, which may find their ways to human. *Candida tropicalis* is prevalent in organically enriched soil and aquatic environments [22] as well as in wild birds [23]. Thus, in the present study, we investigated the presence in soil of these *C. tropicalis* strains exhibiting reduced susceptibility to fluconazole, especially DST98, DST140, and DST149.

Results

Candida tropicalis isolates

A total of 56 *C. tropicalis* isolates (Table 1) analyzed in the present study were recovered from 26 of the 477 soil samples collected from 2006 to 2008 around Taiwan. The positive culture rates of *C. tropicalis* from various types of soil differed significantly. It was 85.7% (6/7) for petroleum-contaminated soil samplings, 44.8% (13/29) for agricultural fields, 38.5% (5/13) for sludge soil, and 0.5% (2/427) for forest soil. To identify whether multiple strains were present in a sample, three individual colonies (if present) of each representative morphotype in every soil sample were picked for subsequent workup. In the present study, 10 samples had only one *C. tropicalis* isolate recovered, 8 had 2, 5 had 3, 2 had 4, and 1 had 7(Table 1). All isolates were susceptible to amphotericin B. Among the 18 isolates with fluconazole MICs≥64 mg/L after 48 h incubation, 17 had voriconazole MICs≥4 mg/L as well. Nevertheless, they all exhibited trailing growth phenotype.

Molecular typing

To determine whether the *C. tropicalis* isolates with reduced susceptibility to fluconazole recovered from soil samples were genetically related to those from human, we applied pulsed field gel electrophoresis (PFGE) method to analyze their genetic relatedness. The 18 isolates that were investigated included 11 soil and 7 human isolates. The soil isolates comprised 6 isolates exhibiting reduced susceptibility to fluconazole (NHUE10, 28, 33, 40, 48, and 56) and 5 susceptible ones (NHUE11, 27, 36, 42, and 52). The 7 clinical isolates (YM990275, 490, 537, 579, 592, 649, and 659) all exhibited reduced susceptibility to fluconazole. As expected, all 6 DST140 isolates from human shared an indistinguishable PFGE pattern (Fig. 1). In addition, 2 isolates exhibiting reduced susceptibility to fluconazole from soil, NHUE48 and NHUE56, shared the same PFGE pattern with those human isolates. Among the remaining 4 soil isolates exhibiting reduced susceptibility to fluconazole, 3 (NHUE28, NHUE33, NHUE40), from different samples, shared at least 97% similarity. Furthermore, fluconazole susceptible isolates recovered from soil were not closely related to the ones exhibiting reduced susceptibility to fluconazole from the same areas, i.e. NHUE 11 vs. NHUE 10, NHUE 36 vs. NHUE 33, and NHUE 42 vs. NHUE 40. The fact that YM990579, a DST149 human isolate, also shared the indistinguishable PFGE pattern with the six DST140 human isolates, suggests that MLST has a higher discriminatory

Table 1. Characteristics of *Candida tropicalis* isolates from soils (*N* = 56).

Isolate	Sample	DST	minimum inhibitory concentration (mg/L)			
			Amp 48 h	Flu 24 h	Flu 48 h	Vor 48 h
NHUE1	16	ND	0.5	0.25	0.5	0.0313
NHUE2	16	ND	0.5	0.25	0.5	0.0313
NHUE3	17	ND	0.5	0.25	0.5	0.0313
NHUE4	17	ND	0.5	0.25	0.5	0.0313
NHUE5	17	ND	0.25	0.25	0.5	0.0313
NHUE6	18	ND	0.5	0.25	0.5	0.0625
NHUE7	18	ND	0.5	0.25	0.5	0.0313
NHUE8	18	ND	0.5	0.25	0.5	0.0625
NHUE9	19	ND	1	0.25	0.5	0.0625
NHUE10	20	140	0.5	0.25	>64	>8
NHUE11	20	231	0.5	0.25	0.5	0.0313
NHUE12	21	ND	0.5	0.25	0.5	0.0313
NHUE13	21	ND	0.5	0.25	0.5	0.0313
NHUE14	21	ND	0.5	0.25	0.5	0.0313
NHUE15	26	ND	0.5	0.25	0.5	0.0313
NHUE16	27	ND	1	0.25	0.5	0.0625
NHUE17	28	149	0.5	0.5	>64	>8
NHUE18	28	149	0.5	0.5	>64	>8
NHUE19	28	149	0.5	0.5	>64	>8
NHUE20	7	ND	0.5	0.25	0.5	0.0625
NHUE21	4	226	0.5	0.25	64	0.0313
NHUE22	4	226	0.5	0.25	8	8
NHUE23	5	168	0.5	0.5	>64	>8
NHUE24	5	ND	0.5	0.25	0.5	0.0156
NHUE25	5	ND	1	2	4	0.0625
NHUE26	5	ND	0.5	2	2	0.0625
NHUE27	23	229	0.5	0.5	1	2
NHUE28	24	149	0.5	0.5	>64	>8
NHUE29	24	149	0.5	4	>64	>8
NHUE30	24	149	0.5	0.5	>64	>8
NHUE31	24	ND	0.5	0.25	0.5	0.0313
NHUE32	22	ND	0.5	0.125	1	0.25
NHUE33	22	149	0.5	2	>64	>8
NHUE34	22	149	0.5	0.125	>64	>8
NHUE35	25	ND	0.0313	0.125	0.125	ND
NHUE36	25	230	0.5	0.25	0.25	0.0156
NHUE37	6	183	0.5	0.5	>64	4
NHUE38	6	ND	0.0313	0.25	0.25	1
NHUE39	6	187	0.5	0.5	>64	>8
NHUE40	6	149	0.5	0.5	>64	>8
NHUE41	6	227	1	0.5	>64	>8
NHUE42	6	139	0.5	0.125	0.5	0.125
NHUE43	6	227	0.5	1	>64	>8
NHUE44	8	ND	0.5	0.25	0.5	0.0625
NHUE45	8	ND	0.5	0.25	0.5	0.0625
NHUE46	9	ND	0.5	0.25	0.5	0.0313
NHUE47	11	ND	0.5	0.25	0.5	0.0313

Table 1. Cont.

Isolate	Sample	DST	minimum inhibitory concentration (mg/L)			
			Amp 48 h	Flu 24 h	Flu 48 h	Vor 48 h
NHUE48	13	140	0.5	0.5	>64	>8
NHUE49	13	ND	1	0.125	0.25	0.0156
NHUE50	14	227	0.5	0.5	1	>8
NHUE51	15	ND	1	0.25	0.5	0.0625
NHUE52	15	191	0.25	0.25	1	0.5
NHUE53	12	ND	0.5	0.25	1	0.125
NHUE54	12	ND	0.5	0.25	0.5	0.0313
NHUE55	10	ND	0.5	0.25	0.5	0.0313
NHUE56	3	140	0.5	0.25	>64	4
ATCC 6258		ND	1	8	32	0.25
ATCC 90028		ND	0.5	1	2	0.0313
ATCC 22019		ND	0.5	0.125	0.125	0.0156
ATCC13803		ND	ND	0.5	0.5	ND

DST, diploid sequence type; h, hours; ND, not determined; amp, amphotericin B; flu, fluconazole; vor, voriconazole;

power than PFGE among *C. tropicalis* isolates. Therefore we further employed MLST to compare the genetic relatedness among all 18 soil isolates exhibiting reduced susceptibility to fluconazole. Among them, 9 from 4 soil samples (#6, 22, 24, and 28) were DST149 and 3 from 2 soil samples (#3 and 20) were DST140 (Table 1). In addition to DST140 and DST149, 10 additional DST patterns were identified from 13 isolates in the present study. They were DST139 (allele combination, 1, 4, 22, 23, 36, and 9), DST168 (allele combination, 1, 3, 3, 17, 57, and 3), DST183 (allele combination, 1, 20, 12, 17, 68, and 3), DST187 (allele combination, 1, 63, 2, 30, 69, and 25), DST191 (allele combination, 9, 54, 2, 10, 70, and 9), DST226 (allele combination, 1, 80, 27, 10, 44, and 9), DST227 (allele combination, 5, 54, 1, 47, 48, and 22), DST229 (allele combination, 1, 82, 3, 17, 57, and 3), DST230 (allele combination, 1, 4, 12, 10, 3, and 9), and DST231 (allele combination, 27, 9, 1, 43, 23, and 7). DST140 and DT168 are closely related. They are single locus variants and differ by only one SNP (nt 242) in the *XYR1* gene. Among the 6 tested fluconazole susceptible isolates, NHUE22 (MIC = 8 mg/L) was less susceptible than the other 5 (MIC $\leq$ 1 mg/L) and they all had their own unique DST. Interestingly, NHUE22 and NHUE21 were both from sample 4 and belonged to DST226.

Growth curves

All 18 soil isolates with fluconazole MICs$\geq$64 mg/L after 48 h incubation, such as NHUE40, NHUE48, YM9900579, and YM990649, were able to grow in the presence of 64 mg/L fluconazole (Fig. 2). In contrast, the growth of the susceptible ones

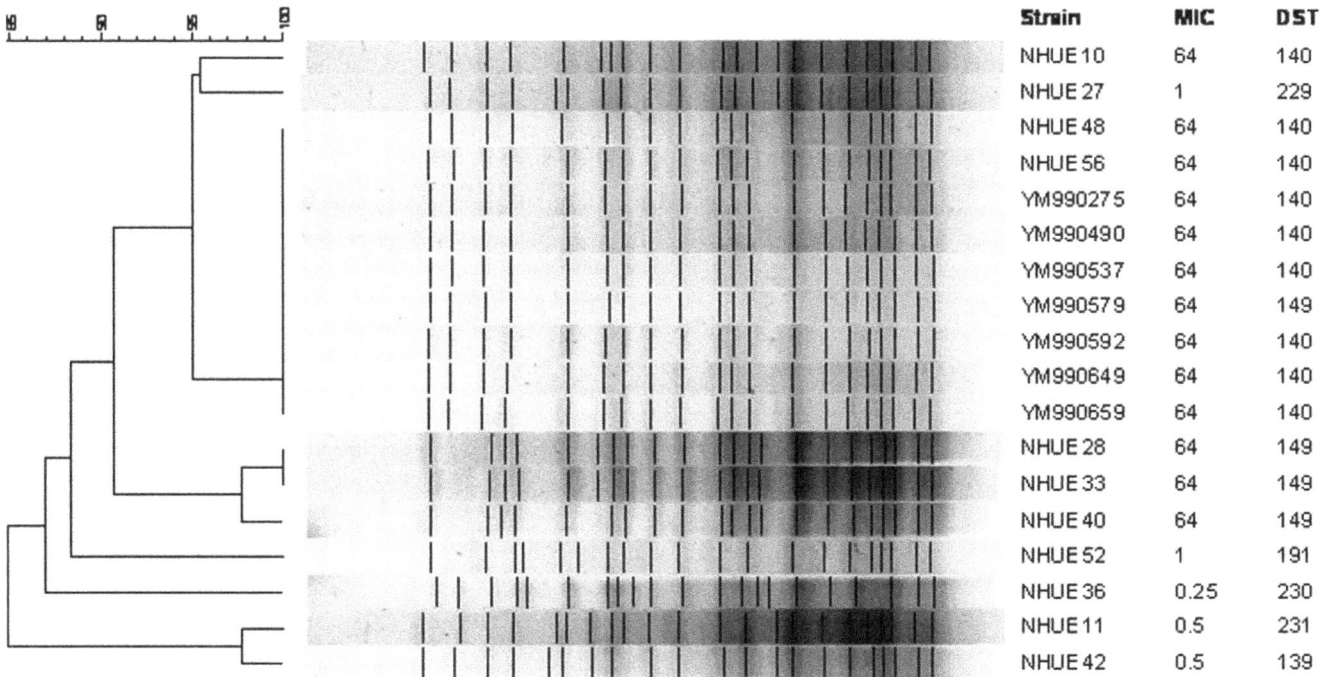

Figure 1. Dendrogram of 18 *Candida tropicalis* isolates from human and soil. The Dice coefficient was used to analyze the similarities between the PFGE band patterns. UPGMA was used for the cluster analysis. The position tolerance and optimization were set at 1%. A total of 7 isolates were recovered from human (YMs) and 11 from soil (NHUEs). MIC refers to fluconazole MIC (in mg/L). An MIC of 64 indicates isolates with reduced susceptibility.

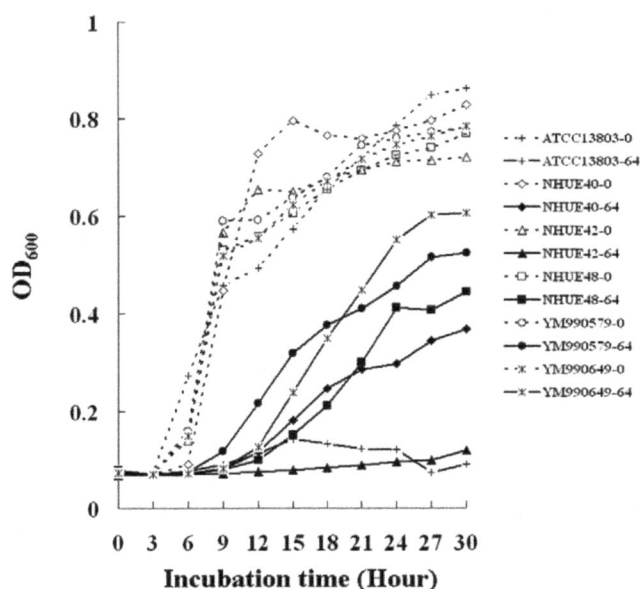

Figure 2. Growth curves of different *C. tropicalis* isolates with different fluconazole susceptibilities. Six isolates, ATCC13803 (rectangles), NHUE40 (diamonds), NHUE42 (triangles), NHUE48 (circles), YM990579 (crosses), and YM990649 (stars) were grown in the RPMI medium 1640 (Gibco BRL31800-022) in the absence (dot lines) or in the presence of 64 mg/L (solid lines) fluconazole.

was inhibited, such as NHUE42 (Fig. 2, triangles solid line) and ATCC13803 (Fig. 2, crosses solid line). Hence, those isolates with fluconazole MICs≥64 mg/L definitely had reduced susceptibility to fluconazole than those isolates with lower fluconazole MICs.

Susceptibilities to drugs used in agriculture

Since 17 of the 18 soil isolates exhibiting reduced susceptibility to fluconazole also had high voriconazole MICs, we selected 6 isolates to investigate whether isolates with reduced susceptibility to fluconazole from human and soil also had reduced susceptibility to other azole drugs used in agriculture. The relative growth of 6 isolates after incubation for 48 h is showed in Table 2. In the presence of 64 mg/L fluconazole, isolates exhibiting fluconazole reduced susceptibility, NHEU40, NHUE48, YM990579, and YM990649 grew better (66–100% of growth) than the fluconazole

susceptible ones, ATCC13803 and NHUE42 (27–32% of growth), confirming our previous results obtained from broth microdilution method. All tested isolates grew poorly (10–11% of growth) in the presence of 64 mg/L penconazole and tebuconazole. Interestingly, the 4 isolates exhibiting fluconazole reduced susceptibility grew better than fluconazole susceptible ones in the presence of 4 mg/L penconazole (36–60% vs. 23–24% of growth), 4 mg/L tebuconazole (54–83% vs. 12–15% growth), 16 mg/L fluquinconazole (71–87% vs. 41–59% of growth), or 64 mg/L tridimenol (24–35% vs. 15–17% growth). Hence, isolates exhibiting fluconazole reduced susceptibility were also less susceptible than the fluconazole susceptible ones to other azole drugs used in agriculture.

Discussion

Candida tropicalis is one of the most frequently isolated non-albicans *Candida* species from patients [8,10,24,25]. Furthermore, *C. tropicalis* develops drug resistance in the presence of fluconazole much more rapidly than other *Candida* species [26,27]. Our findings in the present study suggest that soil may be a potential source of *C. tropicalis* with reduced susceptibility to azole type of antifungal drugs.

The observation that 3 *C. tropicalis* isolates (NHUE28-30) with reduced susceptibility to fluconazole recovered from the same soil sample, #24, belonged to the same DST strain (DST149) suggests that they may be the progenies of the same strain. Nevertheless, one isolate (NHUE31) recovered from the same #24 soil sample was susceptible to fluconazole. On the other hand, of the 7 isolates (NHUE37-43) recovered from another soil sample, #6, 5 exhibited reduced susceptibility to fluconazole and belonged to 4 different DSTs (DST183, 187, 149, and 227). These results indicated that multiple strains of *C. tropicalis* exist in a single soil sample.

Long duration of drug exposure and high numbers of reproducing microorganisms contribute to the selection and/or enrichment of drug-resistant individuals. In fungi, drug resistance is more likely to be the outcome of sequential accumulation of adaptive mutations in the chromosomes [28,29]. Azole-resistant isolates have been shown to emerge following microbial exposure to the drugs in patients and in agricultural settings. The observation that azole-resistant *Aspergillus fumigatus* isolates recovered from soil and compost were genetically related to clinical resistant isolates suggests the existence of an environmental route for developing drug resistance in fungi [30,31,32]. Similar to some antibiotic resistance in bacteria [33,34,35], fluconazole resistance in fungi also involves step-wise mutations affecting more than one

Table 2. Relative growth of six *C. tropicalis* strains in the presence of different drugs.

Drug	none	Fluconazole		Penconazole		Tebuconazole		Tridimenol		Fluquinconazole	
Concentration (mg/L)		4	64	4	64	4	64	4	64	4	16*
ATCC13803	100[#]	34[§]	27	23	11	15	10	49	15	55	41
NHUE40	100	71	71	36	10	58	10	71	34	68	77
NHUE42	100	30	32	24	11	12	10	60	17	75	59
NHUE48	100	87	100	54	11	83	11	78	35	94	87
Ym990579	100	71	66	40	10	59	10	75	35	83	72
Ym990649	100	92	81	60	11	54	11	65	24	100	87

*Due to the solubility of fluquinconazole, the highest concentration of this drug in the present study was 16 mg/L instead of 64 mg/L.
#The growth of each isolate in the absence of drug was defined as 100% and the relative growth of each isolate in the presence of drugs was normalized accordingly.
§Drugs capable of reducing the growth of cells by more than 50% are in bold, considered as susceptible to the concentrations of drugs.

genes [36,37,38]. Therefore, the *C. tropicalis* isolates from soil exhibiting trailing growth phenomena in the present study are prone to be resistant.

In the present study, we have found that *C. tropicalis* isolates with fluconazole MICs≥64 mg/L recovered from human and from soil, albeit from different geographic regions of Taiwan, shared indistinguishable PFGE patterns and belonged to the same DSTs. Whether the original development of *C. tropicalis* exhibiting reduced susceptibility to fluconazole occurred in human hosts or in the environments needs further investigation. However, our observation is consistent with the idea that medically important drug resistant isolates, not only bacteria but also fungi, exist in the environments.

Previous study has shown that in human immunodeficiency virus infected patients, *C. albicans* and other yeast species were cross-resistant to medical and agricultural azole drugs [39]. However, no *C. tropicalis* was recovered in that study. The observation that isolates with reduced fluconazole susceptibility were also more resistant than the susceptible ones to fluquinconazole and tridimenol but not to penconazole and tebuconazole (Table 2) may explain why the later two but not the former two are still used in agriculture in Taiwan. According to the data of "Domestic Manufacturers Production & Sale of Pesticides" published by the Taiwan Crop Protection Industry Association, there is an increased amount of azole-type compounds used in the agriculture in Taiwan, from approximately 100 tons in 2005 to 145 tons in 2009 (about 45% increase). Since both efficient reproduction and spreading of resistant fungi in the environment are to be anticipated, we are confronted with the major challenge of drug-resistance development in environmental setting on a global scale. Hence, to reduce and prevent patients from acquiring azole-resistant *C. tropicalis* and other fungal species, it is advisable to take precaution against unnecessary use of azoles in not only clinical but also agricultural settings. The mechanisms contributing to the reduced susceptibility to azole drugs are under investigation. Further study should be performed to compare *C. tropicalis* isolates recovered from soils in the fields as well as those from farmers for genetic relatedness and susceptibility to various antifungal drugs.

Materials and Methods

Strain isolation

To isolate yeasts from the soil, we transferred approximately one gram of soil from each sample into a tube containing nine ml of sterile water and vortex-mixed. Diluted 0.2 ml of suspensions were plated onto acidified YMA (1% glucose, 0.5% peptone, 0.3% yeast extract, 0.3% malt extract, 1.5% agar, pH3.5) or Dichloran Rose Bengal Chloramphenicol agar (Merck, Darmstadt, Germany). After incubation at 24°C for 3 days, three (if present) representative colonies of each morphotype were picked for subsequent workup. All yeast isolates were identified by the sequences of the D1/D2 domain of the LSU rRNA genes as described by Lee et al. [40].

Drug susceptibility testing

The MICs of amphotericin B (0.0313–16 mg/L), fluconazole (0.125–64 mg/L), and voriconazole (0.0156–8 mg/L) were determined by the same *in vitro* antifungal susceptibility testing established in our laboratory [41,42] according to the guidelines of M27-A3 recommended by the Clinical and Laboratory Standards Institute [43]. RPMI medium 1640 (31800-022, Gibco BRL) was used for the dilution and growth of the yeast culture. Strains from American Type Culture Collection (ATCC),

including *C. albicans* (ATCC 90028), *C. krusei* (ATCC 6258), and *C. parapsilosis* (ATCC 22019), were used as the standard controls. Growth of each isolate was measured by the Biotrak II plate spectrophotometric reader (Amersham Biosciences, Biochrom Ltd., Cambridge England) after incubation at 35°C for 24 h and 48 h.

The MIC of amphotericin B was defined as the minimum inhibitory concentrations of the drug capable of inhibiting cell growth. For susceptibility to amphotericin B, isolates with MIC ≥ 2 mg/L were considered to be resistant and those with MIC≦1 mg/L were susceptible. The MIC for fluconazole and voriconazole was defined as the lowest concentration capable of reducing the turbidity of cells by more than 50%. Isolates with fluconazole MICs≥64 mg/L were considered as to have reduced susceptibility and ≤8 mg/L susceptible. The isolates with MICs in the range of 16–32 mg/L were referred to as susceptible-dose dependent (SDD). Isolates with voriconazole MICs≥4 mg/L were considered as with reduced susceptibility and ≤1 mg/L susceptible. The isolates with MIC 2 mg/L were referred to as SDD.

Pulsed field gel electrophoresis (PFGE)

A total of 18 isolates were examined by PFGE, including 11 soil and 7 human isolates. The soil isolates comprised 6 (NHUE10, 28, 33, 40, 48, and 56) exhibiting fluconazole reduced susceptibility and 5 best matched susceptible ones isolated either from the same sample or from the same type of soil in the same area, NHUE11, 27, 36, 42, and 52 (vs. NHUE10, 28, 33, 40, and 48, respectively). The 7 human isolates, YM990275, 490, 537, 579, 592, 649, and 659, all exhibited reduced susceptibility to fluconazole. Pulsed field gel electrophoresis (PFGE) was performed based on our previous report [44] with some modifications. Briefly, organisms were grown on sabouraud dextrose agar (SDA) plate at 30°C overnight, after which the organisms were embedded in 1% agarose gel (SeaKem Gold agarose, FMC BioProducts). The plugs were subjected to cell lysis by lyticase at 37°C for 2 h, and then treated with proteinase K in 50 mM EDTA buffer at 50°C for 16 h. The genomic DNA was digested with 4 units of *Bss*HII at 37°C for 16 h. Electrophoresis was then performed with a CHEF MAPPER system (Bio-Rad, Hercules, Calif., USA) in 0.5×TBE buffer at 6 V/cm at 14°C with alternating pulses at an angle of 120 degrees in a 5–50 s pulse-time gradient for 20 h. A ladder of *Salmonella choleraesuis* subsp. *choleraesuis* (Smith) Weldin serotype Braenderup (ATCC BAA664) was used as the molecular weight marker. The PFGE pattern was analyzed by BioNumerics software (Version 4.5, Applied Maths, Kortrijk, Belgium). The position tolerance was set at 1% and optimization was set at 1%. The Dice coefficient was used to analyze the similarities of the band patterns. The unweighted pair group method using arithmetic averages (UPGMA) was used for cluster analysis.

Multilocus sequence typing

A total of 18 isolates exhibiting reduced susceptibility to fluconazole and 5 best matched susceptible ones (either from the same sample or from the same type of soil in the same area), NHUE11, 27, 36, 42, and 52 (vs. NHUE10, 28, 33, 40, and 48, respectively) as well as NHUE 22 with MIC of fluconazole 8 mg/L, were analyzed. Based on our previous report [19], the DNA fragments of six genes: *ICL1*, *MDR1*, *SAPT2*, *SAPT4*, *XYR1*, and *ZWF1a* were sequenced for the analyses. The resulted sequences were aligned with BioNumerics 3.0 (Applied Maths, Kortrijk, Belgium) and compared with those in the database of *C. tropicalis* (http://pubmlst.org/website) to obtain the identifiers (DST).

Growth curves

The Bioscreen C analyzer (Oy Growth Curves AB Ltd, Helsinki, Finland) was used to re-assess the growth of different isolates. Approximately 1×10^4 cells in 200 μl of RPMI medium 1640 (Gibco BRL31800-022) were grown in the absence or in the presence of different drugs, including fluconazole, fluquinconazole, penconazole, tebuconazole, and tridimenol, at 35°C. According to the record from the Taiwan Corp Protection Industry Association, penconazole and tebuconazole but not fluquinconazole and tridimenol were currently in use for agriculture in Taiwan. The growths of cells were determined every 15 minutes. Due to the solubility of fluquinconazole, the highest concentration of this drug in the present study was 16 mg/L instead of 64 mg/L. The experiments were repeated twice with similar results.

Acknowledgments

We would like to thank Pfizer for supplying fluconazole and Mission Biotech for performing sequencing. We would like to express our gratitude toward Dr. Tsung-Chain Chang for providing soil samples from petroleum-contaminated areas. We would like to thank Drs. Ferric Fang and Lawrence C. McDonald for their helpful suggestions on the manuscript.

Author Contributions

Conceived and designed the experiments: YLY HJL. Performed the experiments: CCL TPC HTC CWH PCC. Analyzed the data: CCL TPC YLY TLL CFL HJL. Wrote the paper: YLY TLL HJL.

References

1. Iatta R, Caggiano G, Cuna T, Montagna MT (2011) Antifungal Susceptibility Testing of a 10-Year Collection of *Candida* spp. Isolated from Patients with Candidemia. J Chemother 23: 92–96.
2. Pfaller MA, Diekema DJ (2007) Epidemiology of invasive candidiasis: a persistent public health problem. Clin Microbiol Rev 20: 133–163.
3. Pfaller MA, Diekema DJ, Mendez M, Kibbler C, Erzsebet P, et al. (2006) *Candida guilliermondii* , an opportunistic fungal pathogen with decreased susceptibility to fluconazole: geographic and temporal trends from the ARTEMIS DISK antifungal surveillance program. J Clin Microbiol 44: 3551–3556.
4. Warnock DW (2007) Trends in the epidemiology of invasive fungal infections. Nippon Ishinkin Gakkai Zasshi 48: 1–12.
5. Yang YL, Cheng MF, Wang CW, Wang AH, Cheng WT, et al. (2010) The distribution of species and susceptibility of amphotericin B and fluconazole of yeast pathogens isolated from sterile sites in Taiwan. Med Mycol 48: 328–334.
6. Bard M, Bruner DA, Pierson CA, Lees ND, Biermann B, et al. (1996) Cloning and characterization of *ERG25* , the *Saccharomyces cerevisiae* gene encoding C-4 sterol methyl oxidase. Proc Natl Acad Sci U S A 93: 186–190.
7. Beck-Sague C, Jarvis WR (1993) Secular trends in the epidemiology of nosocomial fungal infections in the United States, 1980–1990. National Nosocomial Infections Surveillance System. J Infect Dis 167: 1247–1251.
8. Cheng MF, Yu KW, Tang RB, Fan YH, Yang YL, et al. (2004) Distribution and antifungal susceptibility of *Candida* species causing candidemia from 1996 to 1999. Diagn Microbiol Infect Dis 48: 33–37.
9. Pfaller MA, Boyken L, Hollis RJ, Kroeger J, Messer SA, et al. (2007) In Vitro Susceptibility of Invasive Isolates of *Candida* spp. to Anidulafungin, Caspofungin, and Micafungin: Six Years of Global Surveillance. J Clin Microbiol 46: 150–156.
10. Pfaller MA, Jones RN, Doern GV, Sader HS, Messer SA, et al. (2000) Bloodstream infections due to *Candida* species: SENTRY antimicrobial surveillance program in North America and Latin America, 1997–1998. Antimicrob Agents Chemother 44: 747–751.
11. Pfaller MA, Diekema DJ (2010) Epidemiology of invasive mycoses in North America. Crit Rev Microbiol 36: 1–53.
12. Pfaller MA, Moet GJ, Messer SA, Jones RN, Castanheira M (2011) Geographic variations in species distribution and echinocandin and azole antifungal resistance rates among Candida bloodstream infection isolates: report from the SENTRY Antimicrobial Surveillance Program (2008 to 2009). J Clin Microbiol 49: 396–399.
13. Ann Chai LY, Denning DW, Warn P (2010) *Candida tropicalis* in human disease. Crit Rev Microbiol 36: 282–298.
14. Xess I, Jain N, Hasan F, Mandal P, Banerjee U (2007) Epidemiology of candidemia in a tertiary care centre of north India: 5-year study. Infection 35: 256–259.
15. Yang YL, Ho YA, Cheng HH, Ho M, Lo HJ (2004) Susceptibilities of *Candida* species to amphotericin B and fluconazole: the emergence of fluconazole resistance in *Candida tropicalis*. Infect Control Hosp Epidemiol 25: 60–64.
16. Lee MK, Williams LE, Warnock DW, Arthington-Skaggs BA (2004) Drug resistance genes and trailing growth in *Candida albicans* isolates. J Antimicrob Chemother 53: 217–224.
17. Arthington-Skaggs BA, Lee-Yang W, Ciblak MA, Frade JP, Brandt ME, et al. (2002) Comparison of visual and spectrophotometric methods of broth microdilution MIC end point determination and evaluation of a sterol quantitation method for in vitro susceptibility testing of fluconazole and itraconazole against trailing and nontrailing Candida isolates. Antimicrob Agents Chemother 46: 2477–2481.
18. Lo HJ, Ho AH, Ho M (2001) Factors accounting for mis-identification of *Candida* species. J Microbiol Immunol Infect 34: 171–177.
19. Chou HH, Lo HJ, Chen KW, Liao MH, Li SY (2007) Multilocus sequence typing of *Candida tropicalis* shows clonal cluster enriched in isolates with resistance or trailing growth of fluconazole. Diagn Microbiol Infect Dis 58: 427–433.

20. Wang JS, Li SY, Yang YL, Chou HH, Lo HJ (2007) Association between fluconazole susceptibility and genetic relatedness among *Candida tropicalis* isolates in Taiwan. J Med Microbiol 56: 650–653.
21. Li SY, Yang YL, Lin YH, Ko HC, Wang AH, et al. (2009) Two Closely Related Fluconazole-Resistant Candida tropicalis Clones Circulating in Taiwan from 1999 to 2006. Microb Drug Resist 15: 205–210.
22. Vogel C, Rogerson A, Schatz S, Laubach H, Tallman A, et al. (2007) Prevalence of yeasts in beach sand at three bathing beaches in South Florida. Water Res 41: 1915–1920.
23. Lord AT, Mohandas K, Somanath S, Ambu S (2010) Multidrug resistant yeasts in synanthropic wild birds. Ann Clin Microbiol Antimicrob 9: 11.
24. Hung CC, Yang YL, Lauderdale TL, McDonald LC, Hsiao CF, et al. (2005) Colonization of human immunodeficiency virus-infected outpatients in Taiwan with *Candida* species. J Clin Microbiol 43: 1600–1603.
25. Prasad KN, Agarwal J, Dixit AK, Tiwari DP, Dhole TN, et al. (1999) Role of yeasts as nosocomial pathogens & their susceptibility to fluconazole & amphotericin B. The Indian Journal of Medical Research 110: 11–17.
26. Barchiesi F, Calabrese D, Sanglard D, Falconi DF, Caselli F, et al. (2000) Experimental induction of fluconazole resistance in *Candida tropicalis* ATCC 750. Antimicrob Agents Chemother 44: 1578–1584.
27. Calvet HM, Yeaman MR, Filler SG (1997) Reversible fluconazole resistance in *Candida albicans* : a potential in vitro model. Antimicrob Agents Chemother 41: 535–539.
28. Cowen LE (2001) Predicting the emergence of resistance to antifungal drugs. FEMS Microbiol Lett 204: 1–7.
29. Cowen LE, Sanglard D, Calabrese D, Sirjusingh C, Anderson JB, et al. (2000) Evolution of drug resistance in experimental populations of *Candida albicans*. J Bacteriol 182: 1515–1522.
30. Snelders E, Huis In 't Veld RA, Rijs AJ, Kema GH, Melchers WJ, et al. (2009) Possible environmental origin of resistance of *Aspergillus fumigatus* to medical triazoles. Appl Environ Microbiol 75: 4053–4057.
31. Snelders E, van der Lee HA, Kuijpers J, Rijs AJ, Varga J, et al. (2008) Emergence of azole resistance in *Aspergillus fumigatus* and spread of a single resistance mechanism. PLoS Med 5: e219.
32. Verweij PE, Snelders E, Kema GH, Mellado E, Melchers WJ (2009) Azole resistance in *Aspergillus fumigatus*: a side-effect of environmental fungicide use? Lancet Infect Dis 9: 789–795.
33. Chen FJ, Lo HJ (2003) Molecular mechanisms of fluoroquinolone resistance. J Microbiol Immunol Infect 36: 1–9.
34. McDonald LC, Chen FJ, Lo HJ, Yin HC, Lu PL, et al. (2001) Emergence of reduced susceptibility and resistance to fluoroquinolones in *Escherichia coli* in Taiwan and contributions of distinct selective pressures. Antimicrob Agents Chemother 45: 3084–3091.
35. Yang YL, Lauderdale TL, Lo HJ (2004) Molecular mechanisms of fluoroquinolone resistance in Klebsiella. Curr Drug Targets Infect Disord 4: 295–302.
36. Akins RA (2005) An update on antifungal targets and mechanisms of resistance in *Candida albicans*. Med Mycol 43: 285–318.
37. White TC, Marr KA, Bowden RA (1998) Clinical, cellular, and molecular factors that contribute to antifungal drug resistance. Clin Microbiol Rev 11: 382–402.
38. Yang YL, Lo HJ (2001) Mechanisms of antifungal agent resistance. J Microbiol Immunol Infect 34: 79–86.
39. Muller FM, Staudigel A, Salvenmoser S, Tredup A, Miltenberger R, et al. (2007) Cross-resistance to medical and agricultural azole drugs in yeasts from the oropharynx of human immunodeficiency virus patients and from environmental Bavarian vine grapes. Antimicrob Agents Chemother 51: 3014–3016.
40. Lee CF, Yao CH, Liu YR, Young SS, Chang KS (2009) *Kazachstania wufongensis* sp. nov., an ascosporogenous yeast isolated from soil in Taiwan. Antonie Van Leeuwenhoek 95: 335–341.
41. Yang YL, Cheng HH, Lo HJ (2004) In vitro activity of voriconazole against *Candida* species isolated in Taiwan. Int J Antimicrob Agents 24: 294–296.

42. Yang YL, Wang AH, Wang CW, Cheng WT, Li SY, et al. (2008) Susceptibilities to amphotericin B and fluconazole of *Candida* species in TSARY 2006. Diagn Microbiol Infect Dis 61: 175–180.

43. Clinical Laboratory Standards Institute CLSI (2008) Reference method for broth dilution antifungal susceptibility testing of yeasts; approved standard-third edition. CLSI document M27-A3. Wayne, PA.

44. Chen KW, Lo HJ, Lin YH, Li SY (2005) Comparison of four molecular typing methods to assess genetic relatedness of *Candida albicans* clinical isolates in Taiwan. J Med Microbiol 54: 249–258.

9

In Situ Identification of Plant-Invasive Bacteria with MALDI-TOF Mass Spectrometry

Dominik Ziegler[1,3◉]**, Anna Mariotti**[1,2◉]**, Valentin Pflüger**[3]**, Maged Saad**[1]**, Guido Vogel**[3]**, Mauro Tonolla**[1,2]**,
Xavier Perret**[1]*****

1 Department of Botany and Plant Biology, University of Geneva, Geneva, Switzerland, **2** Institute of Microbiology, Bellinzona, Switzerland, **3** Mabritec AG, Riehen, Switzerland

Abstract

Rhizobia form a disparate collection of soil bacteria capable of reducing atmospheric nitrogen in symbiosis with legumes. The study of rhizobial populations in nature involves the collection of large numbers of nodules found on roots or stems of legumes, and the subsequent typing of nodule bacteria. To avoid the time-consuming steps of isolating and cultivating nodule bacteria prior to genotyping, a protocol of strain identification based on the comparison of MALDI-TOF MS spectra was established. In this procedure, plant nodules were considered as natural bioreactors that amplify clonal populations of nitrogen-fixing bacteroids. Following a simple isolation procedure, bacteroids were fingerprinted by analysing biomarker cellular proteins of 3 to 13 kDa using Matrix Assisted Laser Desorption/Ionization Time of Flight (MALDI-TOF) mass spectrometry. In total, bacteroids of more than 1,200 nodules collected from roots of three legumes of the *Phaseoleae* tribe (cowpea, soybean or siratro) were examined. Plants were inoculated with pure cultures of a slow-growing *Bradyrhizobium japonicum* strain G49, or either of two closely related and fast-growing *Sinorhizobium fredii* strains NGR234 and USDA257, or with mixed inoculants. In the fully automatic mode, correct identification of bacteroids was obtained for >97% of the nodules, and reached 100% with a minimal manual input in processing of spectra. These results showed that MALDI-TOF MS is a powerful tool for the identification of intracellular bacteria taken directly from plant tissues.

Editor: Akos Vertes, The George Washington University, United States of America

Funding: XP acknowledges funding from the Swiss National Science Foundation (grant n°3100A0-116591) and the University of Geneva. DZ thanks Mabritec AG for financial support of the research project. The funders had no role in study design, data collection and analysis, decision to publish, or preparation of the manuscript.

Competing Interests: Co-authors Valentin Pflüger and Guido Vogel are employees of Marbitrec AG. Dominik Ziegler is a PhD student in Geneva whose research is partially funded by Mabritec AG. There are no patents or products in development to declare.

* E-mail: xavier.perret@unige.ch

◉ These authors contributed equally to this work.

Introduction

Of the diverse microorganisms that thrive in the rhizosphere of higher plants, only a few can colonize the inner root tissues. Fewer still can enter plant cells and establish persistent intracellular infections. Amongst them, rhizobia form a heterogeneous group of α- and β-proteobacteria capable of establishing beneficial nitrogen-fixing associations with legumes [1]. Rhizobia are responsible for as much as half of the nitrogen fixed each year by terrestrial biological systems [2], and are often used in agriculture to supplement plants with the reduced forms of nitrogen absent in many soils around the world. Reduction of atmospheric nitrogen (N_2) by rhizobia occurs predominantly inside plant cells found in specialized root (occasionally stem) organs called nodules. To colonize the inner tissues of nodules, rhizobia generally follow a trans-cellular path of infection that crosses the root epidermis as well as several layers of cortical cells. This infection process remains mostly under the control of host plants and rhizobial proliferation is restricted to the tips of infection threads (ITs), which ultimately ramify when reaching the newly formed nodule meristem [3,4]. Plants restrict detrimental infections of root tissues by harmful pathogens or poor nitrogen fixers by using sophisti-

cated screens that involve the coordinated exchange of molecular signals between nodulating rhizobia and host plants [5,6]. Flavonoids released by roots, bacterial surface polysaccharides, nodulation factors (Nod-factors) and type-three secreted proteins released by rhizobia are amongst the most discriminating signals involved in the establishment of a proficient symbiosis. Ultimately, rhizobia are released from ITs into the cytoplasm of nodule cells where they differentiate into nitrogen-fixing bacteroids. In return for reduced nitrogen, plants provide bacteroids with ample carbon supply mostly in the form of dicarboxylates. To protect rhizobial nitrogenase from its irreversible inactivation by traces of free oxygen, plants have evolved a number of mechanisms including a cortical oxygen diffusion barrier that surrounds the infected nodule tissue and the synthesis of leghemoglobin, which together regulate the oxygen tension within infected cells. Leghemoglobin is particularly abundant inside N_2-fixing nodules where it accounts for as much as 30% of all proteins [7]. As a single rhizobial cell attached at the root-hair tip is sufficient to initiate the formation of an infection thread, a population of bacteroids inside a nodule is mostly clonal. In fact nodules can be regarded as small bioreactors in which each of the infected nodule cells contains several hundred bacteroids. Occasionally infection threads can accommodate distinct bacteria, and thus may lead to nodules that house a mixed population of rhizobia [3,4].

Matrix-assisted laser desorption/ionization time of flight (MALDI-TOF) mass spectrometry (MS) has become a method of choice for the identification of bacteria [8]. In such procedures, isolated bacteria are first grown as micro-colonies on solid media prior to MALDI-TOF MS analysis. Then, small numbers of cells are lysed to release the intracellular proteins, which in turn will be ionized and separated according to their mass to charge ratio (m/z). Protein masses are recorded as distinct peaks that together form a complex spectrum or fingerprint, which is characteristic of a bacterial sample (see Fig. 1). Each spectrum of protein masses is subsequently matched against reference spectra that are stored in dedicated databases, and were initially established using well characterized reference strains and defined growth conditions [9]. The Spectral Archive and Microbial Identification System (SARAMIS^{TM} of database management reduces the complexity of mass spectra by restricting datasets to subsets of peaks from the most abundant proteins or peptides found in the 2 to 20 kDa range. These selected peaks together form the so called Super-Spectra^{TM} (SSp) that predominantly include ribosomal proteins as selected biomarkers [10]. Depending on the number of SSp that are stored in a reference database for a given group of bacteria, a sampled strain can be reliably assigned to a known family, genus, species or more rarely a particular subspecies [11,12,13,14,15].

Except for a few protocols in which infectious bacteria from urine or blood samples were concentrated by centrifugation prior to cell typing [16,17,18], the identification of environmental or pathogenic bacteria by MALDI-TOF MS analysis requires preliminary steps of bacterial isolation and purification. Unlike most animal pathogens that form isolated micro-colonies in a matter of hours, plant-interacting bacteria require longer incubation times that, as in the case of bradyrhizobia, may extend over several days. To prevent growth of soil contaminants during such long incubation periods, standard methods for the identification of endosymbiotic rhizobia include a surface-sterilization of root nodules prior to sampling serial dilutions of their contents on plates [19]. Recently, MALDI-TOF MS technology was applied to the identification of fast–growing rhizobia, but only to strains cultivated on solid media [20]. To bypass the initial and time-consuming steps of bacterial isolation and subsequent cultivation, we explored the possibility of applying MALDI-TOF MS analysis directly to populations of endosymbiotic bacteroids found in root nodules. For this study, three symbiotic strains were used in combination with the three host plants: *Vigna unguiculata* (cowpea), *Macroptilium atropurpureum* (siratro) and *Glycine max* (soybean). With a generation time close to 3 h., the fast-growing *Sinorhizobium fredii* strain NGR234 was selected for its unique ability to induce nodule formation on more than 120 genera of legumes, and fix nitrogen with a subset of at least 150 plants [21]. The other fast-grower *S. fredii* strain USDA257 shares with NGR234 many of its symbiotic genes [22,23], but unlike NGR234 forms N$_2$-fixing symbioses with selected soybean cultivars [21]. In contrast, the slow-growing micro-symbiont *Bradyrhizobium japonicum* strain G49 is genetically distant from NGR234 and USDA257 and fixes nitrogen in association with diverse legumes, including *G. max*, *M. atropurpureum* and *V. unguiculata*. Using these distinct strains, we show here that MALDI-TOF MS can allow for the rapid and reliable identification of nodule bacteria, including at the subspecies level, thus paving the way for large-scale population studies of rhizobia isolated from root nodules found on crops of agronomical importance.

Results

MALDI-TOF MS fingerprinting of free-living rhizobia

Reliable identification of bacterial strains requires the selection as biomarkers of protein masses that appear consistently in spectra, irrespective of the growth conditions or age of the cultures. Thus, cells of strains G49, NGR234, and USDA257 were grown at 28°C on solid media for 2, 3, 5, 7, 10, and 14 days. Initially, bacterial samples were subsequently prepared using either (i) the direct smear method, (ii) a suspension in 25% formic acid, or (iii) a pre-purification step for cellular proteins via an extraction with acetonitrile and 70% formic acid. For each strain, 18 averaged spectra of good quality (generated using the 3 methods of sample preparation for each of the 6 incubation periods) were compiled into a single representative SuperSpectrum^{TM} (SSp) comprised of 39 to 42 biomarker masses that are listed in Table S1. Because of their abundance, small size and ubiquitous distribution, ribosomal proteins are often selected as biomarkers [10,24] that, in some cases, are sufficiently discriminatory to allow for an accurate typing at the strain level [25,26]. Accordingly, 35 of the 108 biomarkers selected for the SSp of G49, NGR234 and USDA257 were tentatively assigned to ribosomal proteins with corresponding molecular weights that were calculated using the corresponding genomic sequences (see Table S1). As befits strains that belong to the same *Sinorhizobium* genus, the reference spectra of NGR234 and USDA257 shared 12 biomarkers of which 7 were identified as ribosomal proteins. In contrast, none of the 39 biomarkers specific to G49 were found in NGR234 or USDA257, and none of the 108 masses listed in Table S1 were common to all three rhizobial SSp. Ultimately, the sets of biomarker masses were matched against each other and against the whole SARAMIS database (ver. 4.09, system ver. 3.4.1.2, with >70,000 spectra and >2,600 SSp). This procedure showed that an unequivocal strain assignment was reached when at least 60% of the biomarkers of a given spectrum of G49, NGR234 or USDA257 matched those of the reference SSp. Thus, the identification thresholds were set at 24 (from a total of 39), 25 (of 40) and 26 (of 42) matching biomarkers for the spectra of respectively, strains G49, NGR234 and USDA257.

Generating reproducible spectra of root-nodule bacteria

To establish a reliable protocol of rhizobial identification that was both robust and able to be applied to many symbioses, a number of plant-microsymbiont combinations were tested. Batches of 16 plants of *G. max* cv. Davis, *M. atropurpureum* (DC.) Urb., or *V. unguiculata* cv. Red Caloona were inoculated with either strain G49, NGR234 or USDA257, corresponding to a total of 144 plants in 72 Magenta jars for each legume species. Except for NGR234 that forms uninfected pseudonodules on *G. max* and USDA257 that fixes poorly with *M. atropurpureum*, all remaining plant-bacterium combinations lead to proficient nitrogen-fixing associations. To avoid potential problems linked to senescence of bacteroids, nitrogen-fixing nodules of cowpea, siratro, and soybean were harvested before flowering time at respectively, 42, 49 and 35 days post-inoculation (dpi) and stored at −60°C until further use. Following surface sterilisation, nodules were processed as described in the materials and methods section to remove contaminating leghemoglobin from the samples. Enriched bacteroid pellets were subsequently treated using either of the three methods described above for free-living bacteria, and two replicates for each nodule. A total of 420 nodules were examined in this way, including 60 nodules for each of the following microsymbiont/host-plant combinations G49/cowpea, G49/soybean, NGR234/cowpea, NGR234/siratro, USDA257/cowpea and USDA257/soybean, and 30 nodules for the G49/siratro

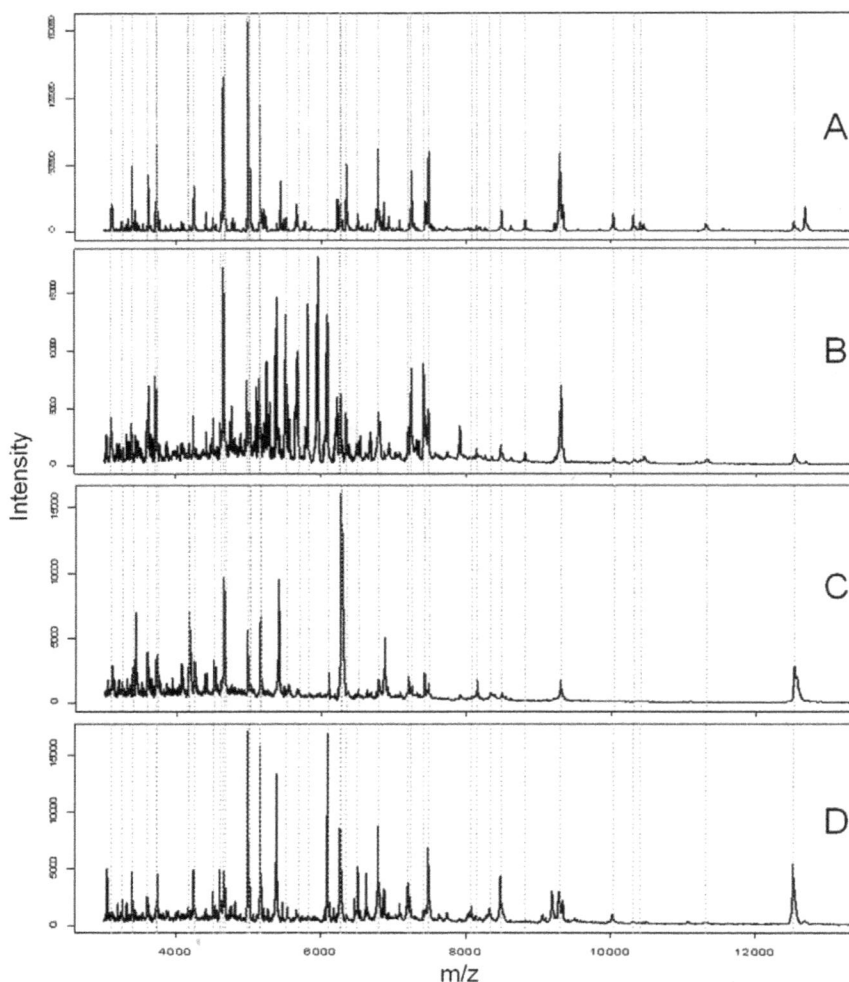

Figure 1. Typical MALDI mass spectra between *m/z* 3000 and *m/z* 13000 of *S. fredii* strain USDA257 grown on TY for 5 days at 28°C (A) or found inside nitrogen-fixing nodules of *G. max* (B), *M. atropurpureum* (C), and *V. unguiculata* (D) collected respectively 35, 49, and 42 days post-inoculation. Sample preparation included washing steps to remove leghemoglobin, and suspension of bacterial pellets in 25% formic acid. Positions of biomarkers selected for the USDA257 SSp are showed with dashed lines.

and USDA257/siratro combinations. Representative spectra of free-living or endosymbiotic cells of USDA257 are shown in Figure 1. Using the threshold of 60% biomarkers in common that was established earlier for G49, NGR234 or USDA257 SSp, a positive strain-identification was reached for 389 of the 420 nodules (92%). When the 30 small and poorly infected nodules formed by USDA257 on siratro were excluded from the analysis, the success rate for the identification of nodule bacteria increased to 97% (379 out of 390 nodules). However, nodule bacteria from the remaining 11 unassigned nodules (3%) were identified by matching manually each of the unknown spectra against the representative spectra selected to generate the SSp of G49, NGR234 or USDA257. Thus, with a minimal operator input, bacteroids from all nodules were successfully identified. The circular dendogram presented in Figure 2 shows the unambiguous clustering in distinct G49, NGR234 or USDA257 groups of 410 individual spectra obtained from free-living bacteria or nitrogen-fixing bacteroids, using *Agrobacterium tumefaciens* as an outgroup. Ultimately, as spectra of optimal quality were produced using the preliminary treatment of isolated bacteria in 25% formic acid, all subsequent samples were prepared in this way.

A competition assay using MALDI-TOF MS to identify nodule bacteria

In the soil, competition between rhizobia to gain access to the nodule environment is widespread and often limits attempts to use improved strains for field inoculations [27,28]. To assess the reliability of the MALDI-TOF MS protocol for identifying root-nodule bacteria, a competition assay was undertaken. Cowpea was selected as a host plant because of its low selectivity for rhizobia [29]. Sinorhizobia NGR234 (Rif[R], Km[S]) and USDA257 (Rif[S], Km[R]) were chosen as microsymbionts because both strains (i) are genetically closely related, (ii) have orverlapping host-ranges [21], and (iii) possess distinct antibiotic resistances that facilitate their identification using appropriate selective media. Inoculants of 2×10^8 rhizobia per plant were prepared using mixtures of NGR234 and USDA257 cells at ratios of 100/1, 50/50 or 1/100. Plants were harvested 42 dpi and a total of 113 to 120 nodules from up to 11 plants were examined for each of the three inoculants (Table 1). Nodule occupancy was determined using both MALDI-TOF MS spectra of bacteroids and by plating isolated nodule bacteria on selective media. In all conditions tested, NGR234 outcompeted USDA257 for nodulation on *V.*

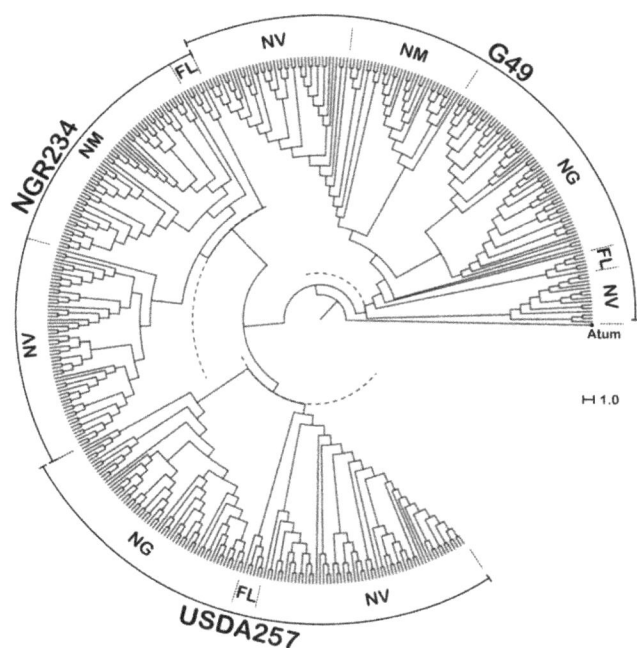

Figure 2. Unsupervised hierarchical cluster analysis (Dice algorithm) of G49, NGR234 and USDA257 MALDI-TOF MS mass spectra comprised in the size range of *m/z* 3000 to 15000. Bacteria were isolated from free-living cultures (FL), or nodules of *G. max* (NG), *M. atropurpureum* (NM), or *V. unguiculata* (NV). The respective positions of the 60% identity level used as a threshold to assign strains are shown as dotted semi-circles. *Agrobacterium tumefaciens* (Atum) was selected as the outgroup.

Table 1. Identification of bacteria competing for nodulation on *V. unguiculata*.

		NGR234 vs. USDA257		
	Cell ratio	100/1	50/50	1/100
	Nodules/plants analyzed	120/10	120/11	113/11
Nodule occupancy	NGR234	115 (95.9)	102 (85.0)	64 (56.6)
	USDA257	0 (0.0)	5 (4.2)	20 (17.7)
	mixed – NGR234 spectra	4 (3.3)	4 (3.3)	5 (4.4)
	mixed – USDA257 spectra	1 (0.8)	9 (7.5)	24 (21.2)
Numbers of manual inputs in MALDI analysis		11 (9.2)	3 (2.5)	2 (1.8)

% of total nodules are shown in brackets.

unguiculata including when cells of NGR234 were outnumbered by a factor of 100. Although MALDI-TOF MS analysis systematically resulted in the identification of a single nodule strain (most probably the strain with the most abundant bacteroid population), growth on selective media confirmed that 4 to 25% of the nodules were infected with both NGR234 and USDA257. As no bacterial growth was observed on plates on which surface-sterilized nodules were rolled, the sterilization procedure was efficient and cases of nodule occupancy documented as mixed were unlikely to result from contaminating bacteria found on the nodule epidermis. In a similar assay using G49 and NGR234 as competitors, the analysis of an additional 484 nodules showed that G49 outcompeted NGR234 for nodulation on *V. unguiculata* irrespective of the cell ratios used as inoculants (data not shown). These competition assays also showed that a correct identification of the predominant bacteroid population was reached by SARAMIS™ in the automatic mode for 753 (90%) of the 837 nodules examined, while strain identification for the remaining 84 nodules required only a minimal operator input.

Discussion

That NGR234 so efficiently outcompeted USDA257 for nodulation on *V. unguiculata* was surprising since both strains are genetically closely related, secrete Nod-factors of similar structures, and have largely overlapping host-ranges [6]. The absence of functional copies of the *nodSU* genes [30] is probably not the reason for the lower competitiveness of USDA257, as a mutant of NGR234 deleted in *nodSU* nodulates cowpea with the same efficiency as the wild type [31]. It seems likely that competitiveness

of NGR234 and USDA257 on *V. unguiculata* is not mediated by Nod-factors, but rather due to specific alterations in the composition and structure of their respective cell surface components. Both NGR234 and USDA257 activate type three protein secretion systems (T3SS) in response to flavonoid stimulation [32], yet the presence or absence of a functional T3SS has no measurable effect on the nodulation of cowpea by NGR234 [33]. Instead, flavonoid-inducible modifications in the *O*-antigen portion of lipopolysaccharides were shown to affect the ability of NGR234 to fix nitrogen on *V. unguiculata* [34]. As genes responsible for rhamnan synthesis are absent from USDA257 [23], surface polysaccharides are candidate determinants of the NGR234 competitivity on cowpea.

More importantly, the competition assays confirmed that our protocol for bacteroid identification via MALDI-TOF MS was capable of discriminating between closely related strains such as NGR234 and USDA257. The possibility to bypass the time-consuming step of cultivating individual colonies on plates to reach a reliable identification of nodule bacteria by MALDI-TOF MS opens up new perspectives. According to our data, bacteroids found within a single mature nitrogen-fixing nodule are in sufficient numbers to generate reproducible spectra of good quality. Except for the abundant leghemoglobin that needs to be washed away to prevent the saturation of spectra, nodule cells or their proteins do not seem to interfere with the process of bacteroid identification. Although spectra of USDA257 isolated from nodules of distinct plants appeared to differ (see Fig. 1), a correct identification was ultimately reached because only the presence or absence (and not the respective signal intensities) of specific biomarker masses were used as discriminating quanta. That one SSp generated from spectra of free-living rhizobia was sufficient to reach a correct strain identification, irrespective of the plant from which the bacteroids were collected, facilitates further developments. The need to improve the SARAMIS database with SSp's of both free-living and endosymbiotic rhizobia would have been impractical, given that some promiscuous strains may nodulate more than one hundred genera of legumes [21]. Currently, a rhizobia-specific database is being established using well-characterized fast- and slow-growing strains belonging to the α- and ß-proteobacteria. Once established, such a database is bound to facilitate the implementation of MALDI-TOF MS as a tool to study the great diversity in populations of symbiotic rhizobia in agricultural or natural soils. To test whether our protocol for strain identification was efficient for such a purpose, 24 nodules were collected in a field of mature soybeans in the Swiss state of Ticino. MALDI-TOF MS spectra of the bacteroids found within these nodules were matched against those of cells of G49, NGR234 and USDA257 isolated from nodules of plants grown in laboratory conditions. Figure 3 shows that field-nodule bacteria clustered together with G49 bacteroids isolated from nodules of *G. max* cv. Davis while free-living cells and bacteroids of G49 found in cowpea or siratro formed distinct subclusters. This result is consistent with the use of *B. japonicum* strains as commercial inoculants for soybean cultures (including in Switzerland), and confirms that direct identification via MALDI-TOF MS of strains found in environmental samples is possible.

However, nodules come in many forms and shapes in nature, and the maturation of bacteroids and their fate once nodules senesce vary considerably in a host-dependent manner [1]. For example, nodules of cowpea, siratro and soybean, are all of determinate type with populations of bacteroids being regarded as relatively homogenous and to some extent capable of resuming growth once outside of plant cells [35,36]. In contrast, nodules of the indeterminate type are characterized by the presence of a persistent meristem, and several developmental zones within which bacteroids display distinct differentiation states [37]. Legumes of the inverted repeat-lacking clade (IRLC) such as

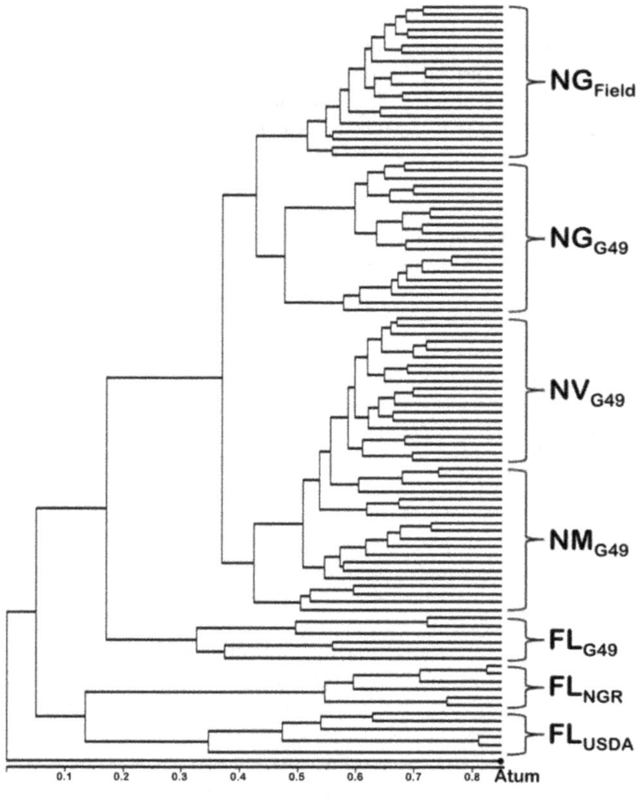

Medicago sativa, form indeterminate nodules in which nitrogen-fixing bacteroids are terminally differentiated [38]. Terminal differentiation results from the action of abundant nodule-specific and cystein-rich peptides (NCRs) of 3 to 5 kDa that induce the endoreduplication of the bacteroid genome and a concomitant cell enlargement [39,40]. It is unclear whether the endoreduplicated state of *Sinorhizobium meliloti* bacteroids in indeterminate nodules of *M. sativa* may affect the process of strain identification by MALDI-TOF MS. In this respect, further studies will also confirm whether NCRs that are ultimately targeted to bacteroids, can be identified by comparing the spectra of free-living and endosymbiotic cells of *S. meliloti*. Nevertheless, our current protocol of strain identification suits many agronomically important legume crops such as soybean, bean (*Phaseolus vulgaris*) or pigeon pea (*Cajanus cajan*), all of which develop determinate nodules on their roots.

Materials and Methods

Characteristics of bacteria and growth conditions

Derivatives of *S. fredii* strains NGR234 resistant to rifampicin (Rif^R) [41] and USDA257 resistant to kanamycin (Km^R) [42] were grown at 27°C in/on TY [43], supplemented with antibiotics at final concentrations of 50 µg ml^{-1}. *B. japonicum* strain G49 was isolated from a batch of the commercial inoculum HiStick Soybean (Becker Underwood, Littlehampton, West Sussex, UK) and grown in/on yeast-mannitol (YM) medium. Strain G49 was sensitive to both, rifampicin and kanamycin (Rif^S, Km^S).

Plant assays

Surface sterilized seeds of *G. max*, *M. atropurpureum* and *V. unguiculata* were dispersed on agar plates and incubated to germinate in the dark, at 27°C. Once germinated, seedlings were planted in Magenta jars containing vermiculite [31], and watered using nitrogen free B&D solution [44]. Two to three days after their transfer to Magenta jars, each plantlet was inoculated with 200 µl of a solution of 2×10^8 bacteria. For competition assays, *V. unguiculata* plantlets were inoculated with either 1/100, 50/50 or 100/1 titters of the NGR234/USDA257 or NGR234/G49 combinations. All plants were grown 6 to 8 weeks post inoculation at a day temperature of 27°C, a night temperature of 20°C and a light phase of 12 hours.

Sample preparation for MALDI-TOF MS identification of free-living bacteria

Prior to spotting MALDI steel target plates, bacteria of G49, NGR234 or USDA257 were prepared using either of the three protocols: (i) direct smear method using a disposable loop, (ii) suspension in 25% formic acid, and (iii) extraction with acetonitrile (ACN) and formic acid (70%). Then, bacterial samples were spotted twenty times on MALDI steel target plates after growth periods at 28°C of 2, 3, 5, 7, 10 and 14 days on plates. Spots were overlaid with 1 µl of matrix consisting of a saturated solution of alpha-cyano-4-hydroxycinnamic acid (Sigma-Aldrich, Buchs, Switzerland) in 33% acetonitrile (Sigma-Aldrich), 33% ethanol and supplemented with 3% trifluoroacetic acid, and air-dried within minutes at room temperature.

Preparation of nodule bacteria

Once harvested, nodules were frozen and kept at −60°C until further use. Prior to isolation of endosymbiotic bacteria, nodules were surface sterilized in 70% ethanol and crushed into 400 µl of sterile ddH$_2$O. For replica plating of nodule bacteria, 20 µl of these homogenates were transferred into independent wells of 96 MicroWell Plates Round Bottom of Nunc (Nunc GmbH & Co.

Figure 3. Typing of bacteroids found in nodules of a soybean field. Unsupervised hierarchical cluster analysis (Dice algorithm) of complex MALDI-TOF MS spectra (*m/z* 3000 to 15000) of rhizobia found inside nodules of *G. max* from a field in the Swiss state of Ticino (NG$_{Field}$), or of G49 isolated from nodules of soybean (NG$_{G49}$), siratro (NM$_{G49}$), and cowpea (NV$_{G49}$). Spectra from free-living cells of G49 (FL$_{G49}$), NGR234 (FL$_{NGR}$), and USDA257 (FL$_{USDA}$), were also included. *A. tumefaciens* (Atum) was used as an outgroup.

KG, Langenselbold, Germany), and aliquots were subsequently plated on selective media. The rest of the crude supernatants were transferred free of nodule debris into new microfuge tubes and were centrifuged at 20,000 g to collect bacteroids. The bacterial pellets were washed three times with 200 μl of sterile ddH$_2$O to remove plant leghemoglobin. Bacteroids were prepared for MALDI-TOF MS fingerprinting as described above.

MALDI-TOF fingerprinting

Protein mass fingerprints were obtained using a MALDI-TOF Mass Spectrometry Axima™ Confidence machine (Shimadzu-Biotech Corp., Kyoto, Japan), with detection in the linear positive mode at a laser frequency of 50 Hz and within a mass range from 2,000–20,000 Da. Acceleration voltage was 20 kV, and the extraction delay time was 200 ns. A minimum of 10 laser shots per sample was used to generate each ion spectrum. For each bacterial sample, a total of 50 protein mass fingerprints were averaged and processed using the Launchpad™ 2.8 software (Shimadzu-Biotech Corp., Kyoto, Japan). For peak processing of raw spectra, the following settings were selected for Launchpad™ 2.8: (i) the advanced scenario from the Parent Peak Cleanup menu, (ii) a peak width set at 80 chans, (iii) the smoothing filter width set at 50 chans, (iv) a baseline filter width of 500 chans and (v) a threshold apex method set as dynamic. The Threshold offset was set at 0.020 mV with a threshold response factor of 1.2. Each target plate was first externally calibrated using spectra of the reference strain *Escherichia coli* DH5α. In addition, the spectra that were acquired to generate the Superspectra were internally calibrated using strain-specific sets of marker masses at mass-to-charge-ratio (m/z) of 4990, 5090, 6188, 6848, 7035, 7480, 7840, 8511 and 8950 for *B. japonicum* strain *G49*; m/z 4979, 5153, 7255, 7484, 8484, 8816, 9239, 10007, 10313 and 11333 for *S. fredii* strain NGR234; and m/z 4979, 5153, 7240, 7484, 8144, 8484, 8815, 9305, 10034, 10312, and 11334 for *S. fredii* strain USDA257. The processed spectra were exported as ASCII datasets with each peak being characterized by an m/z value and a signal intensity.

Generation of Superspectra and cluster analysis

Fingerprints of protein masses were analyzed with SARA-MIS™ (Spectral Archive and Microbial Identification System, AnagnosTec, Potsdam-Golm, Germany). For each preparation, lists of peaks were imported and binary matrix was calculated using the SARAMIS™ Superspectrum tool with an error window of 800 ppm. Average spectra were generated with Excel macros in which noise masses found in less than half of the twenty replicates were eliminated, resulting in 18 representative averaged spectra (for three preparations and 6 cultivation periods) per strain. For each of the three selected strains, a SuperSpectrum™ consisting in a specific pattern of biomarker masses was calculated using the SARAMIS™ Superspectrum tool. For cluster analysis of protein mass fingerprints, binary matrices were generated using the SARAMIS™ Superspectra tool, and exported as text files. Data was imported into the PAST software (Natural History Museum, Oslo University, Norway) and multivariate neighbor joining cluster analysis was performed using the Dice algorithm to calculate distance matrices. Dendograms were drawn using the FigTree v1.3.1 software (http://tree.bio.ed.ac.uk/software/figtree/) and distance matrices formatted as nexus file.

Supporting Information

Table S1 Selected biomarkers for SuperSpectra of strains G49, NGR234 and USDA257.

Acknowledgments

We thank Natalia Giot for her help in many aspects of this work, as well as Gérard Hopfgartner and Franco Widmer for their insightful inputs. We would like to acknowledge Julie Ardley for her critical reading of the manuscript.

Author Contributions

Conceived and designed the experiments: XP MT AM DZ. Performed the experiments: AM DZ MS. Analyzed the data: AM DZ VP GV XP MT. Contributed reagents/materials/analysis tools: XP MT GV. Wrote the paper: XP DZ AM MT.

References

1. Masson-Boivin C, Giraud E, Perret X, Batut J (2009) Establishing nitrogen-fixing symbiosis with legumes: how many rhizobium recipes? Trends Microbiol 17: 458–466.
2. Herridge DF, Peoples MB, Boddey RM (2008) Global inputs of biological nitrogen fixation in agricultural systems. Plant Soil 311: 1–18.
3. Gage DJ (2002) Analysis of infection thread development using Gfp- and DsRed-expressing *Sinorhizobium meliloti*. J Bacteriol 184: 7042–7046.
4. Gage DJ (2004) Infection and invasion of roots by symbiotic, nitrogen-fixing rhizobia during nodulation of temperate legumes. Microbiol Mol Biol Rev 68: 280–300.
5. Jones KM, Kobayashi H, Davies BW, Taga ME, Walker GC (2007) How rhizobial symbionts invade plants: the *Sinorhizobium-Medicago* model. Nat Rev Microbiol 5: 619–633.
6. Perret X, Staehelin C, Broughton WJ (2000) Molecular basis of symbiotic promiscuity. Microbiol Mol Biol Rev 64: 180–201.
7. Baulcombe D, Verma DP (1978) Preparation of a complementary DNA for leghaemoglobin and direct demonstration that leghaemoglobin is encoded by the soybean genome. Nucleic Acids Res 11: 4141–4155.
8. Sauer S, Kliem M (2010) Mass spectrometry tools for the classification and identification of bacteria. Nat Rev Microbiol 8: 74–82.
9. Cherkaoui A, Hibbs J, Emonet S, Tangomo M, Girard M, et al. (2010) Comparison of two matrix-assisted laser desorption ionization-time of flight mass spectrometry methods with conventional phenotypic identification for routine identification of bacteria to the species level. J Clin Microbiol 48: 1169–1175.
10. Fenselau C, Demirev PA (2001) Characterization of intact microorganisms by MALDI mass spectrometry. Mass Spectrom Rev 20: 157–171.
11. Stephan R, Ziegler D, Pfluger V, Vogel G, Lehner A (2010) Rapid genus- and species-specific identification of *Cronobacter* spp. by matrix-assisted laser

desorption ionization-time of flight mass spectrometry. J Clin Microbiol 48: 2846–2851.
12. Stephan R, Cernela N, Ziegler D, Pfluger V, Tonolla M, et al. (2011) Rapid species specific identification and subtyping of *Yersinia enterocolitica* by MALDI-TOF mass spectrometry. J Microbiol Met 87: 150–153.
13. Freiwald A, Sauer S (2009) Phylogenetic classification and identification of bacteria by mass spectrometry. Nat Protoc 4: 732–742.
14. Hahn D, Mirza B, Benagli C, Vogel G, Tonolla M (2011) Typing of nitrogen-fixing *Frankia* strains by matrix-assisted laser desorption ionization-time-of-flight (MALDI-TOF) mass spectrometry. Syst Appl Microbiol 34: 63–68.
15. Rezzonico F, Vogel G, Duffy B, Tonolla M (2010) Application of whole-cell matrix-assisted laser desorption ionization-time of flight mass spectrometry for rapid identification and clustering analysis of *Pantoea* species. Appl Environ Microbiol 76: 4497–4509.
16. Ferreira L, Sanchez-Juanes F, Gonzalez-Avila M, Cembrero-Fucinos D, Herrero-Hernandez A, et al. (2009) Direct identification of urinary tract pathogens from urine samples by matrix-assisted laser desorption ionization-time of flight mass spectrometry. J Clin Microbiol 48: 2110–2115.
17. Prod'hom G, Bizzini A, Durussel C, Bille J, Greub G (2010) Matrix-assisted laser desorption ionization-time of flight mass spectrometry for direct bacterial identification from positive blood culture pellets. J Clin Microbiol 48: 1481–1483.
18. Stevenson LG, Drake SK, Murray PR (2010) Rapid identification of bacteria in positive blood culture broths by matrix-assisted laser desorption ionization-time of flight mass spectrometry. J Clin Microbiol 48: 444–447.
19. Somasegaran P, Hoben HJ (1985) Methods in Legume-*Rhizobium* technology. . 510 p.

20. Ferreira L, Sánchez-Juanes F, García-Fraile P, Rivas R, Mateos PF, et al. (2011) MALDI-TOF mass spectrometry is a fast and reliable platform for identification and ecological studies of species from family *Rhizobiaceae*. PLoS One 6: e20223.

21. Pueppke SG, Broughton WJ (1999) *Rhizobium* sp. strain NGR234 and *R. fredii* USDA257 share exceptionally broad, nested host ranges. Mol Plant-Microbe Interact 12: 293–318.

22. Perret X, Broughton WJ (1998) Rapid identification of *Rhizobium* strains by Targeted PCR Fingerprinting. Plant Soil 204: 21–34.

23. Perret X, Fellay R, Bjourson AJ, Cooper JE, Brenner S, et al. (1994) Subtraction hybridisation and shot-gun sequencing: a new approach to identify symbiotic loci. Nucleic Acids Res 22: 1335–1341.

24. Sun LW, Teramoto K, Sato H, Torimura M, Tao H, et al. (2006) Characterization of ribosomal proteins as biomarkers for matrix-assisted laser desorption/ionization mass spectral identification of *Lactobacillus plantarum*. Rapid Commun Mass Spectrom 20: 3789–3798.

25. Hotta Y, Teramoto K, Sato H, Yoshikawa H, Hosoda A, et al. (2010) Classification of genus *Pseudomonas* by MALDI-TOF MS based on ribosomal protein coding in *S10-spc*-alpha operon at strain level. J Proteome Res 9: 6722–6728.

26. Teramoto K, Sato H, Sun L, Torimura M, Tao H, et al. (2007) Phylogenetic classification of *Pseudomonas putida* strains by MALDI-MS using ribosomal subunit proteins as biomarkers. Anal Chem 79: 8712–8719.

27. Denison RF, Kiers ET (2004) Lifestyle alternatives for rhizobia: mutualism, parasitism, and forgoing symbiosis. FEMS Microbiol Lett 237: 187–193.

28. Kuykendall LD, Weber DF (1978) Genetically marked *Rhizobium* identifiable as inoculum strain in nodules of soybean plants grown in fields populated with *Rhizobium japonicum*. Appl Environ Microbiol 36: 915–919.

29. Lewin A, Rosenberg C, Meyer zAH, Wong CH, Nelson L, et al. (1987) Multiple host-specificity loci of the broad host range *Rhizobium* sp. NGR234 selected using the widely compatible legume *Vigna unguiculata*. Plant Mol Biol 8: 447–459.

30. Krishnan HB, Lewin A, Fellay R, Broughton WJ, Pueppke SG (1992) Differential expression of *nodS* accounts for the varied abilities of *Rhizobium fredii* USDA257 and *Rhizobium* sp. strain NGR234 to nodulate *Leucaena* spp. Mol Microbiol 6: 3321–3330.

31. Lewin A, Cervantes E, Wong C-H, Broughton WJ (1990) *nodSU*, two new *nod* genes of the broad host range *Rhizobium* strain NGR234 encode host-specific nodulation of the tropical tree *Leucaena leucocephala*. Mol Plant-Microbe Interact 3: 317–326.

32. Deakin WJ, Broughton WJ (2009) Symbiotic use of pathogenic strategies: rhizobial protein secretion systems. Nat Rev Microbiol 7: 312–320.

33. Viprey V, Del Greco A, Golinowski W, Broughton WJ, Perret X (1998) Symbiotic implications of type III protein secretion machinery in *Rhizobium*. Mol Microbiol 28: 1381–1389.

34. Broughton WJ, Hanin M, Relić B, Kopcińska J, Golinowski W, et al. (2006) Flavonoid-inducible modifications to rhamnan O antigens are necessary for *Rhizobium* sp. strain NGR234-legume symbioses. J Bacteriol 188: 3654–3663.

35. Oke V, Long SR (1999) Bacteroid formation in the *Rhizobium*-legume symbiosis. Curr Opin Microbiol 2: 641–646.

36. Patriarca EJ, Taté R, Federova E, Riccio A, Defez R, et al. (1996) Down-regulation of the *Rhizobium ntr* system in the determinate nodule of *Phaseolus vulgaris* identifies a specific developmental zone. Mol Plant-Microbe Interact 11: 243–251.

37. Sprent JI, James EK (2007) Legume evolution: where do nodules and mycorrhizas fit in? Plant Physiol 144: 575–581.

38. Kereszt A, Mergaert P, Kondorosi E (2011) Bacteroid development in legume nodules: evolution of mutual benefit or of sacrificial victims? Mol Plant Microbe Interact 24: 1300–1309.

39. Mergaert P, Uchiumi T, Alunni B, Evanno G, Cheron A, et al. (2006) Eukaryotic control on bacterial cell cycle and differentiation in the *Rhizobium*-legume symbiosis. Proc Natl Acad Sci USA 103: 5230–5235.

40. Van de Velde W, Zehirov G, Szatmari A, Debreczeny M, Ishihara H, et al. (2010) Plant peptides govern terminal differentiation of bacteria in symbiosis. Science 327: 1122–1126.

41. Stanley J, Dowling DN, Broughton WJ (1988) Cloning of *hemA* from *Rhizobium* sp. NGR234 and symbiotic phenotype of a gene-directed mutant in diverse legume genera. Mol Gen Genet 215: 32–37.

42. Heron DS, Ersek T, Krishnan HB, Pueppke S (1989) Nodulation mutants of *Rhizobium fredii* USDA257. Mol Plant-Microbe Interact 2: 4–10.

43. Beringer JE (1974) R factor transfer in *Rhizobium leguminosarum*. J Gen Microbiol 84: 188–198.

44. Broughton WJ, Dilworth MJ (1971) Control of leghaemoglobin synthesis in snake beans. Biochem J 125: 1075–1080.

10

Dynamics of Seed-Borne Rice Endophytes on Early Plant Growth Stages

Pablo R. Hardoim[1,2]*, **Cristiane C. P. Hardoim[1¤]**, **Leonard S. van Overbeek[2]**, **Jan Dirk van Elsas[1]**

1 Department of Microbial Ecology, University of Groningen, Centre for Ecological and Evolutionary Studies, Groningen, The Netherlands, **2** Plant Research International, Wageningen, The Netherlands

Abstract

Bacterial endophytes are ubiquitous to virtually all terrestrial plants. With the increasing appreciation of studies that unravel the mutualistic interactions between plant and microbes, we increasingly value the beneficial functions of endophytes that improve plant growth and development. However, still little is known on the source of established endophytes as well as on how plants select specific microbial communities to establish associations. Here, we used cultivation-dependent and -independent approaches to assess the endophytic bacterrial community of surface-sterilized rice seeds, encompassing two consecutive rice generations. We isolated members of nine bacterial genera. In particular, organisms affiliated with *Stenotrophomonas maltophilia* and *Ochrobactrum* spp. were isolated from both seed generations. PCR-based denaturing gradient gel electrophoresis (PCR-DGGE) of seed-extracted DNA revealed that approximately 45% of the bacterial community from the first seed generation was found in the second generation as well. In addition, we set up a greenhouse experiment to investigate abiotic and biotic factors influencing the endophytic bacterial community structure. PCR-DGGE profiles performed with DNA extracted from different plant parts showed that soil type is a major effector of the bacterial endophytes. Rice plants cultivated in neutral-pH soil favoured the growth of seed-borne *Pseudomonas oryzihabitans* and *Rhizobium radiobacter*, whereas *Enterobacter*-like and *Dyella ginsengisoli* were dominant in plants cultivated in low-pH soil. The seed-borne *Stenotrophomonas maltophilia* was the only conspicuous bacterial endophyte found in plants cultivated in both soils. Several members of the endophytic community originating from seeds were observed in the rhizosphere and surrounding soils. Their impact on the soil community is further discussed.

Editor: Scott E. Baker, Pacific Northwest National Laboratory, United States of America

Funding: This study was supported by a collaborative agreement between the University of Groningen and Plant Research International. The funders had no role in study design, data collection and analysis, decision to publish, or preparation of the manuscript.

Competing Interests: The authors have declared that no competing interests exist.

* E-mail: phardoim@gmail.com

¤ Current address: Microbial Ecology and Evolution Research Group, Algarve University, Centre of Marine Sciences (CCMar), Faro, Portugal

Introduction

Endophytes can be defined as microbial communities (bacteria and fungi) that are found inside plant tissues without causing any apparent harm to the host. Microbial endophytes have been reported to occur in virtually all tissues of the host plant, including aseptically regenerated meristematic tissues of micropropagated plants [1,2]. The concept that seeds may serve as the sources of endophytes or pathogens was first launched by Baker et al. [3]. The presence of bacterial endophytes in, and dissemination from, seeds may be considered to represent an atypical event, which is certainly very difficult to demonstrate. However, the presence of bacteria has been documented in ovule tissues (several plants [4]), throughout seed maturing stages of rice [5] and in the endosphere of mature rice seeds [6]. Still, the concept of seeds as important sources of bacterial endophytes has been called controversial until recently [7]. A recent study revealed that a diverse array of endophytes could be obtained from plant tissue that once was considered germ-free, i.e. the callus tissue of micropropagated plants. This community encompassed a total of 11 bacterial and 17 fungal (ascomycete) taxa [8]. Moreover, a core set of seed-borne endophytes has been demonstrated to endure for hundreds

of seed generations, suggesting that select endophytes might establish long relationship with their host thus defeating the boundaries of evolution, human selection and ecology [9]. More recently, the function of seed-borne endophytes that improve seedling development have been demonstrated in a study in which seed-borne *Pseudomonas* sp. SENDO 2, *Acinetobacter* sp. SENDO 1, and *Bacillus* sp. SENDO 6 improved cardon cactus growth by solubilising rock minerals [10]. These results suggest that bacterial endophytes are inherent to plant tissues and may exert more essential functions than is apparent first sight.

The bacterial community inside a plant is obviously prone to influences caused by changing plant physiology [11]. Therefore, many factors that modify plant physiology, e.g. growth stage, soil type, agricultural management regime and even bacterial density, are thought to also promote significant shifts in the endophytic community structure. On the other hand, so-called competent endophytes might thrive in the plant even under adverse conditions [12]. We coined the term 'competent endophyte' for microorganism that successfully colonizes the plant tissues and that has the capacity to incite plant physiology and be selectively favoured, leading to beneficial maintenance of the plant-microbe association [13]. For the great majority of bacterial endophytes, their function or ecology inside the host plant is unknown. However, particular bacterial endophytes might actively influence the physiology of the host as a result of the production of phytohormones and/or the modulation of host ethylene levels. Many other plant-growth-promoting functions, such as fixation of N_2, solubilisation of inorganic phosphate, provision of micronutrients, promotion of photosynthetic activity, induction of the plant defence system, production of antibiotics, biotransformation of heavy metals and biodegradation of organic pollutants, might also enhance host fitness [14]. The effect of these beneficial functions might be drastically improved when plant endophytes establish synergistic interactions with their plant hosts [15–17].

In this study we present a comprehensive analysis of the bacterial endophytes of rice seeds by assessing the culture-dependent and -independent fractions of the bacterial community in two consecutive seed generations. Furthermore, we assessed the development of bacterial endophytes from second-generation seeds up to tiller stage of plants growing in gamma-irradiated soils. To gain insight into how environmental factors affect the bacterial endophytic community, we included different abiotic conditions, i.e. we used two soil types (neutral and low pH) and two water regimes (flooded and unflooded). We also assessed different biotic parameters, i.e. we introduced previously isolated bacterial root endophytes in two densities (low and high bacterial inoculation densities - BID) and compared these with an uninoculated treatment. We then assessed the bacterial communities that emerged in the bulk and rhizosphere soils, and in the root and shoot endosphere. We found that the seed-borne bacterial endophytes were highly diverse. As the plant developed, few of these became dominant while others were suppressed. The endophytic community in plant tissue was largely influenced by soil type, followed by water regime. These results suggest that, under our conditions of reduced soil microbial complexity, rice seeds are important sources of bacterial endophytes that colonize the plant. Furthermore, plant physiology was found to play a major role in shaping the structure and diversity of the endophytic bacterial communities.

Results

Rice seed endophytic communities

The culturable endophytic community of rice seeds was assessed using the seeds from two consecutive generations. Seeds from the first generation showed the highest population density, at 3.5×10^5 CFU g^{-1} fresh weight (FW), whereas the second generation revealed the presence of 4.5×10^3 CFU g^{-1} FW. A total of 16 strains were isolated from internal seed tissues of rice. The 16S rRNA gene identification of these revealed that the endophytes encompassed members of nine genera within the classes *Alpha-* and *Gamma-proteobacteria*, *Flavobacteria*, *Bacilli* and *Actinobacteria* (Table 1). Strains that were closely related to *Stenotrophomonas maltophilia* (R2 and R8), *Mycobacterium abscessus* (R1 and R5) and *Ochrobactrum* spp. (R3 – *O. tritici* and R12 – *O. grignonense*) were observed inside both seed generations. The seed endosphere strains R6, R8, R9, R11, R12, R15 and R16 showed high 16S rRNA gene sequence similarities (>99.0%) to bacteria isolated and/or sequenced from the rice phytosphere, rhizosphere and paddy soil (Table 1), suggesting that these bacteria might be well adapt to rice niche.

PCR-DGGE analysis of the seed and rice tissue (5 days) endophytic communities revealed considerable complexity, with a total of 30 migration positions of the bands (Fig. 1A). Across the samples, the bacterial richness varied between 7 and 15 bands, which included five dominant bands (Fig. 1A bands 3, 9, R13, R14

and one as-yet-unidentified band), which were erratically distributed in the midst of many faint ones. Seeds from the first and second generations revealed a similar endophytic richness with, respectively, nine and seven PCR-DGGE bands. Four PCR-DGGE bands (Fig. 1A bands 11, 12, R13 and one as-yet-unidentified band) were shared in both seed generations, whereas three (9, 10 and one as-yet-unidentified, Fig. 1A) were found in the seeds of the first generation and the remainder in the second seed generation. The endophyte richness assessed from shoot and root tissues of aseptically growing rice seedlings showed slightly higher richness than that observed inside seeds with, respectively, 13 and 11 PCR-DGGE bands on average from both generations. The endophytic community that was shared in both generations of seedling shoot and root tissues encompassed, respectively, 24% (PCR-DGGE bands 9, 12, R13, R14 and one as-yet-unidentified) and 22% (bands 11, 12, R13 and one as-yet-unidentified) of the total community.

We tentatively identified 17 PCR-DGGE bands by sequencing (Table 2) and assigned three additional bands with identical motility behaviour to previously isolated seed endophytes (band identity is preceded by letter R, Fig. 1A). In the PCR-DGGE profile of seed and seedling endophytes, a total of 16 PCR-DGGE bands were identified, of which ten showed high 16S rRNA gene sequence similarity (>99.0%) to bacteria previously assessed from the root endosphere of mature rice plants growing in the Philippines (Fig. 1A, PCR-DGGE bands 1, 2, 3, 4, 5, 7, 9, 10 and 14) and from the rhizosphere of rice plants growing in India (Fig. 1A, band 12; Table 2). PCR-DGGE bands 9, 12 and R13 were the most frequently found bands inside seeds and seedlings of both generations. They were closely related to *S. maltophilia* (99.7% sequence similarity), *Pseudomonas protegens* CHA0^T (100%) and *Plantibacter flavus* DSM 14012^T (99.8%), respectively (Tables 1 and 2). The bands of seed endophyte strains R6 and R8 showed migration behaviour similar to those of PCR-DGGE bands 12 and 9 and were identical 16S rRNA gene sequence, respectively. Two PCR-DGGE bands with identical motility (3 and 4, and 7 and 8) were identified as belonging to different species and these were further analysed as pairs.

We further compared the rice endophytic community against publicly-available endophytic sequences from seeds of rice (*Oryza sativa*) and *Zea* plants. The strains R9, R15 and R16 were closely related to sequences of endophytes that were exclusively found in rice seeds from two independent studies, whereas PCR-DGGE bands 6 and 10 were closely related to strains found in rice and *Zea* seeds (Table 3). The sequences of strains R6 and R8 and of PCR-DGGE bands 2 and 9 were closely related (>99.0% 16S rRNA sequence similarity) to those of endophytic communities found in rice and *Zea* seeds (Table 3).

Dynamics of rice endophytic communities as revealed by plant development

As evidenced by PCR-DGGE, the endophytic bacterial communities inside root and shoot tissues of three- and five-week-old rice plants cultivated in gamma-irradiated Kollumerwaard (K) and Valthermond (V) soils were mainly influenced by soil type (Fig. 1B and C). The richness of endophytes from plants cultivated in the K soil was higher than that found in V soil plants, independent of the plant tissue or time of analysis. The profile of the endophytic community from three-week-old plants cultivated in K soil showed two to eight bands for root and eight to 13 bands for shoot tissues, whereas plants cultivated in V soil harboured between two and four and three and 13 bands, respectively. Plants cultivated in K soil showed dominance of five bacterial communities (Fig. 1B PCR-DGGE bands 7/8, R13, 14, 15, and

Table 1. Identification of isolated seed-borne strains.

Strains[a]	Accession number	Closest type strain (accession number)	Similarity (%)	Closest rice associated bacteria (accession number)	Similarity (%)	Sources[b]
R6*	JN110435	Pseudomonas protegens CHAO[T] (AJ278812)	723/723 (100)	Pseudomonas sp. MDR7 (AM911672)	723/723 (100)	R
R2	JN110431	Stenotrophomonas maltophilia IAM 12423[T] (AB294553)	789/792 (99.6)	Uncultured Stenotrophomonas clone SHCB1148	785/792 (99.1)	RE1
R8*	JN110437	Stenotrophomonas maltophilia IAM 12423[T] (AB294553)	662/663 (99.8)	Uncultured Stenotrophomonas clone SHCB1148	661/663 (99.7)	RE1
R3	JN110432	Ochrobactrum tritici SCII 24[T] (AM114402)	741/741(100)	Ochrobactrum sp. RFNB9 (FJ266319)	727/741 (98.1)	PF
R12	JN110441	Ochrobactrum grignonense OgA9a[T] (AJ242581)	754/755 (99.9)	Ochrobactrum sp. RFNB9 (FJ266319)	749/755 (99.2)	PF
R7	JN110436	Sphingomonas yanoikuyae IFO 15102[T] (D13728)	717/721 (99.4)	Uncultured Sphingomonas clone SHCB0924	696/723 (96.3)	RE1
R11	JN110440	Flavobacterium johnsoniae DSM 2064[T] (AM230489)	608/619 (98.2)	Flavobacterium sp. P-135 (AM412169)	615/620 (99.2)	PS
R4	JN110433	Paenibacillus humicus PC-147[T] (AM411528)	547/590 (92.7)	Paenibacillus sp. RFNB4 (FJ266315)	542/588 (92.2)	PF
R10	JN110439	Agromyces mediolanus DSM 20152[T] (X77449)	674/674 (100)	Curtobacterium sp. Pd-E-(s)-I-D-6(4) (AB242985)	198/204 (97.1)	SE
R9	JN110438	Curtobacterium citreum DSM 20528[T] (NR_026156)	720/721 (99.8)	Curtobacterium sp. Pd-E-(l)-e-D-1(4) (AB291847)	203/203 (100)	LE
R16	JN110445	Curtobacterium herbarum DSM 14013[T] (AM410692)	798/800 (99.7)	Curtobacterium sp. Pd-S-(l)-I-D-3(6) (AB291903)	248/250 (99.2)	LS
R14	JN110443	Frigoribacterium faeni DSM 10309[T] (AM410686)	717/719 (99.7)	Curtobacterium sp. Pd-E-(l)-e-D-3(5) (AB291849)	194/199 (97.5)	LE
R15	JN110444	Microbacterium oleivorans DSM 16091[T] (AJ698725)	791/797 (99.2)	Microbacterium sp. Pd-S-(l)-I-D-6(16) (AB291906)	311/311 (100)	LS
R1	JN110430	Mycobacterium abscessus CIP 104536[T] (AY457071)	574/576 (99.6)	Mycobacterium sp. Pd-E-(r)-m-D-6(5) (AB291833)	329/343 (95.9)	RE2
R5	JN110434	Mycobacterium abscessus CIP 104536[T] (AY457071)	622/623 (99.8)	Mycobacterium sp. Pd-E-(r)-m-D-6(5) (AB291833)	308/322 (95.6)	RE2
R13	JN110442	Plantibacter flavus DSM 14012[T] (AJ310417)	629/630 (99.8)	Microbacterium sp. P-65 (AM411961)	615/631 (97.5)	PS

[a]Rice strains isolated from first (R1-R4) and second (R5-R16) generation of seeds.
*The 16S rRNA gene sequences of strains R6 and R8 were identical to PCR-DGGE products of the bands 12 and 9, respectively.
[b]Source of the closest rice associated bacteria, LE – Leaf Endophyte [21]; LS – Leaf surface [21]; PF – Paddy Field (Islam et al., unpublished); PS – Paddy Soil [28]; R - Rhizosphere [25]; RE1 - Root Endosphere [20]; RE2 - Root Endosphere [21] and SE – Seed endophyte [5].

16) across shoot replicates, whereas the community structure from root tissues was erratically distributed across replicates, with members of the dominant shoot community found in a single replicate (Fig. 1B). One PCR-DGGE band (9) was conspicuously present in all root samples of plants cultivated in V soil, whereas two bands (6 and 7/8) were dominant in the shoot tissues (Fig. 1B).

The PCR-DGGE profiles of the endophytic community from five-week-old plants cultivated in K soil showed four to seven bands in root tissues, of which four (bands 2, 6, 9 and 14, Fig. 1C) were conspicuous. In shoot tissues, 12–16 bands were found, of which six (bands 2, 3/4, 6, 7/8, 9 and 14, Fig. 1C) were conspicuous. The PCR-DGGE profile of plants cultivated in V soil showed five to seven bands in the root tissues, of which two (bands 6 and 13) were conspicuous, and six to 11 were found in shoot tissues, from which five (bands 2, 7/8, 9, 13, 14) were conspicuous.

The endophytic bacterial community of three- and five-week-old rice plants revealed high similarity with types found inside seeds and seedlings, with, respectively, 20 out of 24 and 19 out of 22 PCR-DGGE bands (Fig. 1). Comparison of the endophytic communities during plant growth revealed diverse trends. For instance, in plants cultivated in K soil, the PCR-DGGE bands 3/4 and 9 were erratically found inside seedlings and three-week-old plant tissues, but they became dominant in the shoot tissues of five-week-old plants. Band 6 was also dominant in the five-week samples, however it was never found inside seeds. Other PCR-DGGE bands (5, 10, 12, 13, R13, 15 and 16) found inside the seeds were encountered in the three-week-old plants and not in the five-week samples. Others (bands 11, 17, R14, R16) were only found in the seedlings. Plants cultivated in V soil revealed different patterns, with PCR-DGGE bands 9 and 13 being conspicuously found across the replicates of three-week-old plants (only root tissues) and five-week-old plants (in both tissues), whereas band 1 (found in seeds) was erratically found in five-week-old plants (in both tissues). PCR-DGGE bands 2, 3/4, 10, 14 and 16 were exclusively found in shoot tissues (Fig. 1).

Endophytic bacterial community survey under distinct conditions

To obtain insight into how the endophytic bacterial community in rice evolves in natural conditions, we designed an assay in which

A

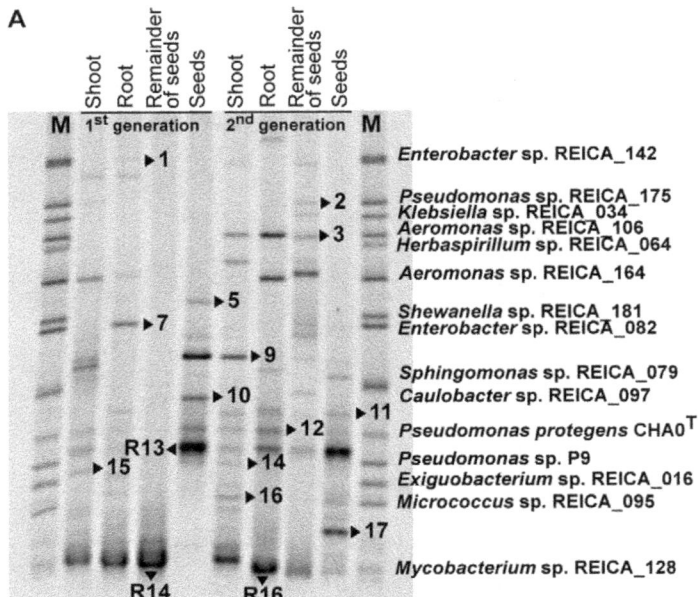

Enterobacter sp. REICA_142

Pseudomonas sp. REICA_175
Klebsiella sp. REICA_034
Aeromonas sp. REICA_106
Herbaspirillum sp. REICA_064

Aeromonas sp. REICA_164

Shewanella sp. REICA_181
Enterobacter sp. REICA_082

Sphingomonas sp. REICA_079
Caulobacter sp. REICA_097

Pseudomonas protegens CHA0^T

Pseudomonas sp. P9
Exiguobacterium sp. REICA_016
Micrococcus sp. REICA_095

Mycobacterium sp. REICA_128

B

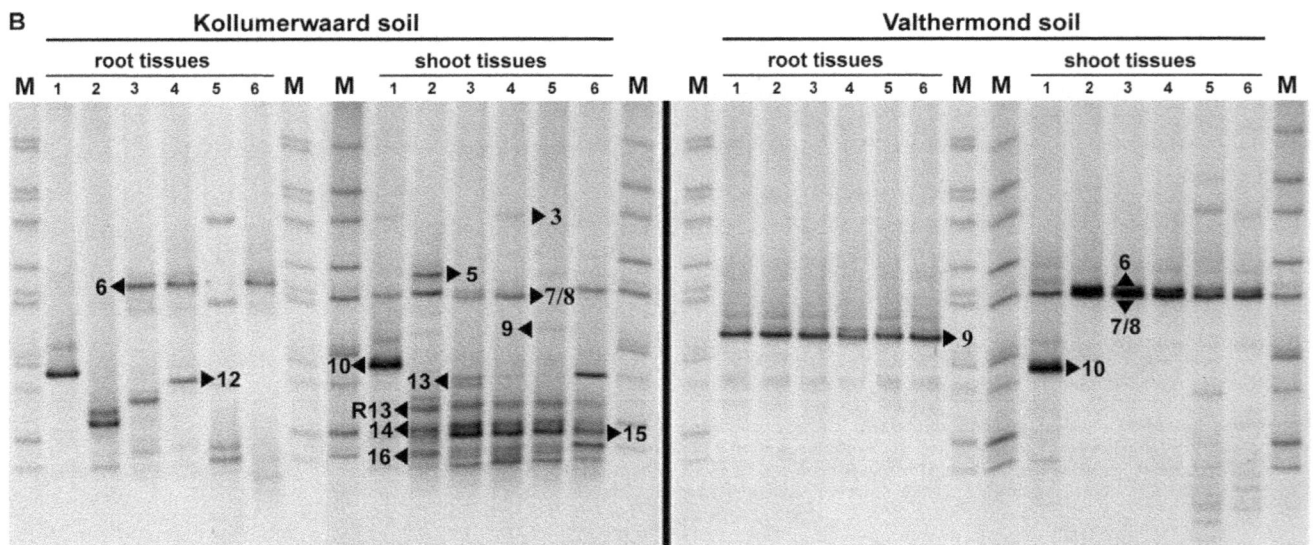

C

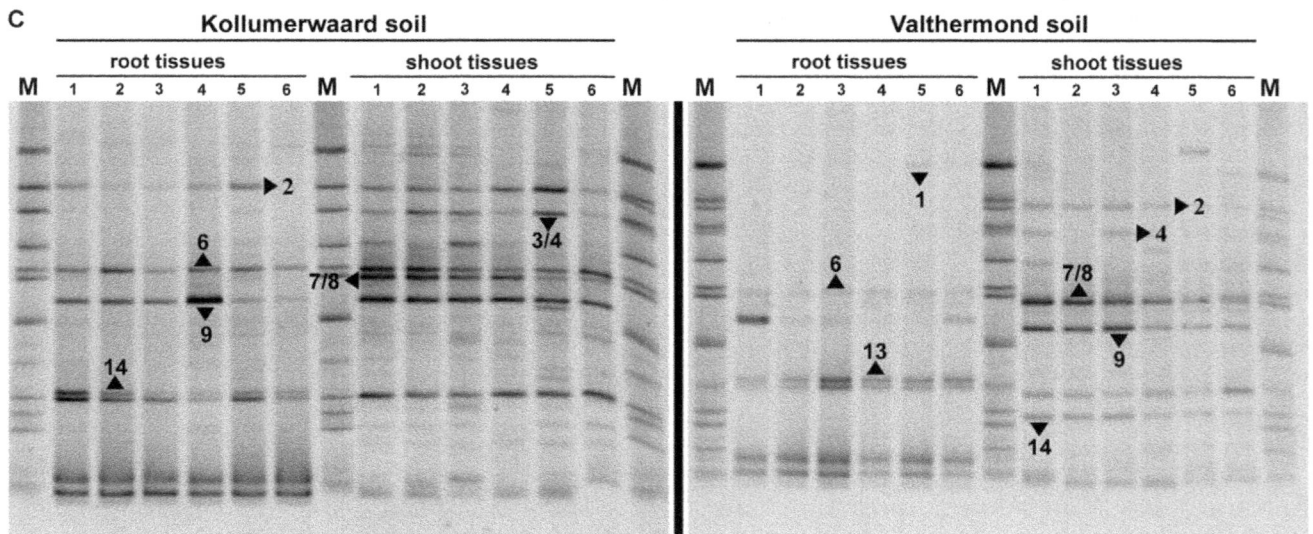

Figure 1. Dynamics of rice endophytes as revealed by PCR-DGGE profiles of seed, three- and five-week-old rice plants. Rice endophyte PCR-DGGE patterns of surface-sterilized dehulled seeds and five-day-old shoot, root and remainder of the seeds from two consecutive generations are shown (panel A). PCR-DGGE patterns of root and shoot endosphere community of three- B) and five- C) week-old rice plants cultivated in two soil types. Six replicates per treatments are shown. Arrow heads indicate identified communities from excised PCR-DGGE bands (only numbers) and strains with identical motility (preceded by letter R; see Table 1 and 2), M – marker with a selection of 15 endophyte ribotypes (panel A).

we reduced the complexity of the system (i.e. rice growing in gamma-irradiated soil inoculated with 'artificial' community encompassed by 18 selected endophytic strains) and then assessed the bacterial community from four distinct microhabitats (i.e. bulk and rhizosphere soil, root and shoot endosphere tissue). In addition to biotic factors, we investigated two abiotic factors, i.e. two soil types (K and V) and two water regimes (flooded and unflooded). As revealed by the PCR-DGGE profiles, soil exerted a major influence on the endophytic bacterial community structure and were analysed separated.

Bacterial distribution on K soil. The seed-borne *Pseudomonas oryzihabitans* (PCR-DGGE band 2) and *Stenotrophomonas maltophilia* (band 9) were observed in all analysed habitats of plant cultivated on

K soil (Fig. 2; Fig. S1). The introduced *Aeromonas* sp. REICA_106 (band 3) were also observed in all habitats, however only for inoculated treatments, whereas *Rhizobium radiobacter* (band 6) was found in the rhizosphere soil, root and shoot tissues, *Pseudomonas putida* (band 14) was conspicuously found in bulk and rhizosphere soils and *Herbaspirillum* sp. REICA_064 (band 4) was restricted to shoot tissues (Fig. 2).

Bacterial distribution on V soil. Plants from V soil selected for members of *Enterobacter* sp. REICA_082 (PCR-DGGE band 7) and *Dyella ginsengisoli* (band 13) for all habitats and *Stenotrophomonas maltophilia* (band 9) mainly in the shoot tissues (Fig. 2; Fig. S2). *Pseudomonas oryzihabitans* (band 2) and *Pseudomonas putida* (band 14) were restricted to shoot tissues, *Enterobacter* sp. REICA_142 (band

Table 2. Identification of excised PCR-DGGE bands.

DGGE band ID	Accession number	Closest type strain or known strain (accession number)	Similarity (%)	Closest rice associated bacteria (accession number)	Similarity (%)	Sources[a]
1	JN110446	*Enterobacter cloacae* subsp. *cloacae* ATCC 13047[T] (AJ251469)	378/382 (99.0)	*Enterobacter* sp. REICA_142	382/382 (100)	RE1
2	JN110447	*Pseudomonas oryzihabitans* IAM 1568[T] (AM262973)	379/380 (99.7)	*Pseudomonas* sp. REICA_175	379/380 (99.7)	RE1
3	JN110448	*Aeromonas hydrophila* subsp. *dhakensis* LMG 19562[T] (AJ508765)	371/373 (99.5)	*Aeromonas* sp. REICA_106	373/373 (100)	RE1
4	JN110449	*Herbaspirillum rubrisubalvicans* ICMP 5777[T] (AF137508)	346/349 (99.1)	*Herbaspirillum* sp. REICA_064	346/349 (99.1)	RE1
5	JN110450	*Acinetobacter beijerinckii* LUH 4759[T] (AJ626712)	382/382 (100)	Uncultured *Acinetobacter* clone SHCB0621	381/382 (99.7)	RE1
6	JN110451	*Rhizobium radiobacter* IAM 12048[T] (AB247615)	378/383 (98.7)	Uncultured *Rhizobium* SHCB0425	369/386 (95.6)	RE1
7	JN110452	*Enterobacter arachidis* Ah-143[T] (EU672801)	374/376 (99.5)	*Enterobacter* sp. REICA_082	376/376 (100)	RE1
8	JN110453	*Escherichia coli* O111:H str. 11128 (AP010960)	382/382 (100)	*Enterobacter* sp. REICA_128	378/382 (98.9)	RE1
9	JN110454	*Stenotrophomonas maltophilia* IAM 12423[T] (AB294553)	382/383 (99.7)	Uncultured *Stenotrophomonas* SHCB1148	382/383 (99.7)	RE1
10	JN110455	*Pantoea agglomerans* DSM3493[T] (AJ233423)	380/380 (100)	Uncultured *Pantoea* SHCB0588	378/380 (99.5)	RE1
11	JN110456	*Neisseria meningitidis* M01-240149 (CP002421)	374/375 (99.7)	Uncultured bacterium clone J-3FECA52 (DQ340883)	291/308 (94.5)	RE2
12	JN110457	*Pseudomonas protegens* CHA0[T] (AJ278812)	378/378 (100)	*Pseudomonas* sp. MDR7 (AM911672)	378/378 (100)	R
13	JN110458	*Dyella ginsengisoli* Gsoil 3046[T] (AB245367)	373/373 (100)	*Dyella* sp. V-6.1 (JF429979)	367/373 (98.4)	PF
14	JN110459	*Pseudomonas putida* BIRD-1 (CP002290)	378/378 (100)	Uncultured *Pseudomonas* SHCB0777	378/378 (100)	RE1
15	JN110460	*Bacillus psychrosaccharolyticus* S156[T] (AY509230)	373/379 (98.4)	*Bacillus* sp. P-150 (AM412171)	367/381 (96.3)	PS
16	JN110461	*Deinococcus ficus* CC-FR2-10[T] (AY941086)	377/379 (99.5)	Uncultured bacterium clone J-3FECC29 (DQ340907)	266/293 (90.8)	RE2
17	JN110462	*Achromobacter spanius* LMG 5911[T] (AY170848)	367/374 (98.1)	Uncultured bacterium clone J-3FECC48 (DQ340912)	365/374 (97.6)	RE2

[a]Source of the closest rice associated bacteria: PF – Paddy Field [65]; PS – Paddy Soil [28]; R - Rhizosphere [25]; RE1 - Root Endosphere [20] and RE2 - Root Endosphere [64].

Table 3. Closest match of sequences obtained in this study against public available rice and *Zea* seed endophyte sequences.

Isolate /DGGE band	Rice						Zea	
	Okunishi et al. [19]	Similarity (%)	Mano et al. [5]	Similarity (%)	Liu et al. unpublished	Similarity (%)	Johnston et al. [9]	Similarity (%)
R2			*Stenotrophomonas* sp. Pd-S-(s)-e-D-1 (4) (AB242927)	301/302 (99.7)			*Stenotrophomonas* sp. DJM1G3 (JF753464)	516/516 (100)
R6							*Pseudomonas* sp. DJM1C10 (JF753430)	513/517 (99.2)
R8			*Stenotrophomonas* sp. Pd-S-(s)-e-D-1(4) (AB242927)	174/174 (100)			*Stenotrophomonas* sp. DJM1G3 (JF753464)	514/515 (99.8)
R9			*Curtobacterium* sp. Pd-E-(s)-I-D-6(4) (AB242985)	241/241 (100)	*Curtobacterium* sp. Fek20 (EU741030)	721/721 (100)		
R15	*Microbacterium* sp. S-(s)-I-D-6(20) (AB178212)	405/408 (99.3)			*Microbacterium* sp. Fek04 (EU741023)	796/797 (99.9)		
R16			*Curtobacterium* sp. Pd-E-(s)-m-D-4(12) (AB242967)	229/231 (99.1)	*Curtobacterium* sp. Fek20 (EU741030)	795/800 (99.4)		
band 2					*Pseudomonas* sp. Fek13 (EU741028)	379/380 (99.7)	*Pseudomonas* sp. DJM1A4 (JF753403)	379/380 (99.7)
band 6					*Agrobacterium* sp. FeL02 (EU741035)	377/382 (98.7)	*Rhizobium* sp. DJM1H4 (JF753477)	381/382 (99.7)
band 9							Uncultured bacterium clone DJM126 (JF753390)	382/383 (99.7)
band 10					*Pantoea* sp. Aek32 (EU741010)	378/380 (99.5)	Uncultured bacterium clone DJM51 (JF753316)	378/380 (99.5)

1) to bulk soil and the introduced *Caulobacter* sp. REICA_097 to bulk and rhizosphere soils (Fig. 2).

Factors affecting the endophytic bacterial community composition of rice

Using the collective data, we performed the Redundancy Analysis (RDA) for each habitat separately, per soil type (Fig. 3 and 4). For both soil types the factors affecting the bacterial community composition shifted from water regime treatments in the shoot and root endosphere to the bacterial inoculation densities (BID) on the soil.

Distribution of bacterial communities inside shoot tissues. On both soil types the rice shoot endosphytes were mainly influenced by water regimes, where the endosphere community of plants subjected to flooded regime differ significantly from those plants conditioned to unflooded treatment (Fig. 3A and B). A total of 76.6 and 69.2% of the RDA diagram distribution was explained by the water regimes of plants cultivated on K and V soils, respectively. The BID treatments were indistinguishable in the K and V soils and only in the K soil the bacterial community from uninoculated treatments differ significantly from inoculated ones. This suggested that the

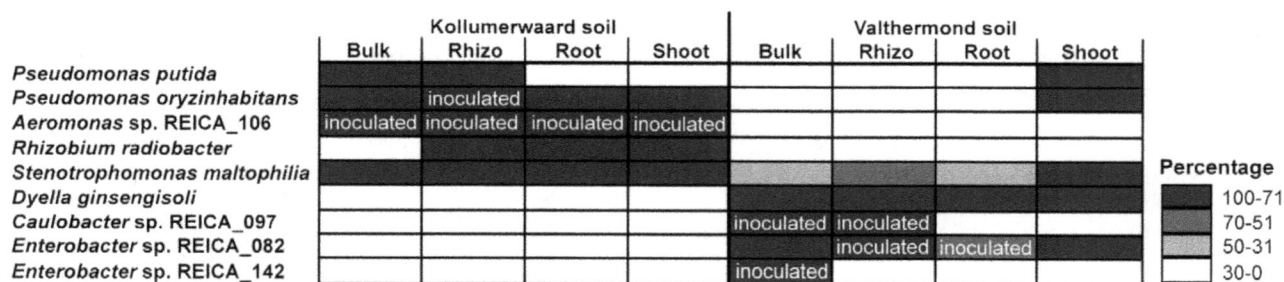

Figure 2. Heat map composition of selected bacterial communities. Distribution of select endophytic bacterial communities (rows) from two soil types (K and V) and four different habitats (root-free and rhizosphere soil, root and shoot endosphere) is shown. Cells are coloured in spectrum of grey that correlates with percentage of observed bacterium in a given habitat. Habitat from which the assessed bacterium was most likely to be originated from 'artificial' soil community is labelled with "inoculated". Unlabelled cells are most likely represented by assessed bacterium originated from rice seeds.

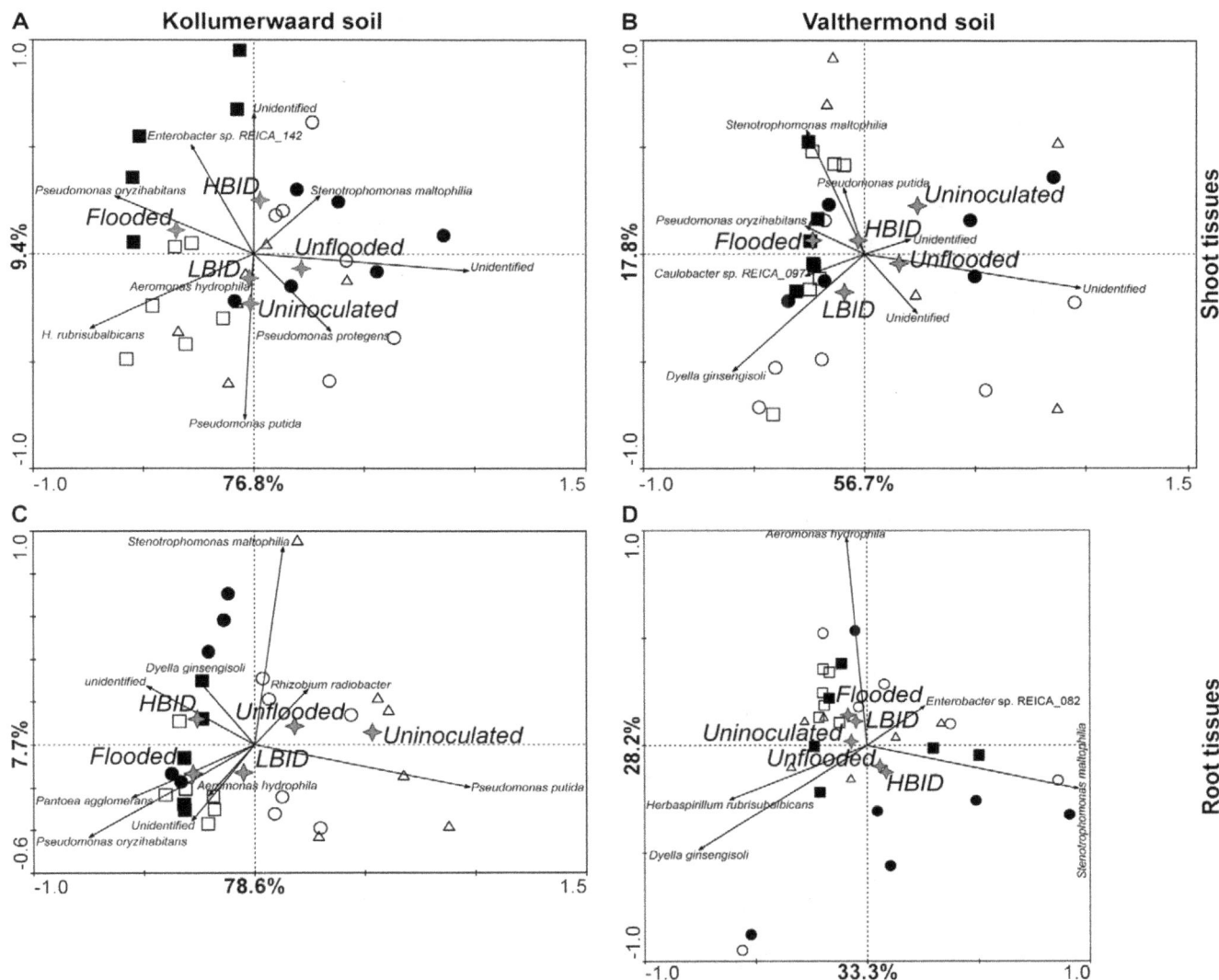

Figure 3. Biplot ordination diagrams of rice shoot and root bacterial endophytes. RDA diagrams generated from PCR-DGGE profiles of endophytic bacterial community sampled from shoot (A and B) and root (C and D) tissues of plants cultivated on K (A and C) and V (B and D) soils are shown. Squares and circle represent PCR-DGGE patterns of bacterial communities from plants submitted to, respectively, flooded and unflooded regimes and exposed to low- (empty symbol) and high- (full symbol) BID. Triangles (control treatment) represent PCR-DGGE patterns of bacterial communities from plants submitted to unflooded regime and cultivated in uninoculated soils. Six replicates of each treatment are shown. Stars represent nominal environmental variables. Arrows represent PCR-DGGE bands in which only the most descriptive communities are shown.

introduced 'artificial' community had exerted a relatively minor effect on the endophytic shoot community for the period investigated.

Distribution of bacterial communities inside root tissues. The distribution of root endophere bacterial community differ on both soils. In the K soil, the endophytic community from uninoculated soil differ significantly from plants exposed to high BID. Both treatments explained 50.2% of the total distribution, while water regimes, which also differ significantly, explained 44.6%. The endophytic bacterial communities from root tissues of plants cultivated on uninoculated soil were placed along the second RDA axis, differing from those of plants cultivated in low- and high-BID soil (Fig. 3C). In contrast to K soil, the distribution of root endophytic communities in V soil seems to be indifferent for bacterial inoculation, where plants of uninoculated soil resembled those from plants of inoculated soil (Fig. 3D). However, the root endophytic community of plants

cultivated under dissimilar water treatments differed significantly, where plants under flooded and unflooded regimes were separated along the diagonal of the RDA diagram. Around 60% of the total distribution was explained by the abiotic factors.

Distribution of bacterial communities in the rhizosphere. As observed on the root tissues, the rhizosphere bacterial communities vary drastically between soil types. In the K soil, most of the treatment was significantly different and the samples from each individual treatment were virtually clustered within one quarter of the RDA diagram. Only the samples from the rhizosphere community of plants cultivated on uninoculated soil were distributed around the centre of the diagram (Fig. 4A). In the V rhizosphere soil, none of the treatments were significantly different, however four out six samples from plants cultivated on uninoculated soil revealed distinct rhizosphere communities and clustered apart from other samples (Fig. 4B).

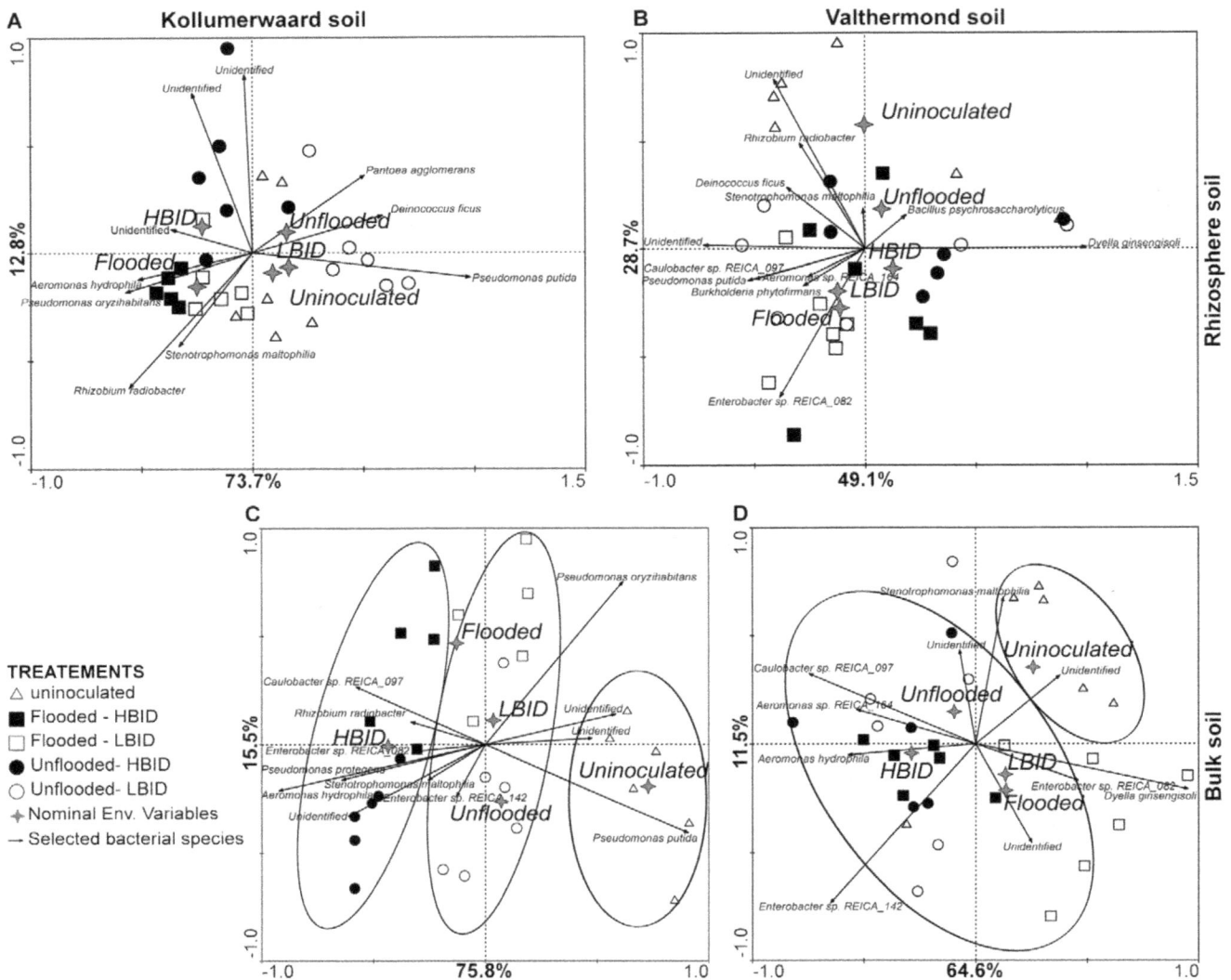

Figure 4. Biplot ordination diagrams of rice rhizosphere and bulk soil bacterial communities. RDA diagrams generated from PCR-DGGE profiles of bacterial community sampled from rhizosphere (A and B) and bulk (C and D) soil of plants cultivated in K (A and C) and V (B and D) soils are shown. See Fig. 3 for symbol description.

Distribution of bacterial communities on root-free soils. The localization of the soil communities in the RDA diagram was mainly influenced by biotic factors for both soil types. In the K soil, the bacterial communities from high-BID, low-BID and uninoculated soils were distributed along the second RDA axis and differed from each other in three main clusters (Fig. 4D). The biotic factors explained 78.8% of the total diagram distribution, while water regimes counted for 17.4%. In the V soil, three clusters were also detected for each BID and uninoculated treatments. The samples were distributed along the second RDA axis, whereas the bacterial communities from flooded and unflooded regimes were distributed along the first axis (Fig. 4D). The biotic factors explained 53.8% of the total bacterial community distribution on root-free soil, while water regimes counted for 33.8%.

Discussion

In this paper, we clearly showed that rice seeds are important sources of the endophytic bacteria that come up in the early rice

growth stages. This was evidenced by experimentation with plants grown in soils deprived of bacterial communities by irradiation. The contention of seed carriage of key endophytes for young plants was supported by three lines of evidence found in this study.

I)   Many (74%) of the rice seed-borne bacterial endophytes found in this study were closely related to bacteria that have previously been isolated from inside maturing and/or mature rice seed tissues [5,18,19] and the endosphere of rice root [20] and leaf tissues [21]. Further, they resembled bacteria from the rhizoplane of rice [22], wheat [23] and sacred fig (*Ficus religiosa*) [24], the rhizosphere of rice [25], the phyllosphere of grasses [26] and rice [20], hay dust [27] and soil in which rice had been cultivated [28].

II)   Throughout plant development, shoot tissues showed higher bacterial endophyte richness than root tissues. Plants cultivated in open fields often reveal the opposite trend, with higher bacterial richness in the root tissues [29]. Mano et al. [21] observed that the endophytic bacterial community in the leaves of rice plants cultivated in the open field was similar to that found in seed tissues, differing drastically from that inside root tissue. The results suggested that rice seed endophytes are generally adapted to plant tissue and rapidly colonize rice shoots, in which there is less competition than in the respective root, which is bathed in rich bacterial communities.

III) The bacterial community from internal plant tissues and the soil surrounding plant roots (cultivated in soil containing an introduced bacterial community or remaining uninoculated) showed similar endophytic bacterial communities, however they differed in the rhizosphere being unrelated to those in the soil.

The bacterial diversity associated with the rice seeds was actually quite astonishing. Recently, two separate studies investigated the correlation of the bacterial community associated with rice seeds across 12 sampling sites [18] and of those with *Zea* seeds across host genotype (i.e. wild ancestor to domesticated maize) [9]. The studies revealed large diversities (284 genomic fingerprint types determined by BOX-PCR from rice seeds and 26 isolated genera from *Zea* seeds) of the bacterial communities associated with the seeds. However, a great majority of the isolates was correlated to the sampling site where the seeds were derived from or to plant genotype, recapitulating the phylogenetic pattern of their *Zea* hosts. Only a few, such as *Enterobacter cloacae*, *Pseudomonas oryzihabitans* (in both rice and *Zea* seeds), *Curtobacterium* spp. (only in rice seeds), *Clostridium beijerinckii*, *Methylobacterium* sp., *Paenibacillus barcinonensis* and *Pantoea agglomerans* (only in *Zea* seeds) were conserved across the sampling sites and host genotypes [9,18]. In addition, strains assigned to *Rhizobium radiobacter*, *Stenotrophomonas maltophilia*, *Acinetobacter* spp., *Herbaspirillum rubrisubalbicans* and *Microbacterium* spp., were isolated from rice seeds collected in more than one (but not all) sampling site [18]. These might be also widespread among rice genotypes. In our study, members of *Rhizobium radiobacter*, *Pantoea agglomerans*, *Stenotrophomonas maltophilia*, *Pseudomonas oryzihabitans*, *Pseudomonas* spp., *Curtobacterium* spp. and *Microbacterium* spp. were also identified. These results suggest that these bacteria are highly adapted to the plant niche.

Many of the aforementioned bacteria are ubiquitous in a range of environment niches, being commonly found in seeds and in the endosphere tissues of rice [5,18,19], graminous (e.g. maize [9]) and leguminous (e.g. soybean [30]) plants, as well as in the soils where these plants had been cultivated. Thus, one might speculate that these organisms form a core microbiota which is conserved across several plant species and that they might use seeds for their own dissemination. For instance, *Stenotrophomonas maltophilia* is an opportunistic bacterium that is often found in soils and in association with plants [31]. It also has a worldwide distribution. Many strains of *Stenotrophomonas maltophilia* have been isolated from the rhizosphere and endosphere of various plants [32]. When inoculated, strains of *Stenotrophomonas* have been shown to enhance plant biomass production in corn [33], sorghum [34], canola [35], potato [36] and poplar [37], all cultivated under greenhouse conditions. Although the genome analysis of *Stenotrophomonas maltophilia* R551-3 has revealed many genes that are dedicated to motility, adaptation to, and colonization of, plant host tissue [38], our results showed that *Stenotrophomonas maltophilia* is transmitted via seeds and can spread out of the host invading the rhizosphere and even surrounding soils. The results suggest that *Stenotrophomonas maltophilia* is highly adapted to niches within the plant and that both dissemination and colonization are two main strategies used in the response to ecological opportunities.

The ecological role of seed endophytes is not thoroughly known. Recently, Puente et al. [10] demonstrated that seed bacterial endophytes are involved in the establishment of giant cardon cactus (*Pachycereus pringlei*) on barren rocks. Cactus seeds disinfected with antibiotics halt seedling development. Plant growth was restored by inoculation of endophytes involved in rock weathering [10]. In another study, introduction of an endophytic consortium composed of *Enterobacter* sp. S_d17, *Pseudomonas* sp. strains S_d12 and S_d13 or of individual strains isolated from surface-sterilized *Nicotiana tabacum* seeds revealed positive effects on plant growth under conditions with and without induced stress (i.e. Cd stress) [39]. The beneficial effects of bacterial endophytes are often more evident in plants cultivated on marginal soils used for phytoremediation or soils conducive to plant disease development [40,41]. Many seed-borne endophytes are involved in plant growth promotion. This is certainly the case for the conserved seed-borne endophytes *Enterobacter cloacae* and *Pseudomonas oryzihabitans*. For instance, *Enterobacter cloacae* strain 501R3 and other unidentified strain are involved in the suppression of damping-off caused by *Pythium ultimum* in many hosts via competitive colonization of the spermosphere and rhizosphere soils, thus reducing the availability of exuded carbohydrate, lipid and amino acid compounds [42,43]. In addition, *Enterobacter cloacae* strain UW5 is involved in the production of IAA [44] and the modulation of plant ethylene levels via 1-aminocyclopropane-1-carboxylate (ACC) deaminase [45]. An extensive assessment of the root endophytic community from mature rice plants cultivated in field soil revealed that members of the genus *Enterobacter* were the most abundant and the most genetically diverse isolated bacteria [20]. Although we have not isolated any *Enterobacter* strain in this particular study, we identified two PCR-DGGE bands from first- and second-generation seed profiles that were identical (at 16S rRNA gene sequence level) to the previously found *Enterobacter* members. Both *Enterobacter* sp. strains REICA_142 and REICA_082 revealed plant-growth-promoting properties such as fixation of N_2, solubilisation of inorganic phosphate and production of ACC deaminase [20]. Members of *Pseudomonas oryzihabitans* containing ACC deaminase (strain Ep4 [46]), or capable of solubilising inorganic phosphate (strain B4M-K [47]), production of IAA, siderophore and fixation of N_2 (strain G6 [48]) have been reported to increase host biomass. In this study, we identified a member closely related to *Pseudomonas oryzihabitans* that extensively colonized plants cultivated in the neutral-pH soil but was almost absent on roots of plants cultivated in the low-pH soil, suggesting pH sensitivity and possibly the importance of plant physiology for community establishment. In addition, we isolated another species, *Pseudomonas* sp. strain R6, that was closely related to the widespread plant-protecting *Pseudomonas protegens* CHA0^T, which is capable of producing the antimicrobial compounds 2,4 diacetyl phloroglucinol and pyoluteorin [49]. The results suggested that selected bacterial communities are hosted by seeds, which might become important when differentially beneficial functions are stimulated in accordance with the local conditions. This may support the development of the new host.

Here, the endophytic bacterial community of rice was shown to be largely influenced by soil type, followed by water regime. The evaluated biotic factors showed minor effect on the diversity and composition of endophytic communities. Rice plants cultivated in K soil (a neutral-pH soil) showed higher richness and were extensively colonized by *Pseudomonas oryzihabitans* and *Rhizobium radiobacter*, whereas plants cultivated in V soil, an acid soil, favoured the growth of *Enterobacter*-like strain REICA_082 and *Dyella ginsengisoli*. Members of these bacteria have been isolated from seeds and/or the phytosphere of various plants

[11,18,30,50], suggesting that they might have a long history of association with diverse host plants. Occasionally, commensalism might come into play, e.g. the plant-associated *Rhizobium radiobacter* (formerly *Agrobacterium tumefaciens*) is the causal agent of crown gall in dicotyledons, however it showed limited pathogenicity towards monocotyledons [51]. The recently-described *Dyella ginsengisoli* has originally been isolated from a ginseng field in South Korea [52]. *Dyella ginsengisoli* strain ATSB10, containing ACC deaminase and with the ability to solubilise inorganic phosphate and to produce β-1,3 glucanase, has been reported to increase the root length of canola seedlings by 145% [50]. The relationship of *Dyella ginsengisoli* with rice plants is unknown and this study is the first documentation that they may be associated.

In summary, seeds from rice plants harbour a great diversity of bacteria that, in response to the plant physiological status, can become competent endophytes. Some organisms might even spread out into rhizosphere and surrounding soil, therefore directly interacting with soil microbial communities [53]. Furthermore, due to their metabolic versatility, seed-borne bacterial endophytes might also increase the fitness of plants, giving the host a competitive advantage over other (indigenous) plant communities [54] and thus might affect whole-ecosystem functioning [55]. Our data suggest that under reduced habitat complexity, this assumption may be met. It remains an open question whether seed-borne endophytes are selected by the host to increase the fitness of the next generations of seeds or whether bacterial endophytes use seeds as vector for dissemination and colonization of new environments.

Materials and Methods

Assessment of endophytic communities from seed endosphere

Rice (*Oryza sativa* L.) seed and five-day-old seedlings from two consecutive generations were analysed. Rice seeds from cultivar APO were obtained from International Rice Research Institute (IRRI, Los Bas, Philippines) and used for seed multiplication in greenhouse conditions at the University of Groningen, Netherlands. Seeds collected from IRRI and Groningen are referred to as first and second generations, respectively. Bacterial communities of the rice seed endosphere from both generations were assessed by culture-dependent and -independent approaches. Under aseptic conditions, the hulls were removed from the rice seeds (1 g) with sterilized forceps and immediately subjected to surface-sterilization with a solution (50 ml) containing 0.12% sodium hypochlorite (NaClO), salts (0.1 and 3% sodium carbonate and sodium chloride, respectively) and 0.15% sodium hydroxide [56] at 30°C for 25 min in orbital shaking (200 rpm). The sterilization procedure was followed by a washing step to remove surface-adhered NaClO in 50 ml 2% sodium thiosulfate [57]. This procedure was repeated twice at 30°C for 10 min under orbital shaking (200 rpm) before the seeds were subjected to rehydration for 1 h at room temperature in 100 ml autoclaved demineralised (demi-)water. In addition, to assess the endophytic communities from early seedling development, 15 surface-sterilized rice seeds from both generations were incubated on R2A medium (DB - Difco) for five days at 28°C and then used to extract DNA from shoot, root and the remainder of the seed tissues.

Endophytic bacterial cells from surface-sterilized seeds and seedlings were released by disrupting the plant tissues with a soft-headed hammer as described [58]. The homogenates (100 μl) were used for serial tenfold dilutions, which were plated onto R2A, after which plates were incubated for one week at 28°C. In addition, homogenates (1 ml) were used for DNA extraction

following the protocol described by Hurek et al. [56]. For each 100 mg of plant material, 1.2 ml cell lysis solution was used, while phenol:chloroform (1:1 v/v) was used for deproteinization. The concentration and quality of the extracted DNA were assessed by electrophoresis in 1% agarose gels, followed by staining with ethidium bromide and visualization under UV light.

Dynamics of rice endophytes

Surface-sterilized rice seeds from the second generation were used to assessed the endophytic bacterial communities from root and shoot endosphere at three and five weeks after seed germination. The plants originating from the germinated seeds were cultivated in two soil types, i.e. Kollumerwaard – K, a clay loam soil with neutral pH (chemical characteristics: pH based on $CaCl_2$ 7.3; total carbon 27.2 g kg^{-1}; organic matter 40.3 g kg^{-1}; dissolved organic matter 86.4 mg kg^{-1}; total nitrogen 1.67 g kg^{-1}; nitrate content 170.12 mg kg^{-1}; and ammonium content 6.37 mg kg^{-1}, soil collected from Groningen, The Netherlands) and Valthermond – V, a loamy sand soil with low pH (chemical characteristics: pH based on $CaCl_2$ 4.5; total carbon 17.8 g kg^{-1}; organic matter 29.2 g kg^{-1}; dissolved organic matter 60.8 mg kg^{-1}; total nitrogen 1.28 g kg^{-1}; nitrate content 123.19 mg kg^{-1}; and ammonium content 10.8 mg kg^{-1}, soil collected from Drenthe, The Netherlands). Both soils were sterilized by applying gamma radiation (minimum 25 kGy, Isotron, Netherlands) and 500 g was aseptically transferred to polyester pots. Sterility of the soil was confirmed by plating, as soil suspensions prepared did not show any colony growth up to 15 days after being plated on R2A medium. Moreover, very faint (residual) bands were observed in PCR-DGGE profiles prepared with soil-extracted DNA.

For the experiment, both soils were watered to a final volume of 70% water holding capacity with filter-sterilized (0.2 μm) 25%-strength Hoagland's nutrient solution [59]. Five-day-old rice seedlings absent of visible microbial outgrowth on R2A medium (at 28°C), were individually transferred to sterile soils. Six replicates for each treatment were used. Rice plants were cultivated in the greenhouse using a day/night cycle of 16/8 h and 25/18°C for light and temperature, respectively. Soil water was replenished daily to holding capacity with freshly prepared filter-sterilized 25%-strength Hoagland's nutrient solution. At weeks three and five, plants were harvested and the bacterial communities in the root and shoot tissues were assessed by PCR-DGGE. Individual rice plants were harvested and roots were carefully washed under running tap water for the removal of adhering soil particles. Root and shoot tissues were segmented with a sterile scalpel and treated as individual sources of endophytes. The surface sterilization procedure was performed in 20-ml tubes filled with 10 ml sterilization solution by exposing rice tissues for 2 min in NaClO solution and manually vortexed at room temperature as described above. Endophytic bacterial DNA was extracted as described above.

Invasion assay

The invasion assay consisted of rice plants cultivated in the greenhouse and subjected to different abiotic and biotic treatments. Surface-sterilized rice seeds from second generation were cultivated in two soil types, i.e. K and V, subjected to two water regimes, i.e. unflooded and flooded, and exposed to three bacterial inoculum densities (BID), i.e., low-, high- and un-inoculated (10^4 and 10^7 bacterial cells g^{-1} soil, respectively). To obtain an 'artificial' community, we used a selection of 15 previously-isolated bacteria, that resembles the community composition found in the root endosphere of mature rice plants [20], i.e. *Enterobacter* sp.

strains REICA_082, REICA_112, REICA_142, *Pseudomonas* sp. REICA_175, *Klebsiella* sp. REICA_034, *Aeromonas* sp. REICA_106 and REICA_164, *Herbaspirillum* sp. REICA_064, *Shewanella* sp. REICA_181, *Exiguobacterium* sp. REICA_016, *Micrococcus* sp. REICA_095, Alphaproteobacterium sp. REICA_149 and *Mycobacterium* sp. REICA_128. In addition three presumably competent endophytes were used as controls, i.e. *Pseudomonas protegens* CHA0[T] [49], *Pseudomonas putida* P9 [60] and *Burkholderia phytofirmans* RG44-4 [61]. Therefore we investigated which bacterium could invade the plant from soil. Each strain was grown separately in R2A broth aerobically at 28°C with shaking (200 rpm). Bacterial cells were harvested in the exponential growth phase by centrifugation and washed twice with sterile PBS buffer. Bacterial cells of each inoculum were combined with their respective amount of cells needed to achieve the final BID. The BID of each treatment was further confirmed using dilution plating on R2A medium. The mixed bacterial cells were diluted in filter-sterilized (0.2 μm) 25% Hoagland's nutrient solution, and added to the soil, establishing 70% of water holding capacity of each soil. Filter-sterilized 25% Hoagland's nutrient solution was used in control treatment (uninoculated). Inoculated and uninoculated soils (500 g pot^{-1}) were covered with aluminium foil and incubated in the greenhouse for one week, for the establishment of the bacterial communities, prior to the placement of five-day old rice seedlings. One seedling per pot and six replicates per treatment were used. Rice plants were then further cultivated in the greenhouse under the aforementioned conditions. At week three, after tiller formation, plants exposed to low- and high-BID were subjected to flooding. At week five, the plants were harvested and the bacterial communities in soil free of roots (denoted bulk soil), rhizosphere soil, the root and shoot tissues were assessed by PCR-DGGE. Individual rice plants were harvested and root-adhering soil particles were removed with a forceps and stored. The bacterial endophytic community of root and shoot tissues were assessed as described above. DNA from bulk and rhizosphere soils were also extracted with the protocol described for seed samples, however DNA from these microhabitats were further purified (twice) using the Wizard DNA clean-up system (Promega).

PCR-DGGE and ordination analyses

For PCR-DGGE analysis, the Chelius-Triplett nested PCR system (799F-1492R followed by 968F-1401R) was the most efficient approach to detect rice endophytic bacteria [62]. DNA amplification conditions and PCR-DGGE analyses were performed as described previously [58]. The denaturing gradient gel was casted with a gradient of 40–55% denaturant (100% denaturant contained 7 M urea and 40% formamide) in a PhorU-2 apparatus, (Ingeny, Goes, Netherlands). The amplicons (150 ng) from each treatment with six replicates were loaded side-by-side in the same gradient gel and were cross-compared. Reference markers containing equal amounts of DNA extracted from the inoculated strains were loaded at both edges and among treatments for normalization purposes. After the run, gels were stained with SYBR gold (Molecular Probes, Leiden, Netherlands) and the DGGE patterns were made visible by illumination with UV. The profiles were digitized using a digital camera and stored as TIFF files.

All PCR-DGGE profiles were analysed using GelCompar II v 4.06 (Applied Maths, Sint-Martens-Latem, Belgium) as described previously [58]. Relative band intensity from each PCR-DGGE profile was exported into matrix. This data combined with the biotic and abiotic factors (assigned as nominal environmental variables) were used to generated the biplot ordination diagrams

by computing the redundancy analysis (RDA) from the package software CANOCO (Biometrics, PRI, Netherlands).

Isolates and PCR-DGGE bands identification

Rice seed endophytes were isolated using R2A at 28°C and replicated on the same medium to obtain pure cultures. Single colonies were used for identification by sequencing the partial 16S rRNA gene as described [63]. For this, the reverse primer 1401R was used in the sequencing reaction. In addition, dominant bands from generated PCR-DGGE profiles were selected for identification. Following excision, band DNA was extracted by incubating the polyacrylamide gel in 50 μl sterile TAE buffer solution for two days at 4°C. From the homogenate, 2 μl was used as DNA template for PCR-DGGE re-amplification. PCR-DGGE bands with identical motility compared with the original PCR-DGGE pattern were subjected to identification by sequencing with reverse primer 1401R. Furthermore, 16S rRNA gene amplicons of rice seed endophyte strains were subjected to PCR-DGGE analysis and PCR-DGGE bands with identical denaturation motility were tentatively assigned to strains. The sequences obtained from this study were assigned to bacterial species by BlastN against NCBI nucleotide database considering only type strains as reference strains. In addition, we compared the generated sequences to publicly available seed-associated (EU741000-EU741045), [5,9,19], rice-associated [20,21,64] and rice paddy soil bacterial sequences (FJ266313-FJ266342), [27,65]. The sequences obtained from the excised PCR-DGGE bands and the partial 16S rRNA gene from strains were deposited in the GenBank under the accession numbers JN110430 to JN110462.

Supporting Information

Figure S1 PCR-DGGE profiles of shoot and root endosphere bacterial community of rice cultivated in Kollumerwaard soil. PCR-DGGE profiles of shoot A) and root B) endosphere community of rice plants cultivated in K soil. Rice plants were subjected to unflooded and flooded regimes and exposed to low-, high- and un-inoculated treatments. Six replicates per treatments are shown. Arrow heads indicate identified communities (see Table 1 and 2).

Figure S2 PCR-DGGE profiles of shoot and root endosphere bacterial community of rice cultivated in Valthermond soil. PCR-DGGE profiles of shoot A) and root B) endosphere community of rice plants cultivated in V soil. Rice plants were subjected to unflooded and flooded regimes and exposed to low-, high- and un-inoculated treatments. Six replicates per treatments are shown. Arrow heads indicate identified communities (see Table 1 and 2).

Acknowledgments

We thank Dr. Darshan Brar at IRRI for providing the rice seeds. This study was supported by a collaborative agreement between the University of Groningen and Plant Research International.

Author Contributions

Conceived and designed the experiments: PRH CCPH LSVO JDVE. Performed the experiments: PRH CCPH. Analyzed the data: PRH CCPH LSVO JDVE. Contributed reagents/materials/analysis tools: PRH CCPH LSVO JDVE. Wrote the paper: PRH CCPH LSVO JDVE. Final approval of the version to be published: PRH CCPH LSVO JDVE.

References

1. Dias ACF, Costa FEC, Andreote FD, Lacava PT, Teixeira MA, et al. (2009) Isolation of micropropagated strawberry endophytic bacteria and assessment of their potential for plant growth promotion. World J Microbiol Biotechnol 25: 189–195.

2. Lucero M, Barrow JR, Osuna P, Reyes I (2008) A cryptic microbial community persists within micropropagated *Bouteloua eriopoda* (Torr.) Torr. cultures. Plant Sci 174: 570–575.

3. Baker KF, Smith SH (1966) Dynamics of seed transmission of plant pathogens. Annu Rev Phytopathol 14: 311–334.

4. Mundt JO, Hinkle NF (1976) Bacteria within ovules and seeds. Appl Environ Microbiol 32: 694–698.

5. Mano H, Tanaka F, Watanabe A, Kaga H, Okunishi S, et al. (2006) Culturable surface and endophytic bacterial flora of the maturing seeds of rice plants (*Oryza sativa*) cultivated in a paddy field. Microbes Environ 21: 86–100.

6. Kaga H, Mano H, Tanaka F, Watanabe A, Kaneko S, et al. (2009) Rice seeds as sources of endophytic bacteria. Microbes Environ 24: 154–162.

7. Mano H, Morisaki H (2008) Endophytic bacteria in the rice plant. Microbes Environ 23: 109–117.

8. Lucero ME, Unc A, Cooke P, Dowd S, Sun S (2011) Endophyte microbiome diversity in micropropagated *Atriplex canescens* and *Atriplex torreyi* var *griffithsii*. PLoS ONE 6(3): e17693. doi:10.1371/journal.pone.0017693.

9. Johnston-Monje D, Raizada MN (2011) Conservation and diversity of seed associated endophytes in *Zea* across boundaries of evolution, ethnography and ecology. PLoS ONE 6(6): e20396. doi:10.1371/journal.pone.0020396.

10. Puente ME, Li CY, Bashan Y (2009) Endophytic bacteria in cacti seeds can improve the development of cactus seedlings. Environ Exp Bot 66: 402–408.

11. Hallmann J, Berg G (2006) Spectrum and population dynamics of bacterial root endophytes. In: Schulz BJE, Boyle CJC, Sieber TN, eds. Microbial Root Endophytes. DorderchtNL: Springer. pp 15–31.

12. Reiter B, Pfeifer U, Schwab H, Sessitsch A (2002) Response of endophytic bacterial communities in potato plants to infection with *Erwinia carotovora* subsp *atroseptica*. Appl Environ Microb 68: 2261–2268.

13. Hardoim PR, van Overbeek LS, van Elsas JD (2008) Properties of bacterial endophytes and their proposed role in plant growth. Trends Microbiol 16: 463–471.

14. Compant S, Clement C, Sessitsch A (2010) Plant growth-promoting bacteria in the rhizo- and endosphere of plants: Their role, colonization, mechanisms involved and prospects for utilization. Soil Biol Biochem 42: 669–678.

15. Glick BR, Todorovic B, Czarny J, Cheng Z, Duan J, et al. (2007) Promotion of plant growth by bacterial ACC deaminase. Crit Rev Plant Sci 26: 227–242.

16. Holland MA (1997) Occam's razor applied to hormonology. Are cytokinins produced by plants? Plant Physiol 115: 865–868.

17. Taghavi S, van der Lelie D, Hoffman A, Zhang YB, Walla MD, et al. (2010) Genome sequence of the plant growth promoting endophytic bacterium *Enterobacter* sp 638. PLoS Genet 6: e1000943. doi:10.1371/journal.p-gen.1000943.

18. Cottyn B, Debode J, Regalado E, Mew TW, Swings J (2009) Phenotypic and genetic diversity of rice seed-associated bacteria and their role in pathogenicity and biological control. J Appl Microbiol 107: 885–897.

19. Okunishi S, Sako K, Mano H, Imamura A, Morisaki H (2005) Bacterial flora of endophytes in the maturing seed of cultivated rice (*Oryza sativa*). Microbes Environ 20: 168–177.

20. Hardoim PR, Sessitsch A, Reinhold-Hurek B, van Overbeek LS, van Elsas (2011) Assessment of rice root endophytes and their potential for plant growth promotion. In: Hardoim PR, ed. Bacterial endophytes of rice – their diversity, characteristics and perspectives Groningen. pp 77–100.

21. Mano H, Tanaka F, Nakamura C, Kaga H, Morisaki H (2007) Culturable endophytic bacterial flora of the maturing leaves and roots of rice plants (*Oryza sativa*) cultivated in a paddy field. Microbes Environ 22: 175–185.

22. Hashidoko Y, Hayashi H, Hasegawa T, Purnomo E, Osaki M, et al. (2006) Frequent isolation of sphingomonads from local rice varieties and other weeds grown on acid sulfate soil in South Kalimantan, Indonesia. Tropics 15: 391–395.

23. Lebuhn M, Achouak W, Schloter M, Berge O, Meier H, et al. (2000) Taxonomic characterization of *Ochrobactrum* sp isolates from soil samples and wheat roots, and description of *Ochrobactrum tritici* sp nov and *Ochrobactrum grignonense* sp nov. Int J Syst Evol Micr 50: 2207–2223.

24. Lai WA, Kämpfer P, Arun AB, Shen FT, Huber B, et al. (2006) *Deinococcus ficus* sp nov., isolated from the rhizosphere of *Ficus religiosa* L. Int J Syst Evol Micr 56: 787–791.

25. Steindler L, Bertani I, De Sordi L, Bigirimana J, Venturi V (2008) The presence, type and role of N-acyl homoserine lactone quorum sensing in fluorescent *Pseudomonas* originally isolated from rice rhizospheres are unpredictable. FEMS Microbiol Lett 288: 102–111.

26. Behrendt U, Ulrich A, Schumann P, Naumann D, Suzuki K (2002) Diversity of grass-associated *Microbacteriaceae* isolated from the phyllosphere and litter layer after mulching the sward; polyphasic characterization of *Subtercola pratensis* sp nov., *Curtobacterium herbarum* sp nov and *Plantibacter flavus* gen. nov., sp nov. Int J Syst Evol Micr 52: 1441–1454.

27. Kämpfer P, Rainey FA, Andersson MA, Lassila ELN, Ulrych U, et al. (2000) *Frigoribacterium faeni* gen. nov., sp nov., a novel psychrophilic genus of the family *Microbacteriaceae*. Int J Syst Evol Micr 50: 355–363.

28. Shrestha PM, Noll M, Liesack W (2007) Phylogenetic identity, growth-response time and rRNA operon copy number of soil bacteria indicate different stages of community succession. Environ Microbiol 9: 2464–2474.

29. Hallmann J, Quadt-Hallmann A, Mahaffee WF, Kloepper JW (1997) Bacterial endophytes in agricultural crops. Can J Microbiol 43: 895–914.

30. Oehrle NW, Karr DB, Kremer RJ, Emerich DW (2000) Enhanced attachment of *Bradyrhizobium japonicum* to soybean through reduced root colonization of internally seedborne microorganisms. Can J Microbiol 46: 600–606.

31. Ryan RP, Monchy S, Cardinale M, Taghavi S, Crossman L, et al. (2009) The versatility and adaptation of bacteria from the genus *Stenotrophomonas*. Nat Rev Microbiol 7: 514–525.

32. Hayward AC, Fegan N, Fegan M, Stirling GR (2010) *Stenotrophomonas* and *Lysobacter*: ubiquitous plant-associated gamma-proteobacteria of developing significance in applied microbiology. J Appl Microbiol 108: 756–770.

33. Mehnaz S, Kowalik T, Reynolds B, Lazarovits G (2010) Growth promoting effects of corn (*Zea mays*) bacterial isolates under greenhouse and field conditions. Soil Biol Biochem 42: 1848–1856.

34. Idris A, Labuschagne N, Korsten L (2009) Efficacy of rhizobacteria for growth promotion in sorghum under greenhouse conditions and selected modes of action studies. J Agr Sci 147: 17–30.

35. De Freitas JR, Banerjee MR, Germida JJ (1997) Phosphate-solubilizing rhizobacteria enhance the growth and yield but not phosphorus uptake of canola (*Brassica napus* L). Biol Fertil Soils 24: 358–364.

36. Sturz AV, Matheson BG, Arsenault W, Kimpinski J, Christie BR (2001) Weeds as a source of plant growth promoting rhizobacteria in agricultural soils. Can J Microbiol 47: 1013–1024.

37. van der Lelie D, Taghavi S, Monchy S, Schwender J, Miller L, et al. (2009) Poplar and its bacterial endophytes: coexistence and harmony. Crit Rev Plant Sci 28: 346–358.

38. Taghavi S, Garafola C, Monchy S, Newman L, Hoffman A, et al. (2009) Genome survey and characterization of endophytic bacteria exhibiting a beneficial effect on growth and development of poplar trees. Appl Env Microbiol 75: 748–757.

39. Mastretta C, Taghavi S, van der Lelie D, Alessio M, Francesca G, et al. (2009) Endophytic bacteria from seeds of *Nicotiana tabacum* can reduce cadmium phytotoxicity. Int J Phytoremediat 11: 251–267.

40. Compant S, Duffy B, Nowak J, Clement C, Ait Barka E (2005) Use of plant growth-promoting bacteria for biocontrol of plant diseases: principles, mechanisms of action, and future prospects. Appl Environ Microb 71: 4951–4959.

41. Weyens N, van der Lelie D, Taghavi S, Newman L, Vangronsveld J (2009) Exploiting plant-microbe partnerships to improve biomass production and remediation. Trends Biotechnol 27: 591–598.

42. Kageyama K, Nelson EB (2003) Differential inactivation of seed exudate stimulation of *Pythium ultimum* sporangium germination by *Enterobacter cloacae* influences biological control efficacy on different plant species. Appl Environ Microb 69: 1114–1120.

43. Roberts DP, McKenna LF, Lohrke SM, Rehner S, de Souza JT (2007) Pyruvate dehydrogenase activity is important for colonization of seeds and roots by *Enterobacter cloacae*. Soil Biol Biochem 39: 2150–2159.

44. Patten CL, Glick BR (2002) Role of *Pseudomonas putida* indoleacetic acid in development of the host plant root system. Appl Environ Microb 68: 3795–3801.

45. Glick BR, Jacobson CB, Schwarze MMK, Pasternak JJ (1994) 1-aminocyclo-propane-1-carboxylic acid deaminase mutants of the plant-growth promoting rhizobacterium *Pseudomonas putida* GR12-2 do not stimulate canola root elongation. Can J Microbiol 40: 911–915.

46. Belimov AA, Safronova VI, Sergeyeva TA, Egorova TN, Matveyeva VA, et al. (2001) Characterization of plant growth promoting rhizobacteria isolated from polluted soils and containing 1-aminocyclopropane-1-carboxylate deaminase. Can J Microbiol 47: 642–652.

47. Collavino MM, Sansberro PA, Mroginski LA, Aguilar OM (2010) Comparison of *in vitro* solubilization activity of diverse phosphate-solubilizing bacteria native to acid soil and their ability to promote *Phaseolus vulgaris* growth. Biol Fert Soils 46: 727–738.

48. Loaces I, Ferrando L, Scavino AF (2011) Dynamics, diversity and function of endophytic siderophore-producing bacteria in rice. Microb Ecol 61: 606–618.

49. Ramette A, Frapolli M, Sauxb MF, Gruffaz C, Meyer J-M, et al. (2011) *Pseudomonas protegens* sp. nov., widespread plant protecting bacteria producing the biocontrol compounds 2,4 diacetylphloroglucinol and pyoluteorin. Syst Appl Microbiol 34: 180–188.

50. Anandham R, Gandhi PI, Madhaiyan M, Sa T (2008) Potential plant growth promoting traits and bioacidulation of rock phosphate by thiosulfate oxidizing bacteria isolated from crop plants. J Basic Microb 48: 439–447.

51. De Cleene M (1985) The susceptibility of monocotyledons to *Agrobacterium tumefaciens*. J Phytopathol 113: 81–89.

52. Jung HM, Ten LN, Kim KH, An DS, Im WT, et al. (2009) *Dyella ginsengisoli* sp nov., isolated from soil of a ginseng field in South Korea. Int J Syst Evol Micr 59: 460–465.

53. Raaijmakers JM, Paulitz TC, Steinberg C, Alabouvette C, Moënne-Loccoz Y (2009) The rhizosphere: a playground and battlefield for soilborne pathogens and beneficial microorganisms. Plant Soil 321: 341–361.

54. Klironomos JN (2002) Feedback with soil biota contributes to plant rarity and invasiveness in communities. Nature 417: 67–70.

55. Himler AG, Adachi-Hagimori T, Bergen JE, Kozuch A, Kelly SE, et al. (2011) Rapid spread of a bacterial symbiont in an invasive whitefly is driven by fitness benefits and female bias. Science 332: 254–256.

56. Hurek T, Reinholdhurek B, Vanmontagu M, Kellenberger E (1994) Root colonization and systemic spreading of *Azoarcus* sp. strain BH72 in grasses. J Bacteriol 176: 1913–1923.

57. Miche L, Balandreau J (2001) Effects of rice seed surface sterilization with hypochlorite on inoculated *Burkholderia vietnamiensis*. Appl Environ Microb 67: 3046–3052.

58. Hardoim PR, Andreote FD, Reinhold-Hurek B, Sessitsch A, van Overbeek LS, et al. (2011) Rice root-associated bacteria: insights into community structures across 10 cultivars. FEMS Microbiol Ecol 77: 154–164.

59. Venema JH, Dijk BE, Bax JM, van Hasselt PR, Elzenga JTM (2008) Grafting tomato (*Solanum lycopersicum*) onto the rootstock of a high-altitude accession of *Solanum habrochaites* improves suboptimal-temperature tolerance. Environ Exp Bot 63: 359–367.

60. Andreote FD, de Araujo WL, de Azevedo JL, van Elsas JD, da Rocha UN, et al. (2009) Endophytic colonization of potato (*Solanum tuberosum* L.) by a novel competent bacterial endophyte, *Pseudomonas putida* strain P9, and its effect on associated bacterial communities. Appl Environ Microb 75: 3396–3406.

61. Sessitsch A, Coenye T, Sturz AV, Vandamme P, Ait Barka E, et al. (2005) *Burkholderia phytofirmans* sp. nov., a novel plant-associated bacterium with plant-beneficial properties. Int J Syst Evol Microbiol 55: 1187–1192.

62. Chelius MK, Triplett EW (2001) The diversity of archaea and bacteria in association with the roots of *Zea mays* L. Microb Ecol 41: 252–263.

63. Stevens P, van Elsas JD (2010) Genetic and phenotypic diversity of *Ralstonia solanacearum* biovar 2 strains obtained from Dutch waterways. Anton Leeuw Int J 97: 171–188.

64. Sun L, Qiu FB, Zhang XX, Dai X, Dong XZ, et al. (2008) Endophytic bacterial diversity in rice (*Oryza sativa* L.) roots estimated by 16S rDNA sequence analysis. Microb Ecol 55: 415–424.

65. Cuong ND, Nicolaisen MH, Sørensen J, Olsson S (2011) Hyphae-colonizing *Burkholderia* sp.—a new source of biological control agents against sheath blight disease (*Rhizoctonia solani* AG1-IA) in rice. Microb Ecol 62: 425–434.

11

Use-Exposure Relationships of Pesticides for Aquatic Risk Assessment

Yuzhou Luo*, Frank Spurlock, Xin Deng*, Sheryl Gill, Kean Goh

Department of Pesticide Regulation, California Environmental Protection Agency, Sacramento, California, United States of America

Abstract

Field-scale environmental models have been widely used in aquatic exposure assessments of pesticides. Those models usually require a large set of input parameters and separate simulations for each pesticide in evaluation. In this study, a simple use-exposure relationship is developed based on regression analysis of stochastic simulation results generated from the Pesticide Root-Zone Model (PRZM). The developed mathematical relationship estimates edge-of-field peak concentrations of pesticides from aerobic soil metabolism half-life (AERO), organic carbon-normalized soil sorption coefficient (KOC), and application rate (RATE). In a case study of California crop scenarios, the relationships explained 90–95% of the variances in the peak concentrations of dissolved pesticides as predicted by PRZM simulations for a 30-year period. KOC was identified as the governing parameter in determining the relative magnitudes of pesticide exposures in a given crop scenario. The results of model application also indicated that the effects of chemical fate processes such as partitioning and degradation on pesticide exposure were similar among crop scenarios, while the cross-scenario variations were mainly associated with the landscape characteristics, such as organic carbon contents and curve numbers. With a minimum set of input data, the use-exposure relationships proposed in this study could be used in screening procedures for potential water quality impacts from the off-site movement of pesticides.

Editor: Guy Smagghe, Ghent University, Belgium

Funding: The authors have no support or funding to report.

Competing Interests: The authors have declared that no competing interests exist.

* E-mail: yluo@cdpr.ca.gov (YL); xdeng@cdpr.ca.gov (XD)

Introduction

As part of the registration process, pesticides are evaluated for their potential to move off-site and impact non-target organisms. Surface runoff and tile flow are significant pathways for pesticides movement to surface waters. Monitoring-based surface water risk assessments of pesticides are usually conducted at the watershed scale using measured concentration data from river sites, especially at watershed outlets. For example, in-stream measurements of pesticides were assessed for U.S. watersheds with spatial scales across 14 orders of magnitude [1,2]. However, water flow from non-application areas and non-agricultural headwaters may significantly dilute pesticide concentrations in the river. For example, in California's Central Valley, one of the most productive agricultural areas in the world, pesticide concentrations are substantially higher in small creeks dominated by irrigation return flows, as compared to main streams where the majority of flow originates in Sierra Nevada mountains [3,4,5,6]. Because of the dilution effects, data in larger streams are associated with great spatial variability and thus not able to provide reliable and comparative information for pesticide management and mitigation. Therefore, assessments of aquatic risk now generally focus on smaller water bodies close to the edge-of-field.

Monitoring data are not always available and adequate for risk assessment, especially for pesticide products with new active ingredients. Environmental fate and transport models may be used to predict likely concentrations and associated risks of pesticides and to determine priorities for monitoring and regulatory

assessments. Water quality modeling is a key component of pesticide management, as in the development of Best Management Practices (BMPs) and Total Maximum Daily Loads (TMDL). Compared to watershed-scale models, field-scale models better account for hydrologic processes within agricultural fields and have the capability to simulate agricultural management practices. Field-scale models, such as the Pesticide Root Zone Model (PRZM) and the Root Zone Water Quality Model (RZWQM) [7], provide dynamic simulation of pesticide fate and transport processes, from pesticide applications to edge-of-field discharge. However, these models usually require a large set of model input parameters for the full descriptions of landscape characteristics, climate conditions, and management practices [8]. Consequently parameterization and simulation using field-scale models could be complicated and time-consuming, particularly when batch simulations and post-data analysis are involved [9]. In addition, during the pesticide registration process, the required input data may be difficult to obtain, especially for new pesticides which have not been applied in field conditions. Therefore, there is a research need to develop simple mathematical relationships to determine the potential aquatic risks of pesticides based on a minimum set of input parameters. Such simple relationships may be used in a screening procedure to identify pesticides that require more refined studies. As an early modeling effort, U.S. Environmental Protection Agency (USEPA) developed the model for Generic Estimated Environmental Concentration (GENEEC) to mimic more sophisticated simulations of pesticide transport from crop field to a standard pond [10]. However, differences in climate, soil, topography or crop are not considered in estimating potential exposure, thus substantially limiting its applications to pesticide evaluation and registration.

This study develops "use-exposure relationships" in the form of linear regression equations that link pesticide application rate and physicochemical properties to a predicted exposure level (such as peak concentration) for specific environmental configurations. The relationship is developed from the results of more detailed field-scale model simulations, but use significantly fewer input parameters. Specific study objectives are [1] to identify governing parameters in pesticide fate and transport processes in canopy-soil system [2]; to establish empirical relationships between those parameters; and [3] to demonstrate the developed model with parameterizations in the crop scenarios of California. The approach enables a quick risk assessment based on limited input data, and yields accuracy comparable to the dynamic simulation of the selected field scale model. The parameterized use-exposure relationship provides useful information for decision making in pesticide registration and management.

Materials and Methods

PRZM Model and Aquatic Exposure Assessment

PRZM is a one-dimensional compartmental model developed by USEPA for predicting pesticide movement in unsaturated soils [11]. It is designed to evaluate the influence of climate, soil properties, and management practices on pesticide transport and transformation processes, e.g., surface runoff, plant uptake, leaching, erosion, and volatilization. PRZM generates daily pesticides fluxes in both dissolved and adsorbed forms at the edge of fields. The resulting fluxes are useful for further analyses, such as aquatic risk assessment [12], loading calculation [13], and water quality evaluation [14]. PRZM has undergone validation and testing to field-scale runoff and leaching studies [15,16]. An enhanced version of PRZM is being used for surface water and groundwater exposure assessments in the European Union [17].

PRZM was selected in this study based on its ability to simulate relevant governing processes of pesticide transport and because of its use by regulatory agencies in their pesticide exposure assessments [10,17]. USEPA has also developed crop scenarios to facilitate the application of PRZM in risk assessment [18]. Those scenarios specify the environmental configurations for typical crops in major agricultural regions of U.S., including weather conditions, landscape characteristics, crop growth parameters, and soil properties.

To assess pesticide risks to aquatic organisms, an exposure index (EI) was defined in this study as follows. First, the estimated environmental concentrations (EEC) of pesticide in surface runoff and soil erosion were predicted as daily time series by PRZM. For dissolved pesticides, the exposure index was then calculated as the peak concentration of 4-day moving averages in the 1-in-3 year return period. This definition is consistent with the current regulatory surface water criteria for two widespread pesticidal surface water contaminates chlorpyrifos and diazinon [19,20].

Simulation Design and Input Data

Stochastic PRZM simulations were conducted to develop crop-scenario-specific "use-exposure relationships", i.e., empirical equations for predicting edge-of-field pesticide runoff concentrations. The simulations were based on crop scenarios developed by USEPA for pesticide risk assessment. A single annual pesticide application, repeated every year during 1961–1990, was simulated for a specific scenario. Annually repeated applications were utilized to incorporate the effects of climatic and hydrologic

variations on pesticide off-site movement. In addition, the 30-year simulation also accounted for the accumulation of persistent pesticides. For those pesticides, residues from previous applications may remain in the soil and add to the newly applied chemicals in the next year. A random application date was assigned to a PRZM simulation and pesticide was applied on the assigned date for each year in that simulation. The random date was generated within the pesticide's application season depending on the actual use pattern of the pesticide, such as dormant-season application, in-season application, and pre-emergent application. On each day of application, pesticide was applied at a fixed rate (a "base rate" or BASE, kg/ha as the active ingredient), which was an arbitrary small application rate for stochastic PRZM runs. A linear relationship was assumed between pesticide application rate and pesticide loadings from the field. A small base rate was used to avoid high predicted concentrations that exceed the water solubility (SOL) of the pesticide during the simulation. In this study, the base rate was set as 0.1 kg/ha. Preliminary simulations showed that, with base rate of 0.1 kg/ha, EECs were always lower than the corresponding SOL in all PRZM runs. Predicted concentrations should be compared to the SOL when applying the developed use-exposure relationships with actual label rates.

The chemical properties of aerobic soil metabolism half-life (AERO), organic carbon-normalized soil adsorption coefficient (KOC), and SOL are the governing factors for pesticide runoff potential in both dissolved and adsorbed phases [21,22]. KOC and SOL are direct input parameters in PRZM, and AERO is used in calculating the model inputs of decay rate constants [23]. In PRZM and most of other field-scale models, SOL is considered only as an upper limit on the dissolved concentration. In addition, significant association between the two properties of KOC and SOL has been reported in several previous studies. For example, linear correlation ($p < 0.001$) was confirmed between log-transformed KOC and SOL [24]. Similar linear relationships were also used to estimate KOC from SOL [11,25]. Therefore, only the independent chemical properties of AERO and KOC were selected in this study for stochastic PRZM simulations. The two selected parameters were also used by other studies for estimating pesticide runoff potentials [21,26,27,28].

The probability distributions for AERO and KOC were derived from a database of physiochemical property and reaction half-life complied by Spurlock [24] for 172 pesticides. Spurlock [24] suggested that log normality is a reasonable assumption for AERO and KOC, and that the two properties are independent ($p = 0.551$). Maximum likelihood estimation (MLE) was applied to estimate the distribution parameters (Table 1). Latin Hypercube Sampling (LHS) was used to generate random input data of

Table 1. Parameters for the log-normal distribution of aerobic soil metabolism half-life (AERO) and organic carbon-normalized soil sorption coefficient (KOC).

Variable	μ	σ	E	SD
AERO	3.44	1.99	226.01	1613.14
KOC	6.51	2.52	1.61e4	3.82e5

Notes:
[1] the parameter estimation was based on the median fate properties derived from registration studies of 172 pesticides [24].
[2] μ and σ are the mean and standard deviation of the data's natural logarithm, respectively; E and SD are the mean and standard deviation of the data, respectively.

AERO and KOC within 95% of cumulative frequency of the corresponding log normal distribution as defined in Table 1. For each PRZM run, the exposure index from the single pesticide application at base rate, denoted as EI_BASE, was calculated from the predicted daily EECs. Finally, the general mathematical relationship between the EI_BASE and input chemical properties of AERO and KOC for the particular crop scenario was developed based on regression analysis. The built-in Monte Carlo simulation in PRZM does not report daily time series of edge-of-field pesticide concentrations. In addition, a deficiency has been reported for the built-in Monte Carlo module in PRZM [29]. Therefore, LHS algorithm was taken from our previous study [30]; and a batch program was developed for stochastic PRZM runs and post-data analysis.

Crop Scenarios in California

Crop scenarios developed by USEPA for California were used for simulations. Available scenarios include standard crop scenarios [18], crop scenarios developed for organophosphate pesticide cumulative risk assessment [31], and crop scenarios developed for effects determinations for California listed endangered and threatened species [32]. Combined, approximately 30 scenarios are available for California, some of which are associated with pesticide use patterns with high runoff potential. These include crops with flood or furrow irrigation, winter rain season application, and pre-emergent herbicide application. These scenarios were selected in this study to demonstrate the development of the pesticide use-exposure screening model. Results of a statewide survey of California irrigation methods [33] indicated that field crops and tomatoes are dominated by flood and furrow irrigation. Scenarios of almond and turf were selected to represent wet season application and pre-emergent herbicide application, respectively. Tables 2 and 3 summarize the selected scenarios for PRZM simulations for California. A non-California scenario (Florida tomatoes) was also included in this study to compare/contrast results with a wetter climate.

The crop scenarios specify the weather conditions, soil properties, and crop growth parameters used in the PRZM simulations. Other model inputs used in this study, including chemical property and pesticide application, are summarized in Table 4. To provide conservative estimation of pesticide residues at the edge of field, pesticides were assumed to be incorporated into the soil at application and all mass loss fluxes by interception, volatilization, and decay on the plant canopy were set as zero.

Results and Discussion

Use-Exposure Relationship for a Single Pesticide Application

The response of EI_BASE to random values of AERO and KOC was evaluated by stochastic PRZM simulation. For each crop scenario, the predicted dissolved or adsorbed EI_BASE were paired with corresponding inputs of AERO and KOC for further regression analysis. The logarithmic transformation was also applied to EI_BASE according to preliminary analyses on pesticide concentrations detected in surface water of California [34]. Finally, an N×3 matrix of (lnAERO, lnKOC, lnEI_BASE), with N denoting the number of stochastic PRZM runs, was generated from Monte Carlo simulation. Demonstrated in Figure 1a is an example plot of the matrix for dissolved pesticides for the standard crop scenario for cotton in California. Significant correlations were identified between dissolved EI_BASE vs. AERO and KOC, especially for pesticides with KOC higher than a certain value (e.g., about 5 for lnKOC as shown in Figure 1a). This correlation reflected the effects of degradation and partitioning of pesticides on the peak concentration predicted at the field edge. For pesticides with lower KOC, EI_BASE was generally invariant with KOC. With low KOC values, pesticides are mainly present in dissolved phase, thus the change in KOC do not have a great effect on the phase partitioning.

Stochastic PRZM simulations for other crop scenarios revealed similar relationships among the predicted EI_BASE and input parameters, i.e., the general linear relationship between EI_BASE and AERO and KOC, and the presence of an approximate lnKOC cutoff, below which EI_BASE is independent of KOC. Therefore, a conceptual model was developed for the use-relationships from a single pesticide application based on that general data structure (Figure 1b). First, a breakpoint for KOC (KOC*) was determined from the trend of dissolved EI_BASE with KOC for the given crop scenario. Multivariate linear regression with logarithmic transformations was conducted between EI_BASE vs. AERO and KOC on the data points with KOC>KOC*. A similar relationship was applied to pesticides with KOC≤KOC* by substituting KOC with KOC*, in order to provide conservative estimation of EI_BASE for those pesticides.

The general use-exposure relationship is:

$$\ln(EI\_BASE)=f(AERO,KOC)=b_1+b_2\ln(AERO)+b_3\ln(KOC) \quad (1)$$

Table 2. Overview of selected California crop scenarios developed by USEPA.

Crop scenario	Represented use pattern	Soil (hydrologic group)	Weather station
Alfalfa (OP)	Pasture, gravity irrigation	Sacramento clay (D)	Fresno
Almond (STD)	Dormant application	Manteca fine sandy loam (C)	Sacramento
Cotton (STD)	Field crop, gravity irrigation	Twisselman Clay (C)	Fresno
Sugar beet (OP)	Field crop, gravity irrigation	Ryde clay loam (C)	Fresno
Tomato (STD)	Tomato, gravity irrigation	Stockton clay (D)	Fresno
Turf (RLF)	Pre-emergent application	CapaySilty Clay Loam (D)	San Francisco
Wheat (RLF)	Grain, gravity irrigation	San Joaquin Loam (D)	Fresno
Tomato_FL (STD)	Tomato scenario in Florida	Riviera Sand (C)	West Palm Beach

Data source: USEPA Tier 2 crop scenarios for PRZM/EXAMS Shell [18,31,32]. "STD" = Standard crop scenarios, "OP" = scenarios developed for the cumulative risk assessment of organophosphate pesticides, and "RLF" = scenarios developed for the effects determinations for the California red-legged frog and other California listed species. "Tomato_FL" denotes the standard USEAP crop scenario for tomato in Florida, provided as an example of the crop scenarios in other states.

Table 3. Landscape characteristics and soil properties of selected California crop scenarios.

Crop scenario	CN	USLE K/LS/P	USLE C	OC1
Alfalfa	90/88/89	0.20/0.30/1.0	0.051–0.217	1.77%
Almond	84/79/84	0.28/0.30/1.0	0.034–0.221	0.81%
Cotton	89/86/89	0.21/0.37/1.0	0.054–0.412	0.29%
Sugar beet	89/86/89	0.28/0.30/1.0	0.015–0.769	3.48%
Tomato	91/87/91	0.24/0.13/1.0	0.035–0.255	0.95%
Turf	80/80/80	0.37/1.80/0.5	0.001	35.6%
Wheat	92/89/90	0.37/0.79/1.0	0.027–0.604	0.44%
Tomato_FL	91/87/91	0.03/0.20/1.0	0.177–0.938	1.16%

Parameters:
CN = Runoff curve numbers of antecedent moisture condition II for fallow, cropping, and residue, respectively;
USLE K = soil erodibility for the universal soil loss equation (USLE);
USLE LS = topographic factor for the USLE;
USLE P = practice factor for the USLE;
USLE C = cover management factor for the USLE;
OC1 = Organic carbon content in the surface soil.

And the relationship for dissolved pesticides is:

$$\ln(EI\_BASE) = f(AERO, KOC)$$
$$= b_1 + b_2 \ln(AERO) + b_3 \ln(\max(KOC, KOC^*))$$
(2)

where b_1, b_2, and b_3 are coefficients derived by regression. KOC* was determined by maximizing the coefficient of determination

(R^2) in the regression analysis. Based on the linear assumption between pesticide application and exposure, the dissolved exposure index (EI, μg/L) from pesticide applications at the actual rate (RATE, kg/ha) was expressed as,

$$EI = \frac{RATE}{BASE} \cdot EI\_BASE$$
(3)

Derived Parameters for Dissolved Pesticides

For each selected scenario, 5 000 stochastic simulations of PRZM (N = 5 000) were conducted for the 30-year period of 1961–1990. Regression coefficients and other statistics for the use-exposure relationship for dissolved pesticides are summarized in Table 5.

Values of lnKOC* varied with different scenarios, ranging from 0.5 to 5.5. With higher organic carbon content in surface soil (OC1), such as for the California turf scenario (35.6%), lower KOC* values were observed relative to other scenarios with smaller OC1 ranging from 0.44% to 3.48% (Table 3). However, the product of KOC* and OC1, equivalent to the corresponding limiting distribution coefficient (KD*), was approximately constant among the scenarios, ranging from 0.5 to 0.7. Since the distribution coefficients indicate pesticide mobility in the soil, the KD* value determined from the PRZM simulations was considered as the critical KD value below which the transport process of dissolved pesticides with surface runoff was insensitive to their KOC values.

The empirical AERO-KOC based use-exposure relationships explained 90–95% of the variances on the predicated EI_BASE of dissolved pesticides. The predictive ability of the relationships was mainly attributable to lnKOC, which solely explained 85–90% of

Table 4. Chemical property and pesticide application in PRZM simulations.

Variable	Description	Values/notes
APPDAY	Application date	Random numbers (uniform) in the application season
APPEFF	Application efficiency	0.99 (ground application) [2]
CAM	Pesticide application method	4 (soil incorporation)
DAIR	Diffusivity in air (cm2/day)	4300 [1]
DEPI	Incorporation depth (cm)	4 [2]
DRFT	Drift fraction	0.01 [2]
DSRATE	Adsorbed phase decay rate (1/d)	= ln2/AERO, LHS sampling
DWRATE	Dissolved phase decay rate (1/d)	= DSRATE [2]
ENPY	Enthalpy of vaporization (kcal/mol)	20 [1]
FEXTRC	Washoff extraction (1/cm)	0.5 [1]
HENRYK	Henry's law constant (g/aq, dimensionless)	0 [3]
IPSCND	Disposition of foliar pesticide after harvest	1 (surface applied) [3]
PLDKRT	Decay rate on foliage (1/d)	0 [2]
PLVKRT	Volatilization rate on foliage (1/d)	0 [2]
KOC	Organic carbon-normalized soil adsorption coefficient (L/kg[OC])	LHS sampling
TAPP	Application rate (kg/ha)	0.1 (base application rate used in this study)
UPTKF	Pesticide uptake	0 [2]

Notes:
[1]Suggested value in the PRZM manual [11].
[2]USEPA-suggested model input parameter value [23].
[3]Assumptions made for conservative evaluation of pesticide exposure.

(a)

(b)

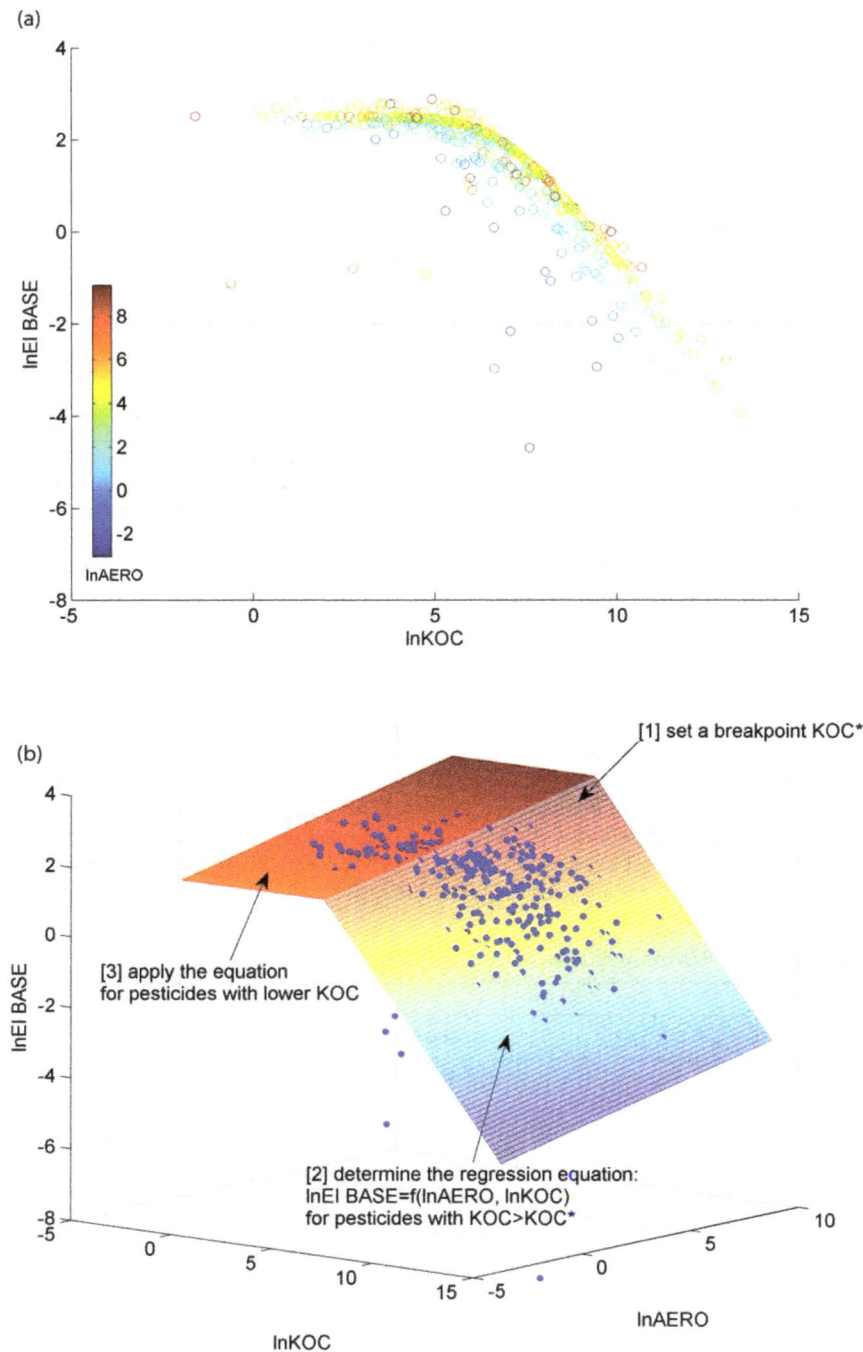

Figure 1. Use-exposure relationship for dissolved pesticides (EI_BASE in µg/L): (a) example results of Monte Carlo simulation and (b) conceptual model.

the total variances in lnEI_BASE, while AERO had only a limited contribution. Generally, the relative magnitudes of dissolved EI_BASE for pesticides with large KOC were mainly related to the regression coefficients of KOC (b3), while those for pesticides with lower KOC were determined by the intercepts (b1). This reflected the competition between phase partitioning and water runoff extraction on the pesticide yields from the applied field. Relatively higher regression coefficients for AERO (b2) were observed for scenarios with higher OC1 such as sugar beet

(OC1 = 3.48%) and turf (OC1 = 35.6%) (Table 3). With elevated OC contents, pesticides are less mobile in the soil and could be accumulated for a longer period. Previous studies indicated that pesticide half-life in the soil is the key parameter in determining the total amount of pesticide residues discharged from fields [21,22,35]. However, the measure of exposure in risk characterization is estimated from peak concentrations, which are usually observed shortly after pesticide applications once surface runoff induced by precipitation or irrigation is available. Thus, soil

Table 5. Use-exposure relationships for dissolved pesticides in selected California crop scenarios.

Scenarios	Coefficients			R2	lnKOC*
	b1	b2	b3		
Alfalfa	5.2156	0.1907	−0.8288	0.9494	3.5
Almond	4.8131	0.1869	−0.7467	0.9335	4.5
Cotton	6.3173	0.1467	−0.7662	0.9102	5.5
Sugar beet	4.9105	0.2412	−0.8377	0.9193	3.0
Tomato	5.9979	0.1785	−0.7844	0.8970	4.0
Turf	3.3647	0.2821	−0.8248	0.9546	0.5
Wheat	6.0764	0.1853	−0.7954	0.9487	5.0
Tomato_FL	4.9362	0.2531	−0.8063	0.9422	4.0

Note: "Tomato_FL" denotes the standard USEAP crop scenario for tomato in Florida, which is provided as an example of the crop scenarios in other states.

metabolism might have only moderate effects on pesticide exposure at the edge of field as measured by peak concentrations.

The regression coefficients for AERO and KOC did not vary much over scenarios. For instance, the maximum regression coefficient for KOC was −0.7467 (almonds), while the minimum value was −0.8377 (sugar beet). The regression coefficients for AERO ranged from 0.1467 (cotton) to 0.2821 (turf) (Table 5). The use-exposure relationship derived for crops in other states, taking tomato in Florida as an example in this study, also showed similar regression coefficients as in California crops. This suggested that, for a specific pesticide, the difference of predicted EI_BASE over scenarios were mainly determined by the intercepts of b1 in Eq. (1). In another words, the effects of chemical fate processes such as partitioning and degradation on pesticide exposure were similar among scenarios, while the spatial variability was related to environmental parameters including climate condition, soil property, and landscape characteristics.

While the California scenarios are developed for areas with similar climate, they are associated with substantial variability in soil type and hydrologic group (Table 3). The intercept in the regression equation for the use-exposure relationship (b1 in Table 5) was significantly correlated to curve numbers for residue surface soil condition (with a p-value, p = 0.008), and moderately correlated to curve numbers for cropping surface condition (p = 0.063). In most of the crop scenarios, curve numbers for residue surface condition were implemented for winter months, or the rainfall season in California. In the use-exposure relationship, the intercept was associated with water runoff generation since the chemical fate processes such as partitioning and degradation were represented by the chemical properties. Therefore, the significant correlation between b1 and curve number for residue surface soil condition indicated that peak concentrations of pesticide at the field edges were most likely observed during the rainfall season in California. The dependence of pesticide runoff potential on curve number in the PRZM simulation has been reported in previous studies [22,36,37]. Findings in this study confirmed the effects of curve number on the predicted pesticide concentrations and loadings from the crop fields.

Derived Parameters for Sediment-Bound Pesticides

At present there are no surface water quality criteria at either federal or state level for sediment-bound pesticides. Water quality assessments for pesticides in sediment, such as those for Clean Water Act Section 303(d) listing [38], are based on 10-day *Hyalella azteca*

sediment toxicity tests [39]. To mimic the sediment toxicity tests, 10-day averages were calculated as adsorbed exposure index from PRZM-predicted daily concentrations of pesticide associated with soil erosion. The same frequency as for dissolved pesticide, i.e., once every three years return period, was used in the development of use-exposure relationship for adsorbed pesticides. Median lethal concentration (LC50) values for sediment toxicity are usually reported on an OC-normalized basis. For example, *Hyalella azteca* 10-day LC50 values for pyrethroids are typically reported at 1% OC, as compiled by Domagalski et al. [40]. To match the toxicity data, PRZM-predicted concentrations of pesticide in eroded sediment were normalized by OC1 defined in each scenario (Table 3).

For pesticide associated with eroded soil, Figure 2 shows the results of stochastic PRZM runs based on the USEPA standard scenario for cotton in California. There was a general increasing trend of EI_BASE with increases of KOC and AERO, especially for pesticides with KOC lower than a certain value. For pesticides with extremely high values, such as those with lnKOC larger than about 11 as shown in Figure 2a, the predicted EI_BASE for adsorbed pesticides were associated with high uncertainty and might not be significantly correlated with their KOC values. For these pesticides, the majority of the residues have been partitioned into the particulate phase. Based on the similar equations for dissolved pesticides, the following use-exposure relationship was developed for adsorbed pesticides,

$$\ln(EI\_BASE) = f(AERO, KOC) \\ = b_1 + b_2 \ln(AERO) + b_3 \ln(\min(KOC, KOC^*)) \quad (4)$$

where EI_BASE (ng/g) is the predicted exposure index associated with eroded soil from a single pesticide application at BASE rate of 0.1 kg/ha, b's are regression coefficients, and KOC* is a threshold value for KOC above which the EI_BASE was assumed to be independent the pesticide's KOC.

Table 6 shows the parameters of the use-exposure relationships for adsorbed pesticides under selected California scenarios. The R^2 values ranged from 60–85%, substantially lower than those for dissolved pesticides (Table 5). Similar to the equations for dissolved pesticides, the majority of the variance in the use-exposure relationships for adsorbed pesticides wasexplained by KOC. Although the KOC* varied greatly, the KD* values as the product of KOC* and OC1 for the modeled scenarios were generally invariant, ranging from 160–240. However, the regression coefficients varied greatly among scenarios. Observed uncertainty in predicted exposure of adsorbed pesticides might be related to soil erosion processes. PRZM simulates soil erosion based on themodified universal soil loss equation (MUSLE) with input parameters. These parameters are usually associated with seasonality and variability in soil properties and management practices (Table 3). With only input parameters of AERO and KOC, therefore, the proposed use-exposure relationship was inadequate for capturing the variability in soil erosion processes.

Use-Exposure Relationship for Multiple Applications

The use-exposure relationships in Eqs. (1)through(4) provide estimates for the exposure index from a single pesticide application. For multiple applications, the exposure could be conservatively estimated from the maximum season application rate. To refined the estimation, especially for pesticides with short field dissipation half-lives (FD), equivalent application rate at any given time, RATE_eq(t), is calculated as the total pesticide amount in the soil available for runoff and soil erosion processes. RATE_eq includes contributions from both applied pesticide on

(a)

(b)

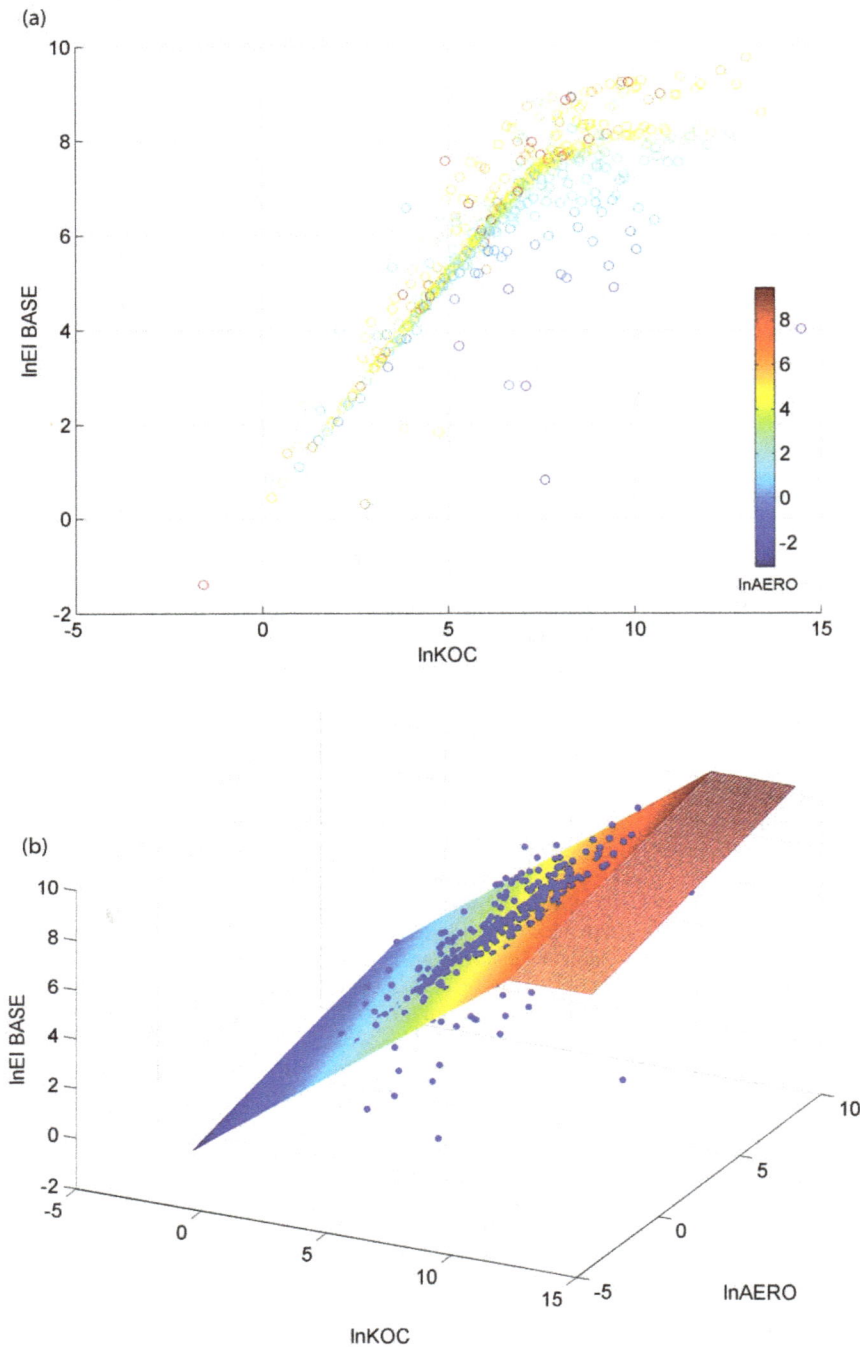

Figure 2. Use-exposure relationshio for adsorbed pesticides (EI_BASE in ng/g): (a) example results of Monte Carlo simulation and (b) conceptual model.

the given day and residues from previous applications, and could be calculated in the form of a convolution,

$$RATE\_eq(t) = \int_{t_0}^{t} [RATE(\tau) \cdot D(t - \tau)] d\tau$$

(5)

$$D(\Delta t) = \exp\left(-\frac{\ln 2}{FD} \Delta t\right)$$

where t0 is the first day of the application season, RATE(t) is the application rate at day t, and D(Δt) is the total fractional decay during Δt. The calculation of RATE_eq accounted for only the pesticide loss by degradation. Losses from surface runoff, soil erosion, and leaching were ignored to provide a conservative estimation of the amount of pesticide remaining in the soil.

Figure 3 presents a schematic of the RATE_eq calculation from multiple applications of carbaryl for tomatoes (maximum single application rate = 2.24 kg/ha, application interval = 7 day, and

Table 6. Use-exposure relationships for adsorbed pesticides in selected California crop scenarios.

Scenarios	Coefficients			R²	ln(KOC*)
	b1	b2	b3		
Alfalfa	1.7756	0.3140	0.4936	0.6896	9.5
Almond	0.1179	0.2116	0.6937	0.7955	10.0
Cotton	0.9213	0.1890	0.7221	0.8466	11.0
Sugar beet	2.7386	0.3254	0.5118	0.6409	8.5
Tomato	3.2070	0.1912	0.6062	0.7770	10.0
Turf	2.7715	0.2832	0.4486	0.6106	6.5
Wheat	1.0782	0.3233	0.5848	0.7210	10.5
Tomato_FL	1.7065	0.4105	0.4809	0.7607	10.0

maximum number of applications = 4) [32]. The exposure index from multiple pesticide applications could be estimated by substituting RATE in Eq. (3) with the maximum value of RATE_eq during a year or a cropping season. For single pesticide application rate and application interval (INTERVAL, day) as fixed values, as usually documented in pesticide labels, the maximum RATE_eq could be directly calculated as,

$$\max[RATE\_eq(t)] = \\ RATE \cdot \sum_{i=1}^{M-1} \exp\left[-\frac{\ln 2}{FD}(M-i)\cdot INTERVAL\right] \quad (6)$$

where M is the maximum number of applications.

Summary and Conclusions

Use-exposure relationships were developed as an alternative approach to field-scale modeling for pesticide runoff and associated aquatic risks. The relationships require a minimum set of input parameters to estimate exposure, which is defined herein as peak pesticide concentrations at the edge of field. The selected input parameters, half-life in the soil, adsorption coefficient, and recommended application rates, are generally available during pesticide registration. Thus the proposed approach is appropriate for quickly screening pesticide products for their potential adverse effects on the environment and human health. While the PRZM model was chosen to parameterize the weighting factors of the selected parameters for this study, the approach could be applied with other field-scale models.

The development of use-exposure relationships was demonstrated using crop scenarios developed by the USEPA for California. The relationships explained 90–95% of variations in the exposure index of dissolved pesticides as predicted by PRZM modeling. Regression coefficients for AERO and KOC for the simulated scenarios varied only in small ranges, suggesting that the effects of chemical property-related fate processes such as partitioning and degradation on the predicted exposure index were similar among scenarios. KOC was the governing factor in predicting pesticide exposures for all scenarios. Since aquatic risk analysis is mainly focused on the peak concentrations of pesticides, and these concentrations are usually observed shortly after pesticide applications, the half-life in the soil had limited influence on the exposure index defined in this study. The results of this study suggested that the selection of evaluation approaches for pesticide exposure could be dependent on the purpose of regulatory assessment and management planning. For instance, total pesticide loadings from agricultural fields might be very sensitive to chemical persistence, while the peak concentrations of a pesticide are mainly related to its mobility. For a particular

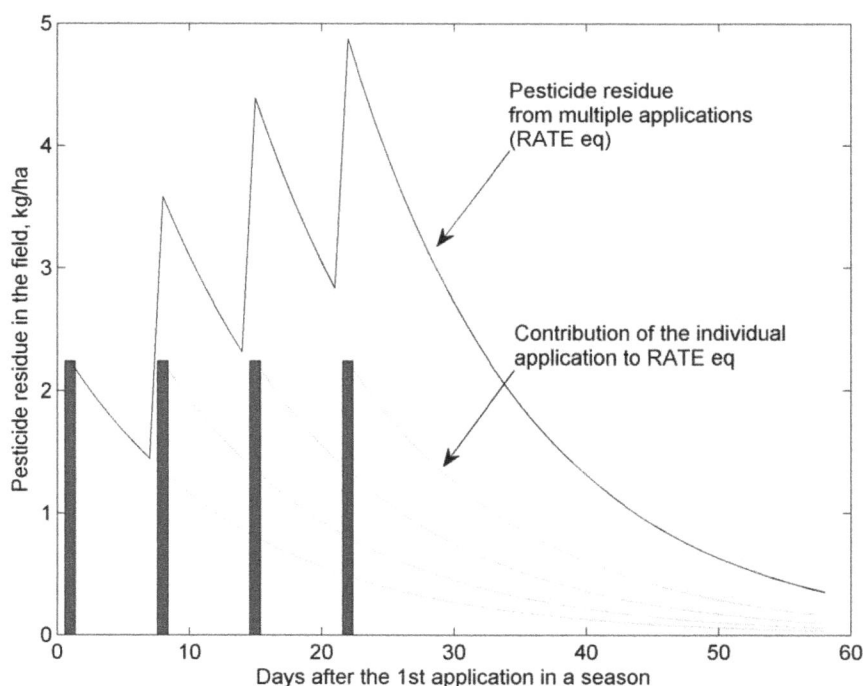

Figure 3. Equivalent application rate from multiple pesticide applications, illustrated with recommended application rates and intervals of carbaryl for tomatoes [32]. Bars represent four applications at 2.24 kg/ha and 7-day intervals.

pesticide, the spatial variability on its exposure indices across various scenarios is associated with landscape characteristics, such as OC and CN values. In addition, significant correlation was observed between intercept constants (b1) in the use-exposure relationships and the curve numbers for residue in the crop scenarios. Those findings indicated the possibility in developing generic equations of use-exposure relationships.

Methods to extend the capability of use-exposure relationship modeling are provided, including applications for assessing sediment-bound pesticides and provisions for multiple pesticide applications within a growing season. Since the use-exposure relationships were parameterized from regression analysis on the simulation results of existing field-scale models, the accuracy of risk assessment based on those relationships is associated with the model itself and the configuration of crop scenarios. Future work issuggested to implement the developed equations in real field conditions and to compare the predictions with measured pesticide

data. An evaluation of measured data with corresponding field conditions would generate instructive information for developing crop scenarios for pesticide exposure assessment and risk characterization.

Acknowledgments

The authors acknowledge Dr. John Sanders, Michael Ensminger, and Keith Starner of California Department of Pesticide Regulation and W. Martin Williams of Waterborne Environmental, Inc. for valuable discussions and critical reviews.

Author Contributions

Conceived and designed the experiments: YL FS XD SG KG. Performed the experiments: YL FS XD. Analyzed the data: YL XD. Contributed reagents/materials/analysis tools: YL FS. Wrote the paper: YL FS XD SG KG.

References

1. Capel PD, Larson SJ, Winterstein TA (2001) The behaviour of 39 pesticides in surface waters as a function of scale. Hydrologic Processes 15: 1251–1269.

2. Capel PD, Larson SJ (2001) Effect of scale on the behavior of atrazine in surface waters. Environmental Science & Technology 35: 648–657.

3. Domagalski JL, Munday C (2003) Evaluation of diazinon and chlorpyrifos concentrations and loads, and other pesticide concentrations, at selected sites in the San Joaquin Valley, California, April to August, 2001. United States Geologic Survey. Water-Resources Investigations Report 03-4088 Water-Resources Investigations Report 03-4088.

4. Kratzer CR, Zamora C, Knifong DL (2002) Diazinon and chlorpyrifos loads in the San Joaquin River Basin, California, January and February 2000. United States Geological Survey Water Resources Investigation Report 02-4103 Water Resources Investigation Report 02-4103.

5. Dubrovsky NM, Kratzer CR, Brown LR, Gronberg JM, Burow KR (1998) Water quality in the San Joaquin-Tulare Basins, California, 1992–95. United States Geological Survey. Circular 1159 Circular 1159.

6. Ross LJ, Stein R, Hsu J, White J, Hefner K (1999) Distribution and mass loading of insecticides in the San Joaquin River, California. Sacramento: Environmental Hazards Assessment Program, Environmental Monitoring and Pest Management Branch, California Department of Pesticide Regulation, EH 99-01 EH 99-01.

7. Ahuja LR, Rojas KW, Hanson JD, Shaffer MJ, Ma L, eds. Root zone water quality model – Modeling management effects on water quality and crop production. Highlands Ranch, CO: Water Resources Publications LLC.

8. USGS (2005) Evaluation of unsaturated-zone solute-transport models for studies of agricultural chemicals, Open-File Report 2005-1196 (http://pubs.usgs.gov/of/2005/1196/ofr20051196.pdf, accessed 02/2011). Reston, VA: U.S. Geological Survey.

9. Nolan BT, Bayless ER, Green CT, Garg S, Voss FD, et al. (2005) Evaluation of Unsaturated-Zone Solute-Transport Models for Studies of Agricultural Chemicals, Open-File Report 2005-1196. Denver, CO: U.S. Geological Survey.

10. USEPA (2010) Water exposure models used by the Office of Pesticide Programs (http://www.epa.gov/oppefed1/models/water/models4.htm, accessed 10/2010). Washington, DC: U.S. Environmental Protection Agency, Office of Pesticide Programs.

11. USEPA (2006) PRZM-3, a model for predicting pesticide and nitrogen fate in the crop root and unsaturated soil zones: users manual for release 3.12.2. Washington, DC: Center for Exposure Assessment Modeling, U.S. Environmental Protection Agency, EPA/600/R-05/111 EPA/600/R-05/111.

12. Luo Y, Zhang M (2009) A geo-referenced modeling environment for ecosystem risk assessment: organophosphate pesticides in an agriculturally dominated watershed. Journal of Environmental Quality 38(32): 664–674.

13. Dasgupta S, Cheplick JM, Denton DL, Troyan JJ, Wiliams WM (2008) Predicted runoff loads of permethrin to the Sacramento River and its tributaries In: Gan J, Spurlock F, Hendley P, eds. Synthetic Pyrethroids, Occurrence and Behavior in Aquatic Environments Oxford University Press.

14. Cryer SA, Fouch MA, Peacock AL, Havens PL (2001) Characterizing agrochemical patterns and effective BMPs for surface waters using mechanistic modeling and GIS. Environmental Modeling and Assessment 6: 195–208.

15. Jones RL, Russell MH (2001) Final Report of FIFRA Environmental Model Validation Task Force (http://femvtf.com/femvtf/Files/FEMVTFbody.pdf, accessed 10/2010). Washington, DC: Federal Insecticide, Fungicide and Rodenticide Act (FIFRA) Environmental Model Validation Task Force, American Crop Protection Association.

16. Singh P, Jones RL (2002) Comparison of pesticide root zone model 3.12: Runoff predictions with field data. Environmental Toxicology and Chemistry 21: 1545–1551.

17. FOCUS (2001) FOCUS surface water scenarios in the EU evaluation process under 91/414/EEC (http://focus.jrc.ec.europa.eu/, accessed 10/2010). European Commission, Forum for the Co-ordination of Pesticide Fate Models and their Use (FOCUS), Document Reference SANCO/4802/2001-rev.2. 245 p.

18. USEPA (2008) USEPA Tier 2 crop scenarios for PRZM/EXAMS Shell (http://www.epa.gov/oppefed1/models/water/index.htm, accessed 09/2010). Washington, DC: U.S. Environmental Protection Agency, Office of Pesticide Programs.

19. Siepmann S, Finlayson B (2000) Water quality criteria for diazinon and chlorpyrifos. Administrative Report 00-3. SacramentoCA: California Department of Fish and Game. 59 p.

20. USEPA (2005) Aquatic Life Ambient Water Quality Criteria, Diazinon (EPA-822-R-05-006). U.S. Environmental Protection Agency, Office of Water, Office of Science and Technology, Washington, DC.

21. Goss EW (1992) Screening procedure for soils and pesticides for potential water quality impacts. Weed Technology 6: 701–708.

22. Luo Y, Zhang M (2010) Spatially distributed pesticide exposure assessment in the Central Valley, California, USA. Environmental Pollution 15: 1629–1637.

23. USEPA (2002) Guidance for selecting input parameters in modeling the environmental fate and transport of pesticides, version II Office of Pesticide Programs, U.S. Environmental Protection Agency.

24. Spurlock F (2008) Distribution and variance/covariance structure of pesticide environmental fate data. Environmental Toxicology and Chemistry 27: 1683–1690.

25. Chapra SC (1997) Surface water-quality modeling. WCB/McGraw-Hill, Boston.

26. Chen W, Hertl P, Chen S, Tierney D (2002) A pesticide surface water mobility index and its relationship with concentrations in agricultural drainage watersheds. Environmental Toxicology and Chemistry 21: 298–308.

27. Kellogg RL, Nehring R, Grube A, Goss DW, Plotkin S (2000) Environmental Indicators of Pesticide Leaching and Runoff from Farm Fields (http://www.nrcs.usda.gov/technical/NRI/pubs/eip_pap.html, accessed 11/2010). Washington, DC: U.S. Department of Agriculture, Natural Resources Conservation Service.

28. Larson SJ, Crawford CG, Gilliom RJ (2004) Development and application of Watershed Regressions for Pesticides (WARP) for estimating atrazine concentration distributions in streams, Water-Resources Investigations Report 03-4047. Sacramento, CA: U.S. Geologic Survey.

29. USEPA (2010) Pesticide Root Zone Model (PRZM) release notes (http://www.epa.gov/ceampubl/gwater/przm3/prz3reln.html, accessed 10/2010). Athens, GA: U.S. Environmental Protection Agency, Center for Exposure Assessment Modeling.

30. Luo Y, Yang X (2007) A multimedia environmental model of chemical distribution: fate, transport, and uncertainty analysis. Chemosphere 66: 1396–1407.

31. USEPA (2006) Organophosphate pesticides: revised cumulative risk assessment (http://www.epa.gov/pesticides/cumulative/rra-op/, accessed 10/2010). Washington, DC: U.S. Environmental Protection Agency.

32. USEPA (2010) Effects Determinations for the California Red-legged Frog and other California Listed Species (http://www.epa.gov/espp/litstatus/effects/redleg-frog/index.html, assessed 02/2011). Washington, DC: U.S. Environmental Protection Agency, Office of Pesticide Programs.

33. CDWR (2002) 2001 Statewide Irrigation Methods Survey (http://www.water.ca.gov/landwateruse/, accessed 09/2010). Sacramento, CA: California Department of Water Resources, Division of Statewide Integrated Water Management.

34. CEPA (2010) Surface Water Database. Sacramento, CA: California Environmental Protection Agency, Department of Pesticide Regulation, (http://www.cdpr.ca.gov/docs/sw/, accessed 10/2010).

35. Villeneuve J-P, Lafrance P, Banton O, Frechette P, Robert C (1988) A sensitivity analysis of adsorption and degradation parameters in the modeling of pesticide transport in soils. Journal of Contaminant Hydrology 3: 77–96.

36. Wolt J, Singh P, Cryer S, Lin J (2002) Sensitivity analysis for validating expert opinion as to ideal data set criteria for transport modeling. Environmental Toxicology and Chemistry 21: 1558–1565.

37. Don DF, Havens PL, Blau GE, Tillotson PM (1992) The Role of Sensitivity Analysis in Groundwater Risk Modeling for Pesticides. Weed Technology 6: 716–724.

38. CEPA (2010) 2010 Integrated Report (Clean Water Act Section 303(d) List/ 305(b) Report (http://www.waterboards.ca.gov/water_issues/programs/tmdl/ integrated2010.shtml, accessed 11/2010). Sacramento, CA: California Environmental Protection Agency, State Water Resources Control Board.

39. USEPA (1999) Methods for Measuring the Toxicity and Bioaccumulation of Sediment-associated Contaminants with Freshwater Invertebrates, Second Edition (EPA-600/R-99/064) U.S. Environmental Protection Agency, Office of Research and Development, Duluth, MI; U.S. Environmental Protection Agency, Office of Water, Washington, DC.

40. Domagalski JL, Weston DP, Zhang M, Hladik M (2010) Pyrethroid insecticide concentrations and toxicity in streambed sediments and loads in surface waters of the San Joaquin Valley, California, USA. Environmental Toxicology and Chemistry 29: 813–823.

12

Temporal and Spatial Profiling of Root Growth Revealed Novel Response of Maize Roots under Various Nitrogen Supplies in the Field

Yunfeng Peng, Xuexian Li, Chunjian Li*

Key Laboratory of Plant-Soil Interactions, Ministry of Education, Department of Plant Nutrition, China Agricultural University, Beijing, China

Abstract

A challenge for Chinese agriculture is to limit the overapplication of nitrogen (N) without reducing grain yield. Roots take up N and participate in N assimilation, facilitating dry matter accumulation in grains. However, little is known about how the root system in soil profile responds to various N supplies. In the present study, N uptake, temporal and spatial distributions of maize roots, and soil mineral N (N_{min}) were thoroughly studied under field conditions in three consecutive years. The results showed that in spite of transient stimulation of growth of early initiated nodal roots, N deficiency completely suppressed growth of the later-initiated nodal roots and accelerated root death, causing an early decrease in the total root length at the rapid vegetative growth stage of maize plants. Early N excess, deficiency, or delayed N topdressing reduced plant N content, resulting in a significant decrease in dry matter accumulation and grain yield. Notably, N overapplication led to N leaching that stimulated root growth in the 40–50 cm soil layer. It was concluded that the temporal and spatial growth patterns of maize roots were controlled by shoot growth and local soil N_{min}, respectively. Improving N management involves not only controlling the total amount of chemical N fertilizer applied, but also synchronizing crop N demand and soil N supply by split N applications.

Editor: Carl J. Bernacchi, University of Illinois, United States of America

Funding: The authors thank the State Key Basic Research and Development Plan of China (No. 2007CB109302), the National Natural Science Foundation of China (No: 30671237) and the Innovative Group Grant of National Natural Science Foundation of China (No. 31121062) for financial support. The funders had no role in study design, data collection and analysis, decision to publish, or preparation of the manuscript.

Competing Interests: The authors have declared that no competing interests exist.

* E-mail: lichj@cau.edu.cn

Introduction

Doubling of the world food production over the past four decades is associated with a seven-fold increase in consumption of synthetic nitrogen (N) fertilizer in agricultural systems [1]. In China, a 71% increase in total annual grain production from 283 to 484 MT (million tons) from 1977 to 2005 was achieved at the cost of 271% increase in synthetic N fertilizer application (from 7.07 to 26.21 MT) over the same period [2]. Maize is one of three major cereal crops in China. Its average grain yield per hectare increased rapidly from 962 kg in 1949 to 5,166 kg in 2007 [3]. The consumption of synthetic N fertilizer in China increased rapidly during the same period, exceeding 32 MT in 2007, accounting for 30% of global N fertilizer production [4]. However, the average maize grain yield per hectare of 5166 kg was much lower than that in Western countries such as the USA, where it was 9359 kg in 2006 [5]. Although the high yield records are more than 15 Mg ha^{-1} in some experimental plots [6,7], and even reached 21 Mg ha^{-1} in Shangdong Province in 2005 [8], this was obtained in small experimental plots and with high input costs. The amount of topdressing N fertilizer applied in the high-yield experimental plots varied from 450 to 720 kg N ha^{-1} [6,8]. The continuous increase in fertilizer supply promotes yield increase on the one hand, and brings serious environmental problems on the other hand. Excessive N fertilization in intensive Chinese agricultural systems does not make significant contributions to grain yield but decreases nitrogen use efficiency (NUE), and increases the risk of N leaching to ground water and soil acidification [2,9–11].

In addition to overapplication, N is often applied incorrectly in China. A survey of chemical N fertilizer application in five major maize-producing provinces in North China during 2001 and 2003 revealed that 31.2–78.3% of the farmers used only a single N application as base fertilizer before sowing [12]. A study of the effects of single N application as base fertilizer on spring maize yield in Jilin province with 110 field experiments in 2004 and 2005 indicated that single N application significantly reduced maize yield compared with optimized N management based on soil mineral N (N_{min}) [13]. In maize 45–65% of the grain N is from pre-existing N in the stover before silking. The remaining 35–55% of the grain N originates from post-silking N uptake [1]. Nitrogen stress at a critical stage may lead to irreversible yield loss. In a greenhouse experiment, Subedi and Ma [14] found that restriction of N supply from seeding to 8-leaf stage could cause an irreparable reduction in maize ear size and kernel yield; however, there was no yield reduction when N was restricted from silking, or 3 weeks after silking, to physiological maturity. Newly developed maize hybrids often show reduced rates of visible leaf senescence, which allows a longer duration of photosynthesis and has a positive effect on N uptake during the grain-filling period [15–18]. Whether N fertilizer application after silking is needed in order to meet the increased N demand of plants in the reproductive growth stage, and how split application of chemical N fertilizer influences root growth and N uptake by plants as well as N movement in the soil, are questions that require addressing.

Chemical N fertilizer applied in the soil is taken up by roots and then assimilated and used by plants. Better root growth and synchronized N supply throughout the crop growing season are beneficial for maximizing fertilizer uptake, optimizing grain yield, and reducing N losses. Many scientists are starting to see roots as central to their efforts to produce crops with a better yield, efforts that go beyond the Green Revolution [19]. However, less attention has been paid to the temporal and spatial dynamics of root growth in the soil profile and how root growth responds to various N supplies [1,20], partially because roots are tangled underground and difficult to study [19]. Few studies have reported root growth plasticity of cereals under different N regimes [1,20]. In a short-term experiment under controlled conditions, N deficiency stimulates root growth, while N oversupply inhibits root growth [21]. Localized nitrate application stimulates lateral root growth (the localized stimulatory effect) [21,22]. Unfortunately, these unsystematic experiments under controlled conditions may not represent real situations in the field. It is interesting to know how roots perform in the soil profile with heterogeneous N distribution in time and space, and whether the responses of root growth to the above-mentioned N applications are repeated in long-term field experiments.

Successful N management requires better understanding of N uptake by roots and synchronized N supply throughout the crop growing season. We hypothesized that N deficiency stimulated early root growth but reduced grain yield. By contrast, N over-application inhibited early root growth and increased potential risk of N leaching without yield increase. Improving N management involved not only controlling the amount of applied N fertilizer, but also synchronizing plant demand and N applications for better root growth and high grain yield. To test this hypothesis and further dissect response strategies of maize roots to various N supplies in the field, comprehensive field studies in three consecutive years (2007–2009) were conducted in the present work to examine temporal and spatial distribution patterns of maize roots, plant N uptake, and N_{min} in the soil profile during the whole growth period, under different chemical N regimes, especially by split application of N fertilizer.

Materials and Methods

Experimental design

The field experiments were conducted in three consecutive years (2007–2009) in three adjacent experimental sites at the Shangzhuang Experimental Station of the China Agricultural University, Beijing. The soil type at the study site is a calcareous alluvial soil with a silt loam texture (FAO classification) typical of the region. The soil N_{min} and related chemical properties of the experimental soils are shown in Table S1. Maize hybrid DH 3719 ('stay-green' cultivar), a popular hybrid in North China, was used in the experiments and sown on 28 April 2007, 27 April 2008, and 27 April 2009, and harvested on 23 September 2007, 19 September 2008, and 21 September 2009. Flooding irrigation before sowing was used to keep the available soil water content above 75%. The amount of rainfall during the maize growing season in the three years was 428 mm, 608 mm, and 216 mm, respectively. In addition, 50 mm and 43 mm of irrigation were applied on 17 June 2007 and 2 July 2009, respectively. The monthly rainfall during the study period is shown in Table S2. Maize was overseeded (three seeds) with hand planters and the

plots were thinned at the seedling stage to a stand of 100,000 plants ha^{-1}. The seeds were sown in alternating 20-cm- and 50-cm-wide rows. The distance between plants was 28 cm in each row. A randomized complete block design with four replicates in each treatment in each year was used. The plot sizes were 56 m^2 (5.6×10 m), 40 m^2 (5×8 m), and 56 m^2 (5.6×10 m) in 2007, 2008, and 2009, respectively.

Fertilization and treatments

There were four (2007 and 2008) or three (2009) N treatments: 1) 0 N as control; no chemical N fertilizer was applied. 2) N topdressing at and after tasseling (TDAT), and 3) N topdressing before tasseling (TDBT). In order to determine the importance of timing of N topdressing, a treatment with delayed N topdressing at and after tasseling was set. In 2007, 175 kg N ha^{-1} as base fertilizer was applied in the TDAT and TDBT treatments, and total amount of N fertilization was 230 and 395 kg N ha^{-1} in TDAT and TDBT, respectively. According to the results of N accumulation in plants and soil N_{min} after the last harvest in 2007, 250 kg N ha^{-1} was set in TDAT and TDBT in 2008 and 2009, in which 60 kg N ha^{-1} was applied as base fertilizer. The remaining N was applied before tasseling (TDBT) at V8 (the eighth leaf emerged with ligule visible) and V12 (the twelfth leaf emerged), or at and after tasseling (TDAT) at VT (tasseling stage) and R2 (grain blister stage), respectively. 4) Traditional N practice (450 N); according to numerous high-yield studies in China, the application rate in the traditional N practice was set at 450 kg N ha^{-1}, in which 175 kg N ha^{-1} was applied as base fertilizer, 50, 170, and 55 kg N ha^{-1} in 2007 and 2008, and 120, 70, 85 kg N ha^{-1} in 2009 were applied in wide interrows by hand as topdressings at the V8, V12 and VT, respectively. Detailed rates and times of N application are shown in Table S3.

The rate and timing of phosphorus and potassium fertilization in each year were the same. In addition, zinc (Zn) was applied in each year as base fertilizer because of the slight Zn deficiency in the experimental region. A total of 135 kg ha^{-1} of P_2O_5 as triple superphosphate [$Ca(H_2PO_4)_2 \cdot H_2O$], 120 kg ha^{-1} of K_2O as potassium sulfate [K_2SO_4], and 30 kg ha^{-1} of $ZnSO_4 \cdot 7H_2O$ were applied. Before sowing, 90 kg ha^{-1} P_2O_5, 80 kg ha^{-1} of K_2O and 30 kg ha^{-1} of $ZnSO_4 \cdot 7H_2O$ were broadcasted and incorporated into the upper 0–15 cm of the soil by rotary tillage. Another 45 kg ha^{-1} of P_2O_5 at V12 and 40 kg ha^{-1} of K_2O at VT were applied in wide interrows by hand as topdressings. Each topdressing (NPK) was applied after harvest.

Harvest 2007. Plants were harvested at 38 (the eighth leaf emerged with ligule visible, V8), 57 (the twelfth leaf emerged, V12), and 74 (tasseling, VT) days after sowing (DAS) before fertilization and at 105 (grain blister stage, R2) and 147 (physiological maturity, R6, when 50% of the plants showed black layer formation in the grains from the mid-portion of the ears) DAS. At harvest, six consecutive plants (three plants each from two narrow rows) were cut at the stem base, chopped to a fine consistency, dried to a constant weight at 60°C and ground into powder to determine aboveground dry weight and N content. To estimate grain yield, ears in the central part of 21 m^2 (2007 and 2009) or 14 m^2 (2008) in each plot were hand-harvested at physiological maturity. Kernels from six randomly selected ears were harvested individually by hand, weighed and calculated to 15.5% moisture content. N content in each plant sample was analyzed by using a modified Kjeldahl digestion method [23]. Briefly, 0.3–0.4 g oven-dried plant tissue was digested with H_2SO_4 (98%)+H_2O_2 at 380°C for 3–4 h in a digestion tube. The digested solution was cooled to room temperature and added deionized water to 100 ml. An aliquot of 5 ml uniform solution was distilled and titrated with standardized 0.01 N sulphuric acid. The total N content was calculated from the concentration of standardized sulphuric acid. After shoot excision at each harvest, three whole root systems were excavated from each plot and washed free of soil with tap water. Two root systems were dried immediately after harvest and used to assess dry weight and N content, and the other root system was stored at $-20°C$ for measuring root length, including embryonic and different whorls of shoot-borne roots [24]. At root harvest, each root system was excavated with a soil volume of 28 cm (14 cm on each side of the plant base in intrarow direction)$\times$35 cm (10 cm in narrow interrow and 25 cm in wide interrow) and a depth of 40 cm. The area of 28 cm$\times$35 cm was the soil surface occupied by each plant at the plant density of 100,000 plants ha^{-1}. In addition, at each harvest, five 2-cm-diameter soil cores per plot were collected and mixed to measure soil N_{min} (auger method, [25,26]). Samples were collected from the 0–90 cm soil layers (in 30 cm increments) in the interrow area. All fresh samples were crushed, sieved through a 3 mm sieve in the field, and extracted immediately after transfer to the laboratory with 0.01 mol L^{-1} CaCl$_2$ solution and analyzed for soil N_{min} (NH_4^+-N+NO_3^--N) by continuous flow analysis (TRAACS 2000, Bran and Luebbe, Norderstedt, Germany) [9,10].

Harvest 2008. Plants were harvested on 53 (V8), 71 (V12), 86 (VT) and 111 (R2) DAS before fertilization and on 130 and 145 (R6) DAS. At each sampling date, shoot harvest and determination of dry weight and N content as well as final dry grain yield were performed as in 2007. In addition, two whole root systems were sampled from each plot at each harvest as in 2007 to determine root dry weight and N content. In order to study the temporal and spatial distribution of maize roots and soil N_{min} during the whole growth period, a different method from that in 2007 was used to obtain root and soil samples at each harvest after shoot excision. Soil samples of 28 cm (width)$\times$35 cm (length)$\times$50 cm (depth) with 10 cm increments in each plot under different treatments were collected. There were five soil blocks of 28 cm$\times$35 cm$\times$10 cm in each plot. All visible roots in each soil block were separated in the field by hand and placed in individual marked plastic bags. These roots were washed free of soil after transfer to the laboratory and then frozen at $-20°C$ until root length analyses were performed [24]. After root harvest, the soil in each block was crushed by hand and sieved through a 3 mm sieve in the field. A representative sample of the mixed soil was placed in a marked plastic bag for N_{min} extraction and analysis as performed in 2007.

Harvest 2009. Plants were harvested on 33, 45 (V8), 61 (V12), 80 (VT), 110 (R2) and 147 (R6) DAS. The methods for shoot and root harvest, dry weight and N content determination, root and soil sampling, and soil N_{min} measurement were identical to those used in 2008. The only difference was that the soil was excavated to a depth of 60 cm.

Statistical analysis

Data were analyzed using analysis of variance with the SAS package (SAS Institute, 1996). Differences between data in all tables were tested with PROC ANOVA. N treatments were treated as fixed effects and means of different N treatments were compared based on least significant difference (LSD) at the significance level of 0.05.

Results

Temporal and spatial distribution patterns of maize roots

The total root length of maize plants increased dramatically after the V8 stage, peaked at the VT stage, and then declined

rapidly until the R6 stage. The dynamic pattern of total root length over the entire growth period was consistent in all three years, irrespective of N regimes (Fig. 1). However, the total root length in the early growth stage was differentially regulated by base N treatments. N Deficiency (0 N) stimulated root growth in the early growth stage (V8 stage), and the total root length peaked before the VT stage, followed by an early decline compared to other treatments with base N fertilizer and N topdressing before tasseling in all three years. Similarly, the treatment with base N fertilizer and delayed N topdressing in 2008 also caused an early decrease in the total root length. In contrast, 175 kg ha^{-1} base N fertilizer (450 N treatment) inhibited root growth in the early growth period. The total root length in the following growth stages under 450 N treatment was comparable with that under TDBT treatment (with 60 kg base N fertilizer) in 2008 and 2009 (Fig. 1).

To further analyze dynamic changes in root structure under different N treatments, the total length of the embryonic roots and each whorl of nodal roots was monitored in 2007 by whole root excavation with a soil volume of 35 cm (length)$\times$28 cm (width)$\times$40 cm (depth) at each growth stage (Fig. 2). The embryonic root began to shorten after the first harvest; therefore, the total root length was largely determined by nodal roots that initiated with root development before VT (except for the 7th whorl of nodal roots of N-deficient plants). The total length of the embryonic root and most nodal roots peaked at the VT stage and then decreased simultaneously until maturity. Although N deficiency (0 N) enhanced embryonic root growth before V12 (57 DAS), it negatively regulated nodal root growth and mortality. In particular, initiation and growth of the 7th whorl of nodal roots of N-deficient plants after the VT stage was almost completely suppressed (Fig. 2).

Monitoring of temporal and spatial root distribution in 2008 and 2009 (Figs. 3, 4) with a different method also confirmed that N deficiency stimulated, while overapplication of base N fertilizer (450 N) inhibited, root growth in the early growth stage. More roots were distributed in upper soil layers than in deep soil layers in all growth stages. Moderate N input promoted root proliferation in the nutrient-rich soil profile during the fast root growth period in 2009 (V8-tasseling; Fig. 4). Maize roots had already grown into the 50–60 cm soil layer one month after sowing in 2009, regardless of N treatment. Interestingly, deep root growth was also stimulated at the reproductive growth stage in 2008, when N was overapplied (Fig. 3).

Dry matter and N accumulation in the shoot and grain yield

The changes in shoot biomass and N accumulation were represented by a typical 'S curve'. However, the increase in shoot biomass and N content was not synchronized. The rapid increase in shoot biomass began at the V8 stage and peaked during the V12–R2 stages, while the N content increased during sowing-V8 and peaked during the V8–VT stages (Tables 1, 2). The shoot biomass and N accumulation patterns were the same in all three years regardless of N regimes. The study site is located on the North China Plain. In this region, total environmental N inputs (atmospheric and irrigation contributions) reach about 104 kg ha^{-1} year^{-1}. Ammonia volatilization and nitrate leaching are the main N loss pathways [2]. Based on N accumulation (Table 2) and soil N_{min} (see below) after the final harvest in 2007, total environmental N inputs in the growing season, and amount of N loss in the region, 250 kg N ha^{-1} fertilizer was applied in delayed N topdressing (TDAT) and moderate N (TDBT) treatments in 2008 and 2009. N deficiency (0 N) significantly reduced shoot biomass, N content, and grain yield; while N

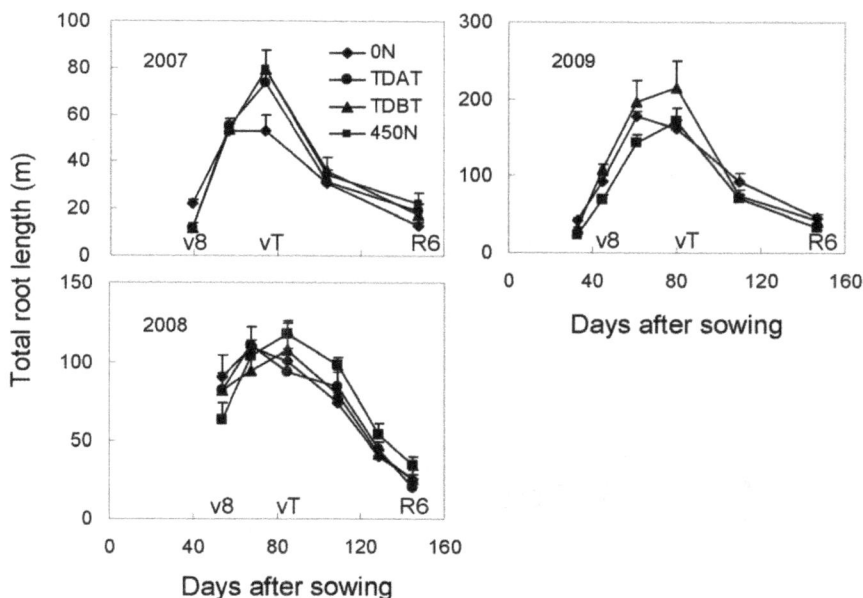

Figure 1. Total root length of maize plants during the whole growth period in response to N fertilization in three consecutive years. In 2007, the whole root system was excavated with a soil volume of 28 cm×35 cm and a depth of 40 cm. In 2008 and 2009, root systems were excavated within a soil volume of 28 cm×35 cm and a depth of 50 cm (2008) or 60 cm (2009) with 10 cm increments. The bars represent the standard error of the mean, $n=4$. TDAT means N top dressing after tasseling. The total amount of N applied in TDAT treatment was 230 and 250 kg ha^{-1} in 2007 and 2008, respectively. TDBT means N top dressing before tasseling. The total amount of N applied in TDBT treatment was 395, 250 and 250 kg ha^{-1} in 2007, 2008 and 2009, respectively, and is the same in the following figures.

overapplication (450 N) failed to increase N uptake, shoot biomass, as well as grain yield, compared with TDBT in all three years. The timing of N topdressing affected shoot growth and N accumulation. Delayed N topdressing decreased N uptake and biomass accumulation during the V8–V12 period, and caused a decrease in final dry matter and grain yield, compared with the TDBT treatment in 2008.

Temporal and spatial dynamics of soil N$_{min}$

In 2007, 0–90 cm soil samples were collected in 30 cm increments in the interrow area at each harvest following the auger method. In 2008 and 2009, soil samples were obtained by digging soil blocks within a soil volume of 35 cm (length)×28 cm (width)×50 (or 60) cm (depth), with 10 cm increments at each time after shoot harvest. Irrespective of methods used for soil sample collection, our three-year experiment consistently demonstrated that the soil N$_{min}$ in each soil layer at each growth stage was positively correlated with the amount of N application (Figs. 5–7). In the 0 N treatment, soil N$_{min}$ was very low and remained relatively steady at each harvest in all three years, in spite of the continuous increase in plant N content with time (Table 2). In the 450 N treatment in all three years, however, high N accumulation in soil profiles was observed, even after the last harvest, indicative of N overapplication. This was also true for the TDBT treatment in 2007 when 395 kg N ha^{-1} was applied. N overapplication led to obvious temporal and spatial fluctuations in soil N$_{min}$ and strong N leaching to the deeper soil layer occurred upon heavy rainfall in 2007 and 2008 growth seasons (Figs. 5, 6). Additionally, late N topdressing also caused N leaching to deep soil layers in 2008, although 250 kg N in total was applied (TDAT, Fig. 6). In comparison, moderate N application (TDBT) ensured plant demand without N leaching.

Discussion

Influence on maize root growth

The root system plays predominant roles in nutrient uptake for plant growth and yield formation [24]. Previous studies indicated that optimized N application is beneficial for maize shoot growth and grain yield [2,10,27,28]. However, little is known about how root development responds to different N supply in the soil profile. The present work showed that N deficiency stimulated early root growth as indicated by increased length of early-initiated nodal roots. This stimulatory effect lasted only for a short time and the total root length began to decrease when maize plants were still in the rapid vegetative growth stage (Fig. 1) with high N uptake activity (Table 2). The early decline in total root length under N deficiency was due to early mortality of the early-initiated nodal roots and growth suppression of the later-initiated nodal roots (Fig. 2). Root growth is closely associated with assimilate supply from the shoot [29]. The stimulated root growth in the early growth stage under N deficiency was achieved at the cost of reduced shoot growth [30], which led to insufficient carbon supply for continuous growth of early-initiated nodal roots and rapid elongation of later-initiated nodal roots. As a result, total plant N content was significantly reduced (Table 2). In contrast, the total root length of maize plants supplied with sufficient N didn't decrease until tasseling (Fig. 1), which favored robust nutrient uptake early in the growing season, and nutrient translocation from roots to reproductive organs later in the season [24,31–34].

Overapplication of base N fertilizer (175 kg ha^{-1} in the 450 N treatment) inhibited early root growth of maize plants, compared with other treatments in all three years (Figs. 1–4). The results suggested the 'systemic inhibitory effect' of high external N concentration on root growth [21] applied in the field. It is envisaged that a single application of total N as base fertilizer in the five major maize producing provinces in North China [12]

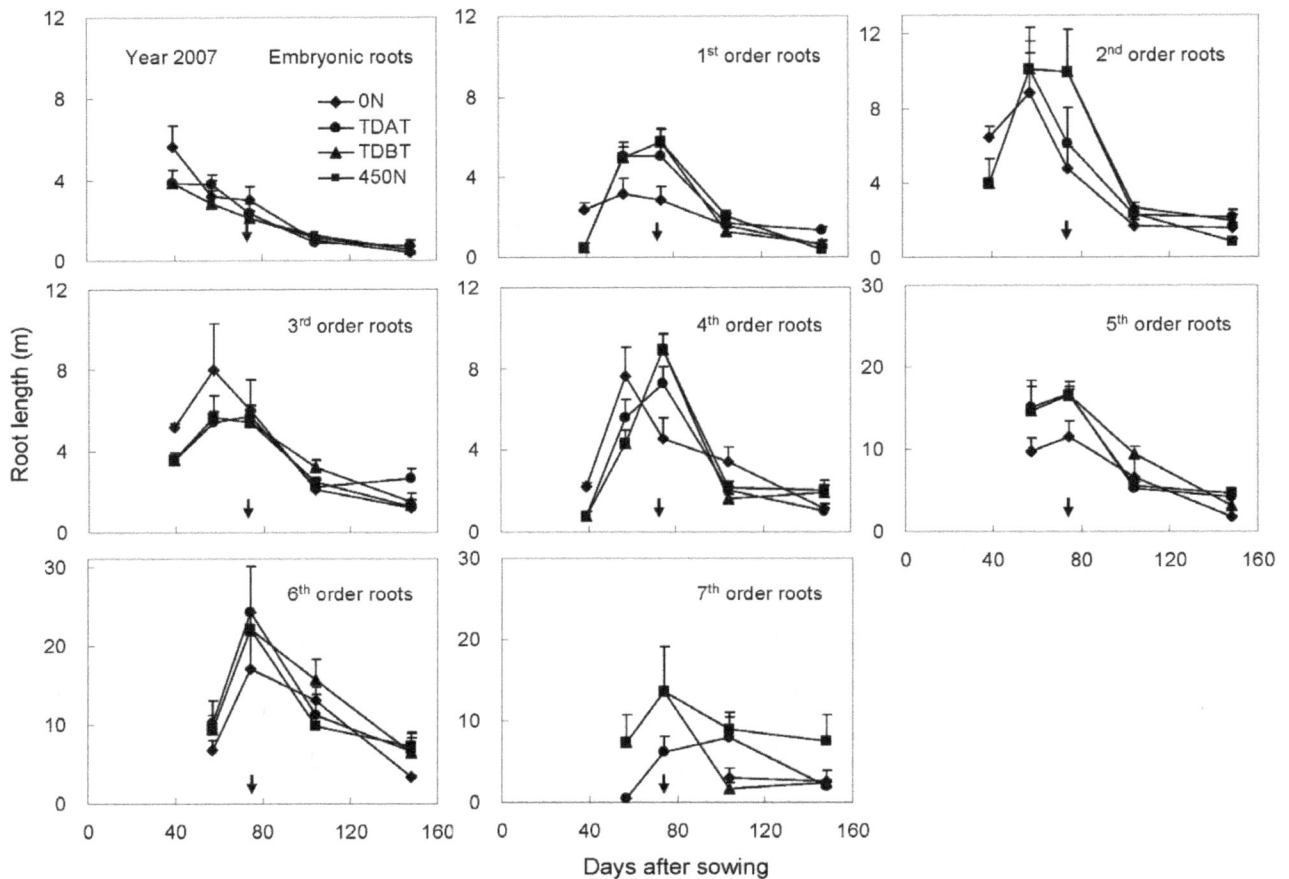

Figure 2. Length of embryonic roots and different whorls (1st to 7th orders) of nodal roots of maize plants in response to N fertilization in 2007. Arrows indicate the time of tasseling. Whole root systems were excavated with a soil volume of 28 cm×35 cm and a depth of 40 cm, and then separated into embryonic roots and different whorls of nodal roots. The bars represent the standard error of the mean, $n = 4$.

would more dramatically inhibit early root growth. A single application of total N before sowing reduced maize yield significantly compared with split applications of chemical N fertilizer based on the soil N_{min} test. However, this reduction of grain yield was not only because of the inhibited root growth in the early growth stage owing to N toxicity, but also because of N deficiency in the reproductive stage owing to N losses by different ways [2,35].

Although maize rooting depth at anthesis varies from around 0.7 m to close to 1 m, approximately 90% of roots grow in the upper 0.3 m soil [36]. Consistently, the present results in 2008 and 2009 (Figs. 3, 4) showed that most of the root length was distributed in the upper 30 cm soil, regardless of N treatments. The decrease in the total root length after tasseling indicates rapid root death that is mainly attributed to lateral root mortality [24], especially in the upper 30 cm soil layer. The root length in deep soil layers (40–60 cm) was quite constant during the whole growth period. Notably, maize roots could sense the changes in soil N_{min}. A localized stimulatory effect of nitrate-N patches on root growth has been reported when the whole root system suffered from N deficiency [21,22]. In this process, nitrate serves as a signal, and the nitrate transporter CHL1 functions as a nitrate sensor [21,37,38]. The majority of the total root length of N-treated plants was distributed in the upper 30 cm soil layer before tasseling in 2008 and during the whole growth period in 2009, because of high soil N_{min} in this soil layer (Figs. 3, 4, 6, 7). However, there was

an obvious increase in root length of maize plants supplied with 450 N in the 40–50 cm soil layer from 109 d after sowing to the last harvest in 2008 (Fig. 3). During the same period increased soil N_{min} owing to N leaching in the same soil profile was also observed (Fig. 6). Together, our study provides direct evidence that N-sufficient maize plants could respond to temporally and spatially heterogeneous soil N_{min} via enhanced root proliferation in the soil profile with higher N_{min}.

Water stress significantly reduces maize N uptake, accelerates leaf senescence, and thus reduces grain yield, compared with the well-watered plants [39,40]. This is supported by the results in the present study that shoot dry weight, N content and grain yield of all treated plants in 2009 were the lowest among the three years (Tables 1, 2), because of the limited precipitation in this year (Table S2). By contrast, the total root length of all treated maize plants in 2009 was extremely high (Fig. 1). Roots are less sensitive to water deficits than leaf, stem or silks growth [41,42]. The results in the present study indicated that maize root growth could be stimulated by low water potential. On the other hand, soil water content influences nutrient availability [30]. Low soil water availability causes low N flux to the root surface. Enhanced root growth (Fig. 1) is beneficial for plants to capture more N under dry soil conditions.

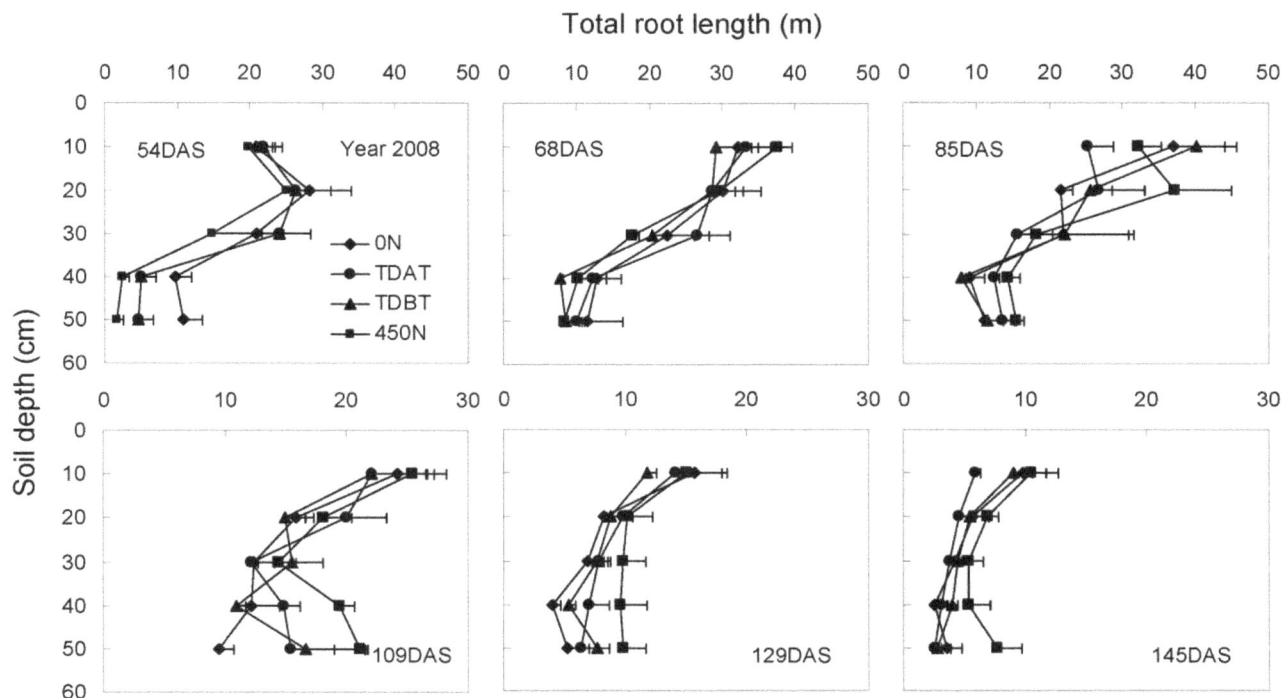

Figure 3. Total root length of maize plants in each soil layer at different growth stages in response to N fertilization in 2008. Root systems were excavated within a soil volume of 28 cm×35 cm and a depth of 50 cm with 10 cm increments. The bars represent the standard error of the mean, $n = 4$.

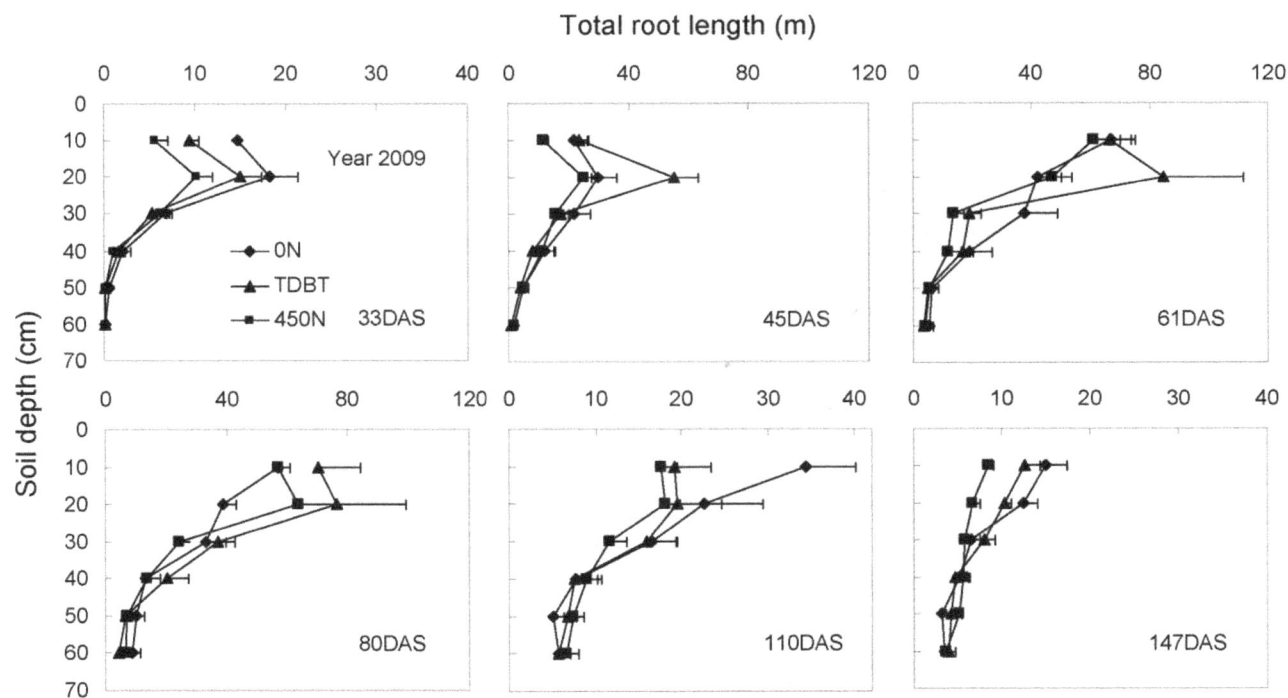

Figure 4. Total root length of maize plants in each soil layer at different growth stages in response to N fertilization in 2009. Roots were excavated within a soil volume of 28 cm×35 cm and a depth of 60 cm with 10 cm increments. The bars represent the standard error of the mean, $n = 4$.

Table 1. Shoot dry matter accumulation (t/ha) in different growth periods, final shoot dry weight (DW) and grain yield (t/ha) of maize plants supplied with different N rates in three years.

Year	Treatments	Growth period					Total DW	Grain yield
		Sowing-V8	V8-V12	V12-VT	VT-R2	R2-R6		
2007	0 N	0.9a	3.8c	3.9b	10.1b	−0.1a	18.6b	9.7b
	TDAT*	0.7a	4.8b	5.0b	13.9a	3.4a	27.8a	12.8a
	TDBT**	0.7a	5.5a	7.3a	8.5b	6.7a	28.7a	12.4a
	450 N	0.7a	5.5a	7.3a	9.2b	6.4a	29.0a	13.3a
2008	0 N	1.2b	2.9b	4.8a	6.8b	4.9b	20.6b	11.0c
	TDAT	1.6a	2.6b	5.2a	8.4ab	5.3ab	23.2ab	12.1b
	TDBT	1.6a	3.8a	4.3a	10.5a	7.0ab	27.2a	13.8a
	450 N	1.4ab	3.5a	4.2a	7.9ab	8.6a	25.6a	13.1a
2009	0 N	1.0a	2.2b	3.6b	4.7b	3.4a	14.9b	6.3b
	TDBT	1.1a	2.9a	5.4a	9.2a	2.0a	20.6a	10.7a
	450 N	1.1a	3.5a	5.4a	9.7a	1.5a	21.2a	11.0a

Values in columns in each year followed by a different letter represent a significant difference between N treatments ($P<0.05$). Values are means $\pm$ SE ($n=4$).
*TDAT, N topdressing after tasseling. The total amount of N applied in the TDAT treatment was 230 and 250 kg/ha in 2007 and 2008, respectively, and is the same in the following tables.
**TDBT, N topdressing before tasseling. The total amount of N applied in the TDBT treatment was 395, 250 and 250 kg/ha in 2007, 2008 and 2009, respectively, and is the same in the following tables.
V8, the eighth leaf emerged with ligule visible; V12, the twelfth leaf emerged; VT, tasseling; R2, grain blister stage; R6, physiological maturity, and they are the same in the following table.

N application, uptake and grain yield

Under N deficiency, grain yield is negatively correlated with early root growth due to competition for N resources [1,43]. In order to obtain high grain yield, fertilizer overapplication in Chinese intensive agricultural systems is very common, since farmers believe that additional fertilizers further improve crop yield [2]. In fact, N overapplication not only inhibited early root growth as discussed above, but also failed to increase shoot dry weight and grain yield of maize plants (Table 1; [44]). N overapplication did not increase total plant N content either, compared with the moderate N treatment (TDBT) in 2008 and 2009 (Table 2). The shoot N concentration of maize plants was the same under 450 N and TDBT treatments at each sampling time. Therefore, excessive N could not be taken up by plants and used to increase grain yield, but instead would increase the risk of N leaching and potential environmental pollution.

Besides quantity control, timing of fertilization is also important. The results in the present study indicated that the increases in shoot biomass and N content were not synchronized. Approximately 60–86% of the total N in maize plants (except the 0 N treatment in 2009) was taken up before tasseling, whereas 53–64% of the dry matter was accumulated after tasseling in all three years (Table 3). Therefore, N topdressing after the V8 stage (TDBT) was necessary to ensure adequate soil N supply for rapid plant growth.

Table 2. Shoot N accumulation (kg/ha) in different growth periods and the final shoot N content of maize plants supplied with different N rates in three years.

Year	Treatments	Growth period					Total N content
		Sowing-V8	V8-V12	V12-VT	VT-R2	R2-R6	
2007	0 N	22a	25c	53b	32ab	−8b	124b
	TDAT	22a	123a	35b	86a	5ab	271a
	TDBT	22a	111b	97a	11b	27ab	267a
	450 N	22a	111b	97a	25b	27a	282a
2008	0 N	36b	35b	37a	41a	27a	176b
	TDAT	52a	33b	53a	52a	26a	218ab
	TDBT	52a	64a	49a	61a	46a	273a
	450 N	44ab	73a	47a	35a	55a	254a
2009	0 N	19a	23c	18b	25a	24a	109b
	TDBT	25a	46b	67a	53a	11a	202a
	450 N	30a	58a	72a	46a	5a	210a

Values in columns in each year followed by a different letter represent a significant difference between N treatments ($P<0.05$). Values are means $\pm$ SE ($n=4$).

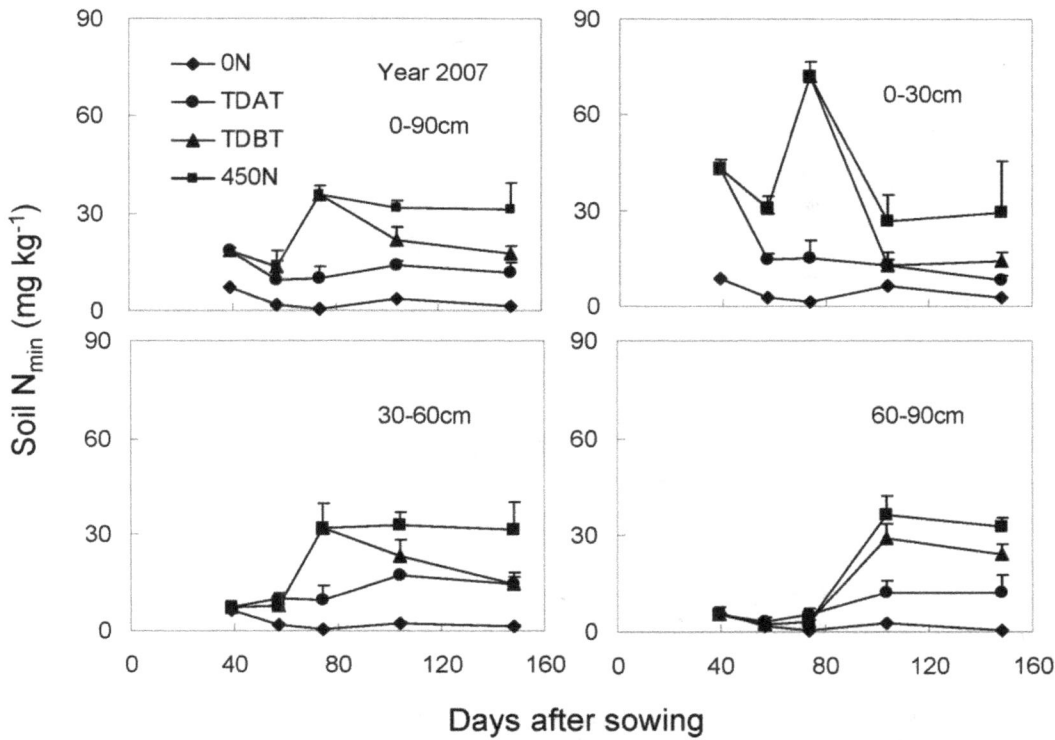

Figure 5. Soil mineral nitrogen (N$_{min}$; NH$_4^+$-N+NO$_3^-$-N) in the 0–90 cm soil profile in response to N fertilization in 2007. The soil samples were obtained using the soil auger method at the same time in each plant harvest. The bars represent the standard error of the mean, $n = 4$.

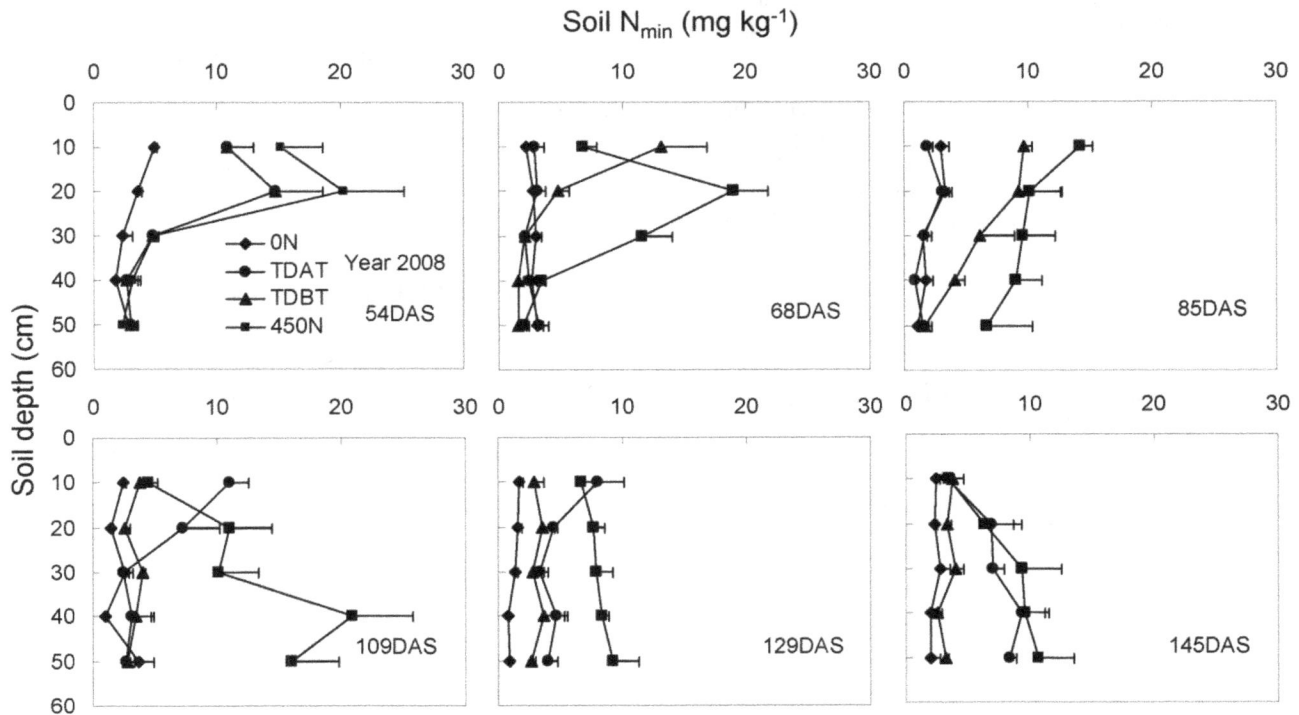

Figure 6. Temporal and spatial distribution of soil mineral nitrogen (N$_{min}$; NH$_4^+$-N+NO$_3^-$-N) in the 0–50 cm soil profile in response to N fertilization in 2008. The soil samples were obtained by excavating soil layers within a soil volume of 28 cm×35 cm and a total depth of 50 cm with 10 cm increments at each time after shoot harvest. The bars represent the standard error of the mean, $n = 4$.

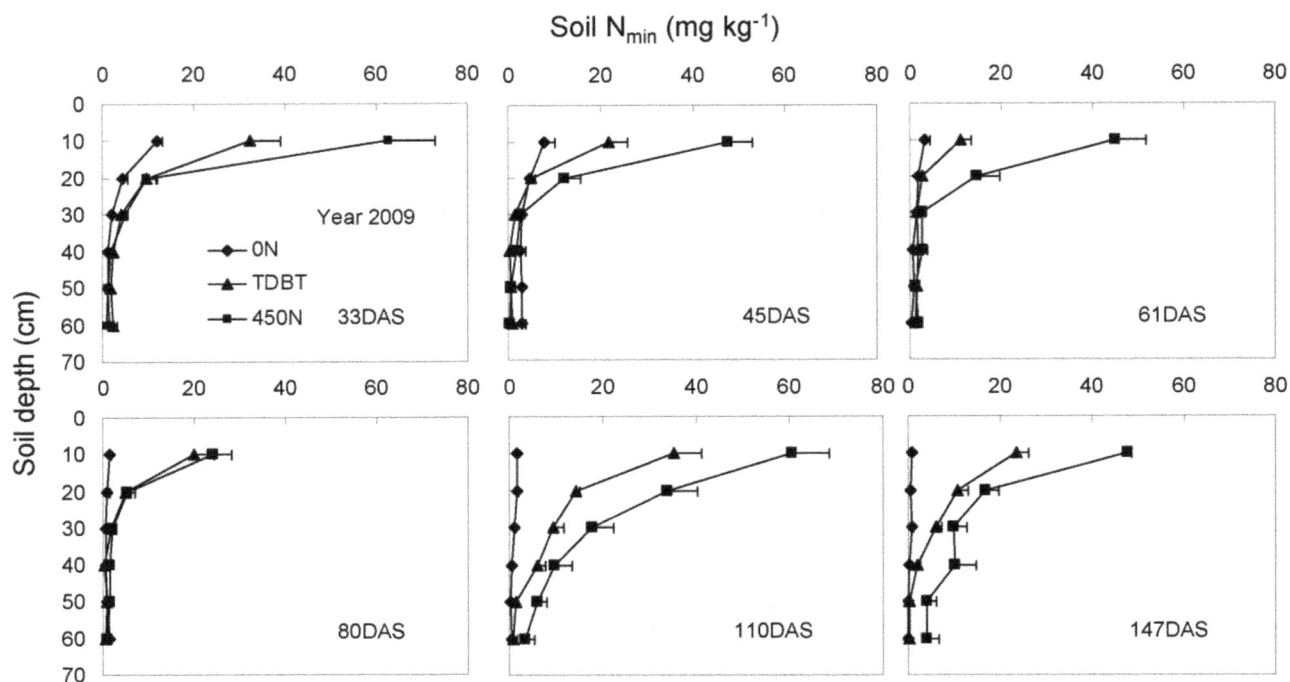

Figure 7. Temporal and spatial distribution of soil mineral nitrogen (N_{min}; NH_4^+-N+NO_3^--N) in the 0–60 cm soil profile in response to N fertilization in 2009. The soil samples were obtained by excavating soil layers within a soil volume of 28 cm×35 cm and a total depth of 60 cm with 10 cm increments at each time after shoot harvest. The bars represent the standard error of the mean, $n = 4$.

Delayed N topdressing (TDAT) decreased not only plant N content and dry matter accumulation during the V8–V12 period, but also caused a decrease in final dry matter and grain yield, compared with the TDBT treatment in 2008 (Tables 1, 2). It is reported that N application rate influences production of maize spikelets and the size of the developing ears before flowering [45]. The period from 1–2 weeks before to 2 weeks after silking is critical for establishment of a large grain sink of maize plants [46]. TDAT could not compensate for the reduction in plant N content and shoot growth caused by insufficient N supply in the vegetative growth stage, although total N supply was sufficient. Moreover,

because of the rapid mortality of the maize root system (Fig. 1) and the slower plant N content increase (Table 3) after tasseling, remaining N fertilizer and mineralized N from the soil were sufficient to meet the plant demand during the reproductive growth stage. Additional unnecessary N application during this stage would increase the risk of N losses. The results indicate that even in intensive agricultural systems with 'stay-green' maize cultivars, N topdressing should be applied before tasseling to maximally synchronize with crop N demand.

Table 3. Ratio of DW and N uptake presilking and total accumulation in maize plants with different N treatments in 2007, 2008 and 2009.

Year	Treatments	Ratio DW pre silking/total accumulation	Ratio N uptake pre silking/total accumulation
2007	0 N	0.46a	0.81a
	TDAT	0.38a	0.66a
	TDBT	0.47a	0.86a
	450 N	0.47a	0.82a
2008	0 N	0.43a	0.61a
	TDAT	0.41ab	0.63a
	TDBT	0.36b	0.60a
	450 N	0.36b	0.65a
2009	0 N	0.46a	0.55b
	TDAT	0.46a	0.68ab
	450 N	0.47a	0.76a

Values in columns represent the significant differences between N treatments, ($P<0.05$). Means ± SE, $n = 4$.

Influence on soil N_{min}

In the 0 N treatment in all three years, soil N_{min} values were very low and remained almost constant in the studied soil profile during the whole maize growth period, except in the 0–10 cm soil layer at the elongation stage (Figs. 5–7). The continuous increase in shoot N content of the 0 N-treated maize plants indicated rapid soil N mineralization in the soil surrounding roots. Studies using the annual grass *Avena barbata* show that the rate of gross N mineralization in rhizosphere soil is about 10-fold higher than that in bulk soil [47]. By contrast, overapplication of chemical N fertilizer in all three years caused obvious N fixation (adsorption in soil lattice and transformation into organic N by soil microorganism) and accumulation in the soil. The more base N fertilizer was applied, the more N was fixed in the soil (Figs. 5–7). High N accumulation in soil profiles not only decreases NUE, but also causes environmental pollution in intensive agricultural systems [48,49]. N leaching is closely correlated to precipitation and soil moisture. With heavy summer rainfall in 2007 and 2008, soil N_{min} moved downward, resulting in high nitrate accumulation in deep soil profiles (Figs. 5, 6; [9]). Even delayed N topdressing in 2008 (TDAT, moderated N input) caused N leaching to deep soil layers (Fig. 6). In comparison with overapplication of chemical N fertilizer, soil N_{min} in the root zone remained at a relatively low level under the TDBT treatment, while maintaining grain yield (Table 1).

Conclusions

With sufficient or very frequently excessive N supply, the highest maize yield reached in North China in recent years was 21 Mg ha^{-1} [8]. The root system has played essential roles in N uptake and dry matter transformation in improving maize grain yield. However, it remained unclear how the root system responded to current fertilization regimes. The present study showed that vast majority of roots grow in the upper 30 cm soil layer. The root length in deep soil layers (40–60 cm) was quite constant during the whole growth period. The decrease in the total root length after tasseling was mainly due to rapid lateral root death in the top 30 cm soil layer. N deficiency stimulated early initiated nodal root growth only for a short time; however, it completely suppressed initiation and growth of the 7th whorl of nodal roots and accelerated older nodal root death, causing an early decrease in the total root length when maize plants were still

in the rapid vegetative growth stage with high N uptake activity. In contrast, 175 kg ha^{-1} base N fertilizer (450 N treatment) inhibited root growth in the early growth period. Importantly, root length increased in the 40–50 cm soil layer in response to N leaching, even under sufficient N supply (450 N) in 2008, indicating that N-sufficient maize plants responded to local N resources via enhanced root proliferation in the soil profile with higher N_{min}. Further, N uptake and accumulation primarily occurred before tasseling, prior to substantial dry matter accumulation and grain formation. Therefore, appropriate N topdressing before tasseling (after the V8 stage) was necessary to ensure adequate N supply for robust plant growth and development. Early N excess, deficiency or delayed N topdressing reduced total N uptake, resulting in a significant decrease in dry matter accumulation and grain yield. Lastly, the soil in the field has great buffering capacity to maintain relatively constant N_{min} in most soil profiles while supporting maize growth under the 0 N treatment in all three years. N overapplication or delayed N topdressing caused N accumulation in the soil and leaching towards deep soil profiles and groundwater upon heavy rainfall in the growth season. TDBT treatment appears to be a good N application strategy because it maintains superior root growth relative to other N application strategies during the maize growing period, and significantly reduces N loss without sacrificing grain yield.

Supporting Information

Table S1 Total soil mineral nitrogen and selected soil chemical properties before maize planting in 2007, 2008 and 2009.

Table S2 Monthly rainfall during the maize growing period in 2007, 2008 and 2009.

Table S3 Rates and times of chemical N application in the field experiments in 2007, 2008 and 2009.

Author Contributions

Conceived and designed the experiments: CL. Performed the experiments: YP. Analyzed the data: YP CL. Contributed reagents/materials/analysis tools: CL. Wrote the paper: YP XL CL.

References

1. Hirel B, Gouis JL, Ney B, Gallais A (2007) The challenge of improving nitrogen use efficiency in crop plants: towards a more central role for genetic variability and quantitative genetics within integrated approaches. J Exp Bot 58: 2369–2387.
2. Ju XT, Xing GX, Chen XP, Zhang SL, Zhang LJ, et al. (2009) Reducing environmental risk by improving N management in intensive Chinese agricultural systems. Proc Nat Acad Sci USA 106: 3041–3046.
3. Li SK, Wang CT (2009) Evolution and development of maize production techniques in China. (In Chinese). Sci Agric Sin 42: 1941–1951.
4. FAO FAOSTAT–Agriculture Database. Available: http://faostat.fao.org/ site/ 291/default.aspx. Accessed 2 October 2010.
5. Liu ZQ, Li WL, Lu LP, Shen HP, Zhou GL, et al. (2007) Revelation of American national maize yield contest in 2006. (In Chinese). J Maize Sci 15: 144–145.
6. Chen GP, Wang HR, Zhao JR (2009) Analysis on yield structural model and key factors of maize high-yield plots. (In Chinese). J Maize Sci 17: 89–93.
7. Chen GP, Yang GH, Zhao M, Wang LC, Wang YD, et al. (2008) Studies on maize small area super–high yield trials and cultivation technique. (In Chinese). J Maize Sci 16: 1–4.
8. Wang YJ (2008) Study on population quality and individual physiology function of super high-yielding maize (Zea mays L.). Shandong Agricultural University. Ph.D. Dissertation.
9. Zhao RF, Chen XP, Zhang FS, Zhang HL, Schroder J, et al. (2006) Fertilization and nitrogen balance in a wheat–maize rotation system in north China. Agron J 98: 938–945.
10. Cui ZL, Zhang FS, Mi GH, Chen FJ, Li F, et al. (2009) Interaction between genotypic difference and nitrogen management strategy in determining nitrogen use efficiency of summer maize. Plant Soil 317: 267–276.
11. Guo JH, Liu XJ, Zhang Y, Shen JL, Han WX, et al. (2010) Significant acidification in major Chinese croplands. Science 327: 1008–1010.
12. Li SK, Wang CT Report of survey on demand for science and technology by farmers in maize production. (In Chinese). Available at http://chinamaize.con. cn/tishengxd/2006ku/2005-77-14/htm.
13. Gao Q, Li DZ, Wang JJ, Bai BY, Huang LH (2007) Studies on the effects of single fertilization on growth and yield of spring maize. (In Chinese). J Maize Sci 15: 125–12.
14. Subedi KD, Ma BL (2005) Nitrogen uptake and partitioning in stay-green and leafy maize hybrids. Crop Sci 45: 740–747.
15. Borrell AK, Hammer GL, Van Oosterom E (2001) Stay-green: a consequence of the balance between supply and demand for nitrogen during grain filling. Ann Appl Biol 138: 91–95.
16. Echarte L, Rothstein S, Tollenaar M (2008) The response of leaf photosynthesis and dry matter accumulation to nitrogen supply in an older and a newer maize hybrid. Crop Sci 48: 656–665.
17. Ma BL, Dwyer ML (1998) Nitrogen uptake and use in two contrasting maize hybrids differing in leaf senescence. Plant Soil 199: 283–291.
18. Rajcan I, Tollenaar M (1999) Source: sink ratio and leaf senescence in maize. II. Nitrogen metabolism during grain filling. Field Crops Res 60: 255–265.
19. Gewin V (2010) An underground revolution. Science 466: 552–553.

20. Amos B, Walters DT (2006) Maize root biomass and net rhizodeposited carbon: an analysis of the literature. SSSAJ 70: 1489–1503.

21. Zhang HM, Jennings A, Barlow PW, Forde BG (1999) Dual pathways for regulation of root branching by nitrate. Proc Nat Acad Sci USA 96: 6529–6534.

22. Drew MC (1975) Comparison of the effects of a localized supply of phosphate, nitrate, ammonium and potassium on the growth of the seminal root system, and the shoot, in barley. New Phytol 75: 479–490.

23. Nelson D W, Somers L E (1973) Determination of total nitrogen in plant material. Agron J 65: 109–112.

24. Peng YF, Niu JF, Peng ZP, Zhang FS, Li CJ (2010) Shoot growth potential drives N uptake in maize plants and correlates with root growth in the soil. Field Crops Res 115: 85–93.

25. Böhm W (1979) Methods of Studying Root Systems. Berlin, New York: Springer-Verlag.

26. Wiesler F, Horst WJ (1993) Differences among maize cultivars in the utilization of soil nitrate and the related losses of nitrate through leaching. Plant Soil 151: 193–203.

27. Chen XP, Zhang FS, Römheld V, Horlacher D, Schulz R, et al. (2006) Synchronizing N supply from soil and fertilizer and N demand of winter wheat by an improved N_{min} method. Nutr Cycl Agroecosys 74: 91–98.

28. Cui ZL, Zhang FS, Chen XP, Miao YX, Li JL, et al. (2008) On-farm evaluation of an in-season nitrogen management strategy based on soil N_{min} test. Field Crops Res 105: 48–55.

29. Ogawa A, Kawashima C, Yamauchi A (2005) Sugar accumulation along the seminar root axis as affected by osmotic stress in maize: A possible physiological basis for plastic lateral root development. Plant Prod Sci 8: 173–180.

30. Marschner P (2011) Mineral Nutrition of Higher Plants, Ed 3. London, UK: Academic Press.

31. Liedgens M, Richner W (2001) Relation between maize (Zea mays L.) leaf area and root density observes with minirhizotrons. Eur J Agron 15: 131–141.

32. Liedgens M, Soldati A, Stamp P, Richner W (2000) Root development of maize (Zea mays L.) as observed with minithizotrons in lysimeters. Crop Sci 40: 1665–1672.

33. Wells CE, Eissenstat DM (2003) Beyond the roots of young seedlings: the influence of age and order on fine root physiology. J Plant Growth Regul 21: 324–334.

34. Niu JF, Peng YF, Li CJ, Zhang FS (2010) Changes in root length at the reproductive stage of maize plants grown in the field and quartz sand. J Plant Nutr Soil Sci 173: 306–314.

35. Gao Q, Li DZ, Wang JJ, Bai BY, Huang LH (2007) Studies on the effects of single fertilization on growth and yield of spring maize. (In Chinese). J Maize Sci 15: 125–12.

36. Dwyer LM, Ma BL, Stewart DW, Hayhoe HN, Balchin D, et al. (1996) Root mass distribution under conventional and conservation tillage. Can J Soil Sci 76: 23–28.

37. Remans T, Nacry P, Pervent M, Filleur S, Diatloff E, et al. (2006) The Arabidopsis NRT1.1 transporter participates in the signaling pathway triggering root colonization of nitrate-rich patches. Proc Nat Acad Sci USA 103: 19206–19211.

38. Ho CH, Lin SH, Hu HC, Tsay YF (2009) CHL1 functions as a nitrate sensor in plants. Cell 138: 1184–1194.

39. Wolfe DW, Henderson DW, Hsiao TC, Alvino A (1988a) Interactive Water and Nitrogen Effects on Senescence of Maize. I. Leaf Area Duration, Nitrogen Distribution, and Yield. Agron J 80: 859–864.

40. Wolfe DW, Henderson DW, Hsiao TC, Alvino A (1988b) Interactive Water and Nitrogen Effects on Senescence of Maize. II. Photosynthetic Decline and Longevity of Individual Leaves. Agron J 80: 865–870.

41. Sharp RE, Davies WJ (1979) Solute regulation and growth by roots and shoots of water-stressed maize plants. Planta 147: 43–49.

42. Sharp RE, Silk WK, Hsiao TC (1988) Growth of the maize primary root at low water potentials. I. Spatial distribution of expansive growth. Plant Physiol 87: 50–57.

43. Gallais A, Coque M (2005) Genetic variation for nitrogen use efficiency in maize: a synthesis. Maydica 50: 531–547.

44. Boomsma CR, Santini JB, Tollenaar M, Vyn TJ (2009) Maize per-plant and canopy-level morpho-physiological responses to the simultaneous stresses of intense crowding and low nitrogen availability. Agron J 101: 1426–1452.

45. Jacobs BC, Pearson CJ (1992) Pre-flowering growth and development of the inflorescences of maize. I. Primordia production and apical dome volume. J Exp Bot 43: 557–563.

46. Cantarero MG, Cirilo AG, Andrade FH (1999) Night temperature at silking affects kernel set in maize. Crop Sci 39: 703–710.

47. Herman DJ, Johnson KK, Jaeger CH, Schwartz E, Firestone MK (2006) Root influence on nitrogen mineralization and nitrification in *Avena barbata* rhizosphere soil. SSSAJ 70: 1504–1511.

48. Halvorson AD, Follett RF, Bartolo ME, Reule CA (2005) Corn response to nitrogen fertilizer in a soil with high residual nitrogen. Agron J 97: 1222–1229.

49. Hong N, Scharf PC, Davis JG, Kitchen NR, Sudduth KA (2007) Economically optimal nitrogen rate reduces soil residual nitrate. J Environ Qual 36: 354–362.

13

Effect of Biocontrol Agent *Pseudomonas fluorescens* 2P24 on Soil Fungal Community in Cucumber Rhizosphere Using T-RFLP and DGGE

Guanpeng Gao, Danhan Yin, Shengju Chen, Fei Xia, Jie Yang, Qing Li, Wei Wang*

State Key Laboratory of Bioreactor Engineering, East China University of Science and Technology, Shanghai, China

Abstract

Fungi and fungal community play important roles in the soil ecosystem, and the diversity of fungal community could act as natural antagonists of various plant pathogens. Biological control is a promising method to protect plants as chemical pesticides may cause environment pollution. *Pseudomonas fluorescens* 2P24 had strong inhibitory on *Rastonia solanacearum*, *Fusarium oxysporum* and *Rhizoctonia solani*, etc., and was isolated from the wheat rhizosphere take-all decline soils in Shandong province, China. However, its potential effect on soil fungal community was still unknown. In this study, the *gfp*-labeled *P. fluorescens* 2P24 was inoculated into cucumber rhizosphere, and the survival of 2P24 was monitored weekly. The amount decreased from 10^8 to 10^5 CFU/g dry soils. The effect of 2P24 on soil fungal community in cucumber rhizosphere was investigated using T-RFLP and DGGE. In T-RFLP analysis, principle component analysis showed that the soil fungal community was greatly influenced at first, digested with restriction enzyme *Hinf* I and *Taq* I. However, there was little difference as digested by different enzymes. DGGE results demonstrated that the soil fungal community was greatly shocked at the beginning, but it recovered slowly with the decline of *P. fluorescens* 2P24. Four weeks later, there was little difference between the treatment and control. Generally speaking, the effect of *P. fluorescens* 2P24 on soil fungal community in cucumber rhizosphere was just transient.

Editor: Jack Anthony Gilbert, Argonne National Laboratory, United States of America

Funding: This work was supported by the National Science Foundation of China (30871664) and Open Project for the State Key Laboratory of Bioreactor Engineering (2060204). The funders had no role in study design, data collection and analysis, decision to publish, or preparation of the manuscript.

Competing Interests: The authors have declared that no competing interests exist.

* E-mail: weiwang@ecust.edu.cn

Introduction

Fungi play important roles in soil ecosystem as major decomposers of plant residues, releasing nutrients that sustain and stimulate plant growth in the process [1,2]. Besides, the phylogenetic diversity of microorganisms can act as natural antagonists of various plant pathogens [3]. A well-developed and diverse rhizosphere community is thought to be critical in the suppression of pathogens [4,5]. Knowledge of the structure and diversity of the fungal community in the plant rhizosphere will lead to a better understanding of pathogen-antagonist interactions [6].

It is suggested that only 17% of the known fungi can be readily grown in culture [7]. As traditional methods have many pitfalls, culture-independent methods show great potential in monitoring shifts or diversity of microbial community in a variety of environmental samples, such as Phospholipid Fatty Acid analysis (PLFA), Fatty Acid Methyl Ester profile (FAME), Terminal Restriction Fragment Length Polymorphism (T-RFLP), Ribosomal Intergenic Spacer Analysis (RISA), Denaturing/Temperature Gradient Gel Electrophoresis (DGGE/TGGE), Single Strand Configuration Polymorphism (SSCP), Amplified Ribosomal DNA Restriction Analysis (ARDRA), etc. Among these, T-RFLP and DGGE are two most widely used and effective methods in analyzing the spatial and temporal shifts of microbial community. T-RFLP method takes advantage in high throughputs, reproducible and web-based RDP database [8], while DGGE has high resolution by separating the same size fragments and sequencing each band [9]. Thus, in this study, the combination of the two methods would give a better understanding of the soil fungal community in cucumber rhizosphere.

Pesticides are widely used in agriculture to improve the yield of crops. However, chemical pesticides have residues and may influence the ecological system, soil fertility and underground water [10,11], thus cause seriously environment pollution. Biological control had been a significant approach to plant health management during the twentieth century and promised through modern biotechnology to be even more significant in the twenty-first century [12]. At present, the global markets of biopesticides become larger and larger especially in North America and Europe [13], and the predicted rate of growth is 10% per year [14].

Pseudomonas spp. commonly inhabits in soil and has been applied for biocontrol, promoting plant growth and bioremediation. 2, 4-diacetylphloroglucinol(DAPG)-producing strains were major groups in biocontrol microorganisms, because of their easy colonization, good competition and broad antimicrobial spectrum. Thus, they were widely used by more and more researchers [15–18]. For example, *P. fluorescens* F113 could inhibit *Erwinia carotovora*, which is the agent of soft rot of potato [19]. It has been also reported that *P. fluorescens* and 2, 4-diacetylphloroglucinol (DAPG) that it produced could prevent *Fusarium oxysporum*, *Septoria tritici*, *Thielaviopsis basicola*, *Rhizoctonia solani* etc [20,21].

P. fluorescens 2P24, which has strong inhibitory on *Rastonia solanacearum*, *F. oxysporum* and *R. solani*, was isolated from the wheat rhizosphere take-all decline soils in Shandong province, China [22]. The root colonization and biocontrol mechanism of it have been studied [23–25] and it has been commercialized. However, the potential effect of *P. fluorescens* 2P24 on agricultural soil fungal community is still unknown, as it is important to address the displacements of indigenous microorganisms by inoculates and assess the potential effects on soil microcosm [26].

This was the first study to investigate the effect of *P. fluorescens* 2P24 on soil fungal community in cucumber rhizosphere. Changes in soil fungal community were detected with T-RFLP and DGGE.

Materials and Methods

1 Bacterial strain and inoculation preparation

The *gfp*-labeled *P. fluorescens* 2P24 was cultured on King's medium B (KB) agar plates containing 100 mg of ampicillin and kanamycin liter^{-1}. The bacteria were growing in liquid KB medium at 28°C, 150 r/min with 100 mg of ampicillin and kanamycin liter^{-1} for 24 h. Bacterial density was measured as the absorbance of the fermentation broth at 600 nm, with reference to a standard curve calibrated by plate enumeration.

2 Experimental site and description

The experimental site was located in the field of National Southern Pesticide Research Centre of Shanghai, China (31.17°N, 121.13°E), where the average annual temperature is 18°C and total rainfall is about 1200 mm per year. Ten plots were established in the experimental area, while each plot contained fifteen cucumber plants. Five plots were treated with *P. fluorescens* 2P24. The fermentation broth of *P. fluorescens* 2P24 was centrifuged to concentrate and then diluted to 2×10^9 CFU/L by water. And then, 1 liter of these diluents was pooled to the root rhizosphere of each cucumber plant in each plot directly. The other five plots were treated with the same volume water as control.

3 Sampling

All soil samples were taken 5–10 cm below the surface and 5 cm away from the plants, the soils were separated by shaking the roots. Soil samples were collected weekly from each treatment and five samples were taken from each plot at each time, mixed and stored at 4°C.

Soil pH and moisture content were immediately determined after sampling. Soil pH was measured using a pH probe and soil moisture was calculated by drying soil at 115°C to a constant dry weight. The soil organic carbon and nitrogen were also measured [27,28].

The soil fungal quantity was calculated by traditional cultivation method. 1 g of each soil sample was mixed with 99 ml sterile water and then diluted to different concentration gradients. 100 µl of these diluents was cultured on PDA plates with 4 days and counted (each with three replicates).

4 Survival of bacterial strain 2P24

Soil samples were dispersed and decimally diluted into sterile water. The dilutions were plated on to KB agar containing 100 mg of ampicillin and kanamycin liter^{-1}. The colonies were calculated after culturing for 48 h.

5 DNA extraction

750 mg of each soil sample and 1.25 g of silica beads were beaten for 5 min with 3 ml TENP washing buffer (50 mM Tris, 20 mM EDTA, 100 mM NaCl, 1% PVPP, pH 8.5), followed by centrifugation for 5 min. 3 ml SDS, 500 µl lysozyme (20 mg ml^{-1}), 500 µl cellulose solution (20 mg ml^{-1}) and 15 µl protease K (20 mg ml^{-1}) were added and vortexed for 10 min. After incubation at 37°C for 30 min, 125 µl of SDS (20%) and 0.15 g of PVPP were added to the mixture and then incubated at 65°C for 2 h, followed by centrifugation for 10 min ($8,000 \times g$). The supernatants were transferred to fresh micro-centrifuge tubes and extracted by mixing an equal volume of phenol-chloroform-isoamyl alcohol (25:24:1; pH 8.0) followed by centrifugation for 10 min ($12,000 \times g$). The aqueous phase was removed by addition of an equal volume of chloroform-isoamyl alcohol (24:1) followed by centrifugation for 10 min ($12,000 \times g$). Ten percent of total volume of NaAC (3 mol l^{-1}, pH 5.2) and sixty percent of total volume of isopropyl alcohol was added, and the total nucleic acids was precipitated at 4°C for 1 h followed by centrifugation for 10 min ($12,000 \times g$). The final nucleic acids were washed in 70% (v/v) ice-cold ethanol and air dried before re-suspension in 100 µl TE buffer (pH 8.0). At last, DNA solutions were stored at -20°C.

6 T-RFLP method

The universal fungal specific primers ITS1-F (5′-CTTGGTCATTTAGAGGAAGTAA-3′) [29] and ITS4 (5′-TCCTCCGCTTATTGATATGC-3′) [30] were used in this study with the forward primer labeled with 6-FAM. PCR was conducted in 25 µl reaction with 12.5 µl Ex Taq (Takara, Japan), 1 µl of extracted DNA, 0.5 µM of each primer and 1% BSA. The thermo cycler reaction conditions were: 5 min initial denaturation at 94°C followed by 35 cycles of 45 s at 94°C, 30 s of annealing at 51°C, and 1 min extension at 72°C. The final extension was 7 min at 72°C. PCR products were purified with PCR purification kits and detected by 1% agarose gel electrophoresis.

Two different restriction enzymes (*Hinf*I, *Taq*I) were used separately. Restriction digests contained 5 U of enzyme, 5 µL of labeled and purified PCR product in a 20-µL total volume. Restrictions were performed with water bath at 37°C for 2 h followed by an inactivation step at 65°C for 15 min.

The samples were separated with GeneScan 1000 Rox (Applied Biosystems) as an internal size standard on an ABI 310 DNA sequencer (Applied Biosystems) using POP6 polymer. Terminal fragments were evaluated by GeneScan Analytical Software.

7 DGGE method

NS1 (5′-GTAGTCATATGCTTGTCTC-3′) [30] and GCFung (5′-CGCCCGCCGCGCCCCGCGCCCGGCCCGCCGCCCC-CGCCCCATTCCCCGTTACCCGTTG-3′) [31] were chosen for amplification of fungal sequences, which had been proved to be the most suitable for detecting fungal diversities in soil using DGGE analysis [32]. A GC-clamp was added to the terminal primer to improve electrophoretic separation amplicons by DGGE. The PCR reactions were carried out in 50 µl volumes containing 25 µl Ex Taq (TAKARA, Japan), 2 µl of extracted soil DNA and 1.0 µM of each primer, 1% DMSO. The thermo cycling program was: 2 min initial denaturation at 94°C, followed by 35 cycles of 30 s at 94°C, 30 s of annealing at 55°C, and 1 min extension at 72°C. The final extension was 5 min at 72°C. Products were checked by electrophoresis in 1% (w/v) agarose gels and ethidium bromide staining. PCR products from each sample were separated by DGGEK-2401 system (C.B.S. Scientific Company, Inc., USA). The PCR products were separated as follows: 8% polyacrylamide gels and denaturing gradient from 25% to 45% were used; gels were electrophoresed in 1 ×TAE buffer at 60°C and 80 V for 16 h.

8 Data analysis

For T-RFLP analysis, profiles in the range of 50–600 bp were used for principal component analysis [33], which was conducted using the Statistical Product and Service Solutions statistics software 17.0 (SPSS Inc.). Further more, the similarity of different TRF clusters was calculated based on Pearson correlation method by SPSS.

For DGGE analysis, the similarity of cluster analysis was calculated based on the density of different bands in different lane. DGGE banding pattern analysis was conducted to compare by cluster analysis via the underweighted pair group method with mathematical averages (UPGMA), using the VisionWorksLS software (UVP, US).

Results

1 Soil characteristics and culturable fungi

The average pH value of soil samples was 5.0, while the average water content was 19%. This kind of acid soil is very typical in south China. The soil total organic carbon content was about 1.8 g/kg, and the total nitrogen content was about 0.19 g/kg.

As can be seen from Figure 1, the amount of soil culturable fungi in cucumber root rhizosphere decreased after inoculation of *P. fluorescens* 2P24 compared to the controls during the following three weeks. However, the discrimination between the treated and control became to be not very obvious. Through the whole experiment, the amount of soil culturable fungi was about 1×10^7 CFU/g dry soil on average.

2 Survival of *P. fluorescens* 2P24 in cucumber rhizosphere soil

The fermentation broth of *P. fluorescens* 2P24 was inoculated into the cucumber root soil directly. The initial amount of organisms in root soil reached at about 2×10^8 CFU/g of dry soil. Survival of *P. fluorescens* 2P24 was detected through cultivation method with gradient dilution by distilled water and then calculated after 48 h. During the following days after inoculation, survival of *P. fluorescens* 2P24 decreased sharply (Figure 2). On the 28th day, populations of

2P24 dropped to 3.6×10^6 CFU/g of dry soil. Then, the survival of 2P24 was not significantly decreased. At the end of this study, the survival of 2P24 still sustained at about 10^5 CFU/g of dry soil.

3 T-RFLP results

T-RFLP was used to detect the fungal community structure in cucumber rhizosphere soil. Although only dominant fungal populations were detected in the T-RFLP method, we assumed that these data represented the total fungal community structure. All of the replicates showed similar results, typical samples were analyzed as follows.

As can be clearly seen from Figures 3 and 4, a substantial change in the T-RFLP pattern was observed as shown by the Principle Component Analysis (PCA) of the TRF data. However, changes between the control and treatments were significantly different as digested by different enzymes.

Digested by H*inf* I (Figure 3), C7 and P7, C14 and P14, C21 and P21, C28 and P28 were far from each other. P7, P14, P21 and C28 cluster together and they could be regarded as one group. Although C7, C14, C21 and P28 were in the same direction, but there was a long distance from them. It suggested that the soil fungal diversity was greatly changed after the inoculation of 2P24. However, the soil fungal diversity of the treatment became close to the control after four weeks. As can be seen in figure 3, the control and treatment on the 35th, 42nd, 49th, 56th, 63rd day got very close to each other.

Digested by T*aq* I (Figure 4), the soil fungal diversity of control and treatment on the 7th day were totally different. But after one week, the distance between them became shorter and shorter. It indicated that the soil fungal diversity was greatly influenced by the inoculation of 2P24 on the 7th day, and then the soil fungal diversity was gradually recovered.

Even with regard to the soil fungal community of controls, there were some changes as the cucumber grew. Whether digested by H*inf* I or T*aq* I, the controls could not cluster together as one group.

Besides PCA analysis, proximity matrix of different treatments digested by H*inf* I and T*aq* I also showed the similar result (Table 1 and Table 2). For example, in table 1, the correlation coefficient between C7 and P7, C14 and P14, C21 and P21, C28 and P28 was less than 0.8, which meant that these treatments had little

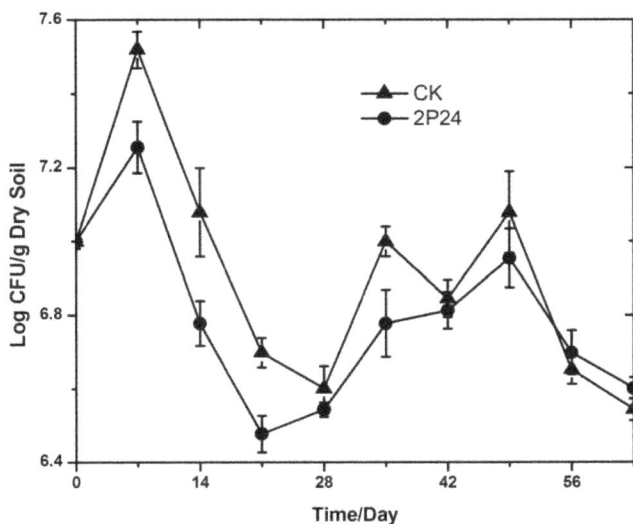

Figure 1. The shifts of soil culturable fungi in cucumber root rhizosphere after inoculation. "CK" was on behalf of the controls amended with water, while "2P24" was on behalf of the treatments amended with *P. fluorescens* 2P24.

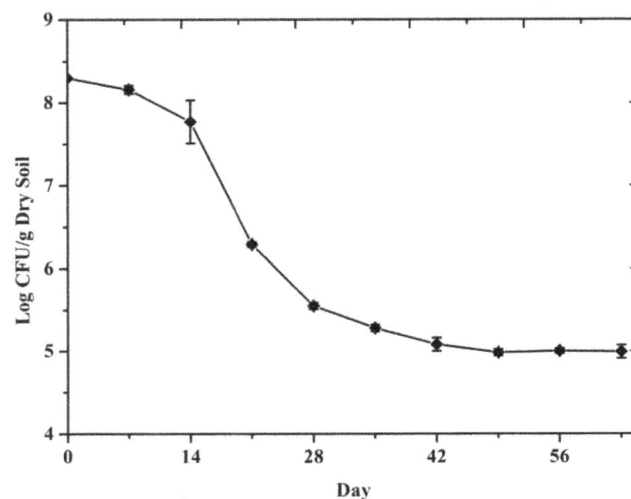

Figure 2. Survival of *P. fluorescens* 2P24 in cucumber rhizosphere soil microcosms after inoculation.

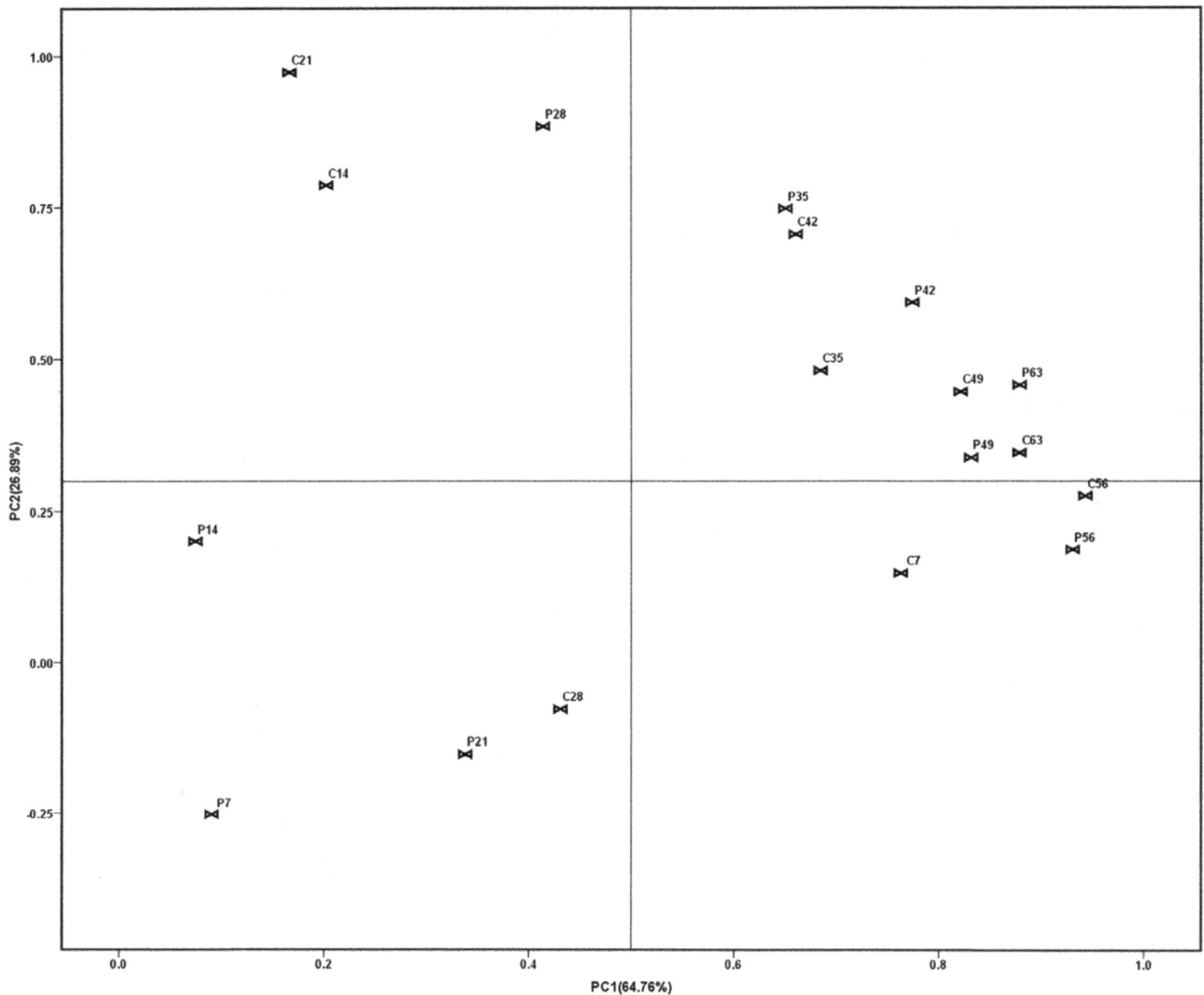

Figure 3. Principal component analysis (PCA) of T-RFLP profiles of soil samples (*Hinf* I). Symbols referred to individual replicates of different treatments. "C" was on behalf of the controls amended with water, and "P" was on behalf of the treatments amended with *P. fluorescens* 2P24. The number (7, 14, 21, 28, 35, 42, 49, 56, 63) following the abbreviation letters "C" and "P" represented the sampling day after inoculation. Numbers in parenthesis were percentage variance explained by each principal component (PC).

relationships. However, the correlation coefficient between control and treatment was between 0.8 and 1, which meant that these treatments had strong relationships. But in table 2, there was no such obvious differences, only the correlation coefficient between C7 and P7 was less than 0.8, while all of the others were more than 0.8, which was similar to the result of PCA analysis.

4 DGGE results

The mixed DNA samples were separated by DGGE fingerprints method. Significant changes between the control and treatment could be observed through the results (Figure 5). All of the replicates showed similar results, typical samples were analyzed as follows.

More than 10 bands could be detected through DGGE analysis. On the 7th day after 2P24 inoculation, bands 3, 4 and 9 of lane P7 almost disappeared compared to lane C7. At the same time, bands 5, 6 and 7 of lane P7 were less concentrated than lane C7. On the

14th day, bands 3, 4 and 9 of lane P14 appeared but were still very dim. Bands 5, 6 and 7 got to be more concentrated than lane P7, but still less than that of lane C14. On the 21st day, the bands of lane P21 came close to the lane C21 except the bands 3 and 4. On the 28th day, the bands of C28 and P28 were mostly similar to each other. After four weeks, the difference between the control and treatment lessened.

The results of cluster analysis by UPGMA method showed that C0, C7, C14, C21, C28, P21, P28 clustered together as one group while others clustered as one group (Figure 6). C0 was divided as a single branch. P7 and P14 were close to the control and treatment of the following five weeks. There was no big difference between the control and treatment after four weeks.

Overall, the soil fungal community was greatly influenced by the inoculation of 2P24 at first. However, this situation lasted about only two weeks. Four weeks later, the effect of biocontrol agent 2P24 had almost vanished.

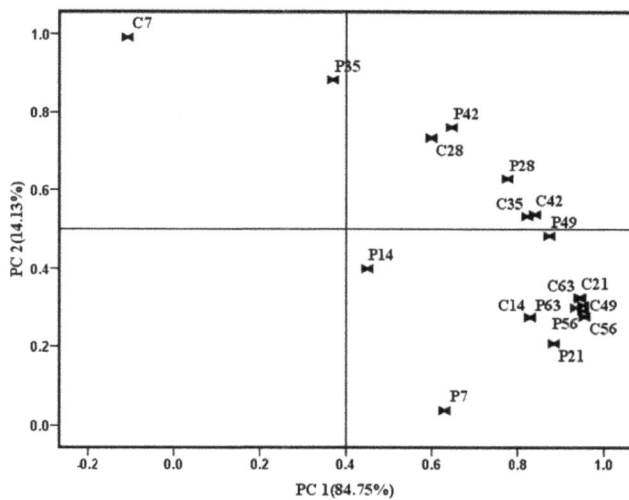

Figure 4. Principal component analysis (PCA) of T-RFLP profiles of soil samples (*Taq* I). Symbols referred to individual replicates of different treatments. "C" was on behalf of the controls amended with water, and "P" was on behalf of the treatments amended with *P. fluorescens* 2P24. The number (7, 14, 21, 28, 35, 42, 49, 56, 63) following the abbreviation letters "C" and "P" represented the sampling day after inoculation. Numbers in parenthesis were percentage variance explained by each principal component (PC).

Discussion

Soil microbial community could be affected by various soil conditions, such as pH, moisture, temperature, CO_2, etc [34]. In this study, the soil was acid soil types, probably because of yearly high temperature and rainfall. However, this kind of soil is a typical agricultural soil both in China and other countries in the world. In general, fungi have been found to be more acid tolerant than bacteria leading to increased fungal dominance in acidic soils [35–37]. The soil organic carbon (SOC) and total soil microorganisms mass correlate with the soil water content [38], and it has been proved that 19% of soil water content was most suitable for plant growth and the activities of soil microorganisms [39].

The amount of soil culturable fungi was about 10^7 CFU/g dry soil. Inoculation of *P. fluorescens* 2P24 decreased the total amount of soil culturable fungi during the following three weeks. It could probably be related with the biocontrol function of *P. fluorescens* 2P24. After that, there was no continued and obvious difference between the treated samples and the controls, which may probably caused by the decreasing of *P. fluorescens* 2P24. Although the amount of total soil fungi was calculated in the experiment, our finally aim was to study the changes of soil fungal diversity, as traditional culture method had a lot of faults.

Many factors could affect the survival of *P. fluorescens* in soil, such as inoculate formulation, soil conditions etc [40,41]. Thus, *P. fluorescens* would decrease quickly after inoculating into soils, just from $10^7 \sim 10^9$ to $10^3 \sim 10^5$ CFU/g dry soil in a month. The difference between variance was mostly dependent on the soil types and initiative inoculation concentrations.

Microorganisms will undergo a large variety of processes following their inoculation, including growth, death, and physio-

Table 1. Proximity matrix of T-RFLP profiles of soil samples (*Hinf* I).

Proximity Matrix (H*inf* I)

	C7	P7	C14	P14	C21	P21	C28	P28	C35	P35	C42	P42	C48	P48	C56	P56	C63	P63
C7	1.000																	
P7	.536	1.000																
C14	.440	.253	1.000															
P14	.154	.180	.704	1.000														
C21	.353	.325	.876	.298	1.000													
P21	.346	.930	.265	.036	.209	1.000												
C28	.245	.898	.168	.010	.116	.992	1.000											
P28	.579	.391	.814	.188	.954	.194	.084	1.000										
C35	.252	.458	.415	.172	.518	.665	.749	.590	1.000									
P35	.536	.047	.658	.128	.815	.207	.318	.910	.860	1.000								
C42	.550	.099	.552	.027	.763	.192	.298	.887	.823	.984	1.000							
P42	.554	.074	.525	.073	.672	.351	.459	.807	.915	.976	.971	1.000						
C48	.469	.276	.410	.091	.522	.537	.635	.660	.967	.909	.896	.973	1.000					
P48	.428	.405	.409	.241	.431	.612	.704	.551	.965	.827	.784	.908	.973	1.000				
C56	.671	.175	.427	.216	.418	.410	.516	.600	.863	.826	.801	.909	.943	.955	1.000			
P56	.580	.313	.358	.247	.321	.522	.618	.490	.875	.759	.722	.861	.930	.971	.987	1.000		
C63	.536	.290	.432	.215	.458	.518	.619	.603	.937	.853	.818	.929	.977	.941	.991	.983	1.000	
P63	.668	.073	.530	.186	.583	.328	.440	.743	.884	.921	.896	.966	.963	.941	.978	.944	.974	1.000

"C" was on behalf of the controls amended with water, and "P" was on behalf of the treatments amended with *P. fluorescens* 2P24. The number (7, 14, 21, 28, 35, 42, 49, 56, 63) following the abbreviation letters "C" and "P" represented the sampling day after inoculation.

Table 2. Proximity matrix of T-RFLP profiles of soil samples (*Taq* I).

Proximity Matrix (T*aq* I)

| | Correlation between Vectors of Values | | | | | | | | | | | | | | | | | |
	C7	P7	C14	P14	C21	P21	C28	P28	C35	P35	C42	P42	C48	P48	C56	P56	C63	P63
C7	1.000																	
P7	.026	1.000																
C14	.221	.906	1.000															
P14	.411	.900	.870	1.000														
C21	.223	.628	.887	.578	1.000													
P21	.144	.879	.993	.815	.915	1.000												
C28	.687	.644	.849	.810	.812	.812	1.000											
P28	.536	.476	.792	.559	.936	.795	.911	1.000										
C35	.422	.410	.742	.436	.944	.761	.833	.983	1.000									
P35	.810	.070	.417	.297	.626	.394	.790	.856	.832	1.000								
C42	.447	.605	.879	.650	.970	.884	.922	.986	.966	.766	1.000							
P42	.686	.448	.752	.606	.857	.735	.954	.980	.938	.911	.954	1.000						
C48	.201	.613	.875	.554	.999	.907	.798	.930	.946	.621	.965	.849	1.000					
P48	.382	.548	.845	.567	.981	.863	.874	.982	.982	.757	.993	.934	.980	1.000				
C56	.176	.635	.887	.578	.993	.922	.793	.911	.925	.584	.955	.830	.996	.972	1.000			
P56	.186	.605	.867	.544	.993	.902	.790	.921	.942	.613	.958	.840	.997	.978	.998	1.000		
C63	.219	.590	.863	.542	.996	.895	.803	.936	.954	.641	.967	.859	.999	.985	.996	.999	1.000	
P63	.184	.469	.777	.413	.976	.823	.735	.921	.959	.655	.936	.833	.983	.971	.978	.987	.988	1.000

"C" was on behalf of the controls amended with water, and "P" was on behalf of the treatments amended with *P. fluorescens* 2P24. The number (7, 14, 21, 28, 35, 42, 49, 56, 63) following the abbreviation letters "C" and "P" represented the sampling day after inoculation.

logical adaption, conversion to nonculturable cells, physical speed and gene transfer [42]. In the view point of biological invasion, the inoculation of a microorganism may break the original ecological balance of soil microbial community. Our results also showed that the inoculation of *P. fluorescens* 2P24 had a significant effect on the soil fungal community at first. But the effect of *P. fluorescens* 2P24 just lasted about one month, after that, the soil fungal community recovered as the control. Some researchers also found that there were only transient effects on soil microbial communities following the inoculation with biocontrol agents, such as *P.fluorescens* [43], *Streptomyces melanosporofaciens* [44] and *Corynebacterium glutamicum* [45].

In T-RFLP analysis, the great change of soil fungal community could be detected. However, the shift of soil fungal community was different as digested by different enzymes either analyzed by PCA method or by proximity matrix method. The soil fungal community was significantly influenced by the inoculation of biocontrol agent at first. But the process of recovering was totally different as digested by different enzyme. The PCA analysis of data digested by H*inf* I showed that the effect of *P. fluorescens* 2P24 on soil fungal community was very strong until 5 weeks later, while there was slight recovery of soil fungal community by PCA analysis of data digested by T*aq* I.

In DGGE analysis, it could be clearly seen that the soil fungal community recovered little by little after the inoculation of *P. fluorescens* 2P24, and the effect of *P. fluorescens* 2P24 lasted about three weeks. This result was mostly close to the result of T-RFLP analysis digested by T*aq* I.

T-RFLP and DGGE methods have already been applied in analyzing many different environmental samples. T-RFLP takes advantage of analyzing quantitative variances, while DGGE is a better choice to discriminate close species. The combination of these two methods would give a better understanding of soil fungal communities. However, Enwall and Hallin [46] showed that DGGE had a higher resolution than T-RFLP and binary data was better for discriminating between samples. But Smalla *et al* [47] showed that DGGE, T-RFLP, and SSCP analysis led to similar findings, although the fragments amplified comprised different variable regions and lengths. Our findings also showed that DGGE and T-RFLP had similar results, in spite of differences between H*inf* I and T*aq* I in T-RFLP analysis.

As PCR-based methods, T-RFLP and DGGE also have some pitfalls. For example, only dominant species can be amplified from soil DNA. Besides, a lot of factors may affect the final results, such as DNA extraction methods, primers, annealing temperature, Taq polymerase, and restriction enzymes etc [48]. For T-RFLP method, although there are specific RDP database, but identification of a TRF profile is usually impossible especially for fungi. Burke *et al.* [49] approved that T-RFLP could be applied to analyze soil fungi, but it could not reflect the real quantity of soil fungi [50]. Furthermore, two or more species may share the same profile, or one species may distribute in different profiles, and it even outputs pseudo-TRFs [51]. For DGGE method, its fragments were less than 500 bp, which was difficult for the following identification and phylogenetic analysis. Furthermore, sometimes a band did not stand for one species, or one species had different bands just as in T-RFLP method.

Conclusions

P. fluorescens 2P24 is a promising biocontrol strain against many fungal pathogens. However, its impact on soil fungal community is still unknown. This is the first study about monitoring the effect of

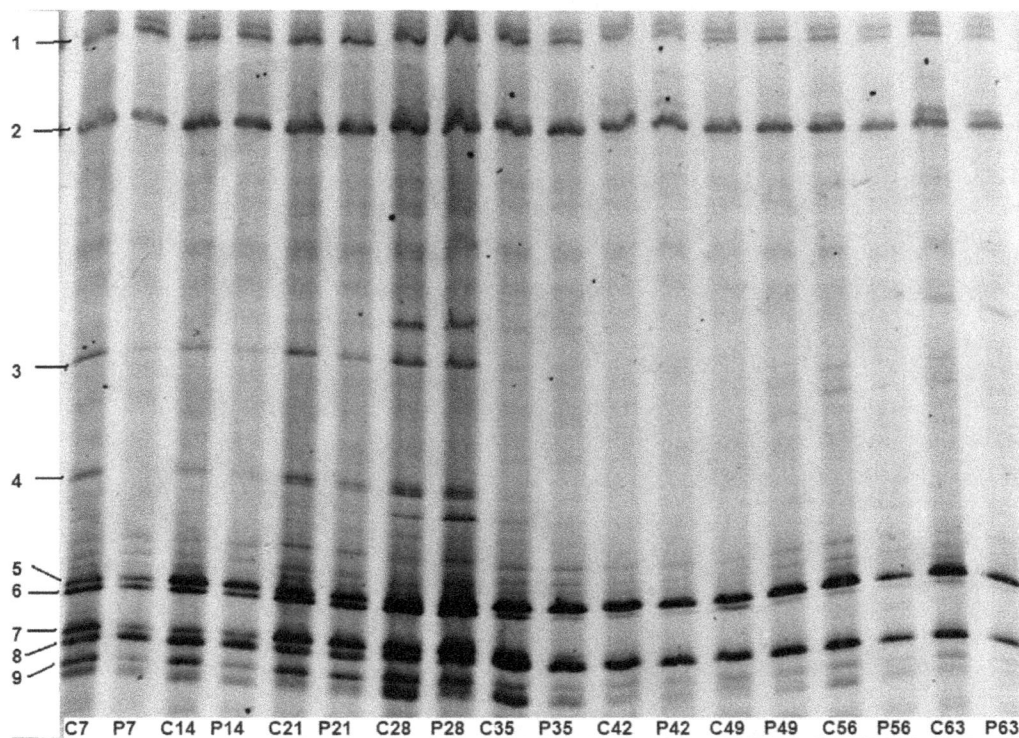

Figure 5. DGGE profiles showed the comparison between the controls and treatments of the soil fungal communities in cucumber rhizosphere after inoculation of *P. fluorescens* **2P24.** The fingerprints of fungal communities were generated by separation of 18S rDNA fragments amplified with primers NS1 and GCfung. "C" was on behalf of the controls amended with water, and "P" was on behalf of the treatments amended with *P. fluorescens* 2P24. The number (7, 14, 21, 28, 35, 42, 49, 56, 63) following the abbreviation letters "C" and "P" represented the sampling day after inoculation.

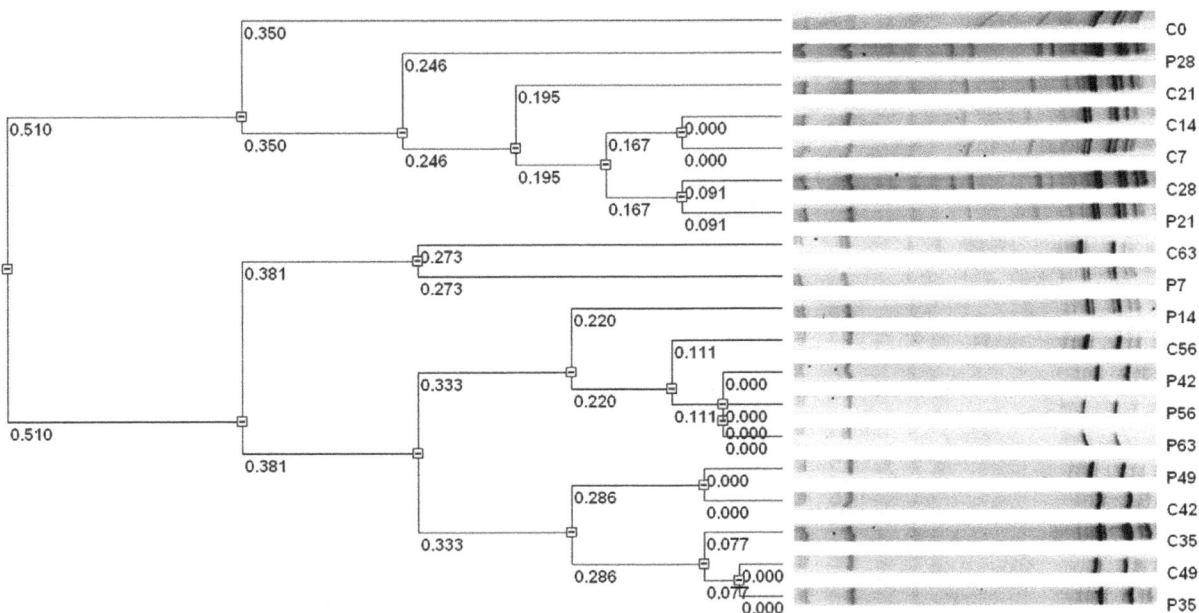

Figure 6. The differences between profiles were indicated by dice similarity. The Dendrogram was based on the RF Values and cluster analysis by the unweighted pair group method analysis (UPGMA) using VisionWorksLS (UVP, US). "C" was on behalf of the controls amended with water, and "P" was on behalf of the treatments amended with *P. fluorescens* 2P24. The number (7, 14, 21, 28, 35, 42, 49, 56, 63) following the abbreviation letters "C" and "P" represented the sampling day after inoculation.

P. fluorescens 2P24 on soil fungal communities in cucumber rhizosphere. After its inoculation, the survival of *P. fluorescens* 2P24 decreased from 10^8 CFU/g dry soil to 10^5 CFU/g dry soil during the whole growth time of cucumber. Thus, the soil fungal community was greatly influenced by its inoculation at the beginning. At the same time, the impact of *P. fluorescens* 2P24 on soil fungal community alleviated slowly weekly. Four weeks later, there was little difference between the control and the treatment.

Generally speaking, there was no significant effect of *P. fluorescens* 2P24 on soil fungal community in cucumber rhizosphere in spite of four-week influence. On the contrary, it suggested that the period of validity of biocontrol agent *P. fluorescens* 2P24 may be less than one month. Besides, our study just focused on the whole fungal community, the relationships between *P. fluorescens* 2P24 and each single fungal species was still unknown.

Acknowledgments

We gratefully acknowledge Dr. Ye Wang for comments on the manuscript.

Author Contributions

Conceived and designed the experiments: WW. Performed the experiments: GPG DHY SJC FX JY. Analyzed the data: GPG. Contributed reagents/materials/analysis tools: FX JY. Wrote the paper: GPG WW. Modified the paper: GPG QL WW.

References

1. Bridge P, Spooner B (2001) Soil fungi: diversity and detection. Plant and Soil 232: 147–154.
2. Vandenkoornhuyse P, Baldauf SL, Leyval C, Straczek J, Peter WY (2002) Extensive fungal diversity in plant roots. Science 295: 2051.
3. Gardener BBM, Fravel DR (2002) Biological control of plant pathogens: research, commercialization, and application in the USA. Plant Health Progress. doi: 10.1094/PHP-2002-0510-01-RV.
4. Alabouvette C (1990) Biological control of *Fusarium wilt* pathogens in suppressive soils. In Biological Control of Soil-Borne Plant Pathogens Hornby D, ed. Wallingford: CAB International. pp 27–43.
5. Jarosik V, Kovacikova E, Maslowska H (1996) The influence of planting location, plant growth stage and cultivars on microflora of winter wheat roots. Microbiological research 151: 177–182.
6. Smit E, Leeflang P, Glandorf B, van Elsas JD, Wernars K (1999) Analysis of fungal diversity in the wheat rhizosphere by sequencing of cloned PCR-amplified genes encoding 18S rRNA and temperature gradient gel electrophoresis. Applied and Environmental Microbiology 75: 2614–2621.
7. Hawksworth DL (1991) The fungal dimension of biodiversity: magnitude, significance and conservation. Mycol Res 95: 641–655.
8. Marsh TL, Saxman P, Cole J, Tiedje J (2000) Terminal restriction fragment length polymorphism analysis program, a web-based research tool for microbial community analysis. Applied and Environmental Microbiology 66: 3616–3620.
9. Muyzer G, Smalla K (1998) Application of denaturing gradient gel electrophoresis (DGGE) and temperature gradient gel electrophoresis (TGGE) in microbial ecology. Antonie van Leeuwenhoek 73: 127–141.
10. Johnsen K, Jacobsen CS, Torsvik V, Sørensen J (2001) Pesticide effects on bacterial diversity in agricultural soils - a review. Biology and Fertility of Soils 33: 443–453.
11. Arias-Estévez M, López-Periago E, Martínez-Carballo E, Simal-Gándara J, Mejuto JC, et al. (2008) The mobility and degradation of pesticides in soils and the pollution of groundwater resources. Agriculture, Ecosystems & Environment 123: 247–260.
12. Cook RJ (2000) Advances in plant health management in the 20th century. Annu Rev Phytopathol 38: 95–116.
13. Bailey KL, Boyetchko SM, Längle T (2010) Social and economic drivers shaping the future of biological control: A Canadian perspective on the factors affecting the development and use of microbial biopesticides. Biological Control 52: 221–219.
14. Bailey KL, Mupondwa EK (2006) Developing microbial weed control products: commercialization, biological, and technological considerations. In: Singh HP, Batish DR, Kohli RK, eds. Handbook of Sustainable Weed Management, The Haworth Press Inc., Binghamton, NY, USA. pp 431–473.
15. Carroll H, Moënne-Loccoz Y, Dowling DN, O'Gara F (1995) Mutational disruption of the biosynthesis genes coding for the antifungal metabolite 2, 4-diacetylphloroglucinol does not influence the ecological fitness of *Pseudomonas fluorescens* F113 in the rhizosphere of sugarbeets. Applied and Environmental Microbiology 61: 3002–3007.
16. Cronin D, Moënne-Loccoz Y, Fenton A, Dunne C, Dowling DN, et al. (1997) Role of 2, 4-diacetylphloroglucinol in the interactions of the biocontrol pseudomonad strain F113 with the potato cyst nematode *Globodera rostochiensis*. Applied and Environmental Microbiology 63: 1357–1361.
17. Saravanakumar1 D, Lavanya N, Muthumeena B, Raguchander T, Suresh S, et al. (2008) *Pseudomonas fluorescens* enhances resistance and natural enemy population in rice plants against leaffolder pest. Journal of Applied Entomology 132: 469–479.
18. Sarniguet A, Kraus J, Henkels MD, Muehlchen AM, Loper JE (1995) The sigma factor sigma s affects antibiotic production and biological control activity of *Pseudomonas fluorescens* Pf-5. PNAS 92: 12255–12259.
19. Cronin D, Moënne-Loccoz Y, Fenton A, Dunne C, Dowling DN, et al. (1997) Ecological interaction of a biocontrol *Pseudomonas fluorescens* strain producing 2, 4-diacetylphloroglucinol with the soft rot potato pathogen *Erwinia carotovora* subsp. *atroseptica*. FEMS Microbiology Ecology 23: 95–106.
20. Bangera MG, Thomashow LS (1996) Characterization of a genomic locus required for synthesis of the antibiotic 2, 4-diacetylphloroglucinol by the biological control agent *Pseudomonas fluorescens* Q2-87. Molecular plant-microbe interactions: MPMI 9: 83–90.
21. Keel C, Schnider U, Maurhofer M, Voisard C, Laville J, et al. (1992) Suppression of root diseases by *Pseudomonas fluorescens* CHA0: importance of the bacterial secondary metabolite 2, 4-diacetylphloroglucinol. Molecular plant-microbe interactions: MPMI 5: 4–13.
22. Wei HL, Wang Y, Zhang LQ, Tang WH (2004) Identification and characterization of biocontrol bacterial strain 2P24 and CPF-10. Acta Phytopathologica Sinica 34: 80–85.
23. Hailei W, Liqun Z (2006) Quorum-sensing system influences root colonization and biological control ability in *Pseudomonas fluorescens* 2P24. Antonie van Leeuwenhoek 89: 267–280.
24. Qing Y, Wei G, Xiaogang W, Liqun Z (2009) Regulation of the PcoI/PcoR quorum-sensing system in *Pseudomonas fluorescens* 2P24 by the two-component PhoP/PhoQ system. Microbiology 155: 124–133.
25. Wu SG, Duan HM, Tian T, Yao N, Zhou HY, et al. (2010) Effect of the hfq gene on 2, 4-diacetyphloroglucinol production and the PcoI/PcoR quorum-sensing system in *Pseudomonas fluorescens* 2P24. FEMS Microbiology Letters 309: 16–24.
26. Elsas JD. van, Duarte GF, Rosado AS, Smalla K (1998) Microbiological and molecular biological methods for monitoring microbial inoculants and their effects in the soil environment. Journal of Microbiological Methods 32: 133–154.
27. Vance ED, Brookes PC, Jenkenson DS (1987) An extraction method for measuring soil microbial biomass C. Soil Bilogy&Biochemical 19: 703–707.
28. Institute of Soil Science, Chinese Academy of Sciences. Soil Physical and Chemical Analysis. Shanghai: Shanghai Scientific and Technical Publishers, 1978. (in Chinese).
29. Bruns TD, Szaro TM, Gardes M, Cullings KW, Pan JJ, et al. (1998) A sequence database for the identification of ectomycorrhizal basidiomycetes by phylogenetic analysis. Molecular Ecology 7: 257–272.
30. White TJ, Bruns TD, Lee S, Taylor J (1990) Analysis of phylogenetic relationships by amplification and direct sequencing of ribosomal RNA genes. PCR protocols a guide to methods and applications Publisher: Academic Press. pp 315–322.
31. May LA, Smiley B, Schmidt MG (2001) Comparative denaturing gradient gel electrophoresis analysis of fungal communities associated with whole plant corn silage. Can J Microbiol 47: 829–841.
32. Hoshino YT, Morimoto S (2008) Comparison of 18S rDNA primers for estimating fungal diversity in agricultural soils using polymerase chain reaction-denaturing gradient gel electrophoresis. Soil Science and Plant Nutrition 54: 701–710.
33. Osborn AM, Moore ERB, Timmis KN (2000) An evaluation of terminal-restriction fragment length polymorphism (T-RFLP) analysis for the study of microbial community structure and dynamics. Environmental Microbiology 2: 39–50.
34. Strickland MS, Rousk J (2010) Considering fungal : bacterial dominance in soils - Methods, controls, and ecosystem implications. Soil Biology and Biochemistry 42: 1385–1395.
35. Högberg MN, Högberg P, Myrold DD (2007) Is microbial community composition in boreal forest soils determined by pH, C-to-N ratio, the trees, or all three? Oecologia 150: 590–601.
36. Joergensen RG, Wichern F (2008) Quantitative assessment of the fungal contribution to microbial tissue in soil. Soil Biology and Biochemistry 40: 2977–2991.
37. Rousk J, Brooks PC, Bååth E (2009) Contrasting soil pH effects on fungal and bacterial growth suggest functional redundancy in carbon mineralization. Applied and Environmental Microbiology 75: 1589–1596.
38. Entry JA, Fuhrmann JJ, Sojka RE, Shewaker GE (2004) Influence of irrigated agriculture on soil carbon and microbial community structure. Environmental Management 33: S363–S373.

39. Wang J, Kang S, Li F, Zhang F, Li Z, et al. (2008) Effects of alternate partial root-zone irrigation on soil microorganism and maize growth. Plant Soil 302: 45–52.

40. Van Elsas JD, Trevors JT, Jain D, Woiters AC, Heijnen CE, et al. (1992) Survival of, and root colonization by, alginate-encapsulated *Pseudomonas fluorescens* cells following introduction into soil. Biol Fertil Soils 14: 14–22.

41. Wessendorf J, Lingens F (1989) Effect of culture and soil conditions on survival of *Pseudomonas fluorescens* R1 in soil. Appl Microbiol Biotechnol 31: 97–102.

42. Prévost K, Couture G, Shipley B, Brzezinski R, Beaulieu C (2006) Effect of chitosan and a biocontrol streptomycete on field and potato tuber bacterial communities. BioControl 51: 533–546.

43. Natsch A, Keel C, Hebecker N, Laasik E, De'fago G (1998) Impact of *Pseudomonas fluorescens* strain CHA0 and a derivative with improved biocontrol activity on the culturable resident bacterial community on cucumber roots. FEMS Microbiol. Ecol 27: 365–380.

44. Prévost K, Couture G, Shipley B, Brzezinski R, Beaulieu C (2006) Effect of chitosan and a biocontrol streptomycete on field and potato tuber bacterial communities. BioControl 51: 533–546.

45. Vahjen W, Munch JC, Tebbe CC (1995) Carbon source utilization of soil extracted microorganisms as a tool to detect the effect of soil supplemented with genetically engineered and non-engineered *Corynebacterium glutamicum* and a recombinant peptide at the community level. FEMS Microbiol Ecol 18: 317–328.

46. Enwall K, Hallin S (2009) Comparison of T-RFLP and DGGE techniques to assess denitrifier community composition in soil. The Society for Applied Microbiology, Letters in Applied Microbiology 48: 145–148.

47. Smalla K, Oros-Sichler M, Milling A, Heuer H, Baumgarte S, et al. (2007) Bacterial diversity of soils assessed by DGGE, T-RFLP and SSCP fingerprints of PCR-amplified 16S rRNA gene fragments: Do the different methods provide similar results? Journal of Microbiological Methods 69: 470–479.

48. Bernboma N, Nørrunga B, Saadbyea P, Mølbakb L, Vogensenc FK, et al. (2006) Comparison of methods and animal models commonly used for investigation of fecal microbiota: Effects of time, host and gender. Journal of Microbiological Methods 66: 87–95.

49. Burke DJ, Martin KJ, Rygiewicz PT, Topa MA (2005) Ectomycorrhizal fungi identification in single and pooled root samples: terminal restriction fragment length polymorphism (TRFLP) and morphotyping compared. Soil Biology and Biochemistry 37: 1683–1694.

50. Lueders T, Friedrich MW (2003) Evaluation of PCR amplification bias by Terminal Restriction Fragment Length Polymorphism Analysis of small-Subunit rRNA and mcrA genes by using defined template mixtures of methanogenic pure cultures and soil DNA extracts. Applied and Environmental Microbiology 69: 320–326.

51. Egert M, Friedrich MW (2003) Formation of pseudo-terminal restriction fragments, a PCR-related bias affecting Terminal Restriction Fragment Length Polymorphism Analysis of microbial community structure. Applied and Environmental Microbiology 69: 2555–2562.

14

Species Accumulation Curves and Incidence-Based Species Richness Estimators to Appraise the Diversity of Cultivable Yeasts from Beech Forest Soils

Andrey M. Yurkov[1]*, **Martin Kemler**[1,2], **Dominik Begerow**[1]

1 Geobotany, Department of Evolution and Biodiversity of Plants, Faculty of Biology and Biotechnology, Ruhr-Universität Bochum, Bochum, Germany, 2 Centre of Excellence in Tree Health Biotechnology, Forestry and Agricultural Biotechnology Institute (FABI), University of Pretoria, Pretoria, South Africa

Abstract

Background: Yeast-like fungi inhabit soils throughout all climatic zones in a great abundance. While recent estimations predicted a plethora of prokaryotic taxa in one gram of soil, similar data are lacking for fungi, especially yeasts.

Methodology/Principal Findings: We assessed the diversity of soil yeasts in different forests of central Germany using cultivation-based techniques with subsequent identification based on rDNA sequence data. Based on experiments using various pre-cultivation sample treatment and different cultivation media we obtained the highest number of yeasts by analysing mixed soil samples with a single nutrient-rich medium. Additionally, several species richness estimators were applied to incidence-based data of 165 samples. All of them predicted a similar range of yeast diversity, namely 14 to 16 species. Randomized species richness curves reached saturation in all applied estimators, thus indicating that the majority of species is detected after approximately 30 to 50 samples analysed.

Conclusions/Significance: In this study we demonstrate that robust species identification as well as mathematical approaches are essential to reliably estimate the sampling effort needed to describe soil yeast communities. This approach has great potential for optimisation of cultivation techniques and allows high throughput analysis in the future.

Editor: Jürg Bähler, University College London, United Kingdom

Funding: This work was funded by the German Research Foundation (DFG) Priority Program 1374 "Biodiversity Exploratories" (BE 2201/9-1) and by the German Academic Exchange Service (DAAD) Program "Sustainable forestry and forest products" (A/07/94549). The funders had no role in study design, data collection and analysis, decision to publish, or preparation of the manuscript.

Competing Interests: The authors have declared that no competing interests exist.

* E-mail: andrey.yurkov@rub.de

Introduction

Soils display a remarkable heterogeneity throughout all climate zones, thus providing a multitude of diverse habitats of different scales and properties. Many studies on the biological diversity associated with particular soil types suggest that soil, in general, may well be a megadiverse habitat dominated by invertebrates, prokaryotes and fungi [1–4]. Fungi living in soils can be divided in two functional groups: filamentous, multicellular fungi and unicellular, yeasts. Yeasts comprise a systematically artificial group of fungi, which includes members of various orders of both Asco- and Basidiomycota [5–6]. The knowledge of yeast species diversity has tremendously increased over the last 50 years with nearly 1500 described species by 2010 [7] and estimations of the total number of yeast species assuming up to to 15,000 species, respectively [8].

Soil yeasts are known from the Polar Regions to the tropics and in total up to 130 species were reported worldwide. However, evidence for a strong association with soil-related substrates is lacking for many of these species [6,9–11] and despite a growing number of studies assessing soil biodiversity using culture-independent techniques [12–14], our knowledge of soil-inhabiting yeasts is mostly derived from cultivation-based approaches [6]. In general, diversity measures rely not only on cultivation success but also on sufficient sampling and reliable species recognition, e.g. [15–17]. Although identification methods based on ribosomal DNA sequencing gained more and more importance in the last decade [5–6,18], most of the studies, which were conducted in forest soils, utilized a combination of morphological and physiological characters for species identification. This is however problematic, because several soil-related yeasts, e.g. *Cryptococcus albidus*, *Cr. humicola*, *Cr. laurentii*, are taxonomically complex and may in fact be difficult to distinguish [19–21]. Therefore, separation of closely related and morphologically similar cryptic species using molecular tools could considerably influence the assessment of the existing microbial diversity.

Yeast species richness studies in forest soils of the temperate zone that were based on conventional phenotypic methods reported 18 to 26 yeast species [22–24]. Surprisingly, the only currently available study attempting species identification on the basis of rDNA sequence data obtained similar species numbers (i.e., 19 and 24 from two different forest sites, respectively) [25]. As molecular techniques are more sensitive in detecting cryptic species this result might indicate some methodological problems (e.g. undersampling), which would lead to a species underestimation.

Because microbial diversity differs largely across habitats, no meaningful comparison of biodiversity assessments can be performed without understanding the ranges of alpha-diversity, i.e. species richness, in a substrate or biotope. We used soil yeast community data to address three basic questions, which have previously not been considered within a single study: (1) Does soil heterogeneity have an effect on the observed species richness? (2) Is cultivation success of different yeast species influenced by different cultivation media and thereby by differences in nutrient availability? (3) What is the expected diversity estimated in relation to sampling effort? Special attention was given to robust species identification by sequencing rDNA and subsequent phylogenetic analysis in order to address species recognition and its impact on estimations of the yeast species richness in soils.

Materials and Methods

Study site, soil sampling and pre-cultivation sample treatment

The study has been performed in the Biodiversity Exploratory Hainich-Dün (http://www.biodiversity-exploratories.de). An overview of the German Biodiversity Exploratories is given by Fischer at al. [26]. Samples were taken in beech (*Fagus sylvatica*) forests of two different management types. The near-natural unmanaged forest consisted of 100 years old beech stands sometimes mixed with *Fraxinus excelsior* and *Acer pseudoplatanus*. The managed forest type consisted of 40-year-old age-class planted forests, with the tree species being *Fagus sylvatica*, and beech selection cutting forest. Details on the study sites, their properties and the field permits are provided in Fischer et al. [26]. Soils were classified in the field according to FAO [27] and KA5 [28] guidelines as Luvisols (Parabraunerde).

During September 2007, soil cores (topsoil, A_h horizon) were collected using a steel ring (volume: 100 cm^3), placed in sterile paper bags, transferred to the laboratory and kept at $-20°C$ before analysis. Soil from one plot (five core samples) was mixed in equal proportion and sieved consequently through 10 mm, 2 mm, 1 mm and 0.065 mm meshes (Figure 1). Roots, stones and woody particles were thereby removed on every step. The soil was afterwards pooled again and five samples were taken to represent one plot. Each sample was inoculated in triplicates, i.e. 15 mixed soil sub-samples and 2 plates per sample, 60 plates in total. Individual samples taken from the same soil cores, were plated in the same way to have the same number of plates per plot in the end (Figure 1). From the 5 core samples taken on a single plot 3 sub-samples were randomly taken (15 sub-samples per plot) and plated on 2 plates each, 60 plates in total. Soils collected in April–May 2008 were studied as mixed samples. Details on soil sampling are given by Yurkov et al. [29].

Isolation of cultures

Soil samples were placed in 50 ml plastic tubes, suspended (w/v) 1:5, 1:10, and 1:20 in sterile water and shaken on an orbital shaker at 200 rpm for 1 hour. An aliquot of 0.15 ml was distributed on the surface of solid media. Glucose-yeast extract-peptone agar (GPYA) acidified with lactate (final pH 4.5), MYP agar [30] supplemented with 0.05% tetracycline, thymine–mineral–vitamin (TMV) agar [31] and modified Browns' nitrogen deficient media

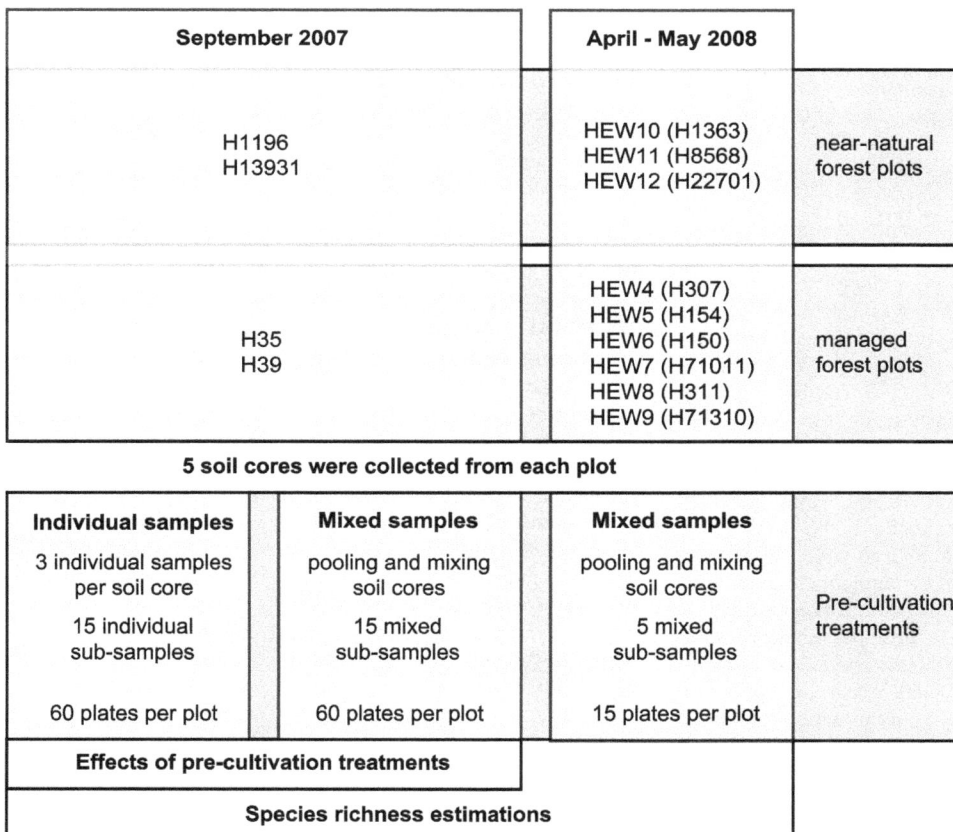

Figure 1. Scheme of the experimental design, sampling dates, analysed forest plots and applied pre-cultivation treatments.

[32] supplemented with imidazol [33] and cycloheximid [34] were used for cultivation experiments. Plates with GPYA and MYP media were incubated at room temperature for 3 days and then at lower temperature (6–10°C) to prevent fast development of moulds. Plates with nitrogen deficient media were incubated at room temperature. All plates were checked after 7, 14 and 21 days of incubation. Colonies were differentiated into macro-morphological types using dissection microscopy, counted, and 1–2 representatives of every colony type per plate were transferred into pure culture.

Identification of cultures

DNA was isolated from 3–4 days old cultures using a technique described by Hoffman and Winston [35], with slight modifications. DNA was precipitated with ethanol and then dissolved in 50 µl TE buffer containing RNAse (10 µg/ml). PCR-fingerprinting with minisatellite-specific oligonucleotides derived from the core sequence of bacteriophage M13 with the sequence given by Sampaio et al. [36] or microsatellite-specific oligonucleotides $(GTG)_5$, $(ATG)_5$ and $(GAC)_5$ as single PCR primer [37] were used to group pure cultures. Strains showing identical electrophoretic profiles were considered as conspecific and only 1–2 representatives of them were chosen for further identification by sequencing of rDNA regions. DNA fragments were amplified by PCR using the primers ITS1f and NL4 [38–39]. Initial denaturation was performed at 96°C for 2 min, followed by 35 cycles of 20 s at 96°C, 50 s at 52°C and 1.5 min at 72°C, respectively. A final extension step of 7 min at 72°C was conducted.

PCR products were purified with the my-Budget Double Pure kit (Bio-Budget Tech., Germany) and sequenced on an ABI3130xl sequencer using the same primers as for PCR amplification. Chromatograms were checked and corrected with Sequencher 4.8–4.10 (Gene Codes Corp., USA). For species identification the obtained nucleotide sequences were compared with sequences deposited in the NCBI (www.ncbi.nih.gov) and CBS (www.cbs.knaw.nl) databases, respectively.

Statistical data analyses

Yeast quantity was calculated as CFU (colony forming units) per gram of soil at natural humidity. The community structure was characterised by frequency of occurrence (incidence) of every observed species in the sample, which was calculated as the relative occurrence of the species to the total number of species in the sample. Statistical evaluations were performed with STATISTICA 8.0 (StatSoft, Inc., Tulsa, USA). The reliability of the different sampling assays was statistically tested by the comparison of results obtained from mixed and individual samples both in near-natural and managed forests using one-way analysis of variance. Only data that passed normality test were used for further analyses. Effects were considered to be statistically significant at the level p<0.05. Significant effects were additionally confirmed with Chi-square test. Species accumulation curves were calculated with EstimateS 8.0 using 50 randomizations, sampling without replacement and default settings for upper incidence limit for infrequent species [40]. As distinct yeast species could form colonies with similar morphology and, thus, make the separation to different types doubtful, we have used only presences/absences (incidence data) in our community matrix. The latter did not depend on the morphological differentiation but relied solely on molecular species identification. Four estimators of species richness were used: Chao 2 richness estimator [41], ICE incidence-based coverage estimator [42], Jackknife 1 first-order Jackknife richness estimator [43], and Bootstrap richness estimator [44]. Of the four species richness estimators, ICE distinguishes between frequent and infrequent species in analyses, Bootstrap does not differentiate the species frequency and the first-order Jackknife richness estimator additionally relies on the number of species only found once. Chao 2 estimator is distinct from the other species estimators as it is an incidence-based estimator of species richness, which relies on the number of unique units and duplicates (species found in only one and two sample units [45]). Species-area (i.e., species-plot) relationship was studied using the procedure described in Ugland et al. [46]. Because yeast communities sampled in September and April differed considerably, we did not use the full randomisation of the areas (plots). Instead, all analysed datasets contained samples from at least one plot sampled at September 2007 and at least one plot sampled in April 2008. For each dataset four species richness estimators were applied. Subsequently, average species richness values were calculated for each combination of samples comprising 2, 3, 4 and 5 plots. The number of samples allocated for the analyses of combinations of plots as well as the species richness values are provided in the Table S2.

Results

Effect of soil heterogeneity on species richness

The total yeast counts varied largely from 10^2 to 10^4 CFU/g. However, there was no significant difference related to the distinct sample treatments. 1.2×10^3 CFU/g of yeasts were observed in mixed and 1.0×10^3 CFU/g in individual samples (Figure S1).

The observed community structures determined from the incidence of eleven shared species were highly similar both in mixed and individual samples (Figure 2 and Table S1). *Trichosporon dulcitum* was found to be the most frequent yeast species irrespective of pre-cultivation sample treatment and was observed in more than half of the analysed samples. *Cryptococcus terricola* and *T. porosum* were observed more frequently in individual samples whereas *Debaryomyces hansenii* and *Kazachstania piceae* were found more frequently in mixed samples. Even though average incidence values of these species differed 3–5 times between individual and mixed samples, confidential interval suggests these differences to be statistically insignificant (ANOVA, p = 0.053). The frequency of appearance of *Trichosporon porosum* was slightly higher in the individual samples. Nevertheless, in both cases it was the second dominating species (Figure 2 and Table S1).

Observed species richness and effect of different media on species richness

We isolated 14 yeast species from near-natural beech forest soils and 12 from managed forest soils. In total we isolated 18 different species, from which 11 species were observed in both forest types (Table S1). The total species number (i.e. 18 species) was recovered by using GPYA and MYP media, and six species with TMV media (*Trichosporon dulcitum*, *Guehomyces pullulans*, *Kazachstania piceae*, *D. hansenii*, *Candida sake*, and *C. vartiovaarae*). Modified Brown's media was the most selective one and yielded isolates of only one species, *G. pullulans* (Table 1).

Dissimilarity in the yeast community composition was observed between soils collected in two different sampling dates (Table 1). Five yeasts, *Aureobasidium pullulans*, *Cryptococcus gastricus*, *Cr. musci*, *Cr. ramirezgomezianus*, and *Trichosporon* cf. *laibachii* were isolated only from the soil samples collected in September 2007, while *Barnettozyma pratensis*, *B. vustinii*, *C. vartiovaarae* and *Lindnera misumaiensis* were detected only in April–May 2008.

Estimated species richness

Because yeast community composition and the frequency of occurrence varied widely among sub-samples (Figure 2), species

frequency of occurence

0.0 0.1 0.2 0.3 0.4 0.5 0.6 0.7 0.8

Trichosporon dulcitum

Trichosporon porosum

Cryptococcus gastricus

Cryptococcus terricola

Cryptococcus ramirezgomezianus

Trichosporon cf. laibachii

Kazachstania piceae

Guechomyces pullulans

Aureobasidium pullulans

Rhodotorula glutinis

Cryptococcus musci

Debaryomyces hansenii

Candida sake

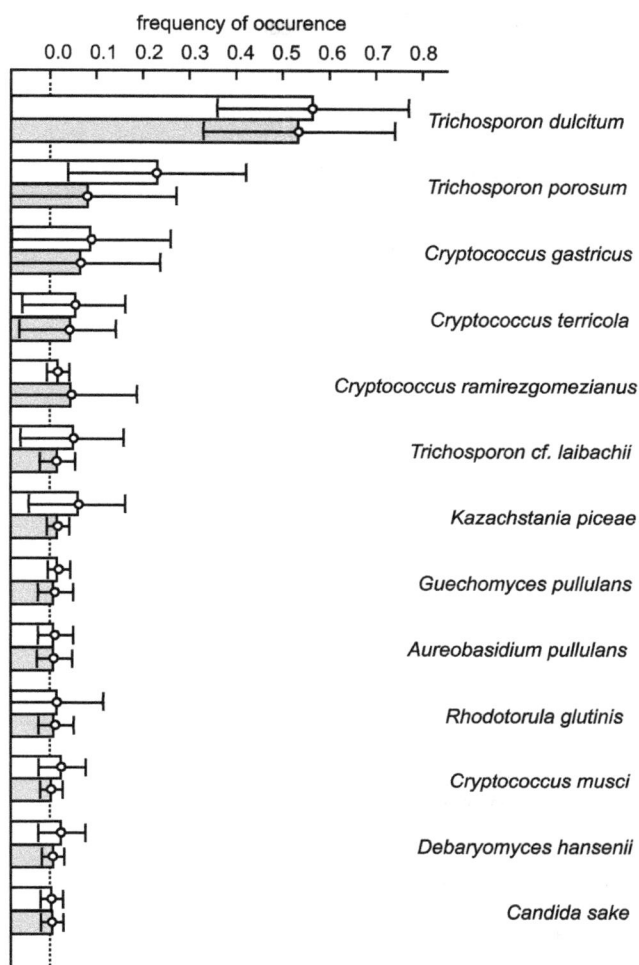

Figure 2. Frequency of appearance of yeasts in individual (white) and mixed (grey) soil samples. Whiskers correspond to confidential interval. Dotted line corresponds to zero frequency of appearance.

richness estimators were applied to obtain reliable numbers of species to be expected with the conventional cultivation technique [17,47]. When a total of $N = 165$ sub-samples were combined into two datasets, for the near-natural ($N = 75$) and managed ($N = 90$) forests, both of the species accumulation curves were close to saturation (Figure 3 and Table 2) showing that our study provides a reliable basis for yeast species richness assessment in a beech forest soils.

Species richness estimators calculated similar values for both forest types (Figure 3 and Table 2). Similarly to the obtained species accumulation curves, estimations were always higher, although not significantly, for natural than managed forests. Depending on the estimator used, means of estimated richness varied only in the range of about 2 species (Table 2). For managed and natural forests the ICE estimator with 15.4 and 16.7 yeasts respectively achieved the highest species richness. The lowest predicted values were 15.2 by Bootstrap in natural and 14.3 species by Chao 2 for managed forests. First order Jackknife estimator predicted relatively high species richness values together with low standard deviations in both forest types. Notably, the estimated species richness values did not differ significantly within a forest type (Table 2).

We estimated the species richness within an area to assess effects of community heterogeneity between the plots on the expected diversity. These estimations were performed for 2, 3, 4 and 5 plots, respectively (Table S2; Figure 4). On average, analysis of 3 managed plots predicts about 14 species to be found. Estimations performed with two additional plots result in 2 more species to be expected in soils under managed forests. Unlike the managed biotopes, near-natural beech forests displayed a contrasting trend, and the curve constructed using the estimations did not reach saturation after 3–4 plots but still increased continuously. In both forest types, the observed number of species followed the same trend as for the estimated species richness.

Discussion

The main purpose of this study was to provide baseline data on the alpha-diversity of yeast fungi in soils under beech forests in central Germany. Although the conventional plating technique has been the preferred method for investigating yeast biodiversity for a long time [48–49], no reliable estimations of yeast species richness in soils were performed to this date. The discrepancy between observed and estimated yeast diversity [8,50] is derived from different approaches of species detection and so far we do not know whether this is due to sampling biases or real limitations of the conventional plating technique. High-throughput dilution-to-extinction cultivation approaches improved the recovery of fungi from plant-related substrates [51–52], but has the same limitations as the applied plating technique. Due to the rise of massive parallel sequencing, culture-independent approaches of soil species diversity have received much attention in the last years, e.g. [3,13–14]. However, yeasts as soil organisms have been often neglected in these studies and reliable data to validate the results from culture independent approaches are lacking. Cultivation from soil suspension enables analysis of soil samples 10–20 times larger than the ones used by culture-independent techniques, e.g. [3,13–14]. In the present study we analysed a considerable amount of soil samples making a total of nearly 1 kg that renders our work the most comprehensive study utilized plating approach followed by molecular identification.

Pre-cultivation sample treatments and scale-dependency of community structure

Analysis of individual and mixed samples was aimed to reveal effects of soil meso- and micro-heterogeneity on estimations of species richness. We hypothesised that patchiness of soil properties, like acidity, water activity, and availability of nitrogen and carbon sources could significantly affect soil yeast communities on the level of soil aggregates. Thus, homogenisations and pooling of soil samples should even the distribution of yeast populations when multiple samples are collected (and mixed) to represent the soil cover of a biotope. Total yeast counts in our study are comparable with the average numbers previously found for soil-related substrates in temperate climate zone [9,22–23], and no significant effect of pre-cultivation sample treatment was observed (Figure 2). As determined from the incidence data, our investigation revealed that soil homogenisation does not significantly affect community structure. Differences in species occurrence were found to be more pronounced between natural and managed forests than between soil of individual and mixed samples (Table S1). Therefore, soil heterogeneity at the level of soil horizon and biotope seems to have no significant effects on the species richness of yeast fungi and the analysis of mixed soil samples provides reliable results and reduces time and costs. Additionally, the absence of pronounced effects of our pre-cultivation treatments suggests that spatial niche separa-

Table 1. Results of cultivation surveys: occurrence of yeasts in soils after different pre-cultivation treatments, in two distinct sampling periods, and with different cultivation media.

Effects:	Pre-cultivation treatments		Sampling date		Media
Species	**Mixed**	**Individual**	**Sept-2007**	**Apr-2008**	**GPYA (G), TMV (T), Brown's agar (B)**
Aureobasidium pullulans	×	×	×	n.o.	G
Barnettozyma pratensis	×	n.o.	n.o.	×	G
B. vustinii	×	n.o.	n.o.	×	G
Candida kruisii	×	×	×	×	G
C. sake	×	×	×	×	G, T
C. vartiovaarae	×	n.o.	n.o.	×	G, T
Cryptococcus gastricus	×	×	×	×	G
Cr. musci	×	×	×	n.o.	G
Cr. ramirezgomezianus	×	×	×	n.o.	G
Cr. terricola	×	×	×	×	G
Debaryomyces hansenii	×	×	×	×	G, T
Guechomyces pullulans	×	×	×	×	G, T, B
Kazachstania piceae	×	×	×	×	G, T
Lindnera misumaiensis	×	n.o.	n.o.	×	G
Rhodotorula glutinis	×	×	×	×	G
Trichosporon dulcitum	×	×	×	×	G, T
Trichosporon cf. laibachii	×	×	×	n.o.	G
Trichosporon porosum	×	×	×	×	G

Abbreviation used: GPYA, glucose-yeast-peptone agar; TMV, thimine-mineral-vitamine medium; n.o., not observed.

tion of soil yeast species might occur at a lower level of soil heterogeneity resembling bacterial communities distribution patterns [53].

Effect of different media on observed species richness

Four different media, representing a gradient of nutrient availability were used in the cultivation experiment. Although various authors have suggested using selective media in order to obtain higher diversity [54–56], our results demonstrate that the use of just one nutrient-rich medium can reveal the majority of cultivable yeasts from soil. Specifically, both of the nutrient-rich media, GPYA and MYP, resulted in recovering the highest number of yeast species, namely isolation of all observed species. We did not observe any significant effect between MYP and GPYA media, which could be explained by the fact that most of the soil-inhabiting species are basidiomycetes that are known to have wide assimilation spectra [6]. By using two more oligotrophic and nitrogen-depleted media (TMV and Brown's agar), only up to six species were be isolated. Nitrogen deficient media were previously used for the isolation of the members of the genus *Lipomyces* because these soil yeasts have the rare ability among yeasts to utilize nitrogen from heterocyclic compounds, such as imidazole, pyrimidine, and pyrazine [33,57]. These compounds were consequently included as a nitrogen source in selective media to isolate *Lipomyces* yeasts from soils [31,58]. Members of the genus *Lipomyces* and their anamorphs *Myxozyma* (Lipomycetaceae, Saccharomycetales, Ascomycota) display a typical oligotrophic behaviour, slow growth on standard nutrient-rich media coupled with the ability to assimilate complex substrates. The ability of other yeasts to grow on medium containing heterocyclic compounds as the only source on nitrogen has been recorded only once [31]. We could not isolate any members of

Lipomycetaceae with nitrogen-depleted media, but six other yeasts (Table S1). Thus, our results suggest that the ability to grow in the absence of external nitrogen sources and utilisation of heterocyclic compounds is more common among soil yeasts than has been assumed previously.

In addition to lower observed diversity, modified Browns' agar and TMV media seems not suitable for soil diversity studies in contrast to nutrient-rich media, because yeast colonies of different species on nitrogen-depleted media looked very similar, either extremely mucous or dimorphic (with developed substrate mycelium). After transfer from these media to GPYA they often split in several morphologically and taxonomically distinct cultures. Therefore, it is very difficult to make assumptions about the observed species abundance using these media. Additionally, we often observed strong development of moulds on our samples although TMV agar has been reported to prevent overgrowth of yeast colonies by filamentous fungi [31].

Estimated species richness and community structure

Rarefaction curves were close to saturation and we applied species richness estimators to predict the number of species to be expected from the soils using the traditional plating technique (Figure 3). All of the applied richness estimators predicted 14 to 16 species per forest type, with a slightly higher (but not significant) diversity to be expected in natural forests. Estimator curves reached saturation starting from 30% to 50% of the analysed sub-samples depending on the estimator applied. In other words, the number of plates for the analysis can be significantly reduced. This result apparently reflects the soil yeast community structure, which is characterised by a few autochthonous species (e.g. *Trichosporon dulcitum*, *T. porosum*, *Cryptococcus terricola*) accounting for the majority of isolates, and a large number of minor or rare species.

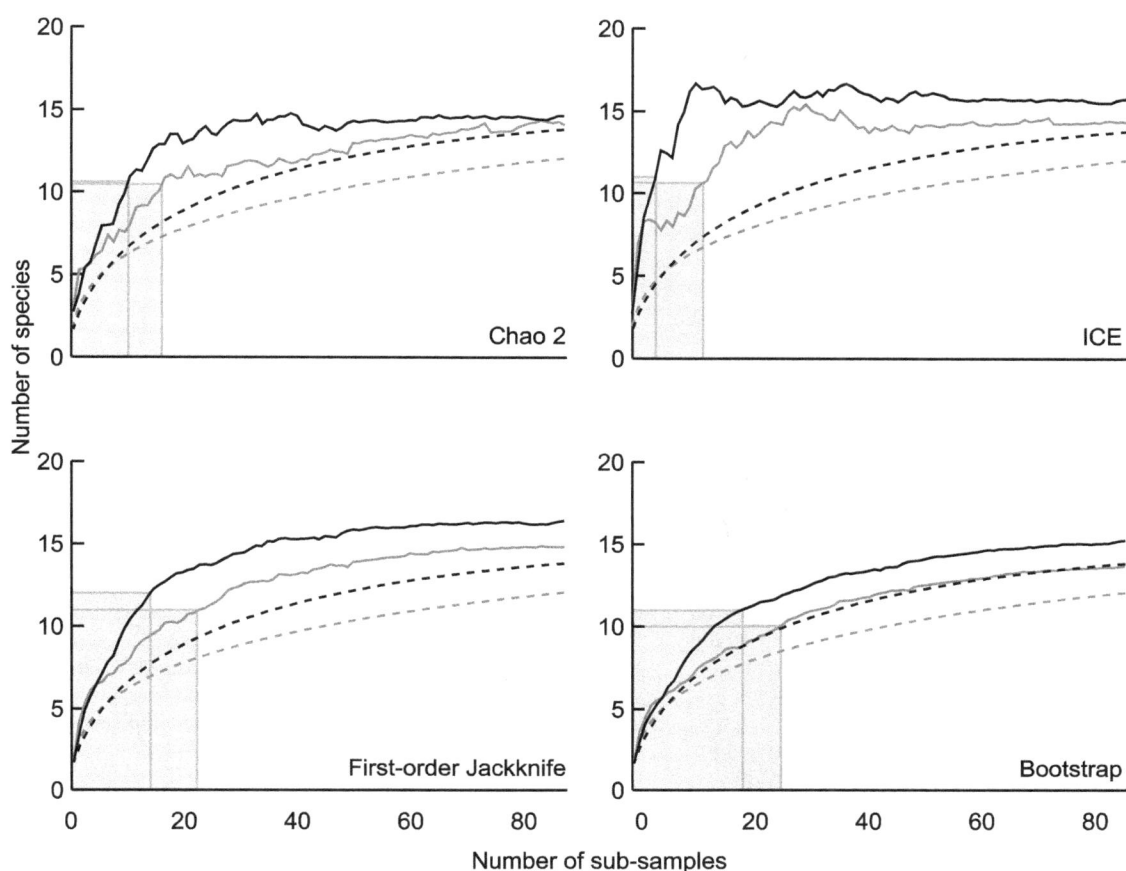

Figure 3. Estimator-based (solid line) and randomised (dashed line) species accumulation curves for near-natural (black) and managed (grey) beech forests obtained with incidence-based coverage (ICE), Chao 2, first-order Jackknife (Jack 1) and bootstrap richness estimators. Shadowed areas correspond to the 75% expected diversity cut-off.

Interestingly, the same pattern, a species-poor yeast community with uneven distribution of species, is known for yeasts inhabiting floral nectar [17]. Unlike soils, nectar is nutrient-rich and is characterised by high cell densities. Nevertheless, in both cases environmental conditions strongly affect the yeast population and select towards a few specialist species. Nearly one third of 18 yeast species isolated in our study was rare (either unique species or duplicates) and could be detected in less than in 3% of the sub-samples (Table S1). In soils, a number of physico-chemical properties of which oligotrophy, water availability and complex organic acids are of the main importance [6] considerably shape the community and, as a result, a limited number of yeasts were common among the two forest management types (Table S1).

Species-plot relationship reflects the importance of rare species for species richness estimations. In the studied areas, three remote forest sites separated by 20–25 km, the yeast community inhabiting soils of managed forests could be assessed by analysing 3–4 plots (Table S2; Figure 4). In contrast, near-natural forests (one forest site, approx. 2×5 km) harbour a larger number of rare and unique species even over a smaller range.

Although all randomised species accumulation curves reached saturation with our sampling effort, ICE and Chao 2 richness estimators approach the plateau earlier than first-order Jackknife or Bootstrap (Figure 3). Our study agree with earlier results obtained for wood-inhabiting fungi [15] and suggests that species richness estimators sensitive to singletons and doubletons, like

Table 2. Estimations of species richness obtained with different models (Mean and standard deviation).

Forest type	N of samples	Observed	ICE	Chao 2	Jackknife 1	Bootstrap
Natural	75	14	16.7±4.3	14.7±3.8	16.4±1.6	15.2±0.4
Managed	90	12	15.4±6.3	14.3±2.6	14.8±1.5	13.6±1.6
Total	165	18	18.6±0.4	19.0±2.3	20.0±1.4	19.0±0.02

Abbreviations used: ICE, incidence-based coverage estimator; Chao 2, Chao 2 richness estimator; Jackknife 1, first-order Jackknife richness estimator [38–41].

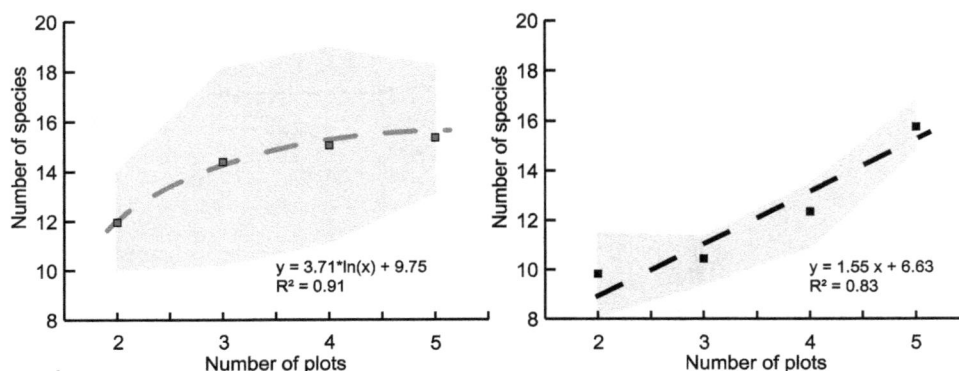

Figure 4. The average estimator based species accumulation curve for one to five plots of near-natural (black) and managed (grey) beech forests. Shadowed areas correspond to the standard deviation.

Chao 2 [41] are promising for the analysis of communities with uneven structure found in heterogenic habitats, like yeasts in a soil environment. In summary, our results suggest that the culture-based approach to estimate the diversity of soil yeasts could be optimised by using 50 mixed soil samples plated in triplicates on a single nutrient-rich medium. Still it must be kept in mind that the number of soil samples to be used as replicates from our estimations might only be applicable to a beech forest located in central Europe and further experiments are needed to evaluate the sampling effort in other soil biotopes.

Implications for diversity assessments of yeasts

Molecular studies on yeast diversity have mainly replaced physiological assimilation tests to identify species as commonly used assimilation tests are not necessarily able to distinguish between closely related species. Additionally, the application of molecular markers has also vastly increased the species numbers in this informal group of fungi [59–61]. Numerous yeast genera were identified containing closely related and morphological cryptic species that seem to have evolved via reproductive isolation [62–64]. The occurrence of cryptic species will also affect biodiversity assessments in a given habitat and therefore studies based only on morphology might underestimate diversity [65]. Indeed, several authors have suggested that next to the application of different media to culture rare species, the exact identification of cryptic diversity by molecular markers is an important factor that will counter underestimation of yeast diversity found in a given area [66–67].

In this study, the used plating strategy resulted in an average of 4.7 colonies per plate, including plates with no yeast colonies. This enabled a reliable differentiation of colonies and adequate isolation of representative strains. Consequently, two new species, described as *Clavispora reshetovae* and *Barnettozyma vustinii*, were found during this survey [68–69]. Out of 18 isolated different phylogenetic species in our study, four belong to the anamorphic yeasts genus *Cryptococcus*. However, only *Cryptococcus ramirezgomezianus* and *Cr. musci* (Trichosporonales) can be interpreted as real cryptic species, as they are difficult to distinguish morphologically [20]. Whether they occur in the same or a similar ecological niche, as could be assumed from their similar assimilation profile [20], still needs further studies. The species pair *Cryptococcus terricola* - *Cr. gastricus* can pose only a minor problem in identification. Although both species belong to the Filobasidiales they are not closely related to each other and exhibit distinct physiological profiles, e.g. assimilation of nitrogen sources [19]. Similarly, *Trichosporon dulcitum*

and *T. porosum* differ phenotypically as they belong to two different clades within the order Trichosporonales, which were suggested to be reclassified into distinct anamorphic genera [20,70]. Based on our results we propose that cryptic species pose no real problem for morpho-physiological biodiversity assessments in temperate forest soils, as we never observed closely related species in the same sample. Nevertheless, this does not eliminate the problem of mislabelling soil-borne yeasts by inappropriate identification tools.

Conclusion

Proper identification of soil fungi using molecular markers is important for diversity estimations and provides a solid background for reliable comparison of results achieved in different surveys. Our study demonstrated that conventional culture-based experiments can be successfully optimised by using mixed soil samples and a single nutrient-rich medium in order to reduce sampling effort in a given substrate or biotope. Additionally, our results demonstrate the usefulness of quantitative estimations to calculate optimal sampling effort in biodiversity- and monitoring-orientated studies. All species richness estimations performed in this study were based on solid species identification, which should be considered in future high throughput analyses as well. The dissimilarities in the yeast community composition observed at the two different sampling dates might hint to seasonal variation in species occurrence and this observation should be addressed in additional studies with more time points.

Supporting Information

Figure S1   Yeast quantity (CFU/g) observed after different pre-cultivation soil treatments. Whiskers correspond to confidential interval and asterisks to outliers.

Table S1   Average frequency of occurrence of yeasts in samples after different pre-cultivation treatments (soils collected in September 2007, see Figure 1) and in relation to the two different management types (all mixed samples, see Figure 1).

Table S2   Species-plot relationship: the number of samples allocated for the analysed combination of plots and the species richness values (Mean and standard deviation).

Acknowledgments

Field work permits were issued by the responsible state environmental offices of Thüringen (according to § 72 BbgNatSchG). We thank the managers of the three exploratories, Swen Renner, Sonja Gockel, Andreas Hemp and Martin Gorke and Simone Pfeiffer for their work in maintaining the plot and project infrastructure, and Markus Fischer, Elisabeth Kalko, Eduard Linsenmair, Dominik Hessenmöller, Jens Nieschulze, Daniel Prati, Ingo Schöning, François Buscot, Ernst-Detlef Schulze and Wolfgang W. Weisser for their role in setting up the Biodiversity Exploratories project. Authors are grateful to Sonja Gockel, Nadine Herold, Marion Schrumpf, Kathrin Henkel, Enrico Weber and other framework participants for exchanging soil samples. We thank Wolfgang Maier and Martin Unterseher for their critical reading and their valuable suggestions on the manuscript. Also we thank Ilse Weßel for the assistance in the laboratory.

Author Contributions

Conceived and designed the experiments: AMY DB. Performed the experiments: AMY. Analyzed the data: AMY. Contributed reagents/materials/analysis tools: AMY MK. Wrote the paper: AMY MK DB.

References

1. Bridge P, Spooner BM (2001) Soil fungi: diversity and detection. Plant Soil 232: 147–154.
2. Fitter AH, Gilligan CA, Hollingworth K, Kleczkowski A, Twyman RM, Pitchford JW (2005) Biodiversity and ecosystem function in soil. Funct Ecol 19: 369–377.
3. Roesch LFW, Fulthorpe RR, Riva A, Casella G, Hadwin AKM, et al. (2007) Pyrosequencing enumerates and contrasts soil microbial diversity. The ISME J 1: 283–290.
4. Fierer N, Strickland MS, Liptzin D, Bradford MA, Cleveland CC (2009) Global patterns in belowground communities. Ecol Lett 12: 1238–1249.
5. Barnett JA (2004) A history of research on yeasts 8: taxonomy. Yeast 21: 1141–1193.
6. Botha A (2011) The importance and ecology of yeasts in soil. Soil Biol Biochem 43: 1–8.
7. Kurtzman CP, Fell JW, Boekhout T (2010) The Yeasts, a Taxonomic Study, 5th edn. Elsevier. 2354 p.
8. Lachance M-A (2006) Yeast biodiversity: how many and how much? In: Peter G, Rosa C, eds. Biodiversity and Ecophysiology of Yeasts, Springer-Verlag, Berlin. pp 1–11.
9. Babjeva IP, Chernov IYu (1995) Geographical aspects of yeast ecology. Physiol Gen Biol Rev 9: 1–54.
10. Chernov IYu (2005) The latitude-zonal and spatial-successional trends in the distribution of yeasts. Zh Obshch Biol 66: 123–135. (English abstract).
11. Vishniac HS (2006) A multivariate analysis of soil yeasts isolated from a latitudinal gradient. Microb Ecol 52: 90–103.
12. Lynch MDJ, Thorn RG (2006) Diversity of basidiomycetes in Michigan agricultural soils. Appl Environ Microbiol 72: 7050–7056.
13. Buée M, Reich M, Murat C, Morin E, Nilsson RH, et al. (2009) 454 Pyrosequencing analyses of forest soils reveal an unexpectedly high fungal diversity. New Phytol 184: 449–456.
14. Lim YW, Kim BK, Kim C, Jung HS, Kim BS, et al. (2010) Assessment of soil fungal communities using pyrosequencing. J Microbiol 48: 284–289.
15. Unterseher M, Schnittler M, Dormann C, Sickert A (2008) Application of species richness estimators for the assessment of fungal diversity. FEMS Microbiol Lett 282: 205–213.
16. Unterseher M, Jumpponen A, Opik M, Tedersoo L, Moora M, Dormann CF, Schnittler M (2011) Species abundance distributions and richness estimations in fungal metagenomics - lessons learned from community ecology. Mol Ecol 20: 275–285.
17. Pozo MI, Herrera CM, Bazaga P (2011) Species richness of yeast communities in floral nectar of southern Spanish plants. Microb Ecol 61: 82–91.
18. Begerow D, Nilsson H, Unterseher M, Maier W (2010) Current state and perspectives of fungal DNA barcoding and rapid identification procedures. Appl Microbiol Biotechnol 87: 99–108.
19. Fonseca A, Scorzetti G, Fell JW (2000) Diversity in the yeast Cryptococcus albidus and related species as revealed by ribosomal DNA sequence analysis. Can J Microbiol 46: 7–27.
20. Takashima M, Sugita T, Shinoda T, Nakase T (2001) Reclassification of the Cryptococcus humicola complex. Int J Syst Evol Microbiol 51: 2199–2210.
21. Takashima M, Sugita T, Shinoda T, Nakase T (2003) Three new combinations from the Cryptococcus laurentii complex: Cryptococcus aureus, Cryptococcus carnescens and Cryptococcus peneaus. Int J Syst Evol Microbiol 53: 1187–1194.
22. Sláviková E, Vadkertiová R (2000) The occurrence of yeasts in the forest soils. J Basic Microb 40: 207–212.
23. Maksimova IA, Chernov IYu (2004) Community structure of yeast fungi in forest biogeocenoses. Microbiology 73: 474–481.
24. Golubtsova YuV, Glushakova AM, Chernov IYu (2007) The seasonal dynamics of yeast communities in the rhizosphere of soddy-podzolic soils. Eurasian Soil Sci 40: 875–879.
25. Wuczkowski M, Prillinger H (2004) Molecular identification of yeasts from soils of the alluvial forest national park along the river Danube downstream of Vienna, Austria ("Nationalpark Donauauen"). Microbiol Res 159: 263–275.
26. Fischer M, Bossdorf O, Gockel S, Hansel F, Hemp A, et al. (2010) Implementing large-scale and long-term functional biodiversity research: The Biodiversity Exploratories. Basic Appl Ecol 11: 473–485.
27. FAO (2006) Guidelines for Soil Description. Food and Agriculture Organisation of the United States, Rome.
28. Ad-hoc-AG Boden (2005) Bodenkundliche Kartieranleitung (KA5). Bundesanstalt für Geowissenschaften und Rohstoffe, Hannover.
29. Yurkov AM, Kemler M, Begerow D (2011) Assessment of yeast diversity in soils under different management regimes. Fungal Ecol;doi: 10.1016/j.funeco.2011.07.004.
30. Sampaio JP, Gadanho M, Bauer R, Weiss M (2003) Taxonomic studies in the Microbotryomycetidae: Leucosporidium golubevii sp. nov., Leucosporidiella gen. nov. and the new orders Leucosporidiales and Sporidiobolales. Mycol Prog 2: 53–68.
31. Cornelissen S, Botha A, Conradie WJ, Wolfaardt GM (2003) Shifts in community composition provide a mechanism for maintenance of activity of soil yeasts in the presence of elevated copper levels. Can J Microbiol 49: 425–432.
32. Brown ME, Burlingham SK, Jackson RM (1962) Studies on Azotobacter species in soil. I. Comparisons of media and techniques for counting Azotobacter in soil. Plant Soil 17: 309–319.
33. LaRue TA, Spencer JFT (1967) The utilization of imidazoles by yeasts. Can J Microbiol 13: 789–794.
34. Danielson RM, Jurgensen MF (1973) The propagule density of Lipomyces and other yeasts in forest soils. Mycopathologia 51: 191–198.
35. Hoffman CS, Winston F (1987) A ten-minute DNA preparation from yeast efficiently releases autonomous plasmids for transformation of Escherichia coli. Gene 57: 267–272.
36. Sampaio JP, Gadanho M, Santos S, Duarte FL, Pais C, et al. (2001) Polyphasic taxonomy of the basidiomycetous yeast genus Rhodosporidium: Rhodosporidium kratochvilovae and related anamorphic species. Int J Syst Evol Microbiol 51: 687–697.
37. Gadanho M, Sampaio JP (2002) Polyphasic taxonomy of the basidiomycetous yeast genus Rhodotorula: Rh. glutinis sensu stricto and Rh. dairenensis comb. nov. FEMS Yeast Res 2: 47–58.
38. Gardes M, Bruns TD (1993) ITS primers with enhanced specificity for basidiomycetes - application to the identification of mycorrhizae and rusts. Molec Ecol 2: 113–118.
39. O'Donnell K (1993) Fusarium and its near relatives. In: Reynolds DR, Taylor JW, eds. The Fungal Holomorph: Mitotic, Meiotic and Pleomorphic Speciation in Fungal Systematics, CAB International, Wallingford, UK. pp 225–233.
40. Colwell RK (2006) Biota: The biodiversity database manager, Version 2. Sinauer Associates, Sunderland, MA.
41. Chao A (1987) Estimating the population size for capture-recapture data with unequal catchability. Biometrics 43: 783–791.
42. Chao A, Lee S-M (1992) Estimating the number of classes via sample coverage. J Am Stat Assoc 87: 210–217.
43. Burnham KP, Overton WS (1979) Robust estimation of population size when capture probabilities vary among animals. Ecology 60: 927–936.
44. Smith EP, Belle G (1984) Nonparametric estimation of species richness. Biometrics 40: 119–129.
45. Chazdon RL, Colwell RK, Denslow JS, Guariguata MR (1998) Statistical methods for estimating species richness of woody regeneration in primary and secondary rain forests of NE Costa Rica. In: Dallmeier F, Comiskey JA, eds. Forest biodiversity research, monitoring and modeling: Conceptual background and Old World case studies, Parthenon Publishing, Paris. pp 285–309.
46. Ugland KI, Gray JS, Ellingsen KE (2003) The species-accumulation curve and estimation of species richness. J Anim Ecol 72: 888–897.
47. Hughes JB, Hellmann JJ, Ricketts TH, Bohannan BJ (2001) Counting the uncountable: Statistical approaches to estimating microbial diversity. Appl Environ Microbiol 67: 4399–4406.
48. Starkey RL, Henrici AT (1927) The occurrence of yeasts in soils. Soil Sci 23: 33–46.
49. di Menna ME (1965) Yeasts in New Zealand soils. New Zeal J Bot 3: 194–203.
50. Boundy-Mills K (2006) Methods for investigating yeast biodiversity. In: Peter G, Rosa C, eds. Biodiversity and Ecophysiology of Yeasts, Springer-Verlag, Berlin. pp 67–100.
51. Collado J, Platas G, Paulus B, Bills GF (2007) High-throughput culturing of fungi from plant litter by a dilution-to-extinction technique. FEMS Microbiol Ecol 60: 521–533.
52. Unterseher M, Schnittler M (2009) Dilution-to-extinction cultivation of leaf-inhabiting endophytic fungi in beech (Fagus sylvatica L.) - different cultivation techniques influence fungal biodiversity assessment. Mycol Res 113: 645–654.

53. Franklin RB, Mills AL (2003) Multi-scale variation in spatial heterogeneity for microbial community structure in an eastern Virginia agricultural field. FEMS Microbiol Ecol 44: 335–346.

54. Golubev VI (2000) Isolation of tremelloid yeasts on glucuronate medium. Microbiology 69: 490–493.

55. Fredlund E, Druvefors U, Boysen ME, Lingsten K-J, Schnürer J (2002) Physiological characteristics of biocontrol yeast *Pichia anomalia* J121. FEMS Yeast Res 2: 395 402.

56. Golubev WI, Sampaio JP (2009) New filobasidiaceous yeasts found in the phylloplane of a fern. J Gen Appl Microbiol 55: 441–446.

57. van der Walt JP (1992) The Lipomycetaceae, a model family for phylogenetic studies. Anton Leeuw Int J G 62: 247–250.

58. Babjeva IP, Gorin SE (1987) Soil yeasts. Moscow: Moscow State University Press. pp 87. (in Russian).

59. Kurtzman CP, Robnett CJ (1998) Identification and phylogeny of ascomycetous yeasts from analysis of nuclear large subunit (26S) ribosomal DNA partial sequences. Anton Leeuw Int J G 73: 331–371.

60. Fell JW, Boekhout T, Fonseca A, Scorzetti G, Statzell-Tallman A (2000) Biodiversity and systematics of basidiomycetous yeasts as determined by large subunit rDNA D1/D2 domain sequence analysis. Int J Syst Evol Microbiol 50: 1351–1371.

61. Scorzetti G, Fell JW, Fonseca A, Statzell-Tallman A (2002) Systematics of basidiomycetous yeasts: a comparison of large subunit D1/D2 and internal transcribed spacer rDNA regions. FEMS Yeast Res 2: 495–517.

62. Sniegowski PD, Dombrowski PG, Fingerman E (2002) *Saccharomyces cerevisiae* and *Saccharomyces paradoxus* coexist in a natural woodland site in North America and display different levels of reproductive isolation from European conspecifics. FEMS Yeast Res 1: 299–306.

63. Sampaio JP, Golubev WI, Fell JW, Gadanho M, Golubev NW (2004) *Curvibasidium cygneicollum* gen. nov., sp. nov. and *Curvibasidium pallidicorallinum* sp. nov., novel taxa in the Microbotryomycetidae (Urediniomycetes), and their relationship with *Rhodotorula fujisanensis* and *Rhodotorula nothofagi*. Int J Syst Evol Microbiol 54: 1401–1407.

64. Marinoni G, Lachance M-A (2004) Speciation in the large-spored *Metschnikowia* clade and establishment of a new species, *Metschnikowia borealis* comb. nov. FEMS Yeast Res 4: 587–596.

65. Schönrogge K, Barr B, Wardlaw JC, Napper E, Gardner MG, et al. (2002) When rare species become endangered: cryptic speciation in myrmecophilous hoverflies. Biol J Linn Soc 75: 291–300.

66. Golubev WI, Sampaio JP, Alves L, Golubeva EW (2005) *Cryptococcus silvicola* nov. sp. from nature reserves of Russia and Portugal. Anton Leeuw Int J G 89: 45–51.

67. Vishniac HS (2006) Yeast biodiversity in the Antarctic. In: Peter G, Rosa C, eds. Biodiversity and Ecophysiology of Yeasts, Springer-Verlag, Berlin. pp 419–440.

68. Yurkov AM, Schäfer AM, Begerow D (2009) *Clavispora reshetovae*. Fungal Planet 35, Persoonia 23: 182–183.

69. Yurkov AM, Schäfer AM, Begerow D (2009) *Barnettozyma vustinii*. Fungal Planet 38, Persoonia 23: 188–189.

70. Okoli I, Oyeka CA, Kwon-Chung KJ, Theelen B, Robert V, et al. (2007) *Cryptotrichosporon anacardii* gen. nov., sp. nov., a new trichosporonoid capsulate basidiomycetous yeast from Nigeria that is able to form melanin on niger seed agar. FEMS Yeast Res 7: 339–350.

15

Diversity in Expression of Phosphorus (P) Responsive Genes in *Cucumis melo* L.

Ana Fita[1]*, Helen C. Bowen[2], Rory M. Hayden[2], Fernando Nuez[1], Belén Picó[1], John P. Hammond[2¤]

1 Centro de Conservación y Mejora de la Agrodiversidad Valenciana, Universitat Politècnica de València, Valencia, Spain, **2** Warwick HRI, University of Warwick, Wellesbourne, Warwick, United Kingdom

Abstract

Background: Phosphorus (P) is a major limiting nutrient for plant growth in many soils. Studies in model species have identified genes involved in plant adaptations to low soil P availability. However, little information is available on the genetic bases of these adaptations in vegetable crops. In this respect, sequence data for melon now makes it possible to identify melon orthologues of candidate P responsive genes, and the expression of these genes can be used to explain the diversity in the root system adaptation to low P availability, recently observed in this species.

Methodology and Findings: Transcriptional responses to P starvation were studied in nine diverse melon accessions by comparing the expression of eight candidate genes (*Cm-PAP10.1, Cm-PAP10.2, Cm-RNS1, Cm-PPCK1, Cm-transferase, Cm-SQD1, Cm-DGD1* and *Cm-SPX2*) under P replete and P starved conditions. Differences among melon accessions were observed in response to P starvation, including differences in plant morphology, P uptake, P use efficiency (PUE) and gene expression. All studied genes were up regulated under P starvation conditions. Differences in the expression of genes involved in P mobilization and remobilization (*Cm-PAP10.1, Cm-PAP10.2 and Cm-RNS1*) under P starvation conditions explained part of the differences in P uptake and PUE among melon accessions. The levels of expression of the other studied genes were diverse among melon accessions, but contributed less to the phenotypical response of the accessions.

Conclusions: This is the first time that these genes have been described in the context of P starvation responses in melon. There exists significant diversity in gene expression levels and P use efficiency among melon accessions as well as significant correlations between gene expression levels and phenotypical measurements.

Editor: John Schiefelbein, University of Michigan, United States of America

Funding: This work was supported by the programs of foreign mobility "Jose Castillejo" [JC2009-00147], Universitat Politecnica de Valencia Professor Mobility Program of the Universitat Politecnica de Valencia [PAID-00-10] and United Kingdom (UK) Department for the Environment, Food and Rural Affairs grant [WQ0119]. The funders had no role in study design, data collection and analysis, decision to publish, or preparation of the manuscript.

Competing Interests: The authors have declared that no competing interests exist.

* E-mail: anfifer@btc.upv.es

¤ Current address: Plant and Crop Sciences Division, University of Nottingham, Loughborough, United Kingdom

Introduction

Phosphorus (P) is major limiting nutrient for plant growth [1], [2]. Therefore, crops are frequently supplied with inorganic phosphate (Pi) fertilizers to maintain yields and quality. However, the application of Pi fertilizers is problematic for both the intensive and extensive agriculture of the developed and developing countries, respectively. In intensive agriculture, excess of soluble inorganic Pi fertilizers is leading to eutrophication and hypoxia of water bodies [3]. In extensive agriculture in the tropics and subtropics, chemically imbalanced soils reduce the availability of P to crops and a lack of infrastructure and purchasing power for fertilizers compound these issues [4]. In addition, since over 85% of mined P is used in food production [5] and consumption of this non-renewable resource will lead to peak phosphorus production (akin to peak oil; [6], [7]), there will be increasing pressures on Pi fertilizer availabilities and costs in the future. These pressures will be exacerbated by increasing demand on food production systems as the human population increases and by fluctuation in oil prices [7]. Sustainable management of P in agriculture requires

developing crops with enhanced P efficiency and management schemes that increase soil Pi availability [8]. Plants have developed adaptive strategies to cope with low soil P availability. These include: i) improvement of Pi-utilization efficiency by enhancing Pi internal remobilization, transport and metabolism, and ii) improvement of Pi-acquisition efficiency by modifying root systems and mobilizing Pi from the soil [1], [9].

Over the past two decades, extensive studies of the response to P starvation using *Arabidopsis thaliana* as a readout phenotype have contributed markedly to the understanding of P signaling and response pathways. In addition, transcriptional profiling of *A. thaliana*, other globally important crops such as *Oryza sativa* (rice), *Lupinus albus* (white lupin), *Phaseolus vulgaris* (common bean), *Brassica rapa* (Chinese cabbage), *Solanum tuberosum* (potato) and *Zea mays* (maize) have extend our knowledge and revealed the complexity of the network of genes necessary for plants to adapt to low soil P availability [10–18]. These studies have revealed a series genes involved in the adaptations to low P, mainly through the regulation of P acquisition, intern remobilization, change in metabolism, and signal transduction [9]. For instance, expression of genes encoding purple acid phosphatases (PAPs) and ribonucleases (RNS) are generally up-regulated under Pi starvation. PAPs are involved in releasing Pi from organic sources, both internally and externally, for efficient transport and subsequent use. They are important in the mobilization of the organic P in soil for root absorption, remobilization of organic P in senescing organs, storage tissues, and intracellular compartments [19,20]. In addition, several studies in arabidopsis, rice and white lupin have demonstrated how Pi starvation affects carbon metabolism [11,21,22], resulting in the accumulation of starch in the leaves and increasing anthocyanin production to protect the photosynthetic machinery. Phosphate starvation also provokes modifications in the lipid composition of plant membranes, including a decrease of phospholipids and an increase of non-phosphorous lipids. Consequently, a typical transcriptional response to Pi starvation is the up regulation of enzymes involved in sulfolipid and galactolipid synthesis [23–26]. In addition, the transcripts and activity of Pi transporters are increased to optimize uptake and remobilization of Pi in Pi-deficient plants [10–12,14,18,27].

Melon (Cucumis melo L.) ranks as the 9th most cultivated horticultural crop in terms of total world production. This species belongs to the botanical family Cucurbitaceae, commonly known as cucurbits. It is an important crop in tropical and subtropical areas, many of which have P-deficient soils. The species is considered to be divided into two subspecies, ssp. melo and ssp. agrestis, each one with several botanical varieties that display a rich morphological diversity [28]. Botanical varieties belonging to the ssp. agrestis are wild or exotic types found in Africa and eastern Asia, from India to Japan, and those belonging to the ssp. melo are mainly cultivated types found from India to Europe and in the Americas. The main ssp. agrestis varieties are conomon (Thunberg) Makino, momordica (Roxburgh) Duthie & Fuller, and the main ssp. melo varieties are dudaim (L.) Naudin, flexuosus (L.) Naudin, cantalupensis Naudin, and inodorus Jacquin, with cantalupensis and inodorus containing most of the commercial varieties [28]. Variability in root morphology and architecture has been described within this species, especially between varieties of both subspecies [29,30]. The relationships between root architecture and response to Pi starvation has also been studied, showing a high variability in the acquisition and use of Pi among melon varieties [31]. However, to our knowledge, there is no information available on the genes involved in the Pi starvation response in melon.

The availability of genetic and genomic resources allows the use of the arabidopsis vast information on P homeostasis related genes on other species [32]. Genomic resources for melon have increased significantly in recent years. These tools include a complete transcriptome, with 53,252 accurately annotated unigenes, assembled from a collection of 125,908 and 689,054 expressed sequence tags (ESTs) [33,34] and 454 sequencing methodologies from a number of accessions belonging to both subspecies [35]. Whole genome sequence of this crop is also in progress [36]. These recent advances make it possible to identify melon orthologues of candidate Pi responsive genes described in other species.

Here we report, diversity in the expression of Pi starvation responsive genes in nine different melon varieties with contrasting phenotypic responses to Pi starvation [31]. Representatives of the two subspecies of melon, the main commercial groups and the exotic types mostly used in melon breeding are included.

Results

Biomass allocation and PUE traits vary significantly between melon accessions

The biomass allocation and PUE traits were analyzed under P replete (control, C) and P starving (NoP) conditions in a genetically diverse set of melon accessions: three accessions belonging to ssp. agrestis (chi-SC and ma-YP of var. conomon, and mo-kha of var. momordica) and six accessions belonging to ssp. melo (cha-PI and flex-Ac of var. flexuosus and four varieties of the main commercial groups, re-Du of var. reticulatus, ca-NC of var. cantalupensis and In-Ps and In-Am of var. inodorus (Table 1).

There were significant differences in root fresh weight (RFW), shoot fresh weight (SFW), root to shoot ratio (RSRa) and root length (RL) among accessions within the control treatment (Figure 1). Accessions belonging to conomon group of ssp. agrestis, chi-SC and ma-YP, in general had low SFW and RFW, and high RSRa, which was not the case of mo-Kha, which also belongs to ssp. agrestis. From the melo ssp., flex-Ac had the highest SFW and RFW, followed by in-Am. Cha-PI, in-PS and re-Du had intermediate values for SFW and RFW, whereas ca-NC displayed morphological traits similar to ma-YP. The lack of P did not produced big reductions in SFW, only mo-kha and flex-Ac showed a significant reduction in SFW between C and NoP treatments (Figure 1A). Despite the tendency of increasing the RFW under NoP condition, only ma-YP, cha-PI and in-Am showed a significant increase in their RFW between C and NoP treatments (Figure 1B). However, the increase of RSRa and RL was significant for all the accessions (except for RL in chi-SC, flex-Ac and re-Du; Figure 1C and D). Although the treatment period was insufficient to significantly reduce SFW, it is clear that the plants are responding to the treatment by altering resource allocation and increasing RSRa. It is interesting to note that flex-Ac had the longest root length in the control treatment, and

Table 1. List of accessions studied and their origin.

Botanical variety[a]	Botanic group[b]	Name[c]	Origin
C. melo ssp.agrestis			
conomon			
chi-SC	chinensis	Songwhan Charmi (PI161375)	Korea
ma-YP	makuwa like	Yamato Purinsu	Japan
momordica			
mo-Kha	momordica	Kharbuja	India
C. melo ssp. melo			
flexuosus			
cha-PI	chate	PI 490388	Mali
flex-Ac	flexuosus	Acuk (PI 167057)	Turkey
cantalupensis			
re-Du	reticulatus	Dulce	USA
ca-NC	cantalupensis	Noir des Carmes	France
inodorus			
in-PS	inodorus	Piel de sapo	Spain
in-Am	inodorus	Amarillo	Spain

[a]Tentative classification according to Munger and Robinson [52].
[b]Tentative classification according to Pitrat [28].
[c]Accessions were kindly supplied by COMAV-UPV, ARS-GRIN-USDA and IPK-Gatersleben germplasm banks.

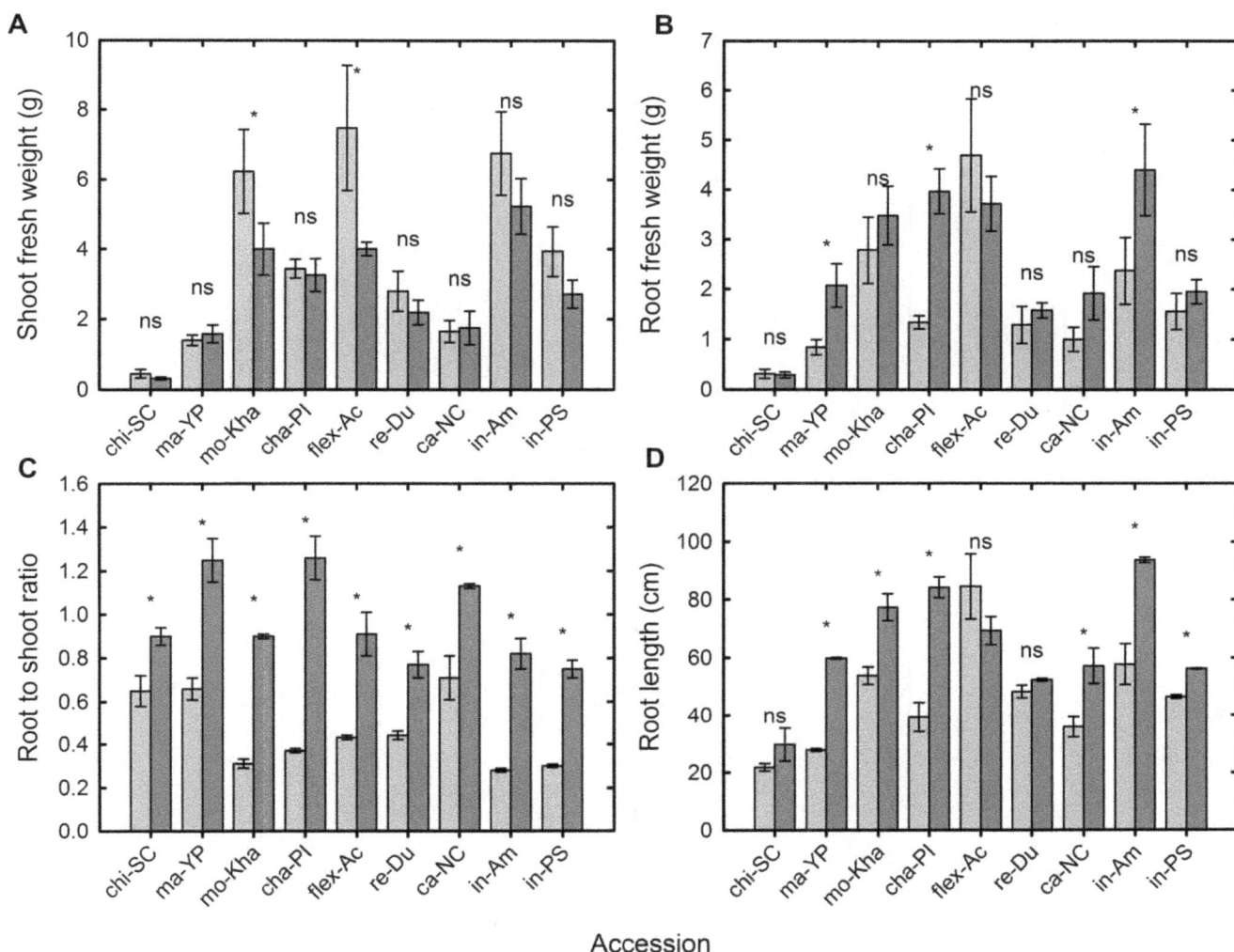

Figure 1. Biomass, root to shoot ratio and root length for nine melon accessions grown hydroponically under P replete and P starved conditions. Mean (a) shoot fresh weight, (b) root fresh weight, (c) root to shoot ratio and (d) root length for melon accessions grown hydroponically with a full nutrient solution (Control, light grey bars) or nutrient solution containing no phosphate (NoP, dark grey) for 21 d. Each bar represents the mean ± se (n = 9). * The mean value in NoP treatment was significantly different at $P<0.05$ from the mean value of the same accession in Control treatment. ns no significant difference.

despite being the only accession to reduce its RL in the NoP treatment still ranked fourth for RL, following in-Am, mo-Kha, and cha-PI, in the NoP treatment (Figure 1D).

There were no significant differences among accessions for shoot P concentration (Shoot-[P], %) in control conditions (Figure 2A), but significant differences were observed for shoot P content (Shoot-P, mg P plant^{-1}), where flex-Ac had the highest Shoot-P (9.48 mg P plant^{-1}; Figure 2B), mo-kha, in-Am, in-PS and cha-PI where intermediate, ranging from 3.28–7.23 mg P plant^{-1}, and chi-Sc, ma-YP and ca-NC had the lowest values (0.98 to 1.97 mg P plant^{-1}). Under NoP conditions, the lowest Shoot-[P] was observed in mo-kha (0.19 P%) and the highest was observed in ca-NC (0.5 P%), followed by in-PS (0.39 P %; Figure 2C). For Shoot-P in NoP conditions, flex-Ac and in-Am had the highest values (1.1 and 1 mg P plant^{-1}, respectively) and chi-SC again had the lowest value (0.11 mg P plant^{-1}; Figure 2D). Taking into account that there was no P in the NoP treatment, these differences in Shoot-P should be considered as differences in

the accumulation of P on the genotypes seeds (all seeds were produced under the same conditions in our greenhouses).

There were significant differences for P use efficiency traits among the melon accessions (Figure 3). Chi-SC, ma-YP and ca-NC had low values for PUpE and PPUE, whereas their values for PUtE were variable, including negative values for ma-YP and a high positive value for Chi-SC. Within the ssp. *agrestis*, mo-Kha had significantly different values for PUpE and PPUE compared with other *agrestis* accessions chi-SC and ma-YP (Figure 3). Within the ssp. *melo*, Flex-Ac had the highest values for PUpE, PUtE and PPUE under control conditions and the second highest for PPUE under the NoP treatment. The *inodorus* accessions also had moderate to high values for the different measures of PUE, with in-Am having the highest value for PPUE under the NoP treatment (Figure 3D). Re-Du and cha-PI, had intermediate values for PUE traits, between *inodorus* and *conomon* groups, cha-PI also had a negative value for PUtE as consequence of its higher SDW under P starvation.

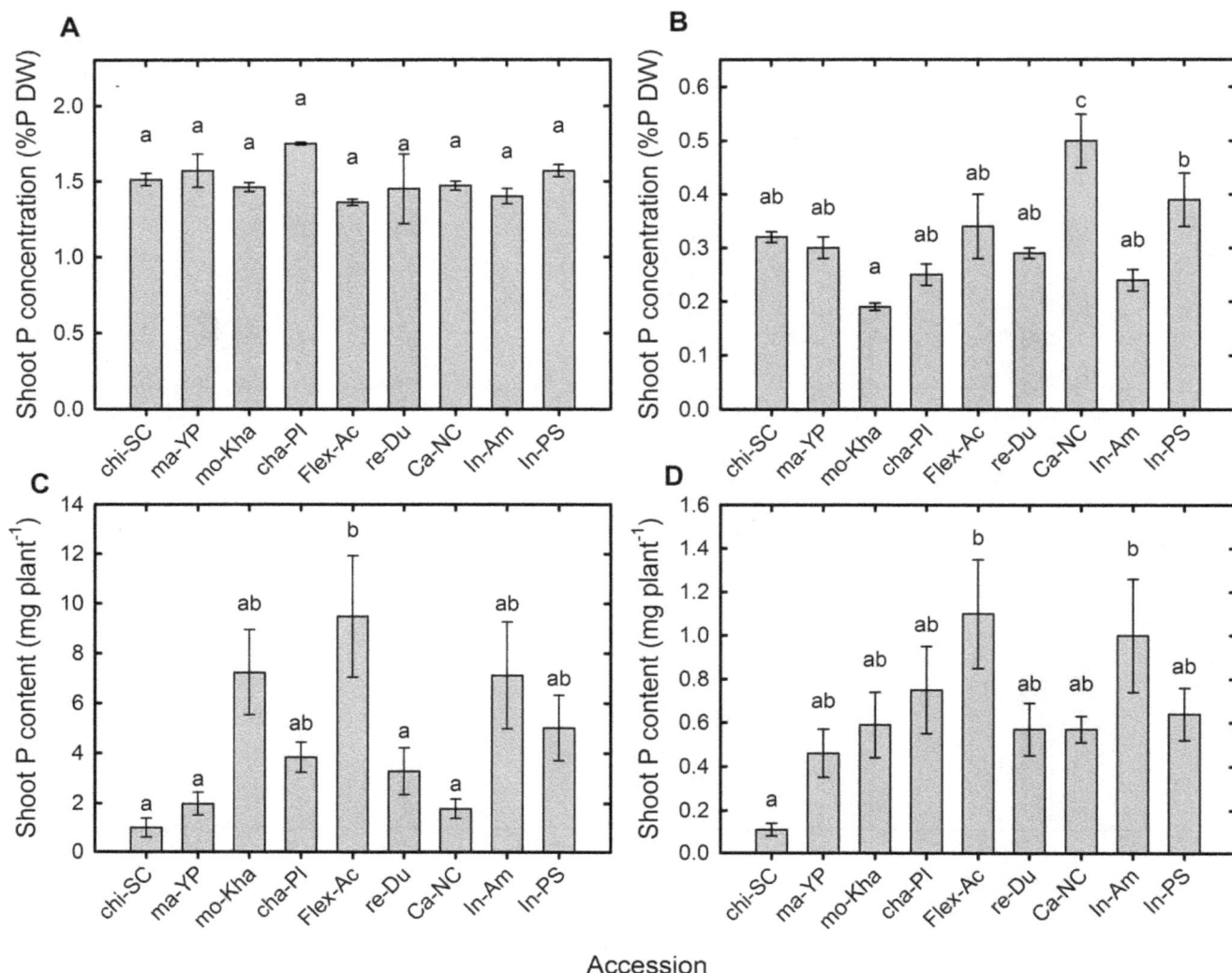

Figure 2. Phosphorus content in shoots of nine melon accessions grown hydroponically under P replete and P starved conditions. Mean (a,b) shoot P concentration and (c,d) shoot P content for melon accessions grown hydroponically with a full nutrient solution (Control, a,c) or nutrient solution containing no phosphate (NoP, b,d) for 21 d. Each bar represents the mean ± se (n = 9). Bars with the same letter are not significantly different at $P<0.05$ by Newmans Keuls multiple range test.

Expression of genes involved in response to P starvation

The expression of genes involved in plant responses to P starvation (phosphorus uptake and mobilization, carbon and secondary metabolism, alteration of membrane lipids composition and P transport) were analyzed under control and NoP conditions in a genetically diverse set of melon accessions.

Phosphate mobilization and re-mobilization

Of the two putative purple acid phosphatases profiled (*Cm-PAP10.1* and *Cm-PAP10.2*, ICUGI unigenes MU46092 and MU50216), the relative expression of *Cm-PAP10.1* was lower than that of *Cm-PAP10.2* in control conditions (Figure 4A and B). Under NoP conditions both genes were induced, with *Cm-PAP10.1* having a greater fold change induction in expression than *Cm-PAP10.2*, with the former induced between 5 to 17-fold, and the latter induced between 1.67 to 6.71-fold (Figure 4B). The relative expression of these two genes in the studied accessions was variable. Chi-SC, re-Du and ca-NC showed low to moderate

relative expression levels in both treatments and genes. Under NoP conditions, mo-Kha, cha-PI, had moderate expression of *Cm-PAP10.1*, but high expression of *Cm-PAP10.2*. In contrast, in-Am and in-PS had high expression of *Cm-PAP10.1* and moderate to low expression of *Cm-PAP10.2* under NoP conditions. Flex-Ac showed the highest relative expression of *Cm-PAP10.1* and a moderate expression of *Cm-PAP10.2* in NoP conditions. Interestingly, despite the low expression of these genes under NoP conditions for chi-SC and ma-YP accessions, they showed a greater induction in expression from control to NoP conditions (Figure 4A and B).

Cm-RNS1 (ICUGI unigen MU47003), which encodes a ribonuclease, was also induced under NoP conditions (Figure 4C). The relative basal expression levels in control conditions were close to one, with a significant increase under NoP conditions. The change of expression was similar among accessions except for chi-SC which showed the lowest increase (7-foldhigher expression in NoP than in the control) and in-PS which showed the highest one (26-fold). The accession with the highest

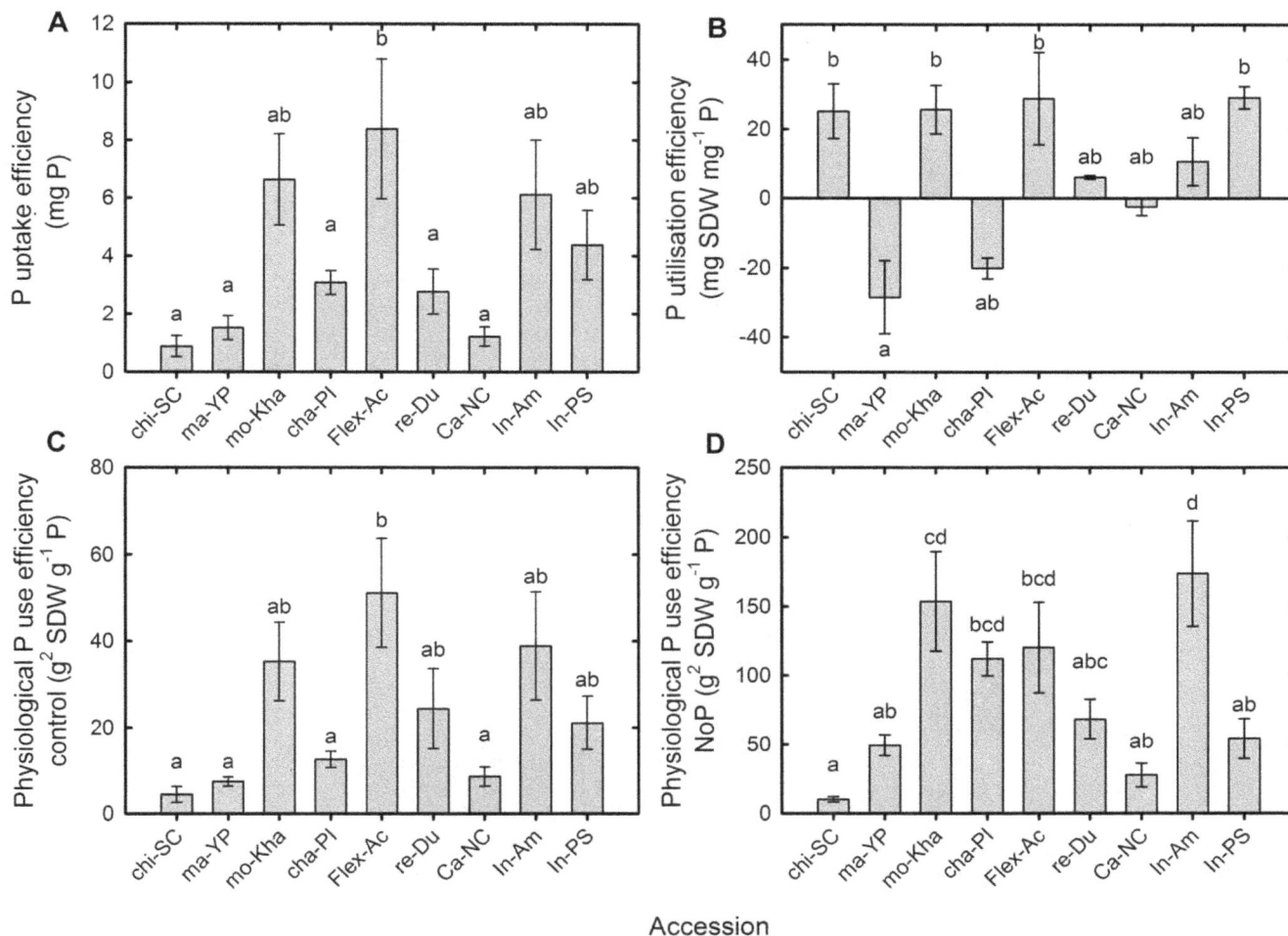

Figure 3. Phosphorus use and uptake measures of nine melon accessions grown hydroponically under P replete and P starved conditions. Mean (a) Phosphorus (P) uptake efficiency, (b) P utilization efficiency, (c) physiological P use efficiency for melon accessions grown hydroponically with a full nutrient solution, and (d) physiological P use efficiency for melon accessions grown hydroponically with a nutrient solution containing no phosphate for 21 d. Each bar represents the mean $\pm$ se (n = 9). Bars with the same letter are not significantly different at $P<0.05$ by Newmans Keuls multiple range test.

relative expression under NoP condition was mo-kha (31.12) followed by flex-Ac and in-Am (Figure 4C).

The expression of these targets putatively involved in P mobilization/remobilization correlated with changes in root structure and measures of PUE (Table 2). The expression of *Cm-PAP10.1* was positively correlated with SFW, RFW and RL in control and NoP treatments, suggesting that higher levels of gene expression are related with higher plant size no matter the treatment. Whereas *Cm-PAP10.2* and *Cm-RNS1* relative expressions showed significant correlations with SFW, RFW and RL, under NoP treatments, indicating that these genes may be associated with improved biomass allocations under NoP conditions. RSRa was negatively correlated with the relative expression of *Cm-PAP10.1* and *Cm-PAP10.2* in control conditions, and accessions with high RSRa, such as chi-SC, ma-YP and ca-NC, had lower basal levels of expression of these genes.

The expression of *Cm-PAP10.1* was also highly positively correlated with shoot P content and PPUE under control conditions but not under NoP conditions. *Cm-PAP10.1* was less expressed in those accessions with small vines and roots in control conditions. These poorly performing accessions had high increases

in the expression of the gene when passing form control to NoP but not great PUpE resulting in a negative correlation between PUpE and *Cm-PAP10.1* induction. Interestingly *Cm-PAP10.2* had different response; it has a negative correlation with [P] and a positive correlation with PPUE under NoP conditions. Therefore, it seems that *Cm-PAP10.1* is correlated with better performance in control conditions and *Cm-PAP10.2* is correlated with higher performance under NoP conditions. *Cm-RNS1* showed similar correlations with other studied traits to *Cm-PAP10.2*, but in this case *Cm-RNS1* was also positively correlated with PUpE, therefore it may contribute to increase uptake in presence of P but also contribute to perform better under NoP conditions.

Carbon and secondary metabolism

The expression of genes known to be involved in carbon and secondary metabolism, were also measured in all nine accessions. These included a phosphoenol pyruvate kinase, *Cm-PPCK1*, and an anthocyanin 5-aromatic acetyltransferase, *Cm-transferase*, (ICUGI unigenes MU52466 and MU47437). Under NoP conditions, the expression of CmPPCK1 increased on average 7-fold across accessions compared with the control treatment

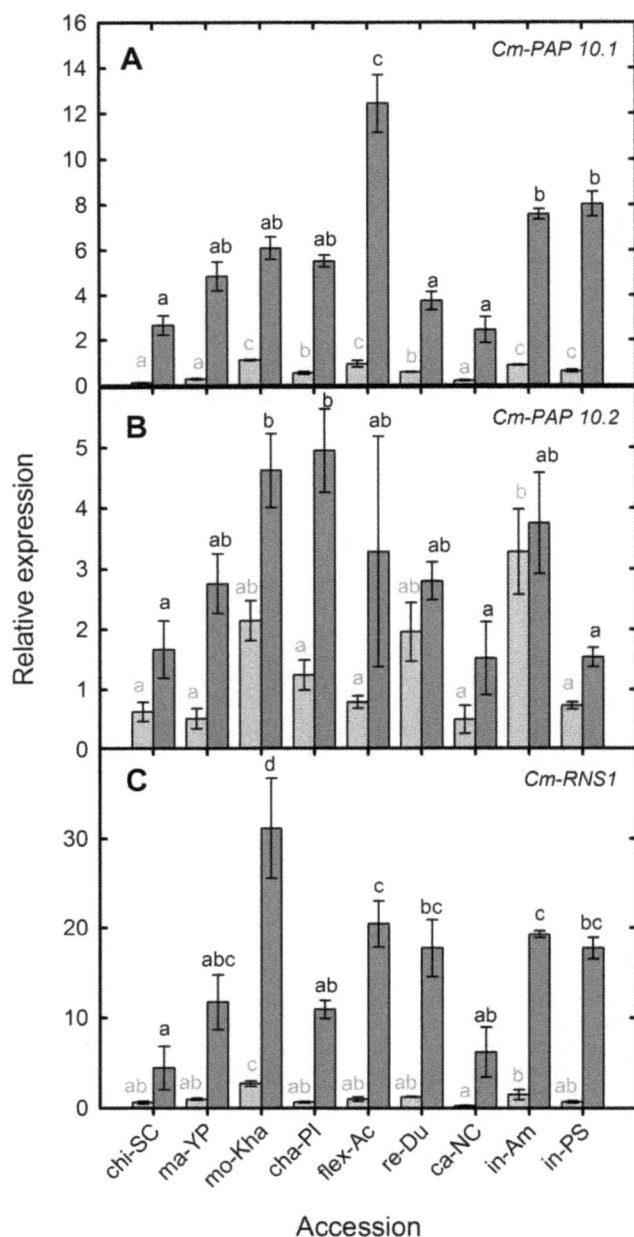

Figure 4. Root transcript abundance for *Cm-PAP10.1*, *Cm-PAP10.2*, *Cm-RNS1* in nine melon accessions under P replete and P starved conditions. Relative transcript abundance for (a) *Cm-PAP10.1* [MU46092], (b) *Cm-PAP10.2* [MU50216], (c) *Cm-RNS1* [MU47003] in different melon accessions grown hydroponically with a full nutrient solution (Control, light grey) or nutrient solution containing no phosphate (NoP, dark grey) for 21 d. Transcript abundance was measured using quantitative PCR (Q-PCR) and expressed relative to that of the housekeeping gene (*Cm-Ubiquitin*). Grey bars with the same grey letter are not significant different at *P*<0.05 by Newmans Keuls multiple range test.

(Figure 5A). Ma-YP and ca-NC showed lower levels of expression under NoP (4.38 and 4.72 respectively) and re-Du had the highest level of expression (12.38) followed by in-PS. Under the NoP condition the expression of *Cm-transferase* increased on average 7-fold across accessions compared with the control treatment

(Figure 5B). There were no large differences in relative expression among accessions under NoP treatment, except for ma-YP (17.81) with the greatest expression, and cha-PI with the lowest (5.33). The highest induction of expression was achieved by flex-AC with a 9-fold increase in expression (Figure 5B).

The only significant correlation between the relative expressions of *Cm-PPCK* and phenotypical traits was a negative correlation with RSRa in both treatments. The expression of *Cm-transferase* was negatively correlated with SFW, RFW, P content and PPUE, under control conditions, indicating that higher expression of this gene under control conditions was associated with low PUE. However, the increased expression of *Cm-transferase* from control to NoP conditions was positively correlated with PUpE and PUtE, suggesting that this gene might be associated with changes in the efficiencies of P uptake and use under NoP conditions.

Alteration membrane lipid composition and metabolism

Genes involved in the manipulation of lipid membrane composition under Pi starvation showed increases in expression under NoP conditions relative to control conditions. The expression of *Cm-DGD1* (ICUGI unigen MU 51583), a putative digalactosyl diacylglycerol synthase, increased its expression approximately 2-fold under NoP conditions (Figure 6A). The accession with highest expression under NoP was chi-SC (8.6) and the lowest were ma-YP (4.71) and flex-Ac (3.65). *Cm-SQD1* (ICUGI unigene MU52028) encodes for a putative UDP sulfoquinovose synthase (SQD1) and was induced 2 to 5-fold under NoP conditions relative to control conditions (Figure 6B). Re-Du, in-Am and in-PS showed the greatest relative expression in NoP (4–5) and ma-Yp with ca-NC the lowest (2).

Cm-DGD1 and *Cm-SQD1* changes in expression were positively correlated with PUtE, indicating that a higher increase in the expression of these genes may be involved in a higher internal P utilization. *Cm-DGD1* was also negative correlated with SFW, RFW, shoot P content and PPUE in control conditions. Indicating that high biomass and PUE accessions under control conditions, have lower basal levels of expression.

P signaling

Cm-SPX2 (ICUGI unigene MU43709) a putative transport protein with a SPX domain was impossible to correctly amplify ca-NC samples. Under NoP conditions, the expression of this gene was 44.5 times higher the basal expression (Figure 7). The higher change of expression was experienced by cha-PI which passed from a basal level of expression of 0.86 to an expression in NoP of 25.78, which in turn was one of the lowest levels of expression in comparison with other accessions. The higher levels of expression in NoP were reached by re-Du and in-PS 44.51 and 45 respectively (Figure 7).

Despite the expression of *Cm-SPX2* was the highest under NoP from all the studied genes, it showed no positive correlations with phenotypical traits, but had a negative correlation with RSRa in both control and NoP conditions.

Discussion

Biomass allocation and PUE traits vary significantly between melon accessions

The responses displayed by the nine melon accessions studied were diverse in terms of plant morphology, P uptake and use. Plants were evaluated in an early stage and, therefore, part of the observed variation might be due to differences in the seed size. Nevertheless, previous studies with these accessions have shown that seed weight affects the initial biomass, especially in NoP

Table 2. Correlation among phenotypical traits and gene expression.

		Cm-PAP10.1[a]	Cm-PAPP10.2	Cm-RNS1	Cm-PPCK1	Cm-transferase	Cm-DGD1	Cm-SQD1	Cm-PHO1
SFW									
	C	0.73***	0.31	0.23	0.29	−039*	−0.40*	0.11	0.35
	NoP	0.51*	0.31	0.65***	0.15	−0.12	−0.36	0.15	0.26
RFW									
	C	0.65***	0.13	0.13	0.12	−0.43*	−0.46*	0.03	0.23
	NoP	0.52*	0.44*	0.53***	−0.03	−0.17	0.32	0.04	0.05
RSRa									
	C	−0.67***	−0.60**	−0.43	−0.68***	−0.05	0.32	−0.28	−0.56*
	NoP	−0.08	0.26	−0.24	−0.54***	−0.05	0.12	−0.55	−0.63*
RL									
	C	0.56***	0.19	0.12	0.21	−0.44	−0.54	0.09	0.26
	NoP	0.43*	0.56*	0.43	−0.11	−0.07	−0.34	−0.13	0.00
P%									
	C	−0.32	−0.30	−0.09	−0.19	0.06	0.11	0.10	−0.11
	NoP	−0.31	−0.77***	−0.54***	0.28	0.02	−0.10	−0.37	−0.46
P									
	C	0.74***	0.26	0.15	0.26	−0.44**	−0.40	−0.1	0.35
	NoP	0.37	−0.12	0.32	0.08	−0.09	−0.52*	−0.02	0.13
PUpE		−0.52***	−0.07	0.61***	−0.22	0.35*	0.22	0.23	−0.32
PUtE		−0.37	−0.26	0.37	0.15	0.37*	0.21*	0.34*	−0.24
PPUE									
	C	0.62***	0.36	0.15	0.28	−0.38*	−0.39*	0.05	0.32
	NoP	0.34	0.51**	0.68***	0.10	−0.06	−0.21	0.15	0.25

Shoot fresh weight (SFW), root fresh weight (RFW), root to shoot ratio (RSRa), root length (RL), P concentration ([P]), shoot P, P uptake efficiency (PUpE), P use efficiency (PUtE) and physiological use efficiency (PPUE) with the relative expression of the studied genes.
[a]Each data is the Pearson correlation of the studied trait with the normalized value of expression of each gene. In PUpE and PUtE the correlation is calculated with the fold-change of expression of the gene from NoP to control treatment.
*,**,***, are significant at P<0.05,0.01 and 0.001 respectively. [ns] non-significant.

conditions, but not other parameters [31]. The studied accessions can be divided into three groups; the first group is composed of chi-SC, ma-YP and ca-NC, which were small plants and had low measures of PUE, but high RSRa; the second group is composed of cha-PI, re-Du and in-PS, which had intermediate values for PUE; the third group is composed by mo-kha, flex-Ac and in-Am, which had greater biomass and higher values of PUE.

Phosphate mobilization and re-mobilization

Transcriptional changes under Pi starvation have been demonstrated in various crop plants [10–17,20,37–44]. Phosphatases and ribonucleases, are important in the mobilization of the organic P in soil for root absorption and remobilization of organic P [19,20,26]. Our results demonstrate an increase in the expression of two purple acid phosphatases and a ribonuclease under Pi starvation in melon (Figure 4). PAPs have different roles in P response, some are secreted and some have internal functions [9,45]. Both melon transcripts studied, Cm-PAP10.1 and Cm-PAP10.2, showed the highest homology to AtPAP10 (At2g16430.2), but they displayed different responses to Pi starvation and different pattern of correlation with other traits, suggesting different roles in Pi nutrition. The fact that Cm-PAP10.1 was associated with improved biomass accumulation and PUE of accessions in control and not under NoP conditions may suggest that it could be a secreted acid phosphatase. Under NoP conditions, the lack of

available Pi or organic sources of P in the hydroponic system would not result in improved acquisition as a consequence of increased expression/secretion of this PAP. In contrast, the correlation between Cm-PAP10.2 expression and PUE, indicate that it could be an internal PAP involved in P efficiency remobilizing Pi under NoP conditions (Table 2). Induction of Cm-RNS1 expression was very intense under NoP conditions (Figure 4C). The orthologue of AtRNS1, encoding a secreted protein up-regulated during Pi starvation, it has been suggested to be involved in Pi mobilization from RNA sources extracellular and intracellular under Pi stress, senescence and wounding [19] [10]. Our correlation results may support a similar role for the melon orthologue described here, but further analysis of RNS and PAPs in melon must be undertaken to define their role in Pi starvation.

Metabolism

Several genes related to glycolysis increase their expression under Pi starvation, increasing the efficiency of this biochemical pathway in terms of Pi [11,21,22]. Cm-PPCK is an orthologue of AtPPCK1 (At1g08650.1). In arabidopsis PPCK genes are rapidly induced by Pi starvation leading to increase phosphorylation of phosphenolpyruvate carboxylase (PEPC). PEPC bypasses the pyruvate kinase step in glycolysis, and the phosphorylation of PEPC reduces its sensitivity to malate inhibition [46]. Increases in the concentration of PPCK have also been reported to be involved

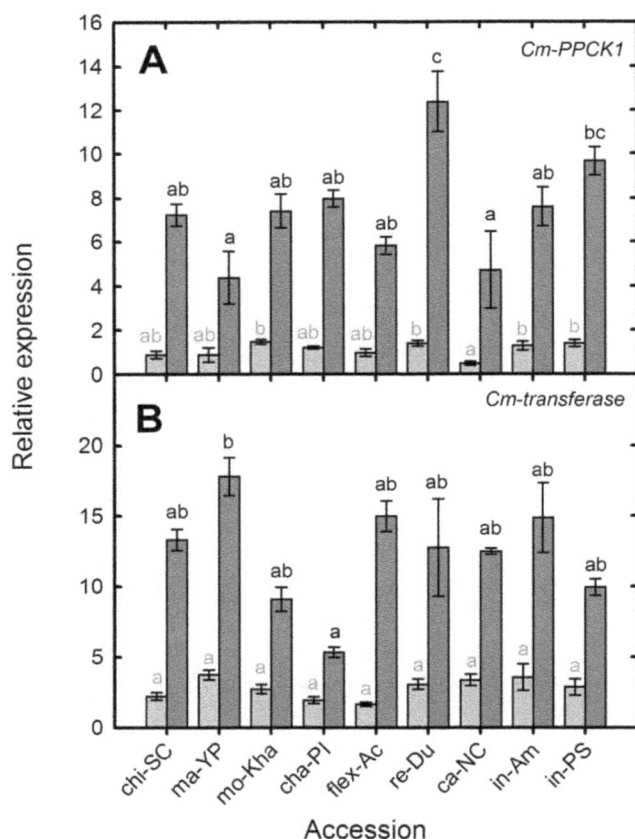

Figure 5. Root transcript abundance for *Cm-PPCK*, *Cm-transferase* in nine diverse melon accessions under P replete and P starved conditions. Relative transcript abundance for (a) *Cm-PPCK1* [MU52466], (b) *Cm-transferase* [MU47437] in different melon accessions grown hydroponically with a full nutrient solution (Control, light grey) or nutrient solution containing no phosphate (NoP, dark grey) for 21 d. Transcript abundance was measured using quantitative PCR (Q-PCR) and expressed relative to that of the housekeeping gene (*Cm-Ubiquitin*). Grey bars with the same grey letter are not significant different at P<0.05 by Newmans Keuls multiple range test. Dark grey bars with the same letter are not significant different at P<0.05 by Newmans Keuls multiple range test.

Figure 6. Relative root transcript abundance for *Cm-DGD1*, (b) *Cm-SQD1* in nine different melon accessions under P replete and P starved conditions. Relative transcript abundance for (a) *Cm-DGD1* [MU51583], (b) *Cm-SQD1* [MU52028] in different melon accessions grown hydroponically with a full nutrient solution (Control, light grey) or nutrient solution containing no phosphate (NoP, dark grey) for 21 d. Transcript abundance was measured using quantitative PCR (Q-PCR) and expressed relative to that of the housekeeping gene (*Cm-Ubiquitin*). Grey bars with the same grey letter are not significant different at P<0.05 by Newmans Keuls multiple range test. Dark grey bars with the same letter are not significant different at P<0.05 by Newmans Keuls multiple range test.

in induction of synthesis and exudation organic acids. Anthocyanin accumulation is a well-documented response to Pi starvation [41], and it is suggested to protect plants against oxidative stress [47]. As expected our transcript *Cm-transferase*, which is an orthologue of the arabidopsis anthocyanin 5-aromatic acetyltransferase gene (At5g39090.1) was upregulated under Pi starvation (Figure 5B). We found differences in the expression of *Cm-PPCK* and *Cm-transferase* genes indicating a different level of response among accessions. These responses were just moderately correlated with P use and efficiency parameters.

Alteration membrane lipid composition and metabolism

Under Pi starvation lipid composition of plant membranes changes drastically, decreasing phospholipids and increasing non-phosphorous lipids [23–26]. In this experiment, we monitored the expression of putative orthologous genes for UDP-sulfoquinovose synthase/SQD1 (At4g33030.1) and UDP-galactosyltransferase/DGD1 (At3g11670.1). Both genes were upregulated under NoP conditions, and as with other genes monitored here there were

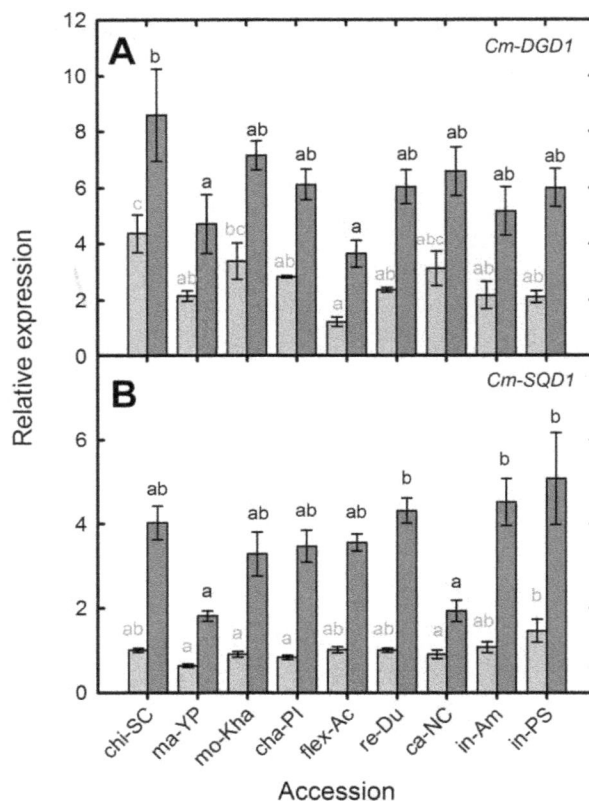

differences in the expression levels among accessions, although they were less intense than for other genes (Figure 6). Nevertheless there was a moderate positive correlation among the best performing accessions between PUtE and expression of *Cm-SQD1* and *Cm-DGD1*.

P sensing and transport

SPX domain containing proteins are involved in Pi transport and sensing [48,49]. The arabidopsis genome contains 20 genes encoding SPX-domain proteins. Our transcript *Cm-SPX2* is orthologue of *AtSPX2* (At2g26660.1), At-SPX2 isoform, is targeted to the nucleus and is weakly induced by Pi starvation [50]. However, it is difficult to drawn inferences about the functional equivalence, if any, among the SPX isoforms between melon and arabidopsis. The *At-PHO1* gene (At3g23430) encodes one SPX protein that is involved in the regulation of Pi homeostasis through the Pi loading to the xylem [48]. In the case of *Cm-SPX2*, it the highest level of induction among the genes studied (Figure 7), but no correlations between its expression and phenotypical traits was observed (Table 2). This may indicate a general response of all the accessions studied despite the different levels of expression.

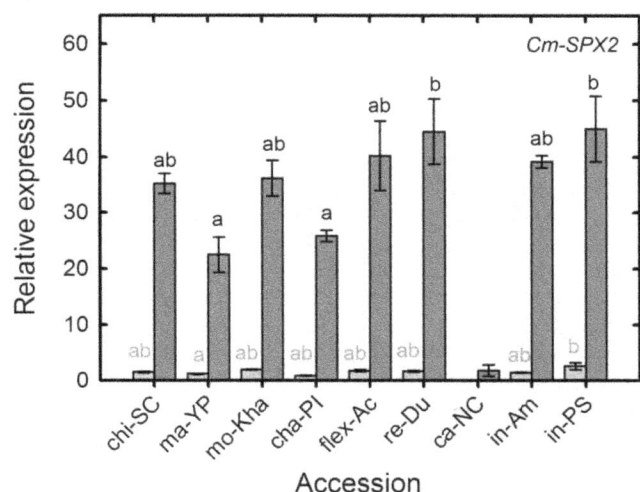

Figure 7. Relative root transcript abundance for *CmSPX2* in nine different melon accessions under P replete and P starved conditions. Relative transcript abundance for *CmSPX2* [MU43709/ MU63649] in different melon accessions grown hydroponically with a full nutrient solution (Control, light grey) or nutrient solution containing no phosphate (NoP, dark grey) for 21 d. Transcript abundance was measured using quantitative PCR (Q-PCR) and expressed relative to that of the housekeeping gene (*CmUbiquitin*). Grey bars with the same grey letter are not significant different at *P*<0.05 by Newmans Keuls multiple range test. Dark grey bars with the same letter are not significant different at *P*<0.05 by Newmans Keuls multiple range test.

Conclusion

The response of plants to Pi starvation is complex involving many genes. Explaining the diversity of Pi starvation responses for different species accessions using nine genes didn't fully account for the phenotypic diversity observed. However, the expression profiles of nine Pi starvation responsive genes in melon have been demonstrated, as well as their differential response among nine diverse accessions. Accessions with higher measures of PUE, such Flex-Ac, in-Am and mo-kha, also had higher *Cm-PAP10.1*, *Cm-PAP10.2* and *Cm-RNS1* expression in NoP than any other accessions, whereas the accessions with lower measures of PUE (chi-SC, ma-YP and ca-NC) had low expression levels for these genes. Therefore higher mobilization and remobilization of P may be a preferential source of diversity in Pi starvation adaptation in melon accessions. The response of other putative Pi responsive genes was also demonstrated to be accession dependent, but they collectively had weaker correlations with measures of biomass accumulation and measures of PUE. Further investigations are required to assess the biochemical impacts of these gene expression changes on their substrates and to characterize additional orthologues of Pi starvation inducible genes for correlations with PUE traits, in the light of new sequence data for melon and related species.

Materials and Methods

Plant material

Nine melon accessions were used in this study. They belong to 5 botanical varieties of both subspecies; three to ssp. *agrestis*: two *conomon* from Japan (chi-SC, and ma-YP), and one *momordica* from India (mo-kha); and six to ssp. *melo*: two *flexuosus* types (cha-PI and flex-Ac), and four varieties of the main commercial groups,

reticulatus (re-Du), *cantalupensis* (ca-NC) and two *inodorus*, Piel de sapo and Amarillo (In-Ps and In-Am; Table 1). These accessions were selected in a previous study as they represent different responses to P starvation in the species [31].

Experimental design

Seeds were germinated on germination paper for 6 d and then transferred to 34 L tanks for hydroponic growth. There were two different treatments with three plants per accession grown in each treatment: i) control nutrient solution (C), composed of: 3 mM KNO_3, 2 mM $Ca(NO_3)_2 \cdot 4H_2O$, 0.5 mM $MgSO_4 \cdot 7H_2O$, 0.5 mM $(NH_4)H_2PO_4$, 25 μM KCl, 12.5 μM H_3BO_3, 1 μM $MnSO_4 \cdot H2O$, 1 μM $ZnSO_4 \cdot 7H_2O$, 0.25 μM $CuSO_4 \cdot 5H_2O$, 1.3 μM $(NH_4)_6MO_7O_{24} \cdot 4H_2O$, and 25 μM Fe-NaEDTA; ii) nutrient solution deficient in Pi (NoP) in which 0.5 mM $(NH_4)_2SO_4$ was added instead of $(NH_4)H_2PO_4$ [31]. The experiment was repeated three independent times.

Plants were grown in a glasshouse compartment at Warwick HRI latitude 52°12′31″N, longitude 1°36′06″W, 46 m above sea level with 16 h light, 26°C day and 20°C night and 60% humidity. Twenty-one days after sowing, plants were harvested and fresh weight of roots and shoots was measured (RFW and SFW respectively) and the root to shoot ratio calculated as RFW/SFW (RSRa). The length of the longest root of the plant was measured (RL). Cotyledons were removed and shoots were oven dried for total P quantification, whereas roots were frozen with liquid nitrogen for RNA extraction.

Total P determination

Shoot dry weight (SDW) was recorded after oven-drying at 60°C for 72 h. Shoot samples were digested by the addition of 2 mL nitric acid to 0.3 g dried, ground material and processed in a closed vessel acid digestion microwave (MARSXpress; CEM Corporation, Matthews, NC, USA). Digested samples were diluted with 23 mL of de-ionised water and analysed using inductively-coupled plasma emission spectrometry (ICP-ES; JY Ultima 2, Jobin Yvon Ltd., Stanmore, Middlesex, UK) to determine shoot P concentrations.

Using the shoot dry weight values and the P concentration in the tissues shoot-[P], several measurements of P-use efficiency (PUE) were calculated [51]:

Total amount of P in the shoot:

Shoot-P (mg P plant^{-1}) = [P]* SDW

P-uptake efficiency (PUpE) was calculated as the increase in total P content:

PUpE (mg P) = ([P_C]*SDW_C)−([P_{NoP}]*SDW_{NoP})

Phosphorus utilization efficiency (PUtE) was calculated as the increase in yield per unit increase in P content:

PUtE (mg SDW mg^{-1} P) = (SDW_C−SDW_{NoP})/ [([P_C]*SDW_C)−([P_{NoP}]*SDW_{NoP})]

Physiological P-use efficiency (PPUE) was calculated as the yield divided by tissue P concentration for either the C or NoP treatment:

PPUE (mg^2 SDW mg^{-1} P) = SDW_C/[P_C] or SDW_{NoP}/[P_{NoP}]

Where SDW is the shoot dry weight for either control (C), or treatments (NoP) and [P] is the tissue P concentration for either control (C), or P starving treatments (NoP).

RNA extraction and reverse transcription

Total RNA was extracted from frozen root tissue using the TRIzol method as described in [10]. To each sample, 1 mL of TRIzol reagent was added, and total RNA was subsequently extracted according to the manufacturer's instructions (Invitrogen, Paisley, UK), with the following modifications: (i) after homogenization with the TRIzol reagent, the samples were centrifuged to remove any remaining plant material and the supernatant was then transferred to a clean Eppendorf tube, and (ii) to aid precipitation of RNA from the aqueous phase, 0.25 mL of isopropanol and 0.25 mL of 1.2 M NaCl solution containing 0.8 M sodium citrate were added. This procedure precipitated the RNA whilst maintaining the proteoglycans and polysaccharides in a soluble form. Extracted total RNA was then purified using the 'RNA Cleanup' protocol for RNeasy columns (Qiagen, Hilden, Germany). RNA yield and purity were determined using a NanoDrop spectrophotometer (Thermo Fisher Scientific Inc, Waltham, MA) and agarose gels.

Reverse transcription was performed on 1 μg of total RNA using the ThermoScript RT-PCR system (Invitrogen, Paisley, UK). The cDNA synthesis reaction was carried out using 0.2 μL random hexamers (50 ng μL^{-1}) and 0.8 μL Oligo(dT)$_{20}$ primer (50 μM) according to the manufacturer's instructions.

Gene identification and primer design

The expression of selected genes, involved in response to P starvation was analyzed by quantitative PCR (Q-PCR). The melon orthologues of genes that are known to be up- or down-regulated under P starvation in *A. thaliana* (Table 3) were identified

in silico using available melon sequence data ([33,35]. BLASTx was used to screen the ICUGI melon EST collection [34].

Primers for quantitative PCR were designed across exon boundaries of the candidate genes with the Primer3, and hairpins and dimer formation were checked with PrimerSelect (DNAS-TAR, Inc., Madison, WI). Primers were also designed for six possible housekeeping genes. The stability of the housekeeping genes was evaluated with samples of all accessions and both treatments with the program GeNorm (http://medgen.ugent.be/~jvdesomp/genorm/). The ubiquitin gene (MU45991) was selected as housekeeping gene because of its stability and good dissociation curve. Primers used for amplifying the selected target genes are listed in Table 3.

Gene expression analysis by quantitative RT-PCR and statistical analysis

Analysis of transcript levels of each gene was determined by quantitative PCR using an ABI Prism 7900 HT (Applied Biosystems, Paisley, UK sequence detector and SYBR Green fluorescent dye (Bioline Ltd., London, UK). For each accession and treatment, cDNA samples from three biological replicates used. Intra-assay variation was evaluated by performing all

Table 3. List of genes analyzed and their function.

Melon Gene	unigene[a]	Primers	Description
Phosphate mobilization			
Cm-PAP10.1	MU46092	aacattctggtttgtcactcctc	AT2G16430.2 PAP10 (Purple acid phosphatase 10); acid phosphatase/protein serine/threonine phosphatase
		tatgcgggtttcgttcgtag	
Cm-PAP10.2	MU50216	catggtcggtcctgatgttc	AT2G16430.2 PAP10 (Purple acid phosphatase 10); acid phosphatase/protein serine/threonine phosphatase
		tgattgtccgcgtaagaaag	
Cm-RNS1	MU47003	gacaggaaaaccaagtgctg	AT2G02990.1 RNS1 (Ribonuclease 1); endoribonuclease/ribonuclease
		tctccatactgctcaccaaatc	
Carbon and secondary metabolism			
Cm-PPCK1	MU52466	cgatgaaactgacaaggaatg	AT1G08650.1 PPCK1 (Phosphoenolpyruvate carboxylase kinase); kinase/protein serine/threonine kinase
		ggtcggaaatcaaacagagg	
Cm-transferase	MU47437	ctttgatttgatgtggctaagg	AT5G39090.1 transferase family protein, similar to anthocyanin 5-aromatic acetyltransferase
		ccagaggaagataatgacgaag	
Alteration membrane lipid composition and metabolism			
Cm-DGD1	MU51583	tcgtaaatggcttgaggaaag	AT3G11670.1 DGD1 (Digalactosyl diacylglycerol deficient 1); UDP-galactosyltransferase/galactolipid galactosyltransferase/transferase, transferring glycosyl groups
		ttggaaggaacaaactgagaag	
Cm-SQD1	MU52028	ccgcacttcatgtttctcag	AT4G33030.1 SQD1; UDPsulfoquinovose synthase/sulfotransferase
		gccatatcaacgtgttccac	
P sensing and transport			
Cm-SPX2	MU43709	gcgagatggttttgttggag	AT2G26660.1 SPX2 (SPX domain gene 2) similar to PHO1 (At-SPX2)
	MU63649	tctgtggtgaagaagggttg	
Control gene			
Cm-Ubiquitin	MU45991	tgtttctaaggtgctgttgtcc	Ubiquitin carrier ligase protein
		cgtgctgttgcttcatacttg	

[a]Sequences available at ICUGI [34].

amplification reactions in technical triplicate. Reactions (10 µL volume) were conducted in 384-well plates consisting of 2-ng cDNA sample, 10 µM forward and reverse primer, and 5 µL of 2× SYBR Green PCR master mix (Bioline Ltd). The quantitative PCR reaction conditions were 50°C (2 min) followed by 95°C (10 min) for 1 cycle, then 95°C (15 s) followed by 60°C (1 min) for 40 cycles. This was followed by a dissociation step of 95°C for 15 s, 60°C for 15 s and 95°C for 15 s. The dissociation step was included to generate data for melting curve analysis so that primer dimers or nonspecific products could be detected in the reaction. A control reaction of each sample was included in each plate using 2 ng of total RNA, to assess the presence of genomic contamination and primer dimers and/or primer contamination

The cycle threshold and normalized fluorescence (ΔR_n) values were determined for each sample during the quantitative PCR cycling reaction using the ABI Prism sequence detector software (v2.0). The cycle threshold value was calculated using a threshold value set at 0.2. A four point standard curve was constructed from 10-fold dilutions of cDNA for each specific gene and the housekeeping gene. Efficiency was calculated from the slope of the linear correlation between Cp (crossing points) values of each dilution and the logarithm of the corresponding amount of RNA,

according to the equation $E = 10^{(-1/slope)} - 1$. Since efficiency for all genes ranged within 95–105, the relative expression of target genes was related to the expression of the housekeeping gene (Cm-$Ubiquitin$) with the equation $R = \dfrac{N_{o,target}}{N_{o,HKG}} = 2^{Cp,HKG - Cp,target}$

Pearson correlations among traits and ANOVA analyses were performed using Statgraphics 5.1 software, Newmans-Keuls multiple Range test was used to separate main effect means when the F-test was significant.

Acknowledgments

The authors thank the COMAV-UPV, ARS-GRIN-USDA, and IPK-Gatersleben germplasm banks for providing many of the accessions studied in this paper.

Author Contributions

Conceived and designed the experiments: AF JH HCB. Performed the experiments: AF HCB RMH. Analyzed the data: AF JH BP. Contributed reagents/materials/analysis tools: JH BP FN. Wrote the paper: AF JH. Critical revision of manuscript: FN.

References

1. Vance CP, Uhde-Stone C, Allan DL (2003) Phosphorus acquisition and use: Critical adaptations by plants for securing a nonrenewable resource. New Phytol 157(3): 423–447.
2. Lynch JP (2011) Root phenes for enhanced soil exploration and phosphorus acquisition: Tools for future crops. Plant Physiol 156(3): 1041–1049.
3. White PJ, Hammond JP (2009) The sources of phosphorus in the waters of great britain RID C-5860-2008. J Environ Qual 38(1): 13–26.
4. Tiessen H (2008) Phosphorus in the global environment. In: White PJ, Hammond JP, eds. The ecophysiology of plant-phosphorus interactions. Dordrecht: Springer. pp 1–7.
5. Heffer P, Prud'Homme MPR, Muirheid B, Isherwood KF (2006) Phosphorus fertilisation:Issues and outlook. In: Anonymous Proceedings. York: International Fertilizer Society.
6. Raven JA (2008) Phosphorus and the future. In: White PJ, Hammond JP, eds. The ecophysiology of plant-phosphorus interactions Springer. pp 271–283.
7. Cordell D, Drangert J, White S (2009) The story of phosphorus: Global food security and food for thought. Global Environmental Change-Human and Policy Dimensions 19(2): 292–305. 10.1016/j.gloenvcha.2008.10.009.
8. Lynch JP (2007) Roots of the second green revolution. Aust J Bot 55(5): 493–512. 10.1071/BT06118 ER.
9. Fang ZY, Shao C, Meng YJ, Wu P, Chen M (2009) Phosphate signaling in arabidopsis and oryza sativa. Plant Science 176(2): 170–180. 10.1016/j.plantsci.2008.09.007 ER.
10. Hammond JP, Bennett MJ, Bowen HC, Broadley MR, Eastwood DC, et al. (2003) Changes in gene expression in arabidopsis shoots during phosphate starvation and the potential for developing smart plants. Plant Physiol 132(2): 578–596. 10.1104/pp.103.020941.
11. Hammond JP, White PJ (2011) Sugar signaling in root responses to low phosphorus availability. Plant Physiol 156(3): 1033–1040.
12. Hammond JP, Mayes S, Bowen HC, Graham NS, Hayden RM, et al. (2011) Regulatory hotspots are associated with plant gene expression under varying soil phosphorus supply in brassica rapa. Plant Physiol 156(3): 1230–1241.
13. Wasaki J, Shinano T, Onishi K, Yonetani R, Yazaki J, et al. (2006) Transcriptomic analysis indicates putative metabolic changes caused by manipulation of phosphorus availability in rice leaves. J Exp Bot 57(9): 2049–2059.
14. Uhde-Stone C, Zinn KE, Ramirez-Yanez M, Li AG, Vance CP, et al. (2003) Nylon filter arrays reveal differential gene expression in proteoid roots of white lupin in response to phosphorus deficiency. Plant Physiol 131(3): 1064–1079. 10.1104/pp.102.016881.
15. Uhde-Stone C, Gilbert G, Johnson JMF, Litjens R, Zinn KE, et al. (2003) Acclimation of white lupin to phosphorus deficiency involves enhanced expression of genes related to organic acid metabolism. Plant Soil 248(1–2): 99–116.
16. Hernandez G, Ramirez M, Valdes-Lopez O, Tesfaye M, Graham MA, et al. (2007) Phosphorus stress in common bean: Root transcript and metabolic responses. Plant Physiol 144(2): 752–767. 10.1104/pp.107.096958.
17. Calderon-Vazquez C, Ibarra-Laclette E, Caballero-Perez J, Herrera-Estrella L (2008) Transcript profiling of zea mays roots reveals gene responses to phosphate deficiency at the plant- and species-specific levels. J Exp Bot 59(9): 2479–2497.
18. Wasaki J, Yonetani R, Kuroda S, Shinano T, Yazaki J, et al. (2003) Transcriptomic analysis of metabolic changes by phosphorus stress in rice plant roots. Plant Cell and Environment 26(9): 1515–1523.
19. Bariola PA, Howard CJ, Taylor CB, Verburg MT, Jaglan VD, et al. (1994) The arabidopsis ribonuclease gene Rns1 is tightly controlled in response to phosphate limitation. Plant Journal 6(5): 673–685.
20. Duff SMG, Sarath G, Plaxton WC (1994) The role of acid-phosphatases in plant phosphorus-metabolism. Physiol Plantarum 90(4): 791–800.
21. Hammond JP, White PJ (2008) Sucrose transport in the phloem: Integrating root responses to phosphorus starvation. J Exp Bot 59(1): 93–109. 10.1093/jxb/erm221.
22. Plaxton WC, Tran HT (2011) Metabolic adaptations of phosphate-starved plants. Plant Physiol 156(3): 1006–1015.
23. Essigmann B, Guler S, Narang RA, Linke D, Benning C (1998) Phosphate availability affects the thylakoid lipid composition and the expression of SQD1, a gene required for sulfolipid biosynthesis in arabidopsis thaliana. Proc Natl Acad Sci U S A 95(4): 1950–1955.
24. Andersson MX, Stridh MH, Larsson KE, Lijenberg C, Sandelius AS (2003) Phosphate-deficient oat replaces a major portion of the plasma membrane phospholipids with the galactolipid digalactosyldiacylglycerol. FEBS Lett 537(1–3): 128–132.
25. Tjellstrom H, Sandelius AS (2008) Lipid asymmetry in root plasma membrane during phosphate limiting conditions. Chem Phys Lipids 154: S30–S30.
26. Andersson M, Larsson K, Tjellstrom H, Liljenberg C, Sandelius AS (2005) The plasma membrane and the tonoplast as major targets for phospholipid- to-glycolipid replacement and stimulation of phospholipases in the plasma membrane RID E-7958-2010. J Biol Chem 280(30): 27578–27586.
27. Hammond JP, Broadley MR, White PJ (2004) Genetic responses to phosphorus deficiency. Annals of Botany 94(3): 323–332. 10.1093/aob/mch156.
28. Pitrat M (2008) Melon (cucumis melo L.). In: Prohens J, Nuez F, eds. Handbook of Plant Breeding. Vegetables I. New York: Springer. pp 287–314.
29. Fita A, Pico B, Nuez F (2006) Implications of the genetics of root structure in melon breeding. J Am Soc Hort Sci 131(3): 372–379.
30. Fita A, Pico B, Monforte AJ, Nuez F (2008) Genetics of root system architecture using near-isogenic lines of melon. J Am Soc Hort Sci 133(3): 448–458.
31. Fita A, Nuez F, Picó B (2011) Diversity in root architecture and response to P deficiency in seedlings of cucumis melo L. Euphytica pp 1–17.
32. Ding G, Liao Y, Yang M, Zhao Z, Shi L, et al. (2011) Development of gene-based markers from functional arabidopsis thaliana genes involved in phosphorus homeostasis and mapping in brassica napus. Euphytica 181(3): 305–322.
33. Gonzalez-Ibeas D, Blanca J, Roig C, Gonzalez-To M, Pico B, et al. (2007) MELOGEN: An EST database for melon functional genomics. BMC Genomics 8: 306.
34. International Cucurbit Genomics Initiative (ICuGI) (2007) Cucurbit genomics database. Available: http://www.icugi.org/ via the Internet. Accessed 2011 Dec 22.
35. Blanca JM, Cañizares J, Ziarsolo P, Esteras C, Mir G, et al. (2011) Melon transcriptome characterization: Simple sequence repeats and single nucleotide polymorphisms discovery for high throughput genotyping across the species. The Plant Genome 4(2): 118–131. 10.3835/plantgenome2011.01.0003.

36. Gonzalez VM, Rodriguez-Moreno L, Centeno E, Benjak A, Garcia-Mas J, et al. (2010) Genome-wide BAC-end sequencing of cucumis melo using two BAC libraries. BMC Genomics 11.

37. Wu P, Ma LG, Hou XL, Wang MY, Wu YR, et al. (2003) Phosphate starvation triggers distinct alterations of genome expression in arabidopsis roots and leaves. Plant Physiol 132(3): 1260–1271. 10.1104/pp.103.021022.

38. Misson J, Raghothama KG, Jain A, Jouhet J, Block MA, et al. (2005) A genome-wide transcriptional analysis using arabidopsis thaliana affymetrix gene chips determined plant responses to phosphate deprivation. Proc Natl Acad Sci U S A 102(33): 11934–11939.

39. Graham MA, Ramirez M, Valdes-Lopez O, Lara M, Tesfaye M, et al. (2006) Identification of candidate phosphorus stress induced genes in *phaseolus vulgaris* through clustering analysis across several plant species. Functional Plant Biology 33(8): 789–797. 10.1071/FP06101.

40. Morcuende R, Bari R, Gibon Y, Zheng W, Pant BD, et al. (2007) Genome-wide reprogramming of metabolism and regulatory networks of arabidopsis in response to phosphorus. Plant Cell and Environment 30(1): 85–112. 10.1111/j.1365-3040.2006.01608.x.

41. Muller R, Morant M, Jarmer H, Nilsson L, Nielsen TH (2007) Genome-wide analysis of the arabidopsis leaf transcriptome reveals interaction of phosphate and sugar metabolism. Plant Physiol 143(1): 156–171. 10.1104/pp.106.090167.

42. Hernandez G, Valdes-Lopez O, Ramirez M, Goffard N, Weiller G, et al. (2009) Global changes in the transcript and metabolic profiles during symbiotic nitrogen fixation in phosphorus-stressed common bean plants. Plant Physiol 151(3): 1221–1238. 10.1104/pp.109.143842.

43. Zheng L, Huang F, Narsai R, Wu J, Giraud E, et al. (2009) Physiological and transcriptome analysis of iron and phosphorus interaction in rice seedlings. Plant Physiol 151(1): 262–274.

44. Li L, Qiu X, Li X, Wang S, Zhang Q, et al. (2010) Transcriptomic analysis of rice responses to low phosphorus stress. Chinese Science Bulletin 55(3): 251–258.

45. Zhang Q, Wang C, Tian J, Li K, Shou H (2011) Identification of rice purple acid phosphatases related to posphate starvation signalling. Plant Biology 13(1): 7–15.

46. Chen Z, Nimmo GA, Jenkins GI, Nimmo HG (2007) BHLH32 modulates several biochemical and morphological processes that respond to P-i starvation in arabidopsis. Biochem J 405: 191–198.

47. Nagata T, Todoriki S, Masumizu T, Suda I, Furuta S, et al. (2003) Levels of active oxygen species are controlled by ascorbic acid and anthocyanin in arabidopsis. J Agric Food Chem 51(10): 2992–2999.

48. Wang Y, Ribot C, Rezzonico E, Poirier Y (2004) Structure and expression profile of the arabidopsis *PHO1* gene family indicates a broad role in inorganic phosphate homeostasis. Plant Physiol 135(1): 400–411.

49. Hamburger D, Rezzonico E, Petetot JMC, Somerville C, Poirier Y (2002) Identification and characterization of the arabidopsis *PHO1* gene involved in phosphate loading to the xylem. Plant Cell 14(4): 889–902.

50. Duan K, Yi KK, Dang L, Huang HJ, Wu W, et al. (2008) Characterization of a sub-family of arabidopsis genes with the SPX domain reveals their diverse functions in plant tolerance to phosphorus starvation. Plant Journal 54(6): 965–975.

51. Hammond JP, Broadley MR, White PJ, King GJ, Bowen HC, et al. (2009) Shoot yield drives phosphorus use efficiency in brassica oleracea and correlates with root architecture traits. J Exp Bot 60(7): 1953–1968. 10.1093/jxb/erp083.

52. Munger HM, Robinson RW (1991) Nomenclature of *cucumis melo* L. Cucurbit Genetics Cooperative Reports 14: 43–44.

16

Changes in N-Transforming Archaea and Bacteria in Soil during the Establishment of Bioenergy Crops

Yuejian Mao[1,2], Anthony C. Yannarell[1,2,3], Roderick I. Mackie[1,2,4]*

1 Energy Biosciences Institute, University of Illinois, Urbana, Illinois, United States of America, **2** Institute for Genomic Biology, University of Illinois, Urbana, Illinois, United States of America, **3** Department of Natural Resources and Environmental Sciences, University of Illinois, Urbana, Illinois, United States of America, **4** Department of Animal Sciences, University of Illinois, Urbana, Illinois, United States of America

Abstract

Widespread adaptation of biomass production for bioenergy may influence important biogeochemical functions in the landscape, which are mainly carried out by soil microbes. Here we explore the impact of four potential bioenergy feedstock crops (maize, switchgrass, *Miscanthus X giganteus*, and mixed tallgrass prairie) on nitrogen cycling microorganisms in the soil by monitoring the changes in the quantity (real-time PCR) and diversity (barcoded pyrosequencing) of key functional genes (*nifH*, bacterial/archaeal *amoA* and *nosZ*) and 16S rRNA genes over two years after bioenergy crop establishment. The quantities of these N-cycling genes were relatively stable in all four crops, except maize (the only fertilized crop), in which the population size of AOB doubled in less than 3 months. The nitrification rate was significantly correlated with the quantity of ammonia-oxidizing archaea (AOA) not bacteria (AOB), indicating that archaea were the major ammonia oxidizers. Deep sequencing revealed high diversity of *nifH*, archaeal *amoA*, bacterial *amoA*, *nosZ* and 16S rRNA genes, with 229, 309, 330, 331 and 8989 OTUs observed, respectively. Rarefaction analysis revealed the diversity of archaeal *amoA* in maize markedly decreased in the second year. Ordination analysis of T-RFLP and pyrosequencing results showed that the N-transforming microbial community structures in the soil under these crops gradually differentiated. Thus far, our two-year study has shown that specific N-transforming microbial communities develop in the soil in response to planting different bioenergy crops, and each functional group responded in a different way. Our results also suggest that cultivation of maize with N-fertilization increases the abundance of AOB and denitrifiers, reduces the diversity of AOA, and results in significant changes in the structure of denitrification community.

Editor: Jack Anthony Gilbert, Argonne National Laboratory, United States of America

Funding: This work was funded by the Energy Biosciences Institute, Environmental Impact and Sustainability of Feedstock Production Program at the University of Illinois, Urbana. The funders had no role in study design, data collection and analysis, decision to publish, or preparation of the manuscript.

Competing Interests: The authors have declared that no competing interests exist.

* E-mail: r-mackie@illinois.edu

Introduction

Bioenergy derived from cellulosic ethanol is a potential sustainable alternative to fossil fuel-based energy, since the energy from green plants is renewable and largely carbon neutral in comparison to fossil fuel combustion. Perennial grasses, such as switchgrass (*Panicum virgatum*) and *Miscanthus×giganteus*, with large annual biomass production potential, are proposed as biofuel feedstocks that can maximize ethanol production without adversely affecting the market for food crops (e.g. maize). However, our knowledge of the impacts of various bioenergy feedstock production systems on the soil microbial ecosystem is still very limited. The chemistry of perennial crop residues and plant root exudates may stimulate or inhibit the growth and activity of different fractions of the soil microbial community, and thus the planting of different crops can result in distinct microbial communities [1,2,3]. Differences in management techniques between traditional row-crop agriculture and perennial biomass feedstocks represent different soil disturbance regimes, altered water use, differing rates of fertilizer application, etc., and these should have a direct impact on soil microbial dynamics, subsequently influencing the terrestrial biogeochemical cycles. In particular, we predict that the cultivation of high nitrogen-use

efficiency perennial grasses will result in altered nitrogen-transforming microbial communities in comparison to those found under N-fertilized maize.

The biological nitrogen cycle is one of the most important nutrient cycles in the terrestrial ecosystem. It includes four major processes: nitrogen fixation, mineralization (decay), nitrification and denitrification. Because many of the microorganisms responsible for these processes are recalcitrant to laboratory cultivation, previous studies of the distribution and diversity of nitrogen-transforming microorganisms have employed cultivation-independent techniques targeting functional genes: *nifH*, *amoA* and *nosZ* genes, which encode the key enzymes in nitrogen fixation, ammonia oxidization and complete denitrification, respectively [4,5,6,7,8,9].

Biological nitrogen fixation, which converts atmospheric N_2 into ammonium that is available to organisms, is an important natural input of available nitrogen in many terrestrial habitats [10]. Although nitrogen fixation in terrestrial ecosystems is thought to be mainly carried out by the symbiotic bacteria in association with plants, free-living diazotrophs in soils can play important roles in N cycling in a number of ecosystems [11,12]. In average, 2–3 kg N ha^{-1} year^{-1} could be imported by free living N-fixers [13]. Various field experiments have shown that the biomass yield of one candidate biofuel feedstock crop, *Miscanthus×giganteus*, is not significantly increased by the addition of mineral N fertilizer [14]. The lack of response to nitrogen fertilization and the high biomass production suggest that biological nitrogen fixation may play an important role in supplying the nitrogen needs of *Miscanthus* [15]. Plant species have previously been shown to have a significant effect on the composition of diazotrophs in the field; for example, diazotroph diversity is higher in soil under *Acacia tortilis* ssp. *raddiana* (a leguminous tree) than *Balanites aegyptiaca* (a non-leguminous tree) [16]. Plant genotype also has a strong effect on the rhizosphere diazotrophs of rice [17]. Agronomic practices can also influence soil diazotrophs, e.g. application of N-fertilization can reduce the diversity of diazotrophs [17]. Therefore, we hypothesize that the cultivation of maize with inorganic N-fertilizer will reduce the abundance and diversity of diazotrophs in the soil ecosystem, while biofuel feedstocks receiving little or no N-fertilizer (e.g. *Miscanthus*) will encourage the development of active diazotrophic communities.

Nitrification, which converts ammonium to nitrate, includes two steps: ammonia oxidation to nitrite, and nitrite oxidation to nitrate. The production of nitrate in soil not only supplies nutrition for plants, but it can also mobilize nitrogen to groundwater through nitrate leaching. Ammonia oxidation is the first and rate-limiting step of nitrification [18]. It is typically thought to be carried out by a few groups in β- and γ- Proteobacteria, referred to as ammonia-oxidizing bacteria (AOB) [18]. However, recent environmental metagenomic analyses revealed that ammonia monooxygenase α-subunit (*amoA*) genes are also present in archaea (AOA) [19], and archaeal *amoA* has been shown to be widespread in many environments, e.g. soils, hot springs and marine water [6,19,20,21,22]. Recent work has found that AOA can be up to 3000 times more abundant in soil than AOB [22,23,24], meaning that AOA are the most abundant ammonia oxidizing organisms in soil ecosystems [25]. The soil ammonia oxidizing community is known to be influenced by plant types and management, but different segments of this community respond differently [24,26,27]. For example, the abundance and composition of AOB is significantly altered by long-term fertilization, but AOA are rarely affected [24,27]. The nitrification activity in soil ecosystems is known to be correlated with the abundances and structures of ammonia oxidizers [24,28,29]. We therefore hypothesize that different biofuel cropping systems, especially those that rely on N-fertilization, will influence the composition of ammonium oxidizers in soil, with potential consequences for nitrification rates.

Denitrification, which reduces nitrate to N$_2$ gas, is carried out by a diverse group of microorganisms belonging to more than 60 genera of bacteria, archaea, and some eukaryotes [30]. Complete denitrification involves four steps: $NO_3^-{\rightarrow}NO_2^-{\rightarrow}NO{\rightarrow}N_2O{\rightarrow}N_2$. The enzyme nitrous oxide reductase (encoded by *nosZ*) that reduces N$_2$O to N$_2$ is essential for complete conversion of NO_3^- to N$_2$. Approximately 17 Tg N is estimated to be lost from the land surface through denitrification every year [31]. It is known that the structure and activity of denitrifiers in the terrestrial ecosystem could be significantly influenced by the plant species [7,29,32]. In a study of a maize-cropped field, it was found that organic or mineral fertilizer applications could affect both the structure and activity of the denitrifying community in the long term, with changes persisting for at least 14 months [33]. The potential denitrification rate was found to be significantly correlated to the denitrifier density, as estimated by the quantification of *nosZ* gene copy numbers [34]. Denitrification releases mineralized nitrogen in the soil ecosystem to the atmosphere, and thus, the balance between denitrification and N-fixation, can determine the biologically available N for the biosphere (Arp, 2000).

It is known that plant species can change the soil microbial community [1]. However, while much previous work has examined the microbial community differences between the established crops [7,29,34,35,36], less is known about how microbial communities in the agricultural soils develop during the transition from one cropping system to another (e.g. annual row crops to perennial biofuel feedstocks). Thus, to improve our knowledge of the effects of bioenergy feedstock production on the complex N-cycling microbial communities of terrestrial ecosystems, we followed the changes in soil microbial communities during a two-year establishment period of maize, switchgrass, *Miscanthus×giganteus*, and mixed tallgrass prairie. We monitored the abundance of key genes for nitrogen fixation, ammonia oxidation and complete denitrification (*nifH*, bacterial/archaeal *amoA* and *nosZ* as well as the structural changes of these N-cycling genes and bacterial/archaeal 16S rRNA genes using real-time PCR and barcoded pyrosequencing methods respectively.

Materials and Methods

Study site and sampling

The experiment was conducted at the Energy Biosciences Institute's Energy Farm located southwest of Urbana, Illinois, USA. Miscanthus (*Miscanthus×giganteus*, MG), switchgrass (*Panicum virgatum*, PV), maize (*Zea mays*, ZM) and restored tallgrass prairie (used as control, NP) were planted in the spring of 2008. Miscanthus was replanted in the spring of 2009 due to its poor growth in 2008. Each crop was planted in a randomized block design, with a 0.7-ha plot of each crop randomly positioned within four blocks (n = 4 for each crop). Samples were collected in April 2008, before planting of these crops, in order to characterize the background soil microbial communities. Bulk soil samples were collected at monthly intervals during the growing seasons (June–September 2008 and 2009). 10 soil cores (0–10 cm depth, 1.8 cm diameter) were collected from each plot and homogenized in an ethanol-sanitized, plastic bucket. About 60 g of the well-mixed soil was then subsampled into a 50 mL tube for each plot, and kept on ice until brought to the lab and stored in a −80°C freezer. In total, 112 soil samples were collected. Following standard agricultural practices, only maize was fertilized (17 g/m^2 in 2008, 20 g/m^2 in 2009) with a mixture of urea, ammonia and nitrate (28% UAN). Herbicides were applied in these crops except the restored tallgrass prairie: 1.56 l/ha of Roundup (only applied in 2008) and 4.13 l/ha of Lumax for maize; 1.37 l/ha of 2,4-Dichlorophenoxyacetic acid for Miscanthus; 1.37 l/ha of 2,4-Dichlorophenoxyacetic acid for switchgrass in 2008.

DNA extraction and purification

Soil samples were freeze-dried overnight until completely dry and then manually homogenized with a sterile screwdriver. DNA was extracted from 0.3 g soil using the FastDNA SPIN Kit For Soil (MP Biomedicals) according to manufacturer's protocol. Extracted DNA was then purified using CTAB. DNA concentrations were determined using the Qubit quantification platform with Quant-iT™ dsDNA BR Assay Kit (Invitrogen). DNA was diluted to 10 ng/μL and stored in −80°C freezer for the following molecular applications.

Real-time PCR

The abundances of *nifH*, archaeal *amoA*, bacterial *amoA* and *nosZ* genes in all the soil samples were quantified using real-time PCR. Quantitative real-time PCR was performed according to the methods modified from previous studies: *nifH* (as a measure of N-fixing bacteria) used primers PolF (5′- TGC GAY CCS AAR GCB

GAC TC-3′) and PolR (5′-ATS GCC ATC ATY TCR CCG GA-3′) [5]; archaeal *amoA* (as a measure of ammonia-oxidizing archaea) used primers Arch-amoAF (5′-STA ATG GTC TGG CTT AGA CG-3′) and Arch-amoAR (5′-GCG GCC ATC CAT CTG TAT GT-3′) [6]; bacterial *amoA* (as a measure of ammonia-oxidizing bacteria) used primers amoA-1F (5′-GGG GTT TCT ACT GGT GGT-3′) and amoA-2R (5′-CCC CTC KGS AAA GCC TTC TTC-3′) [4]; and *nosZ* (as a measure of denitrification bacteria) used primers nosZ-F (5′-CGY TGT TCM TCG ACA GCC AG-3′) [37] and nosZ 1622R (5′-CGS ACC TTS TTG CCS TYG CG-3′) [7]. Purified PCR products from a common DNA mixture (equal amounts of DNA from all samples collected in August of 2008 and 2009) were used to prepare sample-derived quantification standards as previously described [38]. The copy number of gene in each standard was calculated by DNA concentration (ng/µL, measured by Qubit) divided by the average molecular weight (estimated based on the barcoded-pyrosequencing results) of that gene. In comparison to using a clone (plasmid) as standard, this method avoids the difference of PCR amplification efficiency between standards and samples caused by the different sequence composition in the PCR templates (single sequence in a plasmid for the standard vs. mixture of thousands of sequences in a soil sample). The 10 µL reaction mixture contained 5 µL of 2× Power SYBR Green Master Mix (Applied Biosystems), 0.5 µL of BSA (10 mg/mL, New England Biolabs), 0.4 µL of each primer (10 µM) and 5 ng of DNA template. Real-time amplification was performed in an ABI Prism 7700 Sequence Detector with MicroAmp Optical 384-Well Reaction Plate and Optical Adhesive Film (Applied Biosystems) using the following program: 94°C for 5 min; 40 cycles of 94°C for 45 s, 56°C for 1 min (54°C for *nifH* gene), 72°C for 1 min. A dissociation step was added at the end of the qPCR to assess amplification quality. The specificity of the PCR was further evaluated by running twenty randomly selected samples (for each gene) on a 1% (w/v) agarose gel. The corresponding real-time PCR efficiency for each of these genes was estimated based on a two-fold serial dilution of the common DNA mixture described above. The qPCR efficiency (E) was calculated according to the equation $E = 10^{[-1/slope]}$ [39]. Triplicate qPCR repetitions were performed for each of the gene for all the samples. The real-time PCR amplification efficiency of *nifH*, archaeal *amoA*, bacterial *amoA* and *nosZ* genes was 1.90 ± 0.01, 1.90 ± 0.06, 1.76 ± 0.01 and 1.82 ± 0.01, respectively. The R^2 of all these standard curves was higher than 0.99. The detection limit of this real-time PCR assay was 10 copies/µL.

The copy numbers of these genes per gram of dry soil was calculated by the copy numbers of the gene per ng of DNA multiplied by the amount of DNA contained in each gram of dry soil. The quantities of these genes were corrected assuming a DNA extraction efficiency of 30% [40,41].

T-RFLP

The soil samples collected from four replicated (blocks) plots of the four crops prior to planting, and then August of establishment years 1 and 2 (2008 and 2009; 48 samples in total) were analyzed by terminal restriction fragment length polymorphism (T-RFLP). The *nifH*, archaeal *amoA*, bacterial *amoA*, *nosZ* gene were amplified from these samples with FAM-labeled (on forward primer) primers PolF/R, Arch-amoAF/R, amoA-1F/2R and nosZ-F/1622R (see Table S1). The 16S rRNA gene was amplified with 8F (5′-FAM-AGAGTTTGATCMTGGCTCAG-3′) and 1492R (5′-GGTT-ACCTTGTTACGACTT-3′) [42,43]. The PCR reaction mixture (25 µl) contained 5 µl GoTaq Flexi Buffer (5×), 2 µl MgCl₂ (25 mM), 0.25 µl DNA Polymerase (5 U/µl, Promega, Madison, Wis.), 1.25 µl BSA (10 mg/ml), 1 µl forward primer (10 µM), 1 µl

reverse primer (10 µM), 1.25 µl dNTP Mix (10 mM), 2 µl DNA template (10 ng/µl). The PCR reaction was performed in a thermo cycler (BioRad, Hercules, CA) using a 5-min heating step at 94°C followed by 30 cycles of denaturing at 94°C for 1 min, annealing at 60°C (54°C for *nifH*) for 45 s, and extension at 72°C for 1 min, with a final extension step of 5 min at 72°C. The PCR products were purified by QIAquick PCR Purification Kit (QIAGEN) and digested at 37°C overnight in a 20-µl mixture containing 2 µl NEB Buffer (10×), 0.5 µl AluI/HhaI (20 U/µl) and 5 µl PCR product. 5 µl of the digested product was sent to the Core Sequencing Facility (University of Illinois at Urbana-Champaign) for fragment analysis. ROX1000 was used as inner standard. T-RFLP profiles were analyzed by GeneMarker (v 1.85) according to the manufacturer's instruction. Fragments with sizes between 50 bp and the length of the PCR products and peak area >500 were selected for T-RFLP profile statistical analysis. The profile data were normalized by calculating the relative abundance (percentage) of each fragment (individual peak area divided by the total peak area).

Barcoded pyrosequencing

The same samples used in T-RFLP were also used in barcoded pyrosequencing. The four replicated samples collected from the same crop at the same time were combined in to one composite sample for the construction of sequence libraries. Altogether, nine samples (one sample for background soil [mixed from the soils from all the plots before planting these crops], four samples from each of the different crops at Aug 2008, and four from Aug 2009) were obtained. Furthermore, all of these N-transforming genes of the sample collected from MG at the end of the second year (MG2) were sequenced twice in two different lanes to estimate the variation of the sequencing method. The 16S rRNA gene of ZM2 was also sequenced twice. Details of primers and PCR conditions used in the study are listed in Table S1.

The *nifH*, archaeal *amoA*, bacterial *amoA*, *nosZ* and 16S rRNA genes (V4–V5 region) were amplified using the barcode primers PolF/R, Arch-amoAF/R, amoA-1F/2R, nosZ-F/1622R and U519F/U926R, respectively (primers are shown in Table S1). The primers (HPLC purified, Integrated DNA Technologies) were designed as 5′-Fusion Primer+barcode+gene specific primer-3′ (Fusion Primer A, 5′- CGTATCGCCTCCCTCGCGCCAT-CAG-3′; Fusion Primer B, 5′-CTATGCGCCTTGCCAGC-CCGCTCAG-3′). The PCR conditions were optimized and primers with appropriate barcodes (10 bp) were selected. The barcodes used for each primer are described in NCBI SRA, with accession number SRA023700. The 50 µL PCR mixture contained 10 µL of 5× Phusion HF Buffer (Phusion GC Buffer was used for bacterial *amoA* gene amplification, both buffer contains 7.5 mM MgCl₂), 1 µL of 10 mM dNTPs, 2.5 µL of 10 mg/mL BSA, 0.5 µL of 2 U/µL Phusion Hot Start DNA Polymerase (FINNZYMES), 4 µL of 10 µM forward/reverse primer mixture and 4 µL of 10 ng/µL DNA templates. 1 µL of 100% DMSO was supplemented into the PCR mixture in bacteria *amoA* gene (GC rich) amplification. The PCR amplification was performed in a thermal cycler (BioRad, Hercules, CA) using the program 98°C for 2 min; 30 cycles of 98°C for 10 s, 60°C (54°C for *nifH* gene and 56°C for bacterial *amoA* gene) for 30 s, 72°C for 20 s; 72°C for 5 min. The PCR product was first checked on a 1.2% w/v agarose gel, and then purified by QIAquick PCR Purification Kit (QIAGEN). The DNA concentration of the purified PCR product was measured by Qubit Fluorometer using the Quant-iT™ dsDNA BR Assay Kit (Invitrogen) according to the manual. PCR products of the same gene, to be run together in the same lane (1/16 plate) in 454 sequencing, were mixed in equal mole amounts and run on a 2% w/v agarose gel. The target bands were cut from the gel and purified by QIAquick Gel Extraction Kit (QIAGEN). The DNA concentrations of the purified PCR products were measured by Qubit Fluorometer and adjusted to 50 nM. The *nifH*, archaeal *amoA*, bacterial *amoA* and *nosZ* genes PCR products were then mixed in equal mole amounts and sequenced on a Genome Sequencer FLX Instrument (Roche) using GS FLX Titanium series reagents. The 16S rRNA gene was run in a separate lane.

Sequence analysis

Sequences were first extracted from the raw data according the Genome Sequencer Data Analysis Software Manual (Software Version 2.0.00, October 2008) by the sequencing center (Roy J. Carver Biotechnology Center, University of Illinois at Urbana-Champaign). The sequences with low quality (length <50 bp, which ambiguous base 'N', and average base quality score <20, for detail see manual) were removed. The sequences that fully matched with the barcodes were selected and distributed to separate files for each of the different genes, after removal of the barcode, using RDP Pipeline Initial Process (http://pyro.cme.msu.edu/). For each gene, the sequences that didn't match with the gene specific primers or had a read length shorter than 350 bp were removed. The sequences that matched with the reverse primer were converted to their reverse complement counterparts using BioEdit to make all the sequences forward-oriented.

The 16S rRNA gene sequences were aligned by NAST (Greengenes). The sequences with significant matched minimum length <300 and identity <75% were removed. The aligned 16S rRNA gene sequences were used for chimera check using Bellerophon method in Mothur [44]. Distance matrices were calculated by ARB using the neighbor joining method [45]. A lane mask was used in calculating the 16S rRNA gene sequences to filter out the hyper variable regions. Operational Taxonomic Units (OTUs) were then classified using a 97% nucleotide sequence similarity cutoff and rarefaction curves were constructed based on the distance matrices (both of nucleotide and amino acid sequences) using DOTUR [46]. The phylogenetic affiliation of each 16S rRNA gene sequence was analyzed by RDP CLASSI-FIER (http://rdp.cme.msu.edu/) using confidence level of 80%.

The 16S rRNA gene sequences were also processed by QIIME pipeline and denoised by Denoiser V0.91 according to the manual [47,48]. The results were compared to that obtained by RDP pipeline. In total, 26,431 valid sequences were obtained after denoising using QIIME, which is 12.2% less than that obtained by RDP pipeline (without denoising). Using the 97% similarity cutoff, 8,568 OTUs were obtained, which is 4.7% lower than that observed by RDP pipeline. After random re-sampling to the same sequence depth (1789 sequences per sample) using Daisy_chopper (http://www.genomics.ceh.ac.uk/GeneSwytch/Tools.html), the number of OTUs for each sample obtained by two different processing methods (QIIME, denoised and RDP, non-denoised) was compared (Fig. S1). The estimated number of OTUs after denoising was similar to that obtained by RDP pyrosequencing pipeline (without denoising), showing that the denoising process had a very limited influence on our diversity analysis. The data reported in this paper was analyzed using RPD pipeline described in the previous paragraph.

The *nifH*, archaeal *amoA*, bacterial *amoA* and *nosZ* genes sequences were blasted against a non-redundant protein sequence database (download from NCBI) using BLASTX with an *E*-value cutoff of 0.001. The top 10 closest matches of each sequence were estimated using a custom made Perl script to remove possible chimeras and sequences with sequencing errors causing frameshifts. Sequences with different regions matching the same sequence in the database but with different frame positions were considered to be frameshifts. Sequences that matched two or more different origin sequences were classified as chimeras. The nucleotide sequences of *nifH*, archaeal *amoA*, bacterial *amoA* and *nosZ* genes were translated into amino acid by Geneious (http://www.geneious.com/) based on the frame positions obtained from BLASTX. The redundant sequences (identical sequences) were removed using CD-Hit [49], and the representatives with longest length were selected for following phylogenetic analysis. Both of the nucleotide and amino acid sequences of these N-cycling genes were aligned by MUSCLE 3.7 [50] using program default settings. Operational Taxonomic Units (OTUs) were then classified and rarefaction curves were constructed based on the distance matrices (both of nucleotide and amino acid sequences) using DOTUR [46]. Previous studies showed that the amino acid sequences of AmoA and NosZ similarity around 90% is generally relevant to 97% similarity of 16S rRNA gene [51,52]. Thus, all these N-cycling gene sequences were classified into OTUs using a 90% amino acid sequence similarity cutoff, and phylogenetic trees were built in ARB using the neighbor-joining method. Sequences of all the samples and genes were also randomly re-sampled to identical sequencing depth (the smallest sequencing effort) using Daisy_chopper (http://www.genomics.ceh.ac.uk/GeneSwytch/Tools.html) to avoid the potential bias caused by sequencing effort difference [53].

The 454-pyrosequencing data were deposited in NCBI SRA under accession number SRA023700.

Statistical analysis

ANOVA combined with post hoc Tukey B test was used to estimate the difference of archaeal/bacterial *amoA*, *nifH* and *nosZ* genes abundances under different crops based on the quantitative PCR results from the replicated plots. The T-RFLP data from the replicated plots were used to follow the structural changes of soil microbial communities by plant types, and significance tests for these changes were conducted using Analysis of Similarity (ANOSIM) based on Bray–Curtis similarity coefficients. Correspondence analysis (CA) and Canonical correspondence analysis (CCA) were also used to visualize the predominant microbial community changes of archaeal/bacterial *amoA*, *nifH*, *nosZ* and 16S rRNA genes after planting bioenergy crops based on the T-RFLP data. These statistical analyses were done using the free software PAST (http://folk.uio.no/ohammer/past/). Based on our extensive pyrosequencing library, the OTUs/genera that showed monotonic (i.e. continuously increasing or decreasing) trends for each crop treatment over the two year establishment were presumed to be particularly noteworthy in terms of crop impact. The populations with continuously increased or decreased abundance in the two-year period after planting these bioenergy crops were selected using a custom Perl script.

Results

Quantification of *nifH*, archaeal *amoA*, bacterial *amoA* and *nosZ* genes

Quantities of AOA in all of crops fluctuated over the two growing seasons in a similar pattern (Fig. 1), but the abundance of this gene was always higher in MG than NP and PV. The quantity of bacterial *amoA* genes in ZM significantly increased from $1.47 \pm 0.61 \times 10^8$ to $3.26 \pm 0.94 \times 10^8$ during the first three months of establishment and thereafter remained higher in ZM than in the other cropping systems. The nitrification rates under these crops were analyzed in the second year by estimating the accumulation of the nitrate in buried soil bags, and linear regression revealed

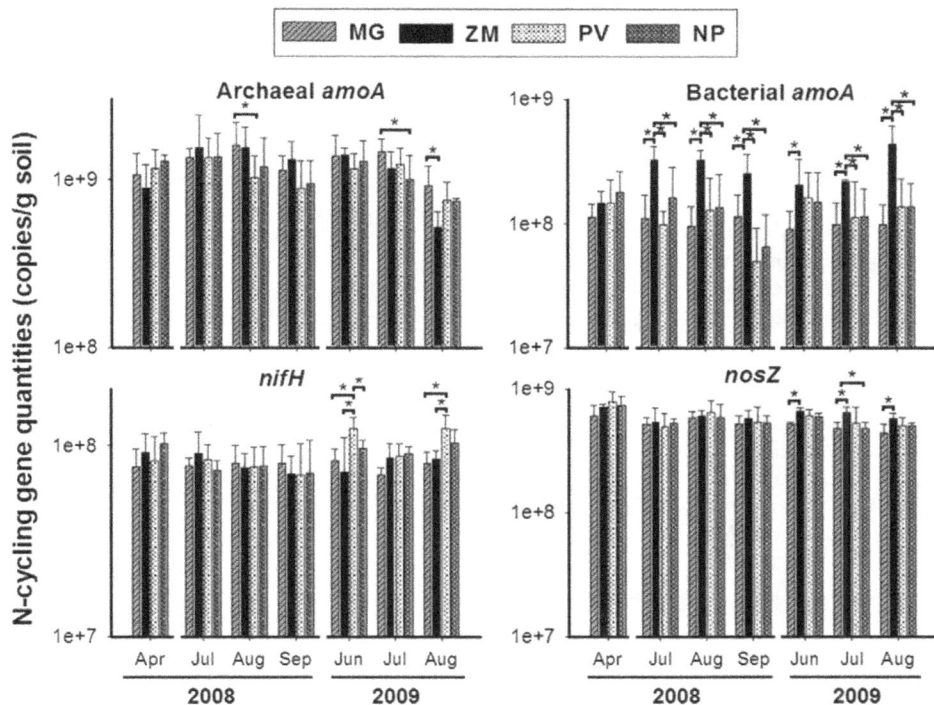

Figure 1. Changes in abundance of *nifH*, archaeal *amoA*, bacterial *amoA* and *nosZ* genes in plots after planting *Miscanthus* × *giganteus* (MG), *Panicum virgatum* (PV), restored prairie (NP) and *Zea mays* (ZM). The copy number of genes in each gram of dry soil was estimated based on the results of real-time PCR (copy number in each ng DNA) and the average amount of extracted DNA (6.23 µg per dry soil) and assuming DNA extraction efficiency was 30% [40]. The R^2 of the standard curve of all these genes was higher than 0.99. The real-time PCR amplification efficiency of *nifH*, archaeal *amoA*, bacterial *amoA* and *nosZ* genes was 1.90±0.01, 1.90±0.06, 1.76±0.01 and 1.82±0.01 respectively. *Represents values that are significantly different (P<0.01).

that the nitrification rates were significantly related to the quantities (log) of archaeal *amoA* genes ($R^2 = 0.61$, $P = 0.03$, $n = 12$), but not to the quantities of bacterial *amoA* genes (Fig. 2). The abundance of *nifH* genes remained stable for all the crops, ranging from 7×10^7 to 9×10^7 copies per gram of dry soil (Fig. 1). No significant differences were observed among the different crops in the first year. In the second year, the population sizes of diazotrophs in PV and NP had significantly increased in comparison to the first year ($P = 0.0001$ and 0.0002). The population size of denitrifiers was less variable in comparison to

the other N-cycling populations; however, the copy number of *nosZ* increased in ZM during the second year of the study and remained higher than in MG (Fig. 1).

Structural changes of N-cycling genes and microbial communities after planting of bioenergy crops

The community structural differentiation of *nifH*, archaeal *amoA*, bacterial *amoA*, *nosZ* and 16S rRNA genes under different bioenergy crops were analyzed by T-RFLP. These analyses used the fully replicated sample set from the randomized block design.

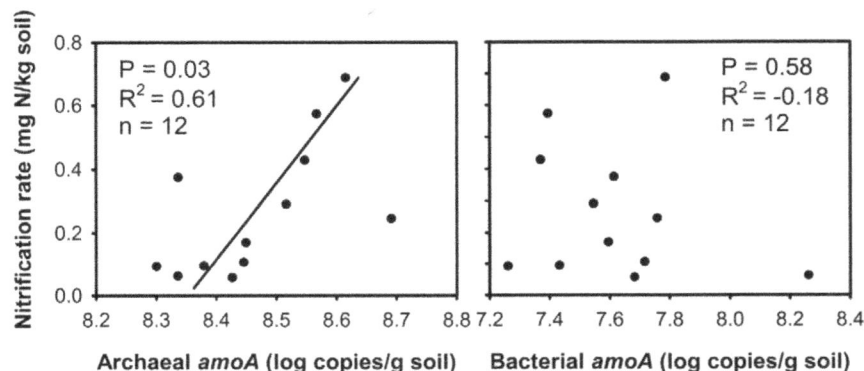

Figure 2. Relationship between the concentration of ammonia-oxidizing archaea/bacteria and nitrification rate. The nitrification rate was determined over the same time as our sample collection in 2009. Nitrification rate was calculated based on the accumulation of nitrate in soil bags incubated in the field (0–10 cm depth) for 15 to 32 days.

Correspondence analysis (Fig. 3) showed that the soil microbial communities in the initial plots did not show any relationship with the crop treatments being applied. During the establishment of these bioenergy crops, the community composition of denitrification bacteria (*nosZ*) under ZM was completely separated (ANOSIM, P<0.05, Table S2) from those under the other crops by the end of the second year along the first-axis, which explained 51.7% of the total variance. None of the other groups showed significant clustering by plant, although the community composition of AOB (bacteria *amoA*) under ZM appeared to be separated from that of MG (ANOSIM, P = 0.17) along the second axis, which explained 14.2% of the total variance. In addition to plant species, the changes of soil microbial communities also could be caused by the variation of environmental conditions. To compare the magnitude of the changes of soil microbial communities related only to plant species, a constrained ordination method was also used. Canonical Correspondence analysis (CCA, Fig. S2) revealed that, at the end of the second year, the microbial communities under ZM were most different from the three cropping systems for bacterial *amoA*, *nosZ*, and 16S rRNA. Archaeal *amoA* was most distinct under MG,

followed by ZM, and *nifH* was equally separated under all cropping systems (Fig. S2).

Diversity of *nifH*, archaeal *amoA*, bacterial *amoA*, *nosZ* and 16S rRNA genes

To further understand the composition of microbial community in the field, the *nifH*, archaeal *amoA*, bacterial *amoA*, *nosZ* and 16S rRNA genes were deeply sequenced using the pyrosequencing approach. In total, 143,487 reads were obtained for these genes. The numbers and qualities of these sequences are described in Table 1, Table S3 and Text S1. The reproducibility of the pyrosequencing result was estimated by comparing the observed microbial composition between repeat sequencing runs for all these genes (Fig. 4). Linear regression analysis indicated a high reproducibility ($R^2 = 0.95$) of our pyrosequencing.

High diversity of *nifH*, archaeal *amoA*, bacterial *amoA* and *nosZ* genes were observed with 10899, 3187, 3945 and 11242 unique nucleotide sequences and 2286, 2246, 3633 and 4208 unique deduced amino acid sequences respectively (Fig. S3). These sequences were then translated to amino acid sequence according

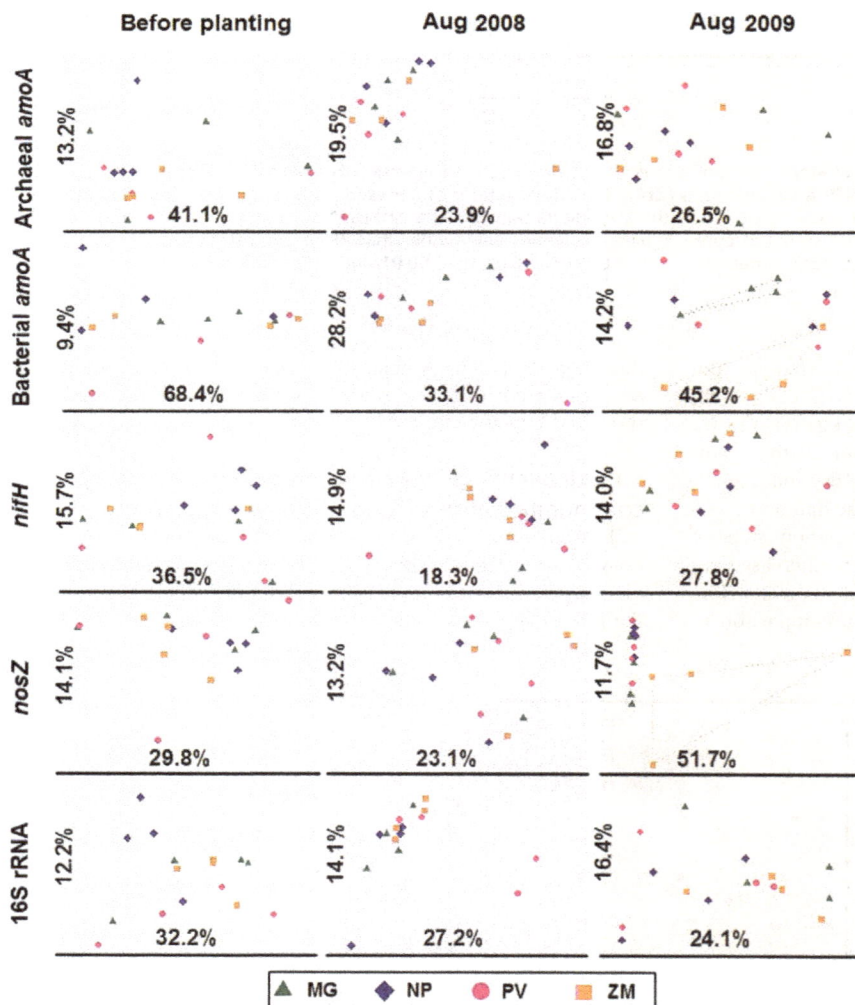

Figure 3. Structural changes of archaeal *amoA*, bacterial *amoA*, *nifH*, *nosZ* and 16S rRNA genes after planting *Miscanthus×giganteus* (MG), *Panicum virgatum* (PV), restored prairie (NP) and *Zea mays* (ZM) revealed by T-RFLP and Correspondence analysis (CA). The number on each axis shows the percentage of total variation explained. The soil samples were collected from four replicate plots for each plant at each time point.

Table 1. Quality of barcoded pyrosequencing reads.

| Genes | Number of sequences | | | |
	Correct barcode and primer	Length >350 bp	[a]Valid	Each sample (range)
nifH	28,334	27,781	21,111	1,312–2,956
Archaeal amoA	16,978	16,226	14,025	697–1,792
Bacterial amoA	28,254	27,874	21,817	1,726–2,569
nosZ	33,838	28,819	22,590	1,600–2,951
16S rRNA	30,487	30,175	30,101	2,034–3,488
Total	137,891 (96.1%)	130,875 (91.2%)	109,644 (76.4%)	697–3,488

Total number of raw reads was 143,487.
[a]Valid sequences of nifH, archaeal amoA, bacterial amoA and nosZ genes were defined as high quality sequences with correct barcode and primer (at 5'-end), length >350 bp and that did not have frameshifts and chimeric structure. The possible sequencing errors causing frameshifts and chimeras were removed based on the BLASTX result. Sequences with different regions matching the same sequence in the database but with different frame positions were considered to be frameshifts. Sequences that matched two or more different origin sequences were classified as chimeras. Valid sequences of 16S rRNA gene were sequences with correct barcode and primer, length >350 bp and passed the chimeric check program in Greengenes with the Bellerophon method. The sequence numbers for each sample are listed in Table S3.

to the BLASTX report. The amino acid sequences of nifH, archaeal amoA, bacterial amoA and nosZ genes were classified into 229, 309, 330 and 331 OTUs, respectively, with a similarity cutoff of 90%. After random re-sampling to the same sequencing depth (697 sequences for each sample), the adjusted total number of OTUs for these genes were 217, 303, 319 and 278, respectively (Table S4). Rarefaction analysis of these genes showed that the diversity of archaeal amoA gene in ZM2 (second year ZM) and MG1 (first year MG) was markedly lower than the others (Fig. S4). The diversity of nifH and nosZ genes slightly decreased in ZM2.

The diversity of bacterial and archaeal 16S rRNA genes was much higher than these N-cycling genes. In total, 19,824 unique 16S rRNA gene sequences and 8,989 species (OTUs classified at 97% similarity cutoff) were observed. RDP classification showed that these sequences covered 16 bacterial and 1 archaeal phyla, including 201 genera (Fig. S5). Proteobacteria and Acidobacteria were the most predominant phyla in the soil (>20%). The

sequences belonging to Proteobacteria were distributed over 86 different genera, while 94.1% of the sequences in Acidobacteria belonged to Family Gemmatimonadaceae, with GP1 as the most predominant genus (accounted 35.9% of the Acidobacteria).

To understand which phylotypes were impacted by vegetation type, the OTUs of nifH, archaeal amoA, bacterial amoA, nosZ and 16S rRNA (genus for 16S rRNA) genes that continuously increased or decreased over the two-year establishment of these bioenergy crops were identified (Fig. 5 and Fig. S6, S7, S8, S9). After planting of these bioenergy crops, 27.5%, 15.4%, 22.7% and 14.5% of the total archaeal amoA, bacterial amoA, nifH, and nosZ phylotypes, respectively, were found to be continuously increasing or decreasing (Table 2). Details of these continuously changed N-cycling OTUs are described in Text S1 and Figure S6, S7, S8, S9.

Pyrosequencing of 16S rRNA gene revealed 19.9% of the bacterial genera (39), spanning six phyla, continuously changed after planting of these bioenergy crops (Fig. 5). Only genus Methylibium was changed in all the crops, with decreased abundance in MG and increased abundance in the other crops. Rhodanobacter only appeared after planting of ZM (7 sequences for both of ZM1 and ZM2), and it was undetectable either in the background soil or in the soil under other crops. Consistent with the changes of Nitrosospira-like bacterial amoA OTU (see above), the abundance of genus Nitrosospira in the 16S rRNA library also increased in ZM. The abundance of genus Nitrospira, which is known as a nitrite-oxidizing bacteria, also increased in ZM. Most of the changed genera in MG decreased or even disappeared, except Terrabacter and Herbaspirillum. All of them were found at low abundance (<1%). Although many genera in Proteobacteria were changed, the total abundance of this most predominant phylum was quite stable under all of the crops (Fig. S5).

Discussion

In this study, we monitored the structural and quantitative changes of the key genes involved in N-cycling as well as the overall bacterial/archaeal community during two-year establishment of four different bioenergy feedstock crops, and analyzed the shifts of specific soil microbial populations in response to different types of crops. We were able to detect significant changes in the abundance of many of these microbial functional groups within 2 years of initial crop establishment. We also found that traditional row-crop agriculture of maize has a larger

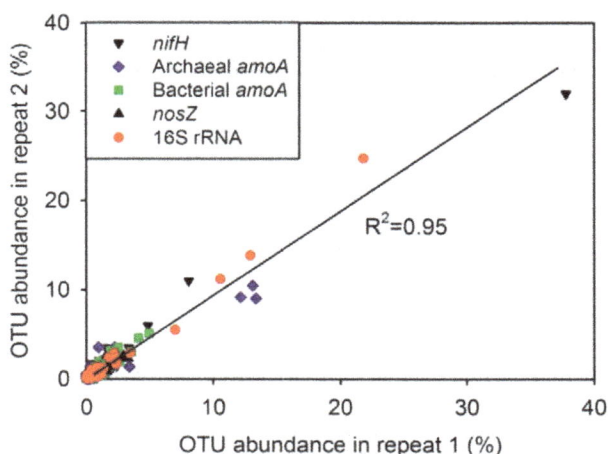

Figure 4. Reproducibility of the pyrosequencing replicates. OTUs of nifH, archaeal amoA, bacterial amoA and nosZ genes were classified at a nucleotide similarity cutoff 90%. The 16S rRNA gene sequences were classified to genus level by RDP classifier. One of the samples was duplicated for each gene.

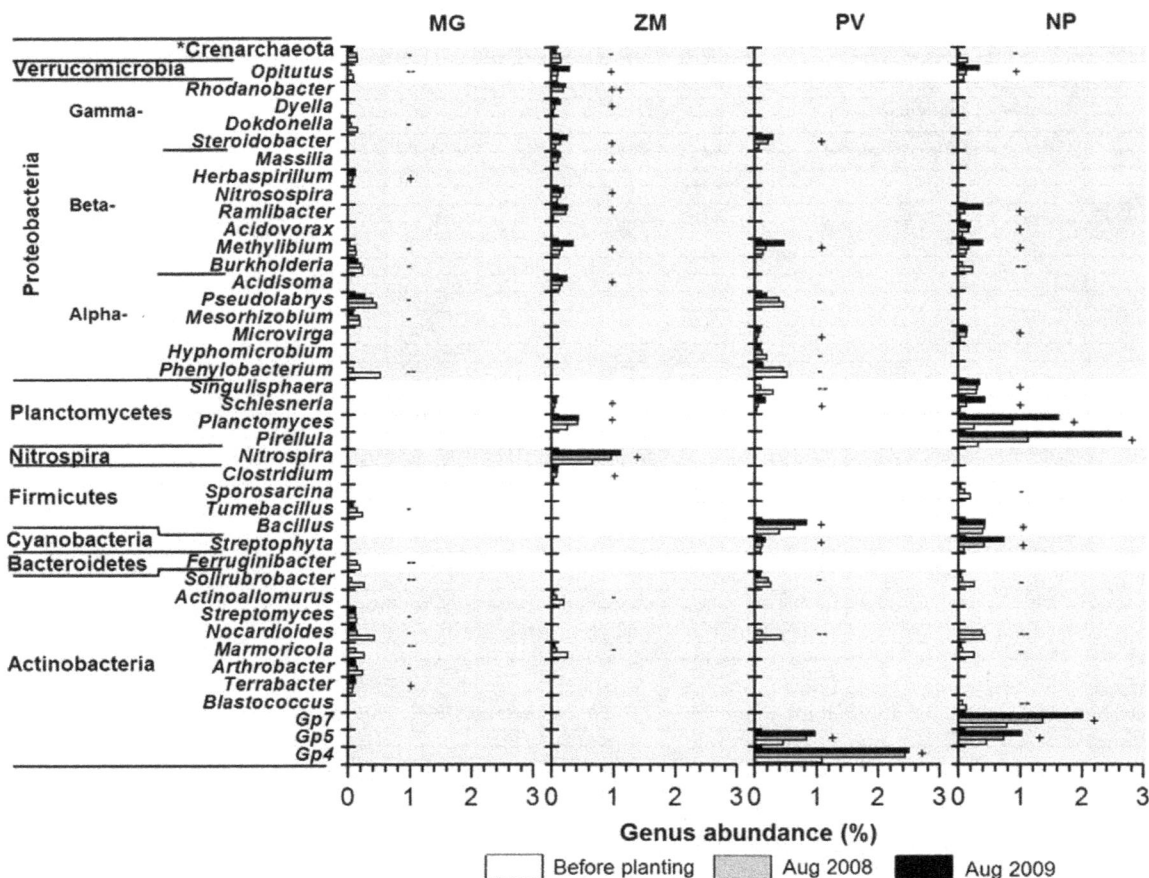

Figure 5. Microbial genera that changed after planting of *Miscanthus×giganteus* **(MG),** *Panicum virgatum* **(PV), restored prairie (NP) and** *Zea mays* **(ZM).** Sequences were classified by RDP Classifier project. −/+ represents the genus continuously decreased/increased after planting the crops; −−/++ represents the genus disappeared/appeared after planting the crops. *sequences belonging to Crenarchaeota, which could not be classified to genus level.

impact on the soil N-cycling community than any of the perennial bioenergy feedstock crops (Figs. 1, S2), while the perennial crops were associated with overall community shifts in the phyla Planctomyces, Firmicutes, and Actinobacteria (Fig. 5).

Table 2. Number of OTUs or genera that changed continuously after planting *Miscanthus×giganteus* (MG), *Panicum virgatum* (PV), restored prairie (NP) and *Zea mays* (ZM).

Genes	MG	PV	ZM	NP	*Total
Archaeal *amoA*	4	16	61	23	85
Bacterial *amoA*	18	18	7	20	51
nifH	8	21	14	23	52
nosZ	12	18	24	16	48
16S rRNA	17	15	14	21	40

OTUs of N-cycling genes were classified based on a cutoff of 90% amino acid sequences similarity. 16S rRNA genes were classified into genus level by RDP Classifier. Details of the abundance changes of the OTUs or genera are shown in Fig. 5 and S6, S7, S8, S9.
*Total number of changed unique OTUs or genera.

Traditional maize cultivation significantly increased the total abundance of ammonia-oxidizing bacteria and denitrifying bacteria (Fig. 1), altered the community composition of denitrifying bacteria (Fig. 3) and decreased the diversity of ammonia-oxidizing archaea, denitrifying bacteria, and diazotrophs (Fig. S4). This may be due to the application of N-fertilizer, which occurred only in ZM plots. Ammonia oxidizers are sensitive to N-fertilizer [24,27], and these responses were manifested in the increased population size of AOB and the high number of markedly changed AOA species. The nitrification rate was significantly correlated with the quantity of archaeal *amoA*, but not bacterial *amoA*, indicating AOA was the major ammonia-oxidizer.

Deep understanding of the structural shifts of key functional genes can help us to better understand changes in microbial activity in the environment. From the present database, we know that the global diversity of the *nifH*, archaeal *amoA*, bacterial *amoA* and *nosZ* genes, as well as the other functional genes of microorganisms, is high. The traditional approaches (e.g. clone library, DGGE and T-RFLP) used in previous studies may largely underestimate the diversity of microbial communities involved in soil nitrogen cycling. Mounier et al. (2004) revealed that even a large library with 713 clones was insufficient to enumerate the diversity *nosZ* gene in maize rhizosphere, showing the high complexity of N-cycling genes [54]. Thus, high-throughput deep sequencing approaches are essential to improve our knowledge of the diversity of these functional genes. In the present study, using barcoded 454-pyrosequencing approach, we found high diversity (ranging from ~3100 to ~11200 unique nucleotide sequences) of nifH, archaeal amoA, bacterial amoA and nosZ genes in the soil ecosystem, which far surpasses the diversity of the N-cycling genes observed in previous studies [54,55,56,57,58,59,60,61]. The rarefaction curves of these genes were close to saturation after sequencing ~1000 for each sample, indicating that such a sequencing effort is sufficient to elucidate the diversity and structure of the complex soil N-cycling communities. The high similarity between repeat runs of these genes (Fig. 4) demonstrates the high reproducibility and reliability of this barcoded pyrosequencing method. This result also indicates that the variation of pyrosequencing, resulting from random sampling of gene targets during emulsion PCR [62], can be greatly reduced by increasing the sequencing depth and library coverage.

During the establishment of these bioenergy crops, about 15%–30% of N-cycling genes and the detected bacterial/archaeal genera were continuously changed, indicating that a large proportion of soil microbes were affected by the transition to different bioenergy feedstock systems. Most of these phylotypes changed uniquely in one of the crops, indicating that the changes were mainly caused by the particular experimental crop treatment (specific plant species or management, such as fertilization) and not due to the environmental conditions that fluctuated in all treatments (e.g. temperature and moisture). Contrary changes of certain populations in different crops also support this conclusion; for example, the abundance of Methylibium (belonging to β-Proteobacteria) decreased in MG but increased in the other crops. The abundance of Bacteroidetes was previously found to be lower in the soil of Miscanthus-dominated grasslands (4%) in comparison to forest soil (6%) [63]. We found that the decrease of Bacteroidetes in MG was mainly due to the disappearance of the genus Ferruginibacter (Fig. 5 and S5).

The structure of nifH gene was completely separated according to vegetation by the end of the second year (Fig. S2), which suggests that the structure of N-fixation bacterial population was particularly sensitive to plant genotype. Tan et al. (2003) has revealed that the structure of diazotrophs was not only different among rice species, but also changed rapidly with fertilization. The diversity of the nifH gene was obviously reduced within 15 days after fertilization [17]. Thus, the decreased diversity of nifH in ZM may be also due to the application of fertilizer not the presence of maize. However, the population size of N-fixing bacteria did not change under ZM, indicating N-fertilization may not change the quantity of soil diazotroph [64]. The N-fixing activity was expected to increase in MG [14,15]. Although the population size of total free-living soil N-fixing bacteria was not significantly increased by growth of MG in the two year period, the abundance of genus Herbaspirillum increased. Herbaspirillum species are known as endophytic diazotrophs that are enriched by C4-prennial grasses including Miscanthus [65,66,67]. Thus, we speculate that the abundance increase of Herbaspirillum in the bulk soil was likely due to the root exudates (e.g. organic carbons) released by Miscanthus, which favored the growth of this population. Our results also suggest that Miscanthus may only selectively enhance the activity of specific diazotrophs, not the whole N-fixing microbial community.

The AOA are thought to be more stable and less responsive to environmental differences than AOB, as revealed by previous quantitative studies [6,68]. However, in the present study we found that, while the population size of AOA was relatively stable, the structure of the AOA community was sensitive to the different cropping systems. The diversity of AOA markedly decreased after planting of maize, with 41 of the AOA phylotypes disappearing (Fig. S6). In contrast, the population size of AOB significantly increased after planting of maize in both the qPCR of bacterial amoA and the 16S rRNA pyrosequencing results. In addition to the increase of genus Nitrosospira (AOB) [69], the abundance of Nitrospira (nitrite-oxidizing bacteria) [70] also increased in N-fertilized maize (Fig. 5). However, the number of changed bacterial amoA phylotypes in ZM was much less than the other crops. Therefore, these two different groups of ammonia-oxidizers respond to the N-fertilization in a very different way. The population size of AOB increased immediately in the first growing season, thus, we hypothesize that the increased AOB abundance in maize was likely due not to the growth of maize plants, but to the application of N-fertilizer, which increased the ammonia content in the soil [24,27]. It has been found that AOB population size increased in seven days after applying of N-fertilizer, and it was still significantly greater than unfertilized soil 8 months after the last application of ammonia [41]. Consistently, we found that the population size of AOB doubled in less than three months and maintained a relatively high level over the two-year study even though measurable NH_4^+ in the soil declined over this time period to levels close to that of the unfertilized plots (Fig. 1 and S10). These results indicate the AOB population size can be quickly increased by N-fertilization and can remain for a long period even after the measurable ammonia has been consumed.

The nitrification rate was found to be significantly correlated with the quantity of archaeal amoA rather than bacterial amoA, indicating AOA rather than AOB may be the major active ammonia-oxidizer in these soils. Contradictory conclusions on the relative importance of AOA and AOB in soil nitrification have been previously reported where nitrification was found to be associated with the changes of archaeal amoA abundance or higher archaeal transcriptional activity in some of the soils [58,71,72]. In contrast, the nitrification kinetics in the other soils were correlated with the growth of AOB [73,74]. It has been found that the ammonia affinity of "Candidatus Nitrosopumilus maritimus" (a marine AOA) is much higher than AOBs, and its growth may be enhanced by relatively low ammonia concentration [75]. Thus, the contradictory conclusions from these studies may be due to the different soil ammonia concentration used in these experiments [58,73]. These results hint that AOB may be more active in soils amended with ammonia, while AOA are more active in soils with low ammonia concentration [58]. The nitrification rate outlier (ZM Aug; Fig. 2) had highest population size of AOB, suggesting that the activity of AOB was enhanced by N-fertilization and also supported the above speculation. In support of this speculation, a recent publication shows that recovery of nitrification potential after disruption was dominated by AOB in cropped soils while AOA were responsible RNP in pasture soils [76].

It is known that nitrogen fertilization can change the structure and activity of denitrifying community, and subsequently affect the N_2O emission [33,34,77,78]. Large amounts (1.3%) of the applied N-fertilizer in maize fields (north Colorado) are converted to N_2O by the combination of nitrification and denitrification [79]. However, it is still unclear how fertilization changes the microbial community, since most of the previous studies are based on the already established fields [7,29,33]. Our study revealed that the structure of denitrifying bacteria in maize soil was significantly differentiated from the other crops at the second year (Fig. 3).

However, the population size of denitrifiers was relatively stable in all the crops in comparison to other N-cycling microbial communities, which only slightly increased in maize at the second year. The high stability of denitrifying population abundance could be explained by the high diversity and functional redundancy of denitrification community [30,80].

In conclusion, our two-year study of transitional agriculture shows that specific N-transforming microbial communities develop in the soil in response to different bioenergy crops. Each N-cycling microbial group responded in a different way after planting with different bioenergy crops. In general, planting of maize has a larger impact on the soil N-cycling community than the other bioenergy crops. Our results also indicate that application of N-fertilizer may not only cause short-term environmental problems, e.g. water contamination, but also can have long-term influence on the global biogeochemical cycles through changing the soil microbial community structure and abundance. Since soil types and other environmental factors may also impact the N-cycling microbial community, the universality of our findings needs to be confirmed by additional study at different sites.

Supporting Information

Figure S1   Number of OTUs obtained by two different processing methods: QIIME (denoised) and RDP pyrosequencing pipeline (non-denoised).

Figure S2   Structural changes of archaeal *amoA*, bacterial *amoA*, *nifH*, *nosZ* and 16S rRNA genes after planting *Miscanthus×giganteus* (MG), *Panicum virgatum* (PV), restored prairie (NP) and *Zea mays* (ZM) revealed by T-RFLP and Canonical correspondence analysis (CCA). The number on each axis shows the explained total variation. The soil samples were collected from four replicated plots for each plant at each time point. * Correspondence analysis was used for the samples collected before planting bioenergy crops.

Figure S3   OTU classification of valid sequences at different distance levels based on nucleotide and deduced amino acid sequences.

Figure S4   Rarefaction analysis of the diversities of *nifH*, archaeal *amoA*, bacterial *amoA*, *nosZ* and 16S rRNA genes in the soil underneath different bioenergy crops. The OTUs of *nifH*, archaeal *amoA*, bacterial *amoA* and *nosZ* genes were classified at 90% similarity cutoff based on amino acid sequences, and 16S rRNA gene was classified at 97% similarity cutoff on nucleotide sequences. BG0 represents the samples collected before planting bioenergy crops. MG, PV, NP, and ZM represent *Miscanthus×giganteus*, *Panicum virgatum*, restored prairie and *Zea mays* respectively. 1 and 2 represent samples collected in the first and second growing seasons.

Figure S5   Phylum level microbial community composition in the soil under different plants before and for two years after transition to bioenergy cropping. * represent significantly changed phylum. MG, *Miscanthus×giganteus*; PV, *Panicum virgatum*; NP, restored prairie; ZM, *Zea mays*.

Figure S6   (a) Phylogenetic tree of and (b) abundance of archaeal *amoA* OTUs that continuously changed after planting *Miscanthus× giganteus* (MG), *Panicum virgatum* (PV), restored prairie (NP) and *Zea*

mays (ZM). OTUs were classified based on a cutoff of 90% amino acid sequence similarity.

Figure S7   (a) Phylogenetic tree of and (b) abundance of bacterial *amoA* OTUs that continuously changed after planting *Miscanthus× giganteus* (MG), *Panicum virgatum* (PV), restored prairie (NP) and *Zea mays* (ZM). OTUs were classified based on a cutoff of 90% amino acid sequence similarity.

Figure S8   (a) Phylogenetic tree of and (b) abundance of *nifH* OTUs that continuously changed after planting *Miscanthus×giganteus* (MG), *Panicum virgatum* (PV), restored prairie (NP) and *Zea mays* (ZM). OTUs were classified based on a cutoff of 90% amino acid sequence similarity.

Figure S9   (a) Phylogenetic tree of and (b) abundance of *nosZ* OTUs that continuously changed after planting *Miscanthus×giganteus* (MG), *Panicum virgatum* (PV), restored prairie (NP) and *Zea mays* (ZM). OTUs of were classified based on a cutoff of 90% amino acid sequence similarity.

Figure S10   Nitrate and ammonia concentration in bulk soil. Soil samples for chemical and microbiological analysis were collected in the same week for each time point, except Sep 2008 when the nitrate and ammonia concentrations were not measured. MG, *Miscanthus×giganteus*; PV, *Panicum virgatum*; NP, restored prairie; ZM, *Zea mays*. These data were measured over the same period of our sample collection by the Biogeochemistry laboratory (C. Smith and M. David).

Figure S11   Pyrosequencing read length based on the raw sequence reads.

Table S1   Primers and annealing temperature for *nifH*, archaeal *amoA*, bacterial *amoA*, *nosZ* and 16S rRNA genes.

Table S2   Comparsion of the microbial community structures between different crops.

Table S3   Number of valid sequences for each gene in each sample.

Table S4   Number of OTUs observed after random re-sampling to the identical sequencing depth (697 sequences/sample.

Text S1

Acknowledgments

The authors gratefully acknowledge help from A. Duong, D. Fieckert, A. Groll, N. Peld, and M. Masters with field sampling and laboratory work. C. Smith and M. David kindly provided denitrification rate estimates for sampled plots. P.Y. Hong provided helpful comments on this manuscript.

Author Contributions

Conceived and designed the experiments: YM ACY RIM. Performed the experiments: YM ACY. Analyzed the data: YM. Contributed reagents/materials/analysis tools: YM. Wrote the paper: YM ACY RIM.

References

1. Garbeva P, van Veen JA, van Elsas JD (2004) Microbial diversity in soil: selection microbial populations by plant and soil type and implications for disease suppressiveness. Annu Rev Phytopathol 42: 243–270.

2. Kowalchuk GA, Buma DS, de Boer W, Klinkhamer PGL, van Veen JA (2002) Effects of above-ground plant species composition and diversity on the diversity of soil-borne microorganisms. Antonie Van Leeuwenhoek International Journal of General and Molecular Microbiology 81: 509–520.

3. Wardle DA, Bardgett RD, Klironomos JN, Setala H, van der Putten WH, et al. (2004) Ecological linkages between aboveground and belowground biota. Science 304: 1629–1633.

4. Rotthauwe JH, Witzel KP, Liesack W (1997) The ammonia monooxygenase structural gene amoA as a functional marker: molecular fine-scale analysis of natural ammonia-oxidizing populations. Appl Environ Microbiol 63: 4704–4712.

5. Poly F, Ranjard L, Nazaret S, Gourbiere F, Monrozier LJ (2001) Comparison of nifH gene pools in soils and soil microenvironments with contrasting properties. Appl Environ Microbiol 67: 2255–2262.

6. Francis CA, Roberts KJ, Beman JM, Santoro AE, Oakley BB (2005) Ubiquity and diversity of ammonia-oxidizing archaea in water columns and sediments of the ocean. Proc Natl Acad Sci U S A 102: 14683–14688.

7. Ruiz-Rueda O, Hallin S, Baneras L (2009) Structure and function of denitrifying and nitrifying bacterial communities in relation to the plant species in a constructed wetland. FEMS Microbiol Ecol 67: 308–319.

8. Bru D, Ramette A, Saby NP, Dequiedt S, Ranjard L, et al. (2011) Determinants of the distribution of nitrogen-cycling microbial communities at the landscape scale. ISME J 5: 532–542.

9. Orr CH, James A, Leifert C, Cooper JM, Cummings SP (2011) Diversity and activity of free-living nitrogen-fixing bacteria and total bacteria in organic and conventionally managed soils. Appl Environ Microbiol 77: 911–919.

10. Zehr JP, Jenkins BD, Short SM, Steward GF (2003) Nitrogenase gene diversity and microbial community structure: a cross-system comparison. Environ Microbiol 5: 539–554.

11. Cleveland CC, Townsend AR, Schimel DS, Fisher H, Howarth RW, et al. (1999) Global patterns of terrestrial biological nitrogen (N-2) fixation in natural ecosystems. Global Biogeochemical Cycles 13: 623–645.

12. Hsu SF, Buckley DH (2009) Evidence for the functional significance of diazotroph community structure in soil. Isme Journal 3: 124–136.

13. Son Y (2001) Non-symbiotic nitrogen fixation in forest ecosystems. Ecological Research 16: 183–196.

14. Schwarz H, Liebhard P, Ehrendorfer K, Ruckenbauer P (1994) The effect of fertilization on yield and quality of Miscanthus sinensis 'Giganteus'. Ind Crop Prod 2: 153–159.

15. Davis S, Parton W, Dohleman F, Smith C, Grosso S, et al. (2010) Comparative biogeochemical cycles of bioenergy crops reveal nitrogen-fixation and low greenhouse gas emissions in a Miscanthus×giganteus agro-ecosystem ecosystems DOI 10.1007/s10021-009-9306-9.

16. Demba Diallo M, Willems A, Vloemans N, Cousin S, Vandekerckhove TT, et al. (2004) Polymerase chain reaction denaturing gradient gel electrophoresis analysis of the N2-fixing bacterial diversity in soil under Acacia tortilis ssp. raddiana and Balanites aegyptiaca in the dryland part of Senegal. Environ Microbiol 6: 400–415.

17. Tan Z, Hurek T, Reinhold-Hurek B (2003) Effect of N-fertilization, plant genotype and environmental conditions on nifH gene pools in roots of rice. Environ Microbiol 5: 1009–1015.

18. Kowalchuk GA, Stephen JR (2001) Ammonia-oxidizing bacteria: a model for molecular microbial ecology. Annu Rev Microbiol 55: 485–529.

19. Schleper C, Jurgens G, Jonuscheit M (2005) Genomic studies of uncultivated archaea. Nat Rev Microbiol 3: 479–488.

20. Konneke M, Bernhard AE, de la Torre JR, Walker CB, Waterbury JB, et al. (2005) Isolation of an autotrophic ammonia-oxidizing marine archaeon. Nature 437: 543–546.

21. Hatzenpichler R, Lebedeva EV, Spieck E, Stoecker K, Richter A, et al. (2008) A moderately thermophilic ammonia-oxidizing crenarchaeote from a hot spring. Proceedings of the National Academy of Sciences of the United States of America 105: 2134–2139.

22. Leininger S, Urich T, Schloter M, Schwark L, Qi J, et al. (2006) Archaea predominate among ammonia-oxidizing prokaryotes in soils. Nature 442: 806–809.

23. He JZ, Shen JP, Zhang LM, Zhu YG, Zheng YM, et al. (2007) Quantitative analyses of the abundance and composition of ammonia-oxidizing bacteria and ammonia-oxidizing archaea of a Chinese upland red soil under long-term fertilization practices. Environ Microbiol 9: 2364–2374.

24. Shen JP, Zhang LM, Zhu YG, Zhang JB, He JZ (2008) Abundance and composition of ammonia-oxidizing bacteria and ammonia-oxidizing archaea communities of an alkaline sandy loam. Environ Microbiol 10: 1601–1611.

25. Wessen E, Soderstrom M, Stenberg M, Bru D, Hellman M, et al. (2011) Spatial distribution of ammonia-oxidizing bacteria and archaea across a 44-hectare farm related to ecosystem functioning. ISME J.

26. Briones AM, Okabe S, Umemiya Y, Ramsing NB, Reichardt W, et al. (2002) Influence of different cultivars on populations of ammonia-oxidizing bacteria in the root environment of rice. Appl Environ Microbiol 68: 3067–3075.

27. Wang Y, Ke X, Wu L, Lu Y (2009) Community composition of ammonia-oxidizing bacteria and archaea in rice field soil as affected by nitrogen fertilization. Syst Appl Microbiol 32: 27–36.

28. Le Roux X, Poly F, Currey P, Commeaux C, Hai B, et al. (2008) Effects of aboveground grazing on coupling among nitrifier activity, abundance and community structure. Isme J 2: 221–232.

29. Patra AK, Abbadie L, Clays-Josserand A, Degrange V, Grayston SJ, et al. (2006) Effects of management regime and plant species on the enzyme activity and genetic structure of N-fixing, denitrifying and nitrifying bacterial communities in grassland soils. Environ Microbiol 8: 1005–1016.

30. Philippot L, Hallin S, Schloter M (2007) Ecology of denitrifying prokaryotes in agricultural soil. Advances in Agronomy, Vol 96 96: 249–305.

31. Schlesinger WH (2009) On the fate of anthropogenic nitrogen. Proc Natl Acad Sci U S A 106: 203–208.

32. Rich JJ, Heichen RS, Bottomley PJ, Cromack K, Jr., Myrold DD (2003) Community composition and functioning of denitrifying bacteria from adjacent meadow and forest soils. Appl Environ Microbiol 69: 5974–5982.

33. Dambreville C, Hallet S, Nguyen C, Morvan T, Germon JC, et al. (2006) Structure and activity of the denitrifying community in a maize-cropped field fertilized with composted pig manure or ammonium nitrate. FEMS Microbiol Ecol 56: 119–131.

34. Hallin S, Jones CM, Schloter M, Philippot L (2009) Relationship between N-cycling communities and ecosystem functioning in a 50-year-old fertilization experiment. Isme J 3: 597–605.

35. Knops JMH, Bradley KL, Wedin DA (2002) Mechanisms of plant species impacts on ecosystem nitrogen cycling. Ecology Letters 5: 454–466.

36. Priha O, Grayston SJ, Pennanen T, Smolander A (1999) Microbial activities related to C and N cycling and microbial community structure in the rhizospheres of Pinus sylvestris, Picea abies and Betula pendula seedlings in an organic and mineral soil. FEMS Microbiol Ecol 30: 187–199.

37. Kloos K, Mergel A, Rosch C, Bothe H (2001) Denitrification within the genus Azospirillum and other associative bacteria. Australian Journal of Plant Physiology 28: 991–998.

38. Chen J, Yu ZT, Michel FC, Wittum T, Morrison M (2007) Development and application of real-time PCR assays for quantification of erm genes conferring resistance to macrolides-lincosamides-streptogramin B in livestock manure and manure management systems. Applied and Environmental Microbiology 73: 4407–4416.

39. Bustin SA (2000) Absolute quantification of mRNA using real-time reverse transcription polymerase chain reaction assays. Journal of Molecular Endocrinology 25: 169–193.

40. Mumy KL, Findlay RH (2004) Convenient determination of DNA extraction efficiency using an external DNA recovery standard and quantitative-competitive PCR. Journal of Microbiological Methods 57: 259–268.

41. Okano Y, Hristova KR, Leutenegger CM, Jackson LE, Denison RF, et al. (2004) Application of real-time PCR to study effects of ammonium on population size of ammonia-oxidizing bacteria in soil. Appl Environ Microbiol 70: 1008–1016.

42. Heuer H, Krsek M, Baker P, Smalla K, Wellington EM (1997) Analysis of actinomycete communities by specific amplification of genes encoding 16S rRNA and gel-electrophoretic separation in denaturing gradients. Appl Environ Microbiol 63: 3233–3241.

43. Lane DJ (1991) 16S/23S rRNA sequencing. InIn E. Stackebrandt, M. Goodfellow, eds. Nucleic acid techniques in bacterial systematics. pp 115–175.

44. Schloss PD, Westcott SL, Ryabin T, Hall JR, Hartmann M, et al. (2009) Introducing mothur: open-source, platform-independent, community-supported software for describing and comparing microbial communities. Appl Environ Microbiol 75: 7537–7541.

45. Ludwig W, Strunk O, Westram R, Richter L, Meier H, et al. (2004) ARB: a software environment for sequence data. Nucleic Acids Research 32: 1363–1371.

46. Schloss PD, Handelsman J (2005) Introducing DOTUR, a computer program for defining operational taxonomic units and estimating species richness. Applied and Environmental Microbiology 71: 1501–1506.

47. Caporaso JG, Kuczynski J, Stombaugh J, Bittinger K, Bushman FD, et al. (2010) QIIME allows analysis of high-throughput community sequencing data. Nat Methods 7: 335–336.

48. Reeder J, Knight R (2010) Rapidly denoising pyrosequencing amplicon reads by exploiting rank-abundance distributions. Nat Methods 7: 668–669.

49. Li W, Godzik A (2006) Cd-hit: a fast program for clustering and comparing large sets of protein or nucleotide sequences. Bioinformatics 22: 1658–1659.

50. Edgar RC (2004) MUSCLE: multiple sequence alignment with high accuracy and high throughput. Nucleic Acids Research 32: 1792–1797.

51. Palmer K, Drake HL, Horn MA (2009) Genome-derived criteria for assigning environmental narG and nosZ sequences to operational taxonomic units of nitrate reducers. Appl Environ Microbiol 75: 5170–5174.

52. Koops H-P, Purkhold U, Pommerening-Röser A, Timmermann G, Wagner M (2006) The lithoautotrophic ammonia-oxidizing bacteria. Prokaryotes 5: 788–811.

53. Gilbert JA, Field D, Swift P, Thomas S, Cummings D, et al. (2010) The taxonomic and functional diversity of microbes at a temperate coastal site: a 'multi-omic' study of seasonal and diel temporal variation. PLoS One 5: e15545.

54. Mounier E, Hallet S, Cheneby D, Benizri E, Gruet Y, et al. (2004) Influence of maize mucilage on the diversity and activity of the denitrifying community. Environ Microbiol 6: 301–312.

55. Teng Q, Sun B, Fu X, Li S, Cui Z, et al. (2009) Analysis of nifH gene diversity in red soil amended with manure in Jiangxi, South China. J Microbiol 47: 135–141.

56. Duc L, Noll M, Meier BE, Burgmann H, Zeyer J (2009) High diversity of diazotrophs in the forefield of a receding alpine glacier. Microb Ecol 57: 179–190.

57. Palmer K, Drake HL, Horn MA (2010) Association of novel and highly diverse acid-tolerant denitrifiers with N$_2$O fluxes of an acidic fen. Appl Environ Microbiol 76: 1125–1134.

58. Zhang LM, Offre PR, He JZ, Verhamme DT, Nicol GW, et al. (2010) Autotrophic ammonia oxidation by soil thaumarchaea. Proc Natl Acad Sci U S A 107: 17240–17245.

59. Reed DW, Smith JM, Francis CA, Fujita Y (2010) Responses of ammonia-oxidizing bacterial and archaeal populations to organic nitrogen amendments in low-nutrient groundwater. Appl Environ Microbiol 76: 2517–2523.

60. Moin NS, Nelson KA, Bush A, Bernhard AE (2009) Distribution and diversity of archaeal and bacterial ammonia oxidizers in salt marsh sediments. Appl Environ Microbiol 75: 7461–7468.

61. Nicol GW, Leininger S, Schleper C, Prosser JI (2008) The influence of soil pH on the diversity, abundance and transcriptional activity of ammonia oxidizing archaea and bacteria. Environ Microbiol 10: 2966–2978.

62. Zhou J, Wu L, Deng Y, Zhi X, Jiang YH, et al. (2011) Reproducibility and quantitation of amplicon sequencing-based detection. ISME J.

63. Lin YT, Lin CP, Chaw SM, Whitman WB, Coleman DC, et al. (2010) Bacterial community of very wet and acidic subalpine forest and fire-induced grassland soils Plant and Soil DOI:10.1007/s11104-010-0308-3.

64. Wakelin SA, Colloff MJ, Harvey PR, Marschner P, Gregg AL, et al. (2007) The effects of stubble retention and nitrogen application on soil microbial community structure and functional gene abundance under irrigated maize. FEMS Microbiol Ecol 59: 661–670.

65. Rothballer M, Eckert B, Schmid M, Fekete A, Schloter M, et al. (2008) Endophytic root colonization of gramineous plants by *Herbaspirillum frisingense*. FEMS Microbiol Ecol 66: 85–95.

66. Miyamoto T, Kawahara M, Minamisawa K (2004) Novel endophytic nitrogen-fixing clostridia from the grass *Miscanthus sinensis* as revealed by terminal restriction fragment length polymorphism analysis. Appl Environ Microbiol 70: 6580–6586.

67. Kirchhof G, Eckert B, Stoffels M, Baldani JI, Reis VM, et al. (2001) *Herbaspirillum frisingense* sp. nov., a new nitrogen-fixing bacterial species that occurs in C4-fibre plants. Int J Syst Evol Microbiol 51: 157–168.

68. Hai B, Diallo NH, Sall S, Haesler F, Schauss K, et al. (2009) Quantification of Key Genes Steering the microbial nitrogen cycle in the rhizosphere of sorghum cultivars in tropical agroecosystems. Applied and Environmental Microbiology 75: 4993–5000.

69. Kowalchuk GA, Stephen JR, De Boer W, Prosser JI, Embley TM, et al. (1997) Analysis of ammonia-oxidizing bacteria of the beta subdivision of the class Proteobacteria in coastal sand dunes by denaturing gradient gel electrophoresis and sequencing of PCR-amplified 16S ribosomal DNA fragments. Appl Environ Microbiol 63: 1489–1497.

70. Ehrich S, Behrens D, Lebedeva E, Ludwig W, Bock E (1995) A new obligately chemolithoautotrophic, nitrite-oxidizing bacterium, *Nitrospira moscoviensis* sp. nov. and its phylogenetic relationship. Arch Microbiol 164: 16–23.

71. Offre P, Prosser JI, Nicol GW (2009) Growth of ammonia-oxidizing archaea in soil microcosms is inhibited by acetylene. Fems Microbiology Ecology 70: 99–108.

72. Tourna M, Freitag TE, Nicol GW, Prosser JI (2008) Growth, activity and temperature responses of ammonia-oxidizing archaea and bacteria in soil microcosms. Environmental Microbiology 10: 1357–1364.

73. Jia Z, Conrad R (2009) Bacteria rather than Archaea dominate microbial ammonia oxidation in an agricultural soil. Environ Microbiol.

74. Di HJ, Cameron KC, Shen JP, Winefield CS, O'Callaghan M, et al. (2009) Nitrification driven by bacteria and not archaea in nitrogen-rich grassland soils. Nature Geoscience 2: 621–624.

75. Martens-Habbena W, Berube PM, Urakawa H, de la Torre JR, Stahl DA (2009) Ammonia oxidation kinetics determine niche separation of nitrifying Archaea and Bacteria. Nature 461: 976–979.

76. Taylor AE, Zeglin LH, Dooley S, Myrold DD, Bottomley PJ (2010) Evidence for different contributions of archaea and bacteria to the ammonia-oxidizing potential of diverse Oregon soils. Appl Environ Microbiol 76: 7691–7698.

77. Kramer SB, Reganold JP, Glover JD, Bohannan BJ, Mooney HA (2006) Reduced nitrate leaching and enhanced denitrifier activity and efficiency in organically fertilized soils. Proc Natl Acad Sci U S A 103: 4522–4527.

78. Wallenstein MD, Peterjohn WT, Schlesinger WH (2006) N fertilization effects on denitrification and N cycling in an aggrading forest. Ecol Appl 16: 2168–2176.

79. Hutchinson GL, Mosier AR (1979) Nitrous oxide emissions from an irrigated cornfield. Science 205: 1125–1127.

80. Wertz S, Degrange V, Prosser JI, Poly F, Commeaux C, et al. (2006) Maintenance of soil functioning following erosion of microbial diversity. Environ Microbiol 8: 2162–2169.

17

Organic and Inorganic Carbon in Paddy Soil as Evaluated by Mid-Infrared Photoacoustic Spectroscopy

Du Changwen[1]*, Zhou Jianmin[1], Keith W. Goyne[2]

1 The State Key Laboratory of Soil and Sustainable Agriculture, Institute of Soil Science Chinese Academy of Sciences, Nanjing, People's Republic of China, **2** Department of Soil, Environmental and Atmospheric Sciences, University of Missouri, Columbia, Missouri, United States of America

Abstract

Paddy soils are classified as wetlands which play a vital role in climatic change and food production. Soil carbon (C), especially soil organic C (SOC), in paddy soils has been received considerable attention as of recent. However, considerably less attention has been given to soil inorganic carbon (SIC) in paddy soils and the relationship between SOC and SIC at interface between soil and the atmosphere. The objective of this research was to investigate the utility of applying Fourier transform mid-infrared photoacoustic spectroscopy (FTIR-PAS) to explore SOC and SIC present near the surface (0–10 µm) of paddy soils. The FTIR-PAS spectra revealed an unique absorption region in the wavenumber range of 1,350–1,500 cm^{-1} that was dominated by C-O (carbonate) and C-H bending vibrations (organic materials), and these vibrations were used to represented SIC and SOC, respectively. A circular distribution between SIC and SOC on the surface of paddy soils was determined using principal component analysis (PCA), and the distribution showed no significant relationship with the age of paddy soil. However, SIC and SOC were negatively correlated, and higher SIC content was observed near the soil surface. This relationship suggests that SIC in soil surface plays important roles in the soil C dynamics.

Editor: Maria Gasset, Consejo Superior de Investigaciones Cientificas, Spain

Funding: This research was supported by the National Natural Science Foundation of China (40871113) and the Knowledge Innovation Program of the Chinese Academy of Sciences (KZCX2-YW-QN411). The funders had no role in study design, data collection and analysis, decision to publish, or preparation of the manuscript.

Competing Interests: The authors have declared that no competing interests exist.

* E-mail: chwdu@issas.ac.cn

Introduction

Surface horizon soils are primarily composed of particles of unbound organic matter and soil aggregates covered by an organic layer, and this organic layer holds some inorganic components (such as metals and carbonates) within the matrix [1,2]. These organic materials comprise the stock of soil organic carbon (SOC), whereas carbonates primarily represent the stock of soil inorganic carbon (SIC) [3]. Through dynamic processes, there is a relationship between soil carbon (C) and atmosphere C; therefore, soil can partly buffer rising atmospheric CO_2 concentrations [4]. Indeed, the pool of C in soil and vegetation is approximately three times higher than in the atmosphere indicating that any increase of C sequestration by soils could significantly offset the rising of atmospheric CO_2 and the resulting in global warming [5,6]. Moreover, elevated CO_2 often stimulates primary production and as a feedback, greater C input is expected to increase C sequestration in soil. Soil organic C, such as humic materials, is relatively stable, but organic C at the outer edge of soil particles and aggregates is more dynamic since it is greatly influenced by water, atmosphere, and biota [7]. Although C at the edge of particles and aggregates occupies a small percentage of total SOC, it is the most active part of soil C and it plays important roles in the soil C fluxes as well as biogeochemical reactions [8,9].

Paddy soils are the largest anthropogenic wetlands on earth, but they are highly modified by anthropogenic activities [10]. Due to their wide extent, paddy soils not only provide food for millions of people but they also play a role in climatic change through fluxes of carbon dioxide (CO_2) and methane [11–13]. Paddy soil research has focused primarily on SOC [14–16], but much less attention has been devoted to SIC even though it also contributes to soil C stocks [17,18]. One of the main reasons that the relationship between SOC and SIC remains understudied is due to the difficulty of differentiating SOC and SIC with conventional extractant-based chemical methods [2,19].

Fourier transform mid-infrared photoacoustic spectroscopy (FTIR-PAS) and has been applied in a wide disciplines [20] and it offers an alternative option for studying SOC and SIC in soils. This novel technique has recently been applied to study soils, and these studies have demonstrated that FTIR-PAS is very suitable for investigating soil samples with varied morphology and particle size [21–23]. One promising feature associated with FTIR-PAS is its depth profiling function [24,25], which permits investigating chemical characteristics at differing depths from the surface of soil particles and aggregates via collection of spectra with different moving mirror velocities. The objectives of this study were to investigate SOC and SIC on the outer surface of paddy soil particles and aggregates using the depth profiling technique of FTIR-PAS and to investigate relationships between SOC and SIC in paddy soils.

Materials and Methods

Paddy Soil Samples

Contemporary paddy soil samples, a total of 739, were collected from eastern China (119.3120–121.0908 E, 30.7955–32.0471 N),

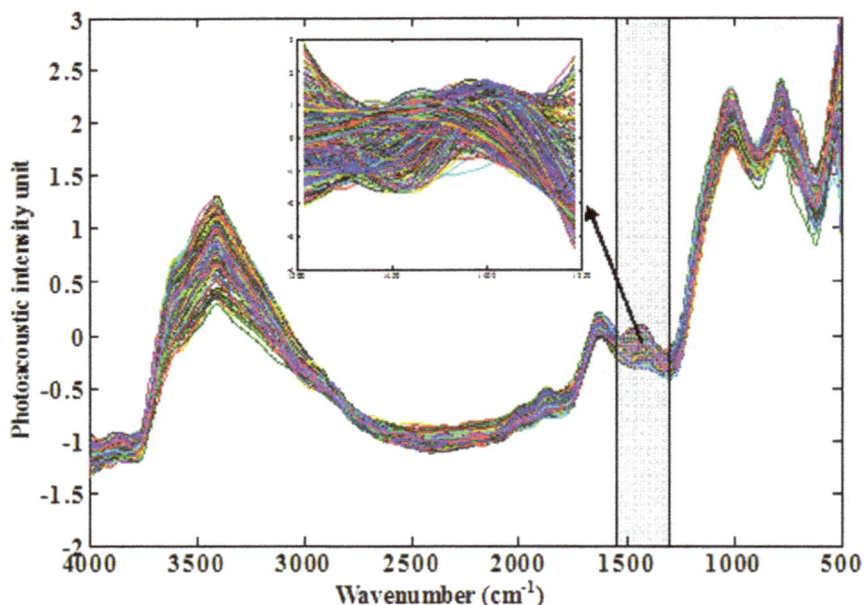

Figure 1. FTIR-PAS spectra of paddy soils (n=739) collected with a moving mirror velocity of 0.64 cm s⁻¹. The shaded frame showed the absorption band in the wavenumber region of 1,350–1,500 cm⁻¹.

and additional 117 historic paddy soil samples were collected from an archaeological site in the city of Cixi also located in eastern China (121.2333 E, 30.1667 N). Contemporary paddy soils were sampled at all sites from 0–15 cm depth. The age of ancient paddy soils ranging from 50 to 2,000 years old was determined by ^{14}C dating in the organic matter and in the carbonized rice [11]. All soil samples were air-dried, passed through a 2 mm sieve, and kept refrigerated at 4°C prior to use.

FTIR-PAS Analysis

Paddy soil sample spectra were collected using a Nicolet 380 spectrophotometer (Thermo Electron, USA) equipped with a

photoacoustic cell (Model 300, MTEC, USA). Samples ($\sim$ 200 mg) were placed in the cell holding cup (5 mm diameter × 3 mm height), after which the cell was purged with dry helium (20 mL min⁻¹) for 30 seconds to minimize interferences due to water vapor and impurities. The samples were then rapid scanned from 500 to 4,000 cm⁻¹ wavenumber with a resolution of 4 cm⁻¹ and moving mirror velocity of 0.64 cm s⁻¹. Following these analyses, 118 paddy soil samples (119.3120–119.4167E, 31.5220–32.0353) randomly selected from 739 contemporary paddy soil samples were scanned at moving mirror velocities of 0.16, 0.32, 0.64 and 1.89 cm s⁻¹. All FTIR-PAS spectra were normalized to a

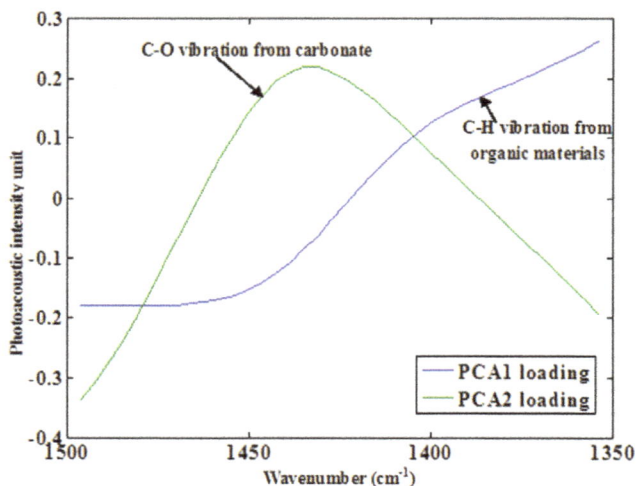

Figure 2. The first two PCA loadings in the wavenumber region of 1,350–1,500 cm⁻¹ for contemporary paddy soil samples (n=739). The explained variances a PCA1 and PCA2 were 57.5% and 30.2%, respectively.

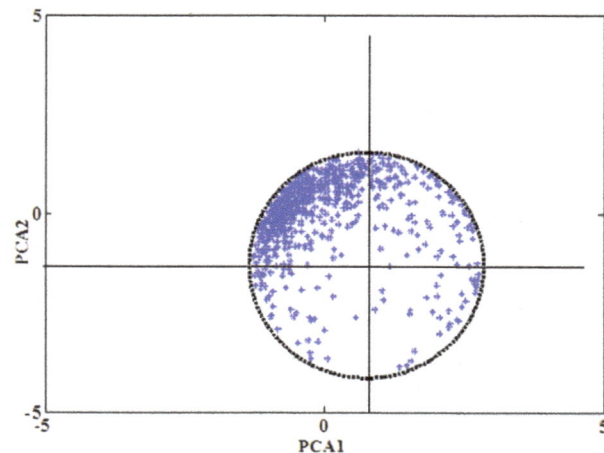

Figure 3. Circular distribution of the first and the second principal components associated with contemporary paddy soil FTIR-PAS spectra (n=739) in wavenumber region of 1,350–1,500 cm⁻¹. Spectra were collected with moving mirror velocity of 0.64 cm s⁻¹. Thickness of the scanning depth was calculated as 6.1 μm; r, defined as soil carbon capacity unit.

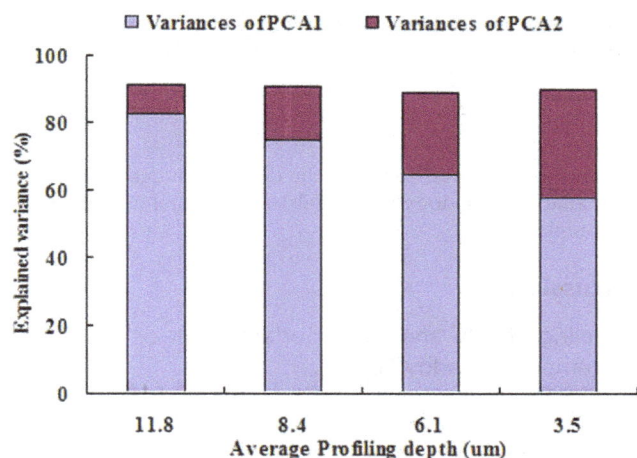

Figure 4. Variance explained by the first two PCA components (PCA1 and PCA2) in samples (n=118) scanned at different profiling depths using FTIR-PAS. Four velocities of the moving mirror (0.16 cm s^{-1}, 0.32 cm s^{-1}, 0.64 cm s^{-1}, 1.89 cm s^{-1}) were used to perform depth profiling of soil samples in the wavenumber region of 1,350–1,550 cm^{-1}. The corresponding average profiling depths were calculated as 11.8 µm, 8.4 µm, 6.1 µm, 3.5 µm, respectively.

carbon black background, and 32 successive scans were recorded and standardized for each sample.

The profiling depths associated with each FTIR-PAS were calculated using the following function [26]:

$$\mu_{th} = \sqrt{\frac{D}{\pi f_M}} \tag{1}$$

where μ_{th} is thermal diffusion length (µm), D is soil thermal diffusion coefficient (cm^2 s^{-1}), f_M is modulated frequency of infrared incident (Hz), which equals moving mirror velocity plus wavenumber in rapid FTIR-PAS scanning. A value of $D \approx 10^{-3}$ cm^2 s^{-1} from reference [27] was chosen for organic materials.

Processing of Spectra Data

The spectra were pre-processed by applying a smoothing filter and normalizing the amplitudes [22]. Data reduction was achieved through use of principal component analysis (PCA), which is commonly used to reduce the dimensionality of infrared spectra and yields a small number of coefficients (so called PCA scores) that retain most of the variability (information) present in the original data.

Modeling of the relationship between soil FTIR-PAS and soil age was performed using partial least square regressions (PLSR) [28]. For PLSR modeling, 87 samples and 30 samples were used in the calibration and validation sets, respectively. The PLS factor was optimized as six [23] and root mean square error (RMSE) was calculated to evaluate model performance. Software of Matlab was used in the processing of spectral data.

Results

Characterization of FTIR-PAS Spectra of Paddy Soils

Figure 1 shows the spectral characteristics of paddy soils from eastern China (n = 739). Generally, three main absorption regions are observed in each spectrum: (1) 2,000–4,000 cm^{-1}; (2) 1,200–

2,000 cm^{-1}; (3) and 500–1,200 cm^{-1}. The 2,000–4,000 cm^{-1} region contains O-H, N-H and C-H vibrations from almost all soil components (i.e., all organic materials and clay minerals), thus the absorption band is relatively broad. For the absorption region of 500–1,200 cm^{-1}, usually called as fingerprint region, two strong absorption bands are present and they represent stretching and bending vibrations from various groups (e.g., C-O, C-H, Si-O-Si, etc.). Strong interferences were also observed in this range [29]. Therefore, the absorption region of 1,200–2,000 cm^{-1} appears to be the candidate region for SOC/SIC related soil analysis. Vibration bands at ~1,600 cm^{-1} and ~1,100 cm^{-1} are primarily associated with soil water and silicon contained within clay minerals, respectively [21]. However, there is relatively little interference in the absorption region of 1,350–1,500 cm^{-1}. The main vibrations in this region are C-H bending vibrations from organic materials (including aliphatic and aromatic C-H) (1,300–1,500 cm^{-1}) and C-O stretching vibrations from carbonates (1,400–1,500 cm^{-1}) [29–31].

A principal component analysis (PCA) was conducted using data extracted from the FTIR-PAS spectra of the contemporary paddy soils. The analysis indicated two main principle components, PCA1 and PCA2, could explain variances of 57.5% and 30.2%, PCA1 and PCA2, respectively. The loadings associated with PCA1 and PCA2 in the candidate region are shown in Figure 2. The loading of PCA1 shows characteristics associated with C-H bending vibrations over a relatively wide wavenumber range (1,350–1,500 cm^{-1}), and the loading of PCA2 shows the typical carbonate absorption band at ~1,450 cm^{-1}. The variance explained by the first two principle components was 87.7%, and the PCA loading spectra show very good separation of C-H from organic materials and C-O from carbonate in this spectral range. Thus, PCA1 and PCA2 were chosen to represent SOC and SIC, respectively.

Relationship between SOC and SIC in Surface Layer of Spatial Contemporary Paddy Soils

PCA distribution of FTIR-PAS spectra of contemporary paddy samples collected with a moving mirror velocity of 0.64 cm s^{-1} were drawn over the wavenumber range of (1,350–1,500 cm^{-1}). The distribution between SOC (PCA1) and SIC (PCA2), shown in Figure 3, exhibits a circular distribution with radius unit of 2.0 and original points of (0.8, −1.5).

From the 739 contemporary paddy soils, 118 samples were randomly selected for FTIR-PAS analysis at different profiling depths (3.5–11.8 um) with varied moving mirror velocities. A principle components analysis was conducted using data extracted from the wavenumber region of 1,350–1,500 cm^{-1} (Figure 4) to investigate the relationship between SOC and SIC as a function of profiling depth. For the profiling FTIR-PAS spectra, there were two main principal components, PCA1 and PCA2, which explained 90% of in the spectral range investigated, thus suggesting that interference from other vibrational bands was relatively low (about 10%). Furthermore, the explained variance of PCA2, increased with decreasing profiling depth, whereas explained variance associated with PCA1 showed the exact opposite trend with profiling depth.

The distribution of SOC and SIC demonstrated as a circle function in the spatial paddy soils, does it work for time series based archaeological soil samples? Temporal ancient paddy soil samples were used to check the C dynamic in surface soil. The occupied variances in the PCA1 and the PCA2 were 57.7% and 30.1%, respectively (very similar to the results of contemporary paddy soils); also, a circular distribution was observed with radius unit of 1.2 and original points of (−0.2, −1.5) (Figure 5a). There

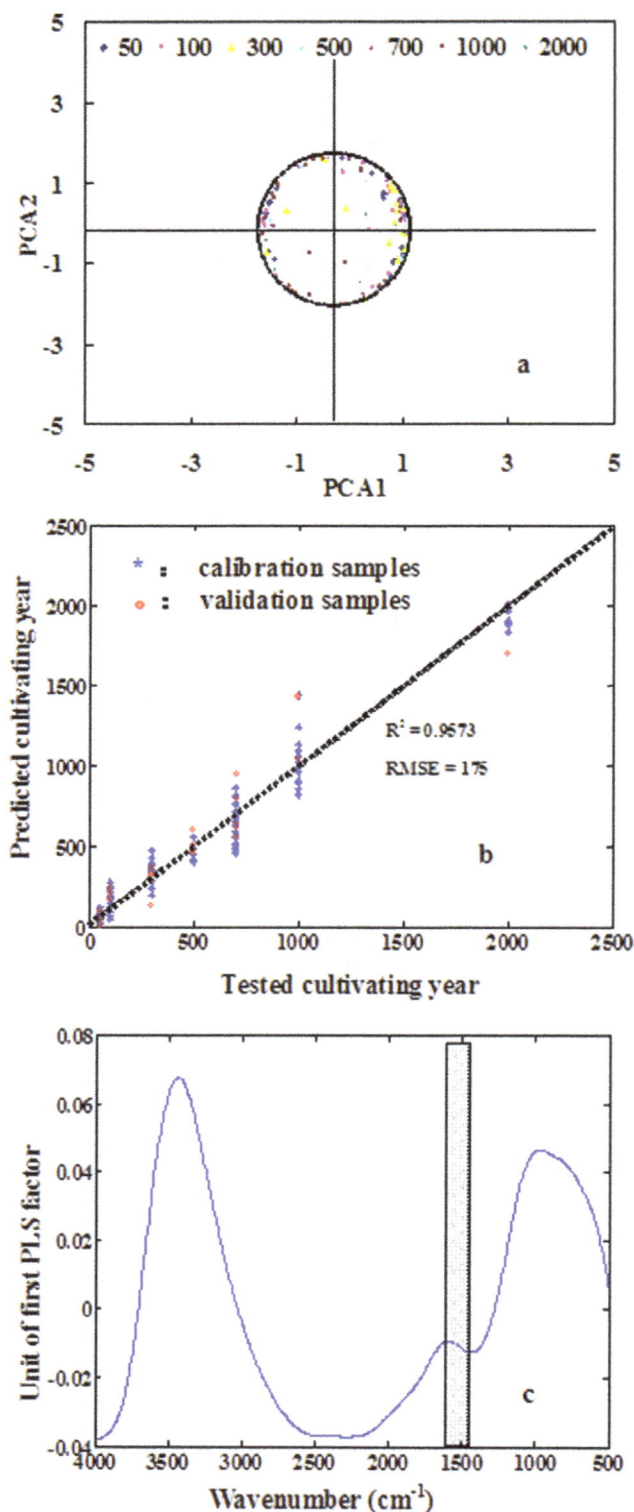

Figure 5. Analysis of FTIR-PAS spectra of historic paddy soils of different age (50–2,000) years using PCA and partial least square analysis (PLS) (*n*=117): (a) distribution of the first (representing SOC) and the second (representing SIC) principal components in the wavenumber region of 1,350–1,500 cm⁻¹; (b) the relationship between tested soil age and predicted soil age using PLSR model (87 soil samples were used in calibration and 30 soil samples were used in validation); (c), the PLSR

loading spectra of first PLSR factor, which shows the main wavenumber region (1,350–1,500 cm⁻¹) related to soil age (shaded frame). The moving mirror velocity during time of collection was 0.64 cm s⁻¹.

was also significant difference in the distribution density between the spatial contemporary paddy soils and temporal ancient paddy soil samples.

Discussion

Modeling of SOC and SIC in Surface Layer of Contemporary Paddy Soils

Using Eq. (1), the profiling depths associated with FTIR-PAS analysis of contemporary paddy samples collected with a moving mirror velocity of 0.64 cm s⁻¹ were calculated to be 5.8 to 6.1 μm over the wavenumber range of (1,350–1,500 cm⁻¹). The distribution between SOC (PCA1) and SIC (PCA2), shown in Figure 4, exhibits a circular distribution, which can be expressed with the following function:

$$(x-a)^2 + (y-b)^2 \leq r^2 \qquad (2)$$

where x (PCA1) denotes SOC, y (PCA2) denotes SIC, and a, b, r are constants. We assume that a, b, r are soil type dependent constants, where a and b might represent soil basic properties, and r represents soil C capacity unit. If the distribution circle is shifted to the origin of the graph (0, 0), the expression is changed to:

$$x^2 + y^2 \leq r^2 \qquad (3)$$

From Figure 2, the distribution density is observed to be significantly different within the circle. The highest density of points is located in upper left-hand corner, near the border of the distribution circle; thus, in most of cases, the expression can be represented as:

$$x^2 + y^2 \approx r^2 \qquad (4)$$

Assuming that r in Eq. (3) represents maximum soil C content, it can be deduced that x (SOC) or y (SIC) is less than r, indicating that the there would be a maximum SOC or SIC content in a specific soil type. It has been observed previously that SOC does not increase continuously even under conditions of great C input [32]. This was also verified in a long-term manure fertilization experiment conducted on a Fluvo-aquic soil [33]. Soil organic carbon increased during the first five to eight years, but thereafter a balance was reached.

From Eq. (2), neither a positive nor negative relationship can be drawn between SIC and SOC in the surface layer of soil particles and aggregates. However, a negative relationship between SOC and SIC can be deduced from Equation (4). This suggests that SOC content could possibly be limited by SIC content when soil C maxima has been reached for a specific soil. The apparent relationship between SIC and SOC could be important for global climatic change, as it may be possible for increases in SIC to compensate for losses in SOC in arable soils.

Differences in explained variances as a function of scanning depth (Figure 4) are likely due to differences in soil conditions favoring SIC accumulation at the edge of particles and aggregates.

These result suggested that the SIC decreased with the profiling depth, while SOC increased with the profiling depth, and a negative relationship between SIC and SOC in this surface layer was observed. Calcite precipitation and dissolution in soils is dependent on a number of environmental and geologic factors, including temperature, concentrations of dissolved Ca^{2+} and CO_2 in the soil pores, and the alkalinity and pH of the soil solution [17]. In general, the precipitation of calcite (as well as other carbonates) in soils is favored as Ca^{2+} concentration increases, CO_2 concentration decreases, and pH and alkalinity are increased. Presumably, CO_2 concentration is greater within organic particles and aggregates, due to diffusion limitations, than soil on or near the surface of particles which is in greater contact with the atmosphere or pores exchanging gases with the atmosphere. Lower CO_2 concentrations at the surface of particles and aggregates as well as greater concentrations of cations, such as Ca^{2+}, Mg^{2+}, could result in increased carbonate mineral formation and precipitation [17], thus resulting in a greater apparent signal from SIC at more shallow scanning depths (e.g., 3.5 μm).

Comparison of SOC and SIC in Contemporary and Historic Paddy Soils

Distribution density for the ancient paddy soil samples was almost entirely along the circle border, thus they were well fitted by Equation (4) and SIC was most likely negatively related with SOC. Because of the significantly larger r value (2.0 versus 1.2), the contemporary paddy soils might demonstrate a larger soil C stock than historic paddy soils, thus the paddy soil has a tendency to stock more C after longer cultivating time, which will contribute to restrain CO_2 in the atmosphere. From this view, paddy soil showed a sustainable capability in agricultural production; however, soil age (cultivating time) showed almost no influence on the distribution between SIC and SOC (Figure 5a). The relationship between soil spectra and the soil age was simulated using PLSR model, and a significant relationship can be found (Figure 5b), but the main PLS loading spectra showed the main

related spectra range included most of the spectral ranges but excluded the range of $1,350–1,500$ cm^{-1} (Figure 5c), which confirmed the result from distribution between PCA1 and PCA2. Therefore, the SIC content in surface layer of paddy soil was more likely linked with climatic change during the cultivating time.

Though there are some feedbacks from the soil C cycle to global warming, the feedback mechanisms are complicated [5,34,35]. Our findings can be the compensation for the feedbacks, which was indicated through the circular distribution between SIC and SOC in soil surface layer; most likely, the paddy soil can stock more inorganic C responding to global warming in soil surface layer, as a result, it may restrict the accumulation of SOC as well as the increase of global temperature.

Conclusions

FTIR-PAS was a useful technique for investigating the spectroscopic profile of paddy soils at the surface of particles and aggregates. The spectral range of $1,350–1,500$ cm^{-1} was found to best represent SOC and SIC signatures due to less interference. Soil organic carbon and SIC in the surface of soil samples studied were negatively correlated, suggesting that SIC may limit SOC content in soils. Additionally, the distribution between SIC and SOC in historic samples was related with climatic changes during cultivating times.

Acknowledgments

We thank Professor Cao Zhihong and Professor Lin Xiangui for their assistance in the collecting of paddy soil samples. Deng Jing and Zhou Huiqin made some contributions to the determination of soil FTIR-PAS spectra.

Author Contributions

Conceived and designed the experiments: DC ZJ. Performed the experiments: DC. Analyzed the data: DC. Contributed reagents/materials/analysis tools: DC ZJ KG. Wrote the paper: DC KG.

References

1. Huang PM, Wang MK, Chiu CY (2005) Soil mineral-organic matter-microbe interactions: impact on biogeochemical processes and biodiversity in soils. Pedobiologia 49: 609–635.
2. Ellerbrock RH, Gerke HH (2004) Characterizing organicmatter of soil aggregate coatings and biopores by Fourier transform infrared spectroscopy. European Journal of Soil Science 55: 219–228.
3. Hirmas DR, Amrhein C, Graham R (2010) Spatial and process-based modeling of soil inorganic carbon storage in an arid piedmont. Geoderma 154: 486–494.
4. Fontaine S, Bardoux G, Abbadie L, Mariotti A (2004) Carbon input to soil may decrease soil carbon content. Ecology Letters 7: 314–320.
5. Davidson EA, Trumbore SE, Amundson R (2000) Soil warming and organic carbon content. Nature 408: 789–790.
6. Lai R (2004) Soil Carbon Sequestration Impacts on Global Climate Change and Food Security. Science 304: 1623–1636.
7. Fang C, Smith P, Moncrieff JB, Smith JU (2005) Similar response of labile and resistant soil organic matter pools to changes in temperature. Nature 433: 57–59.
8. Paustian K, Six J, Elliott ET, Hunt HW (2000) Management options for reducing CO_2 emissions from agricultural soils. Biochemistry 48: 147–163.
9. Davidson EA, Janssens IA (2007) Temperature sensitivity of soil carbon decomposition and feedbacks to climate change. Nature 440: 165–172.
10. Kögel-Knabner I, Amelung W, Cao ZH, Fiedler S, Frenzel P, et al. (2010) Biogeochemistry of paddy soils. Geoderma 157: 1–14.
11. Cao ZH, Ding JL, Hu ZY, Knicker H, Kögel-Knabner I, et al. (2006) Ancient paddy soils from the Neolithic age in China's Yangtze River Delta. Naturwissenschaften 93: 232–236.
12. Redeker KR, Wang N-Y, Low JC. McMillan A, Tyler SC, et al. (2000) Emissions of methyl halides and methane from rice paddies. Science 290: 966–969.
13. Yan XY, Yagi K, Akiyama H, Akimoto H (2005) Statistical analysis of the major variables controlling methane emission from rice fields. Global Change Biology 11: 1131–1141.
14. Xie ZB, Zhu JG, Liu G, Cadisch G, Hasegawa T, et al. (2003) Storage and sequestration potential of topsoil organic carbon in China's paddy soils. Global Change Biology 10: 79–92.
15. Pan GX, Li LQ, Wu LS, Zhang XH (2003) Storage and sequestration potential of topsoil organic carbon in China's paddy soils. Global Change Biology 10: 79–92.
16. Bellamy PH, Loveland PJ, Bradley RI, Murray LR, Kirk G (2005) Carbon losses from all soils across England and Wales. Nature 437: 245–248.
17. Goddard MA, Mikhailova EA, Post CJ, Schlautman MA, Galbraith JM (2008) Continental united states atmospheric wet calcium deposition and soil. Inorganic Carbon Stocks. Soil Sci. Soc. Am. J. 73: 989–994.
18. Wang YG, Li Y, Ye XH, Chu Y, Wang XP (2010) Profile storage of organic/inorganic carbon in soil: From forest to desert. Science of the Total Environment 408: 1925–1931.
19. Jones DL, Willett VB (2006) Experimental evaluation of methods to quantify dissolved organic nitrogen (DON) and dissolved organic carbon (DOC) in soil. Soil Biology & Biochemistry 38: 991–999.
20. Zoltan B, Andrea P, Gabor S (2011) Photoacoustic Instruments for Practical Applications: Present, Potentials, and Future Challenges. Applied Spectroscopy Reviews 46: 1–37.
21. Du CW, Zhou JM (2007) Prediction of soil available phosphorus using Fourier transform infrared-photoacoustic spectroscopy. Chinese Analytical Chemistry 35: 119–122.
22. Du CW, Linker R, Shaviv A, Zhou JM (2008) Identification of agricultural Mediterranean soils using mid-infrared photoacoustic spectroscopy. Geoderma 143: 85–90.
23. Du CW, Zhou JM,Wang HW, Chen XQ, Zhu AN, Zhang JB (2009) Determination of soil properties using Fourier transform mid-infrared photoacoustic spectroscopy. Vibrational Spectroscopy 49: 32–37.
24. Du CW, Zhou JM, Wang HW, Chen XQ (2010) Depth profiling of clay-xanthan complexes using step-scan mid-infrared photoacoustic spectroscopy. Journal of Soils and Sediments 10: 855–862.

25. Drapcho D, Curbelo R, Jiang E, Crocombe RA, McCarthy WJ (1997) Digital signal processing for step-scan Fourier transform infrared photoacoustic spectroscopy Applied Spectroscopy 51: 453–460.

26. McClelland JF, Jones RW, Bajic SJ (2002) Photoacoustic Spectroscopy, In Handbook of Vibrational Spectroscopy, Vol. 2, Chalmers, J.M., and Griffiths, P.R. Eds., Wiley: Chichester, UK, 1231–1251.

27. Zhang WR, Lowe C, Smith R (2009) Depth profiling of coil coating using step-scan photoacoustic FTIR. Progress in Organic Coatings 65: 469–476.

28. Geladi P, Kowalski BR (1986) Partial least-squares regression: a tutorial. Analytica Chimica Acta 185: 1–17.

29. Calderon FJ, Reeves JB, Collins EAP (2011) Chemical differences in soil organic matter fractions determined by diffuse-reflectance mid-infrared spectroscopy. Soil Sci. Soc. Am. J. 75: 568–579.

30. Du CW, Linker R, Shaviv A, Zhou JM (2007) Characterization of soils using photoacoustic mid-infrared spectroscopy. Applied Spectroscopy 61: 1063–1067.

31. Tatzber M, Mutsch F, Mentler A, Leitger E, Englich M, et al. (2010) Determination of organic and inorganic carbon in forest soil samples by mid-infrared spectroscopy and partial least squares regression. Applied Spectroscopy 10: 1167–1175.

32. Gill RA, Polley HW, Johnson HB, Anderson LJ, Maherall H, Jackson RB (2002) Nonlinear grassland responses to past and future atmospheric CO_2. Nature 417: 279–282.

33. Du CW, Lei MJ, Zhou JM, Wang HW, Chen XQ, et al. (2011) Effect of long term fertilization on the transformations of water extractable phosphorus in Fluvo-aquic soil. Journal of Plant Nutrition and Soil Science 117: 20–27.

34. Christian PG, Michael GR (2000) Evidence that decomposition rates of organic carbon in mineral soil do not vary with temperature. Nature 404: 858–861.

35. Melillo JM, Steudler PA, Aber JD, Newkirk K, Lux H, et al. (2002) Soil Warming and Carbon-Cycle Feedbacks to the Climate System. Science 298: 2173–2176.

18

Response of CH$_4$ and N$_2$O Emissions and Wheat Yields to Tillage Method Changes in the North China Plain

Shenzhong Tian[1], Tangyuan Ning[1]*, Hongxiang Zhao[1], Bingwen Wang[1], Na Li[1], Huifang Han[1], Zengjia Li[1], Shuyun Chi[2]*

1 State Key Laboratory of Crop Biology, Shandong Key Laboratory of Crop Biology, Shandong Agricultural University, Taian, Shandong PR, China, 2 College of Mechanical and Electronic Engineering, Shandong Agricultural University, Taian, Shandong PR, China

Abstract

The objective of this study was to quantify soil methane (CH$_4$) and nitrous oxide (N$_2$O) emissions when converting from minimum and no-tillage systems to subsoiling (tilled soil to a depth of 40 cm to 45 cm) in the North China Plain. The relationships between CH$_4$ and N$_2$O flux and soil temperature, moisture, NH$_4^+$-N, organic carbon (SOC) and pH were investigated over 18 months using a split-plot design. The soil absorption of CH$_4$ appeared to increase after conversion from no-tillage (NT) to subsoiling (NTS), from harrow tillage (HT) to subsoiling (HTS) and from rotary tillage (RT) to subsoiling (RTS). N$_2$O emissions also increased after conversion. Furthermore, after conversion to subsoiling, the combined global warming potential (GWP) of CH$_4$ and N$_2$O increased by approximately 0.05 kg CO$_2$ ha^{-1} for HTS, 0.02 kg CO$_2$ ha^{-1} for RTS and 0.23 kg CO$_2$ ha^{-1} for NTS. Soil temperature, moisture, SOC, NH$_4^+$-N and pH also changed after conversion to subsoiling. These changes were correlated with CH$_4$ uptake and N$_2$O emissions. However, there was no significant correlation between N$_2$O emissions and soil temperature in this study. The grain yields of wheat improved after conversion to subsoiling. Under HTS, RTS and NTS, the average grain yield was elevated by approximately 42.5%, 27.8% and 60.3% respectively. Our findings indicate that RTS and HTS would be ideal rotation tillage systems to balance GWP decreases and grain yield improvements in the North China Plain region.

Editor: Ben Bond-Lamberty, DOE Pacific Northwest National Laboratory, United States of America

Funding: This work was financially supported by the Nature Science Fund of China (30900876 and 31101127), the National Science and Technology Research Projects of China (2012BAD14B17), and Special Research Funding for Public Benefit Industries (Agriculture) of China (201103001). The funders had no role in study design, data collection and analysis, decision to publish, or preparation of the manuscript.

Competing Interests: The authors have declared that no competing interests exist.

* E-mail: ningty@163.com (TN); chishujun1955@163.com (SC)

Introduction

CH$_4$ and N$_2$O play a key role in global climate change [1]. The emission of gas from disturbed soils is an especially important contributory factor to global change [2]. N$_2$O is emitted from disturbed soil, whereas CH$_4$ is normally oxidized by aerobic soils, making them sinks for atmospheric CH$_4$ in dry farmland systems [3]. According to estimates of the IPCC [4], CH$_4$ and N$_2$O from agricultural sources account for 50% and 60% of total emissions, respectively. Therefore, it is critical to reduce emissions of greenhouse gases (GHG) from agricultural sources. Many studies have reported that soil tillage has significant effects on CH$_4$ and N$_2$O emissions from farmland because the production, consumption and transport of CH$_4$ and N$_2$O in soil are strongly influenced by tillage methods [5–8].

The North China Plain is one of the most important grain production regions of China. Harrow tillage (HT), rotary tillage (RT) and no-tillage (NT) are frequently used conservation tillage methods in this region because they not only improve crop yield but also enhance the utilization efficiency of soil moisture and nutrients [8–12]. However, successive years of shallow tillage (10–20 cm) exacerbate the risk of subsoil compaction, which not only leads to the hardening of soil tillage layers and an increase in soil bulk density, but also reduced crop root proliferation, limited water and nutrient availability and reduced crop yield [13].

Subsoiling is an effective method that is used to break up the compacted hardpan layer every 2 or 4 years in HT, RT or NT systems [14,15]. Subsoiling significantly increases soil water content and temperature and decreases soil bulk density as well [16,17]. These rotation tillage systems are currently utilized in the North China Plain. Soil moisture and temperature are two factors controlling CH$_4$ and N$_2$O emissions [18–22]. In addition, CH$_4$ and N$_2$O emissions are normally associated with N application (as fertilizer) under wet conditions [23].

Collectively, reasonable soil tillage methods may reduce GHG emissions and may be important for developing sustainable agricultural practices [24]. However, it is unclear how conversion to subsoiling would affect CH$_4$ and N$_2$O emissions and whether subsoiling increases or reduces GHG emissions and the GWP of these agricultural techniques. In addition, there is little information on the soil factors affecting CH$_4$ and N$_2$O emissions after conversion to subsoiling in the North China Plain. The aim of this study was to determine whether conversion to subsoiling can reduce CH$_4$ and N$_2$O emissions.

Materials and Methods

Ethics Statement

The research station of this study is a department of Shandong Agricultural University. This study was approved by State Key Laboratory of Crop Biology, Shandong Key Laboratory of Crop Biology, Shandong Agricultural University.

Study Site

The study was conducted at Tai'an (Northern China, 36°09′N, 117°09′E), which is characteristic of the North China Plain. The average annual precipitation is 786.3 mm, and the average annual temperature is 13.6°C, with the minimum (−1.5°C) and maximum (27.5°C) monthly temperatures in January and July, respectively. The annual frost-free period is approximately 170–220 days in duration, and the annual sunlight time is 2462.3 hours. The soil is loam with 40% sand, 44% silt and 16% clay. The characteristics of the surface soil (0–20 cm) were measured as follows: pH 6.2; soil bulk density 1.43 g cm^{-3}; soil organic matter 1.36%; soil total nitrogen 0.13%; and soil total phosphorous 0.13%. The meteorological data during the experiment are shown in Figure 1.

Experimental Design

The experiment was designed as HT, RT and NT farming methods that started in 2004. In 2008, each plot was bisected, with one half maintained using the original tillage method as the control and the other half converted to subsoiling, resulting in six treatment plots: HT and HT conversion to subsoiling (HTS); RT and RT conversion to subsoiling (RTS); and NT and NT conversion to subsoiling (NTS) in a split-plot design with three

replicates. Each replicate was 35 m long and 4 m wide. After maize was harvested in each plot, straw was returned to the soil by one of the six following tillage operations:

HT - disking with a disc harrow to a depth of 12 cm to 15 cm,
RT - rototiller plowing to a depth of 10 cm to 15 cm,
NT - no tillage,
HTS, RTS, and NTS - plowed using a vibrating sub-soil shovel to a depth of 40 cm to 45 cm,

The experimental site was cropped with a rotation of winter wheat (*Triticum aestivum* Linn.) and maize (*Zea mays L.*). The wheat was sown in mid-October immediately after tilling the soil and was harvested at the beginning of June the following year. The maize was sown directly after the wheat harvest and was harvested in early October. During the wheat growth period, fertilizer was used at a rate of 225 kg N ha^{-1}, 150 kg ha^{-1} P$_2$O$_5$ and 105 kg ha^{-1} K$_2$O, and 100 kg N ha^{-1} was used as topdressing in the jointing stage with 160 mm of irrigation water. During the maize growth period, 120 kg N ha^{-1}, 120 kg ha^{-1} P$_2$O$_5$ and 100 kg ha^{-1} K$_2$O were used as a base fertilizer, and 120 kg N ha^{-1} was used as topdressing in the jointing stage.

CH$_4$ and N$_2$O Sampling and Measurements

CH$_4$ and N$_2$O content was measured using the static chamber-gas chromatography method [25]. The duration of gas sample collection was based on the diurnal variations in this region: the collection of CH$_4$ occurred from 9:00 a.m. to 10:00 a.m., and N$_2$O was collected between 9:00 a.m. and 12:00 p.m. from October 10, 2007, to May 19, 2009 at approximately 1-month intervals [26]. Both CH$_4$ and N$_2$O were sampled at 5 minutes, 20 minutes and 35 minutes after chamber closing. Simultaneously, the atmospheric temperature, the temperature in the static chamber, the land

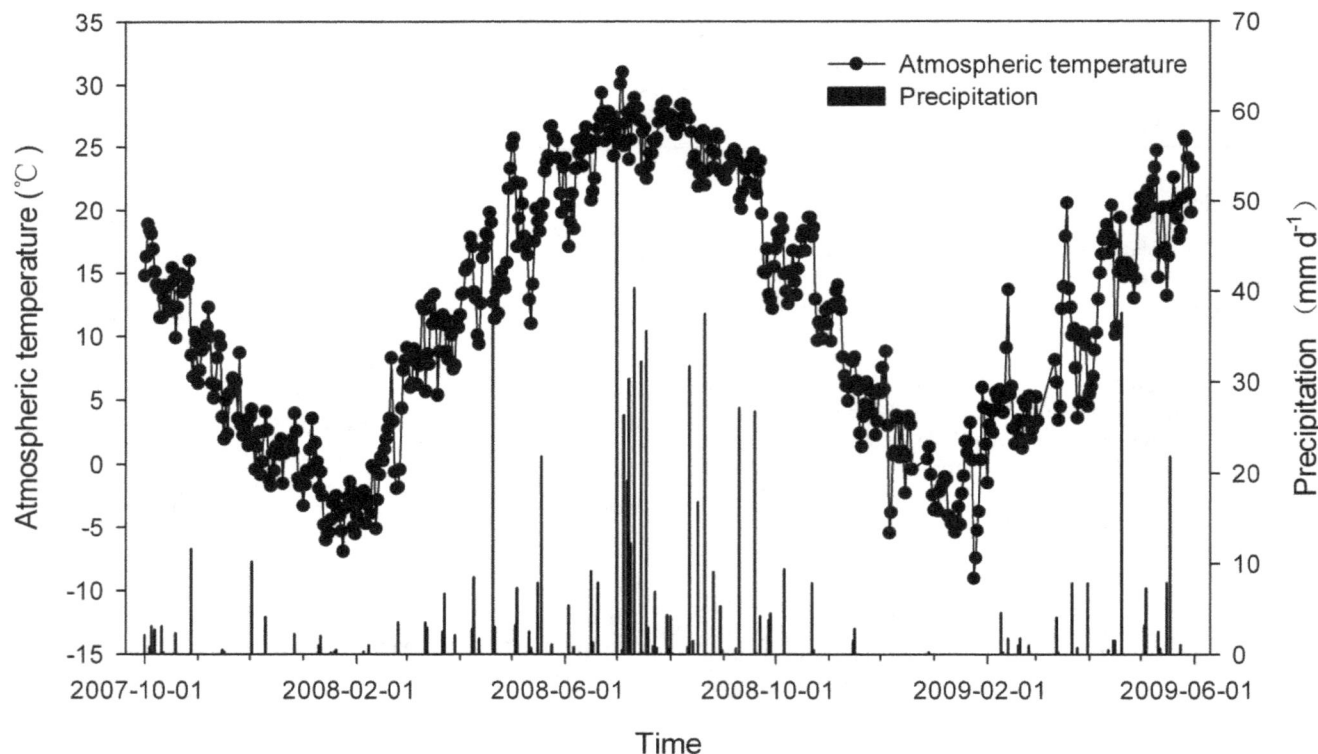

Figure 1. The atmospheric temperature and precipitation at the experiment site. The data were collected by the agricultural meteorological station approximately 500 m from the experiment field.

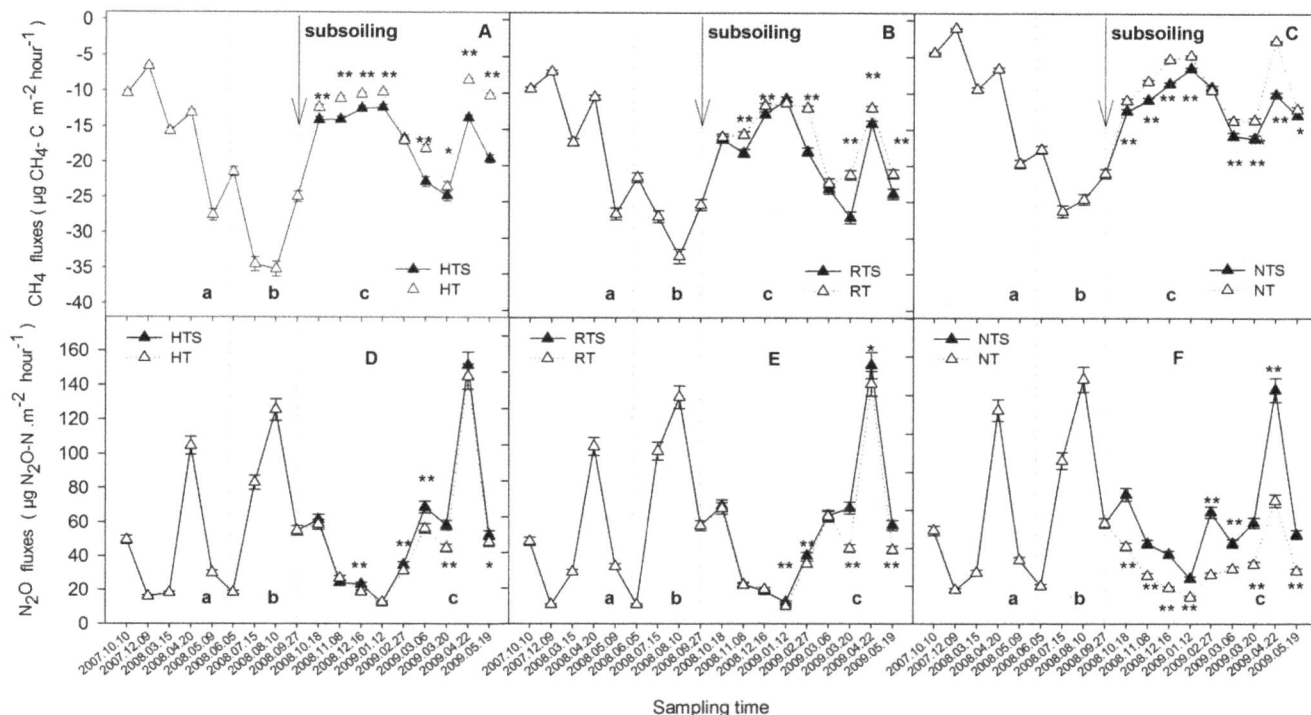

Figure 2. A to C CH$_4$ flux variations of H, R, and N after subsoiling in different periods; D to F N$_2$O flux variations of H, R, and N after subsoiling in different periods. a in Fig. 2 is the wheat growth stage of 2007 to 2008; **b** is the maize growth stage of 2008 to 2009; c is the wheat growth stage of 2008 to 2009. Arrows indicate time of subsoiling. Dotted lines distinguish the growth period of wheat and maize. * indicates $P<0.05$ and **indicates $P<0.01$ between subsoiling and the control.

Table 1. GWP and total changes in CH$_4$ and N$_2$O after subsoiling (2008.10~2009.05).

Treatments	HT	HTS	RT	RTS	NT	NTS
CH$_4$ total emission (kg·ha^{-1})	−0.73	−0.84	−0.64	−0.78	−0.39	−0.52
GWP of CH$_4$ (kgCO$_2$ ·ha^{-1})	−0.17	−0.19	−0.15	−0.18	−0.09	−0.12
N$_2$O total emission (kg·ha^{-1})	2.14	2.42	2.26	2.46	1.46	2.67
GWP of N$_2$O (kgCO$_2$ ·ha^{-1})	0.49	0.56	0.52	0.57	0.35	0.61
Total emissions of CH$_4$ and N$_2$O (kg·ha^{-1})	1.41	1.58	1.62	1.68	1.07	2.15
GWP of CH$_4$ and N$_2$O (kgCO$_2$ ha^{-1})	0.32	0.37	0.37	0.39	0.26	0.49
Increased emissions after conversion (kg·ha^{-1})	–	0.17	–	0.06	–	1.08
Increased GWP after conversion (kgCO$_2$ ·ha^{-1})	–	0.05	–	0.02	–	0.23

Total emissions of CH$_4$ and N$_2$O (kg·ha^{-1}), N$_2$O total emission flux added CH$_4$ total emission flux; **GWP of CH$_4$ and N$_2$O (kgCO$_2$·ha^{-1})**, GWP of N$_2$O added GWP of CH$_4$; **Increased emissions after conversion (kg·ha^{-1})**, difference of total emission of CH$_4$ and N$_2$O before and after conversion; **Increased GWP after conversion (kgCO$_2$·ha^{-1})**, difference of GWP of CH$_4$ and N$_2$O before and after conversion.

surface temperature and the soil temperature at a depth of 5 cm were determined after collecting samples.

The samples were measured using a Shimadzu GC-2010 gas chromatograph. CH$_4$ was measured using a flame ionization detector with a stainless steel chromatography column packed with a 5A molecular sieve (2 m long); the carrier gas was N$_2$. The temperatures of the column, injector and detector were 80°C, 100°C and 200°C, respectively. The total flow of the carrier gas was 30 ml min^{-1}, the H$_2$ flow was 40 ml min^{-1}, and the airflow was 400 ml min^{-1}. N$_2$O was measured using an electron capture detector with a Porapak-Q chromatography column (4 m long); the carrier gas was also N$_2$. The temperatures of the column, injector and detector were 45°C, 100°C and 300°C, respectively. The total flow of the carrier gas was 40 ml min^{-1}, and the tail-blowing flow was 40 ml min^{-1}. The gas fluctuations were calculated by the gas concentration change in time per unit area.

Emission changes in CH$_4$ and N$_2$O were calculated using the following formula [25]:

$$F = \frac{60HMP}{8.314(273+T)}\frac{dc}{dt}$$

where F is the change in gas emission or uptake (µg·m^{-2}·h^{-1}); 60 is the conversion coefficient of minutes and hours; H is the height (m); M is the molar mass of gas (g·mol^{-1}); P is the atmospheric pressure (Pa); 8.314 is the Ideal Gas Constant (J mol^{-1} K^{-1}); T is the average temperature in the static chamber (°C); and dc/dt is the line slope of the gas concentration change over time.

Table 2. Correlation analysis between changes in CH_4 and N_2O with soil temperature and soil moisture per sampling time.

Sampling time	Soil temperature				Soil moisture			
	CH_4		N_2O		CH_4		N_2O	
	R^2	n	R^2	n	R^2	n	R^2	n
2008.10.18	0.6020*	3	0.3832	3	0.5429*	3	0.1020	3
2008.11.08	0.6180*	3	0.0377	3	0.2945	3	0.1241	3
2008.12.16	0.7314**	3	0.0087	3	0.0085	3	0.5142*	3
2009.01.12	0.6490**	3	0.0723	3	0.2988	3	0.5200*	3
2009.02.27	0.6597**	3	0.3053	3	0.5370*	3	0.0914	3
2009.03.06	0.3824	3	0.1461	3	0.0417	3	0.0005	3
2009.03.20	0.2876	3	0.0257	3	0.4966*	3	0.6132*	3
2009.04.22	0.4476*	3	0.3044	3	0.5154*	3	0.6735**	3
2009.05.19	0.8870**	3	0.0503	3	0.4593*	3	0.5027*	3

*$P<0.05$,
**$P<0.01$.

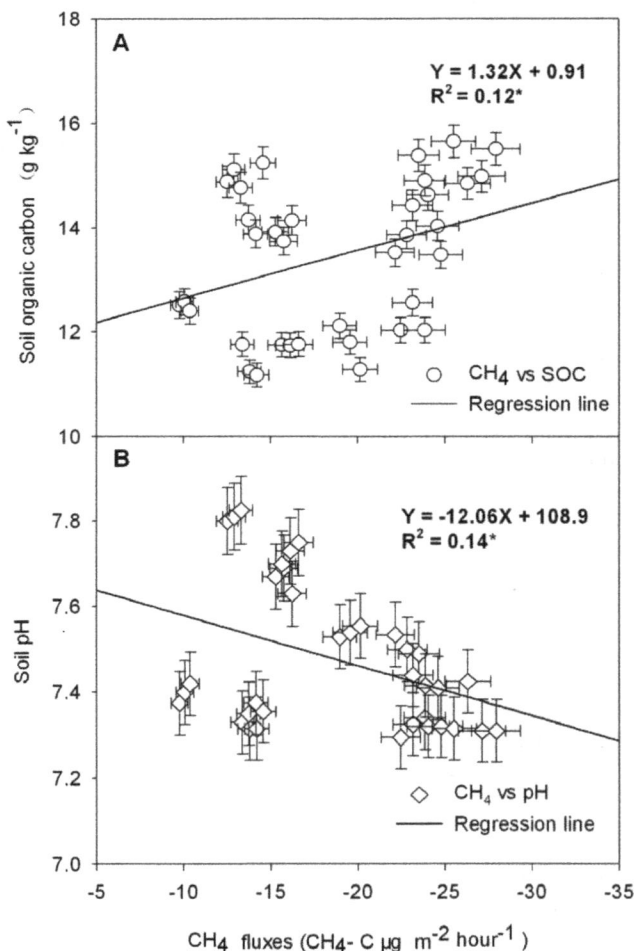

Figure 3. A Linear regression between the CH_4 uptake fluxes and SOC, B Linear regression between the CH_4 uptake fluxes and soil pH. Arrows indicate the regression equation between the CH_4 uptake fluxes and soil organic carbon, soil pH. *indicates $P<0.05$.

GWP of CH_4 and N_2O

The global warming potentials (GWP) were determined by measuring CH_4 and N_2O emissions. The GWP of CH_4 and N_2O are 25 and 298 times higher, respectively, than that of CO_2 (the GWP of CO_2 is 1) [27] and are calculated as follows:

$$GWP(CH_4) = \frac{TF(CH_4) \times 25}{100}$$

$$GWP(N_2O) = \frac{TF(N_2O) \times 298}{100}$$

where $GWP(CH_4)$ is the GWP of CH_4 (kg CO_2 ha^{-1}); $TF(CH_4)$ is the total uptake of CH_4 (kg CO_2 ha^{-1} a^{-1}); 25 is the GWP coefficient of CH_4; 100 is the time scale of climate change (a); $GWP(N_2O)$ is the GWP of N_2O (kg CO_2 ha^{-1}); $TF(N_2O)$ is the total emission of N_2O (kg CO_2 ha^{-1} a^{-1}); and 298 is the GWP coefficient of N_2O.

Soil Factor Measurements

The meteorological data during the experiment were obtained from an agricultural weather station in the experimental area. To evaluate the relation between soil temperature and moisture and CH_4 and N_2O emissions, we measured soil temperature at a depth of 5 cm and the soil moisture in the 0–20 cm soil layers simultaneously using a soil temperature, moisture and electric conductivity instrument (WET brand, made in the UK) as the temperature and moisture data collection tool. The soil samples were collected using a soil sampler with five replicates in each different tillage treatment and were dried and triturated after mixing. This sample was used to determine the SOC, NH_4^+-N and pH using the Potassium Dichromate Heating Method, the UV Colorimetric Method and the Potentiometry Method, respectively [28].

Grain Yield

The grain yield of winter wheat was sampled from the 1.5 m× 6 m portion in the central area of each plot.

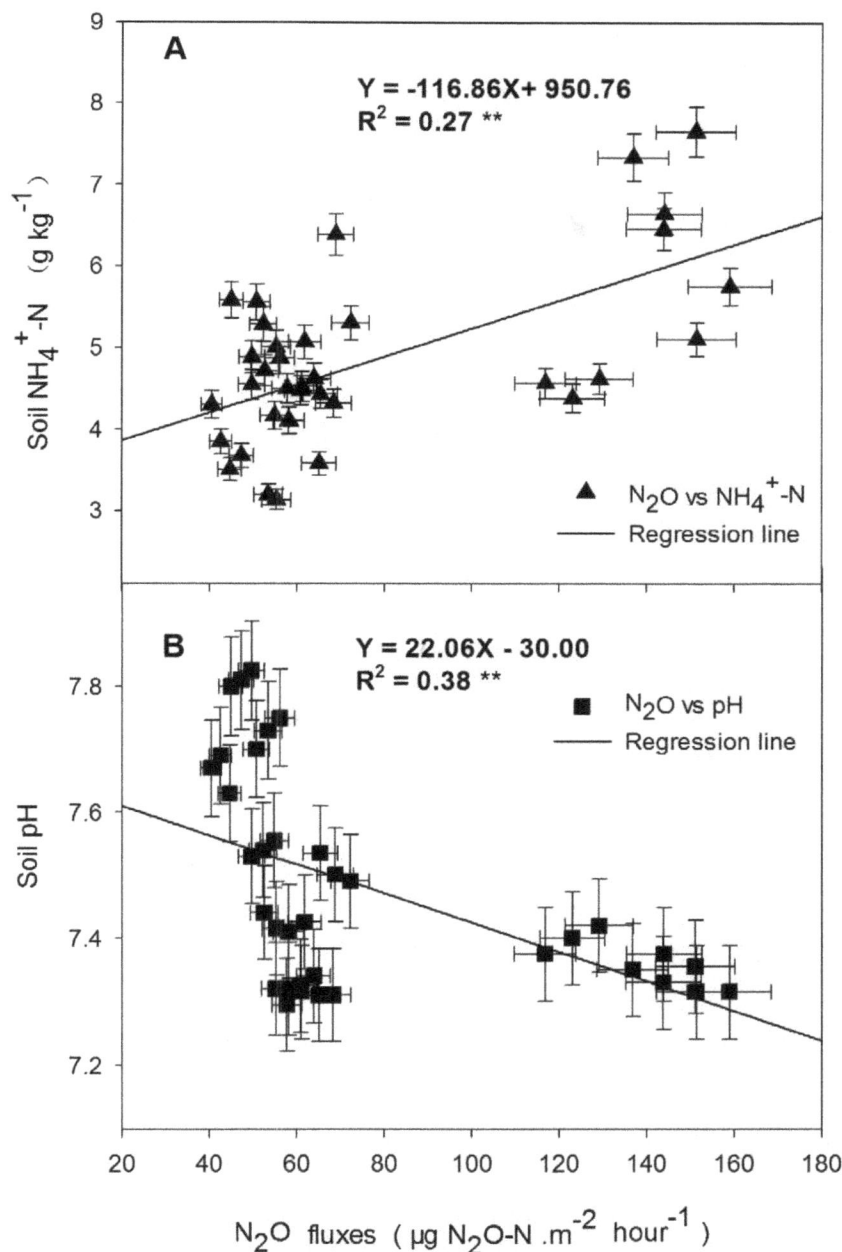

Figure 4. A Linear regression between the N_2O emission fluxes and soil NH_4^+-N, B Linear regression between the N_2O emission fluxes and soil pH. Arrows indicate the regression equation between the N_2O emission fluxes and soil NH_4^+-N, soil pH. **indicates $P<0.01$.

Statistical Analyses

The data were analyzed using analyses of variance and the SPSS 17.0 Statistical Analysis System and were mapped using Sigma Plot 10.0. The mean standard deviation and least significant difference were calculated for comparison of the treatment means.

Results

CH4 and N2O

Differences in CH_4 flux were observed when converting from HT to HTS, from RT to RTS and from NT to NTS (Figs. 2 A to

C). The soil absorption of CH_4 increased in different periods after conversion to subsoiling compared with the control. The soil absorption of CH_4 increased from 13.53 $\mu g \cdot m^{-2} \cdot h^{-1}$ under HT to 16.72 $\mu g \cdot m^{-2} \cdot h^{-1}$ under HTS, from 15.59 $\mu g \cdot m^{-2} \cdot h^{-1}$ under RT to 18.20 $\mu g \cdot m^{-2} \cdot h^{-1}$ under RTS and from 9.01 $\mu g \cdot m^{-2} \cdot h^{-1}$ under NT to 11.36 $\mu g \cdot m^{-2} \cdot h^{-1}$ under NTS, respectively. However, N_2O emission also increased after subsoiling (Fig. 2 D to F), which increased from 49.07 $\mu g \cdot m^{-2} \cdot h^{-1}$ under HT to 54.05 $\mu g \cdot m^{-2} \cdot h^{-1}$ under HTS and from 47.49 $\mu g \cdot m^{-2} \cdot h^{-1}$ under RT to 53.60 $\mu g \cdot m^{-2} \cdot h^{-1}$ under RTS. Compared with the above two treatments, however, the N_2O emissions from the

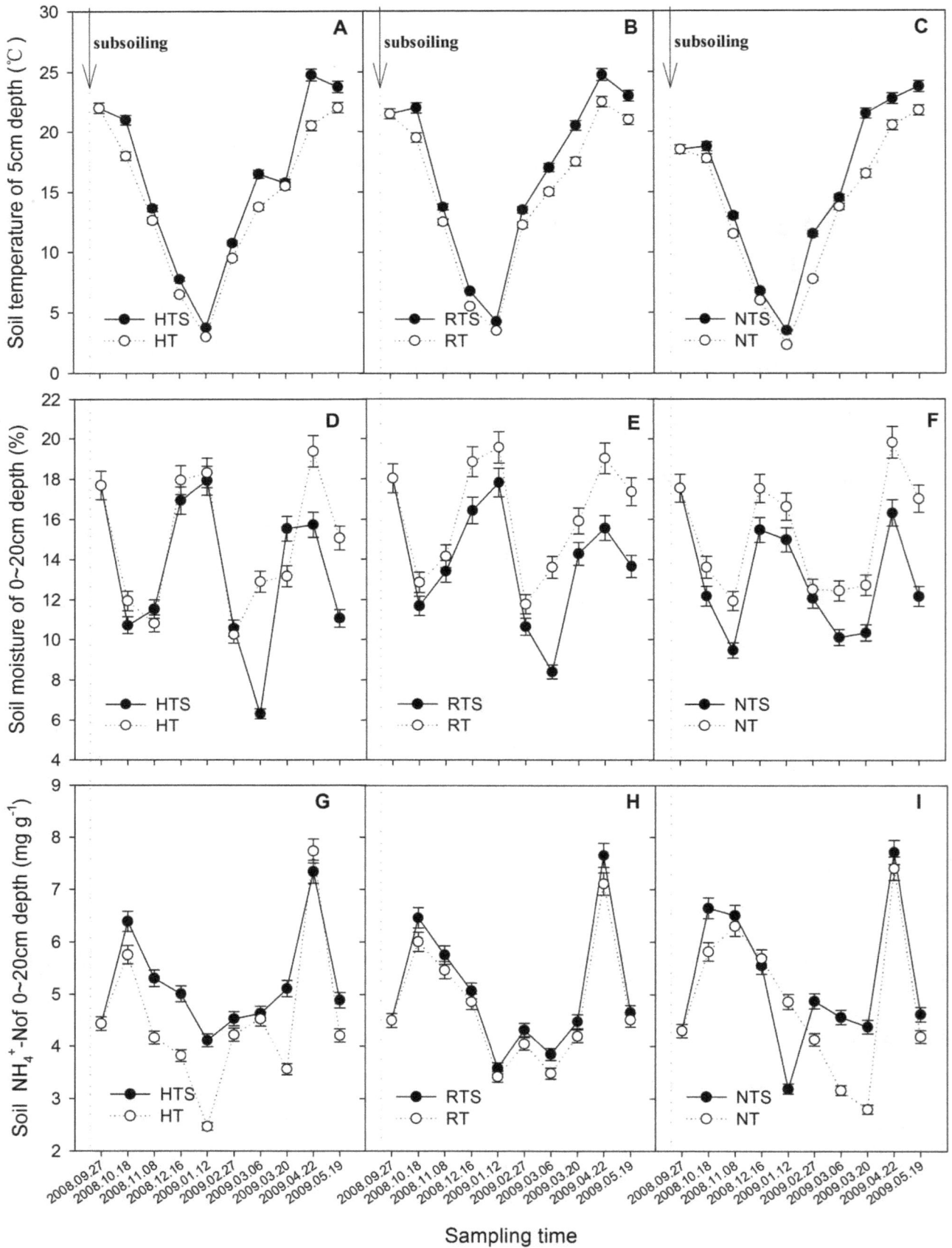

Figure 5. A to C Variation of Soil temperature at a 5 cm depth (°C) after subsoiling; D to F Variation of Soil water content at a 0~20 cm depth (%) after subsoiling; G to I Variation of Soil NH_4^+-N at a 0~20 cm depth (mg·kg^{-1}) after subsoiling. Arrows and the dotted line indicate time of subsoiling.

soil after conversion to NTS increased significantly, from 30.92 µg·m^{-2}·h^{-1} under NT to 55.15 µg·m^{-2}·h^{-1} under NTS.

GWP of CH_4 and N_2O

CH_4 uptake increased under HTS, RTS and NTS; consequently, the GWP of CH_4 decreased using these tilling methods compared with HT, RT and NT. However, the GWP of N_2O increased under HTS, RTS and NTS (Table 1). Overall, therefore, the GWPs of the CH_4 and N_2O emissions taken together increased from 0.32 kg CO_2 ha^{-1} under HT to 0.37 kg CO_2 ha^{-1} under HTS, from 0.37 kg CO_2 ha^{-1} under RT to 0.39 kg CO_2 ha^{-1} under RTS and from 0.26 kg CO_2 ha^{-1} under NT to 0.49 kg CO_2 ha^{-1} under NTS, respectively.

Correlation Analysis between CH_4 and N_2O and Soil Factors

Soil temperature significantly affected the CH_4 uptake in soils, especially in lower (i.e., December, $R^2 = 0.7314$, $P<0.01$; January, $R^2 = 0.6490$, $P<0.01$; February, $R^2 = 0.6597$, $P<0.01$) or higher (i.e., May, $R^2 = 0.8870$, $P<0.01$) temperatures ($P<0.01$) (Table 2). At other sampling times, however, temperature did not affect on CH_4 uptake, and soil moisture became a main influencing factor on the absorption of CH_4 by the soils, especially in wet soil, such as after rain ($R^2 = 0.5154$, $P<0.05$) and irrigation ($R^2 = 0.5154$, $P<0.05$), when CH_4 absorption was significantly limited ($R^2 = 0.5429$, $P<0.05$). Higher soil moisture generally promoted the emission of N_2O ($R^2 = 0.6735$, $P<0.01$), but there was no obvious correlation between soil temperature and N_2O emissions.

In this study, SOC was also correlated with greater CH_4 uptake ($R^2 = 0.12$, $P<0.05$) (Fig. 3 A), whereas higher soil pH limited its absorption in the soil ($R^2 = 0.14$, $P<0.05$) (Fig. 3 B).

The emission of N_2O was correlated with higher soil NH_4^+-N content ($R^2 = 0.27$, $P<0.01$) (Fig. 4 A), while, similar to CH_4, a higher pH in soil strongly limited the emission of N_2O ($R^2 = 0.38$, $P<0.01$) (Fig. 4 B).

Variation of Soil Factors

The soil factors under HTS, RTS and NTS changed after subsoiling. The soil temperature at a depth of 5 cm rose under HTS, RTS and NTS compared with the temperatures under HT, RT and NT (Fig. 5 A to C). Soil temperature variations followed atmospheric temperature changes, but the average soil temperature during sampling period increased from 13.5°C under HT to 15.3°C under HTS, from 14.4°C under RT to 16.2°C under RTS and from 13.1°C under NT to 15.1°C under NTS, respectively. However, soil moisture decreased in the soil at 0–20 cm when converting to subsoiling that in the order of RTS>HTS>NTS (Fig. 5 D to F). The most obvious decrease, by 15.74%, occurred under the NTS treatment, while HTS and RTS decreased by 10.34% and 14.85%, respectively. The soil NH_4^+-N content increased with subsoiling that was NTS>HTS>RTS. Moreover, two peaks occurring on October 18, 2008, and April 22, 2009 (Fig. 5 G to I), due to the application of nitrogenous base fertilizer and topdressing fertilizer.

The CH_4 uptake and N_2O emission were correlated with the content of soil pH and SOC (Table 3). The pH value decreased after conversions, but with the pH under the NTS treatment being higher than that of the HTS and RTS treatments not only at 0~10 cm but also at 10~20 cm. Conversely, SOC content increased under HTS, RTS and NTS, with the highest values was under RTS, followed by NTS and then HTS. SOC was higher in the soil at 0–10 cm than at 10–20 cm.

Grain Yield

The highest wheat yields under RT were 5937.20 kg ha^{-1} in 2009 and 6164.83 kg ha^{-1} in 2010, which were only 3.8% greater than those under HT and NT (Table 4). However, the wheat yields under HTS, RTS and NTS improved significantly ($P<0.01$) than the control, not only in 2009 but also in 2010. The average yield of the two years increased by approximately 2416.25 kg ha^{-1}, 1695.38 kg ha^{-1} and 2804.33 kg ha^{-1} with subsoiling compared with that under HT, RT and NT, respectively. The increases of average yield were not only related to the number of spikes, which increased by 59×10^4 ha^{-1} after conversions as determined by the average of the three conversion treatments, but were also correlated with the grains per ear and 1000-grain weight, which increased by an average of 6.0 grains and 2.8 g, respectively.

Table 3. Soil pH and SOC variations after conversion to subsoiling.

Treatments		pH						SOC					
		HT	HTS	RT	RTS	NT	NTS	HT	HTS	RT	RTS	NT	NTS
0~10 cm	(i)	7.37^c	7.33^d	7.25^e	7.21^f	7.72^a	7.66^b	8.62^f	9.45^e	9.69^d	11.47^b	11.79^a	10.32^c
	(ii)	7.25^d	7.21^e	7.27^c	7.25^d	7.69^a	7.62^b	10.77^d	12.25^a	9.82^f	10.21^e	11.68^c	11.93^b
	(iii)	7.25^e	7.23^f	7.38^a	7.34^c	7.37^b	7.31^d	11.43^d	12.58^b	12.07^c	13.11^a	10.13^e	9.75^f
	(iv)	7.44^{cd}	7.42^d	7.45^c	7.40^e	7.86^a	7.82^b	9.01^f	9.39^e	10.83^b	12.42^a	10.57^c	10.49^d
10~20 cm	(i)	7.71^c	7.67^d	7.52^e	7.46^f	7.77^a	7.75^b	5.93^f	6.29^e	9.10^b	9.44^a	8.09^d	8.34^c
	(ii)	7.46^c	7.43^d	7.36^e	7.35^f	7.85^a	7.83^b	9.22^f	9.97^d	9.45^e	10.07^c	11.35^b	11.77^a
	(iii)	7.44^c	7.40^d	7.39^e	7.37^f	7.56^a	7.52^b	9.76^f	10.62^c	10.11^e	10.40^d	10.88^b	11.76^a
	(iv)	7.71^c	7.68^d	7.43^e	7.43^e	7.83^a	7.81^b	7.63^f	9.90^a	8.26^d	9.55^b	8.31^c	7.84^e

Different small letter means $P<0.01$; (i), (ii), (iv) and (iii) means time of sample collection in 2008.10.18, 2009.03.17, 2009.04.20 and 2009.05.19 respectively.

Table 4. The wheat yield variations of HT, RT and NT after subsoiling from 2008–2010.

Treatments	Number of spikes ($10^4 \cdot ha^{-1}$)	Grains per ear	1000-grain weight (g)	Grain yield ($kg \cdot ha^{-1}$)	Increased ($kg \cdot ha^{-1}$)
2008–2009					
HT	646.50[bc]	30.05[bc]	33.79[b]	5582.83[b]	
HTS	683.50[a]	34.45[a]	34.31[b]	6866.55[a]	+1283.72
RT	655.00[b]	31.45[b]	33.94[b]	5937.20[b]	
RTS	637.50[c]	35.00[a]	36.83[a]	6985.20[a]	+1048.00
NT	583.00[d]	28.60[c]	32.40[c]	4595.87[c]	
NTS	688.50[a]	34.70[a]	33.96[b]	6895.06[a]	+2299.19
2009–2010					
HT	644.67[e]	30.93[e]	33.73[d]	5716.53[e]	
HTS	741.00[b]	38.59[a]	37.70[a]	9161.94[a]	+3548.77
RT	705.00[c]	31.68[d]	32.47[f]	6164.83[d]	
RTS	754.67[a]	35.78[c]	36.77[b]	8439.35[b]	+2342.76
NT	601.67[f]	28.02[f]	32.70[e]	4685.80[f]	
NTS	682.00[d]	37.72[b]	36.13[c]	7898.86[c]	+3309.46

Different small letter means $P<0.05$.

Discussion

Effect of Conversion to Subsoiling on CH_4 Uptake and N_2O Emissions

Long periods of shallow or no-tillage have resulted in an increase in soil bulk density and compacted hardpan in this region, especially in the subsoil [29,30], while subsoiling changed the soil structure, allowing increased gas diffusion in the soil. In this study, soils under HT conversion to HTS, RT conversion to RTS and NT conversion to NTS increased CH_4 absorption and strengthened the sink capacity of the soils (Fig. 2 A to C); however, these conversions also promoted the emission of N_2O (Fig. 2 D to F). This increase may be due to changes in soil conditions as a result of conversion to tillage (Fig. 5). For example, the increase in CH_4 absorption after conversion was mainly correlated with soil temperature, soil moisture, soil pH and SOC content according to the correlation analysis (Fig. 3 and Table 2), which is consistent with some previous studies [31–33]. A higher temperature and greater SOC may be advantageous to increasing the amount of CH_4 absorbed by the soil (Table 2, Fig. 3A) [34,35]. However, soil moisture and pH were two limiting factors in our study (Table 2, Fig. 3B) that had negative effects on CH_4 absorption in the soils [36].

At the same time, subsoiling would reduce subsoil compaction, and some have found improved permeability of soil to increased soil methane sinks [37] and higher bulk density to limit gas diffusion from the soil to the atmosphere, prolonging methane transfer pathways and thereby reducing CH_4 and O_2 diffusion between the soil and the atmosphere [38]. Sometimes, although increased soil tillage may slightly decrease CH_4 uptake [39], this effect is small and can be largely ignored [6,40].

The conditions for the aeration of the soil profile were reduced after irrigation [41,42] that increases emissions of the greenhouse gas N_2O through denitrification in farmland [22], the N_2O emission peaks also coincided with higher moisture and NH_4^+-N content in this study (Fig. 2 D to F, Table 2, Fig. 4A), the emissions of N_2O were significantly affected by soil moisture and NH_4^+-N content in each treatment. Some studies have indicated that there

is a significant linear relationship between N_2O emissions and soil moisture and nitrogenous fertilizer [21,22]. In addition, there was no significant correlation between N_2O emission and soil temperature in this study, and similar results were found by Koponen et al. [43]. In contrast, other studies found that at low temperatures, N_2O emissions may be hindered by soil N and water content [44,45]. However, in different experimental sites, N_2O emission was often related to increased soil temperature [46,47]. These studies demonstrated that when soil moisture and N fertilization were not limiting factors to N_2O emission, the rate of N_2O emission increased as soil temperature increased [22].

Similarly, soil pH also influenced N_2O production in soil (Fig. 4B). N_2 was mainly produced through denitrification when the soil pH was neutral, and the N_2O/N_2 ratio increased when soil pH decreased [48]. In our study, when soil pH values decreased with irrigation, N_2O emissions significantly increased, however, there was no relation to N_2O emission in periods of without irrigation, so soil pH does not directly cause soil GHG emissions [36] but via affected the action of microbes [49]. On the other hand, the predominant form of nitrogen is NO_3-N or NH_4-N after sufficient mixed between soil and straw through tillage, which may produced little N_2O in soil, particularly near the soil surface, with an important influence on N_2O emissions [12].

Therefore, the CH_4 uptake and N_2O emissions under HTS, RTS and NTS were higher than those under HT, RT and NT, respectively, due to the effect of subsoiling. Moreover, the emission differences of CH_4 and N_2O between HTS, RTS and NTS were largely due to the original tillage systems, because they had different background value of soil environment factors, these soil factors change extent after conversion highly affected on CH_4 and N_2O emissions among treatment in this study. Therefore, the variations in CH_4 uptake and N_2O emissions correlated with subsoiling are mainly due to alterations in soil conditions resulting from subsoiling, including soil temperature, moisture, NH_4^+-N, SOC and pH.

GWP of CH_4 and N_2O after Conversion to Subsoiling

Although there was a negative effect on the GWP of N_2O after conversion to subsoiling, the increased CH_4 absorption by soils partially counteracted this negative effect. The total GWP of CH_4 and N_2O increased slightly compare with the original tillage systems, especially under HTS and RTS (Table 1). Some previous studies reported that no-tillage is a better tillage system at mitigating GHG emissions [6,50], and the lowest GWP of CH_4 and N_2O was only measured under NT in this study. However, the GWP of CH_4 and N_2O would increase if NT was converted to NTS.

Yield Variation after Conversion to Subsoiling

In this study, the fields where the HT, RT and NT methods were previously used showed only slight improvements in wheat grain yields between two years (Table 4), possibly due to the subsoil hardpan. However, under HTS, RTS and NTS, the number of spikes, grains per ear and 1000-grain weight significantly increased, which is in agreement with other reports in which subsoiling was found to be an effective method to increase wheat production [51–53].

Conclusions

Significant variations were measured in CH_4 and N_2O emissions after conversion to subsoiling in the North China Plain. While the uptake of CH_4 improved greatly, N_2O emissions also increased after subsoiling. As a result, we demonstrated that the GWP would increase if converted from minimum or no-tillage to subsoiling, especially from no-tillage. Soil temperature, moisture, SOC, NH_4^+-N and pH also varied and were strongly related to CH_4 uptake and N_2O emissions. In addition, the original tillage systems had an important effect on soil factors and GWP variations after conversion to subsoiling. Therefore, the results of our study provide evidence that conversion from rotary tillage to subsoiling (RTS) or harrow tillage to subsoiling (HTS) had a lower GWP for CH_4 and N_2O compared with conversion from no-tillage to subsoiling (NTS), while the grain yields under both RTS and HTS increase. Therefore, we suggest that these two rotation tillage systems be developed in this region.

Author Contributions

Conceived and designed the experiments: ST TN ZL HH SC. Performed the experiments: ST HZ BW NL. Analyzed the data: ST TN. Contributed reagents/materials/analysis tools: ST TN. Wrote the paper: ST TN.

References

1. Forster P, Ramaswamy V, Artaxo P, Berntsen T, Betts R, et al. (2007) Changes in atmospheric constituents and in radiative forcing. In: Solomon S, Qin D, Manning M, Chen Z, Marquis M, et al., eds. Climate Change 2007: The Physical science basis. Contribution of working group I to the fourth assessment report of the intergovernmental panel on climate change. Cambridge University Press, Cambridge, United Kingdom and New York, NY, USA.
2. Bouwman AF (1990) Exchange of greenhouse gases between terrestrial ecosystems and the atmosphere. In: Bouwman AF, eds. Soils and the Greenhouse Effect. Wiley, Chichester, pp. 61–127.
3. Goulding KWT, Hütsch BW, Webster CP, Willison TW, Powlson DS (1995) The effect of agriculture on methane oxidation in the soil. Philip Transaction Royal Society London A 351: 313–325.
4. IPCC (2001) Climate change 2001, The scientific basis–contribution of work group I to the third assessment report of IPCC. Cambridge University Press, Cambridge.
5. Bruce CB, Albert S, John P, Parker (1999) Fields N_2O, CO_2 and CH_4 fluxes in relation to tillage, compaction and soil quality in Scotland. Soil and Tillage Research 53: 29–39.
6. Six J, Ogle SM, Breidt FJ, Conant RT, Mosier AR, et al. (2004) The potential to mitigate global warming with no-tillage management is only realized when practised in the long term. Global Change Biology 10: 155–160.
7. Lee J, Six J, King AP, Van Kessel C, Rolston DE (2006) Tillage and field scale controls on greenhouse gas emissions. Journal of Environment Quality 35: 714–725.
8. Bhatia A, Sasmal S, Jain N, Pathak H, Kumar R, et al. (2010) Mitigating nitrous oxide emission from soil under conventional and no-tillage in wheat using nitrification inhibitors. Agriculture Ecosystem & Environment 136: 247–253.
9. Zhang HL, Gao WS, Chen F, Zhu WS (2005) Prospects and present situation of conservation tillage. Journal of China Agriculture University 10: 16–20.
10. Chatskikh D, Olesen JE (2007) Soil tillage enhanced CO_2 and N_2O emissions from loamy sand soil under spring barley. Soil and Tillage Research 97: 5–18.
11. Elder JW, Lal R (2008) Tillage effects on gaseous emissions from an intensively farmed organic soil in North Central Ohio. Soil and Tillage Research 98: 45–55.
12. Bai XL, Zhang HL, Chen F, Sun GF, Hu Q, et al. (2010) Tillage effects on CH_4 and N_2O emission from double cropping paddy field. Transactions of the CSAE 26: 282–289.
13. Xu YC, Shen QR, Ran W (2002) Effects of no-tillage and application of manure on soil microbial biomass C, N, and P after sixteen years of cropping. Acta Pedologica Sinica 39: 89–96.
14. Bowen HD (1981) Alleviating mechanical impedance. In: Arkin GF, Taylor HM, eds. Modifying the Root Environment to Reduce Crop Stress. Published by the ASAE. St. Joseph, MI, pp. 21–57.
15. Balbuena HR, Aragon A, McDonagh P, Claverie J, Terminiello A (1998) Effect of three different tillage systems on penetration resistance and bulk density. In: Proceedings of the IV CADIR (Argentine Congress on Agricultural Engineering), vol. 1. pp.197–202.
16. Huang M, Li YJ, Wu JZ, Chen MC, Sun JK (2006) Effects of subsoiling and mulch tillage on soil properties and grain yield of winter wheat. Journal of Henan University Science and Technology 27: 74–77.

17. Qin HL, Gao WS, Ma YC, Ma L, Yin CM (2008) Effects of Subsoiling on Soil Moisture under No-tillage 2 Years Later. Science of Agriculture Sinica 41: 78–85.
18. Bradford MA, Ineson P, Wookey PA (2001) Role of CH_4 oxidation, production and transport in forest soil CH_4 flux. Soil Biology & Biochemistry 33: 1625–163.
19. Watanabe T, Kimura M, Asakawa S (2007) Dynamics of methanogenic archaeal communities based on rRNA analysis and their relation to methanogenic activity in Japanese paddy field soils. Soil Biology & Biochemistry 39: 2877–2887.
20. Zheng XH, Wang MX, Wang Y, She R, Shangguan X, et al. (1997) CH_4 and N_2O emissions from rice paddy fields in Southeast China. Scientia atmospherica Sinica 21: 231–237.
21. Merino P, Artetxe A, Castellon A, Menendez S, Aizpurua A, et al. (2012) Warming potential of N_2O emissions from rapeseed crop in Northern Spain. Soil & Tillage Research, 123: 29–34.
22. Gregorich EG, Rochette P, Vandenbygart AJ, Angers DA (2005) Greenhouse gas contributions of agricultural soils and potential mitigation practices in Eastern Canada. Soil and Tillage Research 83: 53–72.
23. Clayton H, Arah JRM, Smith KA (1994) Measurement of nitrous oxide emissions from fertilised grassland using closed chambers. Journal of Geophysical Research 99: 16599–16607.
24. Paustian K, Andren O, Janzen HH, Lal R, Smith P, et al. (1997) Agricultural soil as a C sink to offset CO_2 emissions. Soil Use and Management 13: 230–244.
25. Robertson G (1993) Fluxes of nitrous oxide and other nitrogen trace gases from intensively managed landscapes: a global perspective. In: Harpwr LA, Mosier AR, Duxbury JM, Rolston DE, (eds) Agricultural ecosystem effects on trace gases and global climate change. ASA Special Publication No. 55. ASA, CSSA, SSSA, Madison, wi 95–108.
26. Tian SZ, Ning TY, Chi SY, Wang Y, Wang BW, et al. (2012) Diurnal variations of the greenhouse gases emission and their optimal observation duration under different tillage systems. Acta. Ecol. Sinica. 32, 879–888.
27. IPCC (2007) Climate change 2007: The physical science basis. Contribution of working group I to the fourth assessment report of the intergovernmental panel on climate change. Cambridge University Press, Cambridge, United Kingdom and New York, NY, USA.
28. Bao SD (2000) Soil and Agricultural Chemistry Analysis. China Agriculture Press, Beijing.
29. Han B, Li ZJ, Wang Y, Ning TY, Zheng YH, et al. (2007) Effects of soil tillage and returning straw to soil on wheat growth status and yield. Transactions of the CSAE 23: 48–53.
30. Ahmad S, Li C, Dai G, Zhan M, Wang J, et al. (2009) Greenhouse gas emission from direct seeding paddy field under different rice tillage systems in central China. Soil and Tillage Research 106: 54–61.
31. Qi YC, Dong YS, Zhang S (2002) Methane fluxes of typical agricultural soil in the north china plain. Rural Ecology Environment 18: 56–60.
32. Wu FL, Zhang HL, Li L, Chen F, Huang FQ, et al. (2008) Characteristics of CH_4 Emission and Greenhouse Effects in Double Paddy Soil with Conservation Tillage. Science Agriculture Sinica 419: 2703–2709.
33. Dijkstra FA, Morgan JA, von Fischer JC, Follett RF (2011) Elevated CO_2 and warming effects on CH_4 uptake in a semiarid grassland below optimum soil moisture. Journal of Geophysical Research 116: 1–9.

34. Wang ZP, Han XG, Li LH (2005) Methane emission from small wetlands and implications for semiarid region budgets. Journal of Geophysical Research 110(D13): Art. No. D13304.

35. Bayer CL, Gomes J, Vieira FCB, Zanatta JA, Piccolo MC, et al. (2012) Methane emission from soil under long-term no-till cropping systems. Soil & Tillage Research, 124: 1–7.

36. Ouyang XJ, Zhou GY, Huang ZL, Peng SJ, Liu JX, et al. (2005) The incubation experiment studies on the influence of soil acidification on greenhouse gases emission. China Environment Science 25: 465–470.

37. Dong YH, Ou YZ (2005) Effects of organic manures on CO_2 and CH_4 fluxes of farmland. Chinese Journal of Applied Ecology 16: 1303–1307.

38. Ball BC, Scott A, Parker JP (1999) Field N_2O, CO_2 and CH_4 fluxes in relation to tillage, compaction and soil quality in Scotland. Soil and Tillage Research 53: 29–39.

39. Hütsch BW (1998) Tillage and land use effects on methane oxidation rates and their vertical profiles in soil. Biology and Fertilizer of Soils 27: 284–292.

40. Robertson GP, Paul EA, Harwood RR (2000) Greenhouse gases in intensive agriculture: Contributions of individual gases to the radiative forcing of the atmosphere. Science 289: 1922–1925.

41. Czyz EA (2004) Effects of traffic on soil aeration, bulk density and growth of spring barley. Soil & Tillage Research. 79, 153–166.

42. Berisso FE, Schjønning P, Keller T, Lamande M, Etana A, et al. (2012) Persistent effects of subsoil compaction on pore size distribution and gas transport in a loamy soil. Soil & Tillage Research 122: 41–45.

43. Koponen HT, Flojt L, Martikainen PJ (2004) Nitrous oxide emissions from agricultural soils at low temperatures: a laboratory microcosm study. Soil Biology & Biochemistry 36: 757–766.

44. Conen F, Dobbie KE, Smith KA (2000) Predicting N_2O emissions from agricultural land through related soil parameters. Global Change Biology 6: 417–426.

45. Sehy U, Ruser R, Munch J C (2003) Nitrous oxide fluxes from maize fields: relationship to yield, site-specific fertilization, and soil conditions. Agriculture Ecosystem & Environment 99: 97–111.

46. Groffman PM, Hardy JP, Driscoll CT, Fahey TJ (2006) Snow depth, soil freezing, and fluxes of carbon dioxide, nitrous oxide and methane in a northern hardwood forest. Global Change Biology 12: 1748–1760.

47. Rachphal S, Jassal T, Andrew B, Real R, Gilbert E (2011) Effect of nitrogen fertilization on soil CH_4 and N_2O fluxes, and soil and bole respiration. Geoderma 162: 182–186.

48. Daum N, Schenk MK (1998) Influence of nutrient solution pH on N_2O and N_2 emissions from a soilless culture system. Plant and Soil 203: 279–287.

49. Robertson LA, Kuenen JG (1991) Physiology of nitrifying and denitrifying bacteria. In: Rogers JE and Whitman WBC, (eds) Microbial production and consumption of greenhouse gases: Methane, Nitrogan oxides and Halo methane. American Society for microbiology Washington D. C., 189–199.

50. Lal R (2004b) Soil carbon sequestration impacts on global climate change and food security. Science 304: 1623–1627.

51. He J, Li HW, Gao HW (2006) Subsoiling effect and economic benefit under conservation tillage mode in Northern China. Transactions of the CSAE 22: 62–67.

52. Gong XJ, Qian CR, Yu Y, Zhao Y, Jiang YB, et al. (2009) Effects of Subsoiling and No-tillage on Soil Physical Characters and Corn Yield. Journal of Maize Science 17: 134–137.

53. Huang M, Wu JZ, Li YJ, Yao YQ, Zhang CJ, et al. (2009) Effects of different tillage management on production and yield of winter wheat in dryland. Transactions of the CSAE 25: 50–54.

19

How Early Can the Seeding Dates of Spring Wheat Be under Current and Future Climate in Saskatchewan, Canada?

Yong He[1,2], Hong Wang[1]*, Budong Qian[3], Brian McConkey[1], Ron DePauw[1]

1 Semiarid Prairie Agricultural Research Centre, Agriculture and Agri-Food Canada, Swift Current, Saskatchewan, Canada, **2** Department of Soil and Water Sciences, China Agricultural University, Beijing, China, **3** Eastern Cereal and Oilseed Research Centre, Agriculture and Agri-Food Canada, Ottawa, Ontario, Canada

Abstract

Background: Shorter growing season and water stress near wheat maturity are the main factors that presumably limit the yield potential of spring wheat due to late seeding in Saskatchewan, Canada. Advancing seeding dates can be a strategy to help producers mitigate the impact of climate change on spring wheat. It is unknown, however, how early farmers can seed while minimizing the risk of spring frost damage and the soil and machinery constraints.

Methodology/principal findings: This paper explores early seeding dates of spring wheat on the Canadian Prairies under current and projected future climate. To achieve this, (i) weather records from 1961 to 1990 were gathered at three sites with different soil and climate conditions in Saskatchewan, Canada; (ii) four climate databases that included a baseline (treated as historic weather climate during the period of 1961–1990) and three climate change scenarios (2040–2069) developed by the Canadian global climate model (GCM) with the forcing of three greenhouse gas (GHG) emission scenarios (A2, A1B and B1); (iii) seeding dates of spring wheat (*Triticum aestivum* L.) under baseline and projected future climate were predicted. Compared with the historical record of seeding dates, the predicted seeding dates were advanced under baseline climate for all sites using our seeding date model. Driven by the predicted temperature increase of the scenarios compared with baseline climate, all climate change scenarios projected significantly earlier seeding dates than those currently used. Compared to the baseline conditions, there is no reduction in grain yield because precipitation increases during sensitive growth stages of wheat, suggesting that there is potential to shift seeding to an earlier date. The average advancement of seeding dates varied among sites and chosen scenarios. The Swift Current (south-west) site has the highest potential for earlier seeding (7 to 11 days) whereas such advancement was small in the Melfort (north-east, 2 to 4 days) region.

Conclusions/significance: The extent of projected climate change in Saskatchewan indicates that growers in this region have the potential of earlier seeding. The results obtained in this study may be used for adaptation assessments of seeding dates under possible climate change to mitigate the impact of potential warming.

Editor: Alex J. Cannon, Pacific Climate Impacts Consortium, Canada

Funding: This study was funded by Agriculture and Agri-Food Canada (Abase project #96). The authors gratefully acknowledge support for this research from the Visiting Fellowships in Canadian Government Laboratories Program, managed by the Natural Science and Engineering Research Council of Canada. The funders had no role in study design, data collection and analysis, decision to publish, or preparation of the manuscript.

Competing Interests: The authors have declared that no competing interests exist.

* E-mail: hong.wang@agr.gc.ca

Introduction

The date of seeding is an important decision for wheat (*Triticum aestivum* L.) cultivation in Saskatchewan, Canada, since it has a significant impact on the timing of certain stages of phenological development, such as heading and ripening. This can have a profound impact on the damage the plants experience from adverse weather conditions during the growing season, or late season events such as a killing frost (in fall). Khan et al. [1] pointed out that the seeding date of spring wheat is more important than seeding rate for achieving greatest yields. For achieve high yields, therefore, growers must seed as early as possible during suitable weather conditions in the spring when the whole crop is not at sensitive stages when frost may occur. In addition to weather

conditions, the soil conditions can also be a key factor for seeding operations. In particular, soil water content and snow cover are assumed to be the main factors affecting seeding date [2,3], as wet soil may limit the access of seeding equipment to the field.

Using a four year seeding date trial, Gan et al. [4] found that seeding 10 to 12 days earlier than normal could increase grain yield of spring wheat near Swift Current Saskatchewan, Canada. This may be due to lower water stress during anthesis period. By using the concept of "seeding eras" classified by Major et al. [5] and the recent adoption of no-till and continuous cropping, there is a huge potential for early seeding under current and future conditions. With equipment and acreages getting larger, the need for earlier seeding becomes a necessity to ensure that all the land is seeded in a timely manner. These studies offer an opportunity to seek the potential of earlier seeding dates especially considering climate change. However, it is still unknown how early the farmers can seed with minimal risk of spring frost damage and have the land still trafficable while maximizing yields. To confirm the possibility of earlier seeding, long term analysis of a consistent source of seeding dates is required. However, such a requirement is hard to satisfy due to lack of consistency in data collection and tabulation methods.

To obtain a feasible seeding period, many empirical models were developed based on observed seeding dates. Bootsma and Suzuki [6] built a single linear model using average daily mean and mean maximum air temperature as variables to estimate the optimum seeding period. Bootsma et al. [7] further refined the model of Bootsma and Suzuki [6] to obtain a better regression relationship between predicted and observed seeding dates. Major et al. [5] provided a best-fit model of nonlinear relationships through the use of neural networks to predict seeding dates more close to the observed ones. These are very complex with numerous inputs and little user control, making it too difficult to be widely used. Another method of modelling seeding date, proposed by Bootsma and De Jong [3] was to estimate seeding dates of spring wheat with selected environmental criteria and then compare it with observed data. This method considers five criteria and requires that they all be met for 10 days (not necessarily consecutive) prior to seeding: ① Daily precipitation <2.5 mm; ② The snow cover <10 mm for the day; ③ ($\frac{3}{4}$Tmax+$\frac{1}{4}$Tmin) >7°C, where Tmax and Tmin are daily maximum and minimum air temperature; ④ The soil moisture at the top 5% of the soil profile must not exceed 90% Available Water Holding Capacity (AWHC) and ⑤ the next 7.5% of the profile must not exceed 95% AWHC. McGinn et al. [8] found that the method developed by Bootsma and DeJong [3] underpredicted the seeding dates at Swift Current, Saskatchewan. In order to get closer to the observed data, they adjusted the criteria such that the daily average temperature ($\frac{1}{2}$Tmax+$\frac{1}{2}$Tmin) must exceed 10°C and soil water in the top two soil zones must be less than 90% AWC. The adjusted air temperature and soil moisture criteria, combined with the same daily precipitation and snow cover criteria as Bootsma and DeJong [3] were to be met for 10 days for seeding to occur.

The cited studies all had similar conclusions, namely, that to create a relationship between these criteria and seeding dates, air temperature must exceed a certain value and precipitation and soil moisture must be below certain values for a period of days before seeding occurs. The predicted seeding dates of these models provided considerable information on seeding practices. However, those models aimed to match the observed seeding dates, not to achieve higher yields. The recorded seeding dates only indicate days when seeding occurred as other factors could affect the actual seeding date in a farm. Emphasising a significant relationship between actual and predicted seeding dates eliminates other earlier days that may have been suitable for seeding. In addition, the seeding dates estimated by these empirical models were not connected with yield response.

Inability to control and manipulate other contributing factors in the field makes using traditional field experiments and these empirical models difficult to investigate the effects of earlier seeding. Our previous field phenology study with the widely well-used crop model DSSAT-CSM (Decision Support System for Agrotechnology Transfer - Cropping System Model) [9,10] provided a basis in this study to model the early seeding dates in different agricultural districts across Saskatchewan. Since previous studies have shown that the DSSAT-CSM model could accurately predict measurements of crop growth in the region, we expect that this model can be used to investigate the effects of early seeding on

crop growth and yield. Important attributes of this approach are the ability to freely manipulate seeding periods as desired and to observe the response, which can be difficult to measure experimentally. The objective of this study is to estimate the earliest seeding dates under baseline climate and projected future climate in Saskatchewan, Canada.

Materials and Methods

Study sites and crop

Swift Current, Saskatoon and Melfort, were selected for this study (Fig. 1) and are representative of the brown, dark brown and grey-black soil zones of Saskatchewan, respectively. These sites span an increasing moisture gradient from south-west to northeast, with growing season (May to August) precipitation totals increasing from 188 mm at Swift Current, 202 mm at Saskatoon and 233 mm at Melfort [11]. The selection of these sites takes both the soil and climate into consideration.

Because spring wheat is grown in all crop districts and also because it accounts for almost 70% of the total wheat production in this study area [12], it was selected as the reference crop.

The DSSAT-CSM Model

The DSSAT-CSM, a widely used process-based modelling package [13], was selected for simulating a wheat production system. This model simulates wheat grain and biomass yield reasonably well in western Canada [10,14,15]. The model requires input parameters describing crop and soil characteristics as well as daily weather data. The latter include maximum and minimum temperature, global solar radiation and precipitation, which in the present case were obtained from Meteorological Service of Canada (MSC) [11].

Genetic coefficients of Canadian Prairie Spring wheat class were calibrated with the data collected by Jame and Cutforth [16] and tested using data from the long term New Rotation experiment near Swift Current [17]. In order to predict the long-term effect we used the Sequence Analysis option in the DSSAT-CSM to run the model. Soil property inputs for DSSAT-CSM (organic carbon, clay and silt in percent, pH, soil lower limit of plant extractable soil water, drained upper limit, saturated hydraulic conductivity, saturated upper limit, and bulk density) were observed on site (Table 1). The management used for the simulations was a continuous wheat rotation under no-till with the seeding depth of 5 cm. Nitrogen fertilizer was applied at a rate of 0.1 t ha^{-1} at planting time.

Seeding dates modelling

Four selected environmental criteria (shown in Table 2) were used to restrict the seeding date. The criterion of daily precipitation, snow on ground and soil moisture (top 5 cm) was kept the same as Bootsma and De Jong [3] and McGinn et al. [8]. However, a more practical temperature criterion for the Canadian Prairies (Tmax>10°C) (Stewart Brandt, personal communication) was selected instead of criteria that was used to fit observed seeding dates. Only the top soil (5 cm) moisture was used as a restriction of equipment access. This is not only because soil moisture in many parts of the Prairies in early spring is well below field capacity [18] but also because of better seeding equipment that can allow seeding even if the soil moisture at the deeper depth is equal to 95% soil water capacity (Stewart Brandt, personal communication). Seeding was predicted when conditions met the entire selected criterion simultaneously for four consecutive days.

The daily maximum air temperature and precipitation were obtained from MSC. The soil moisture was simulated by DSSAT

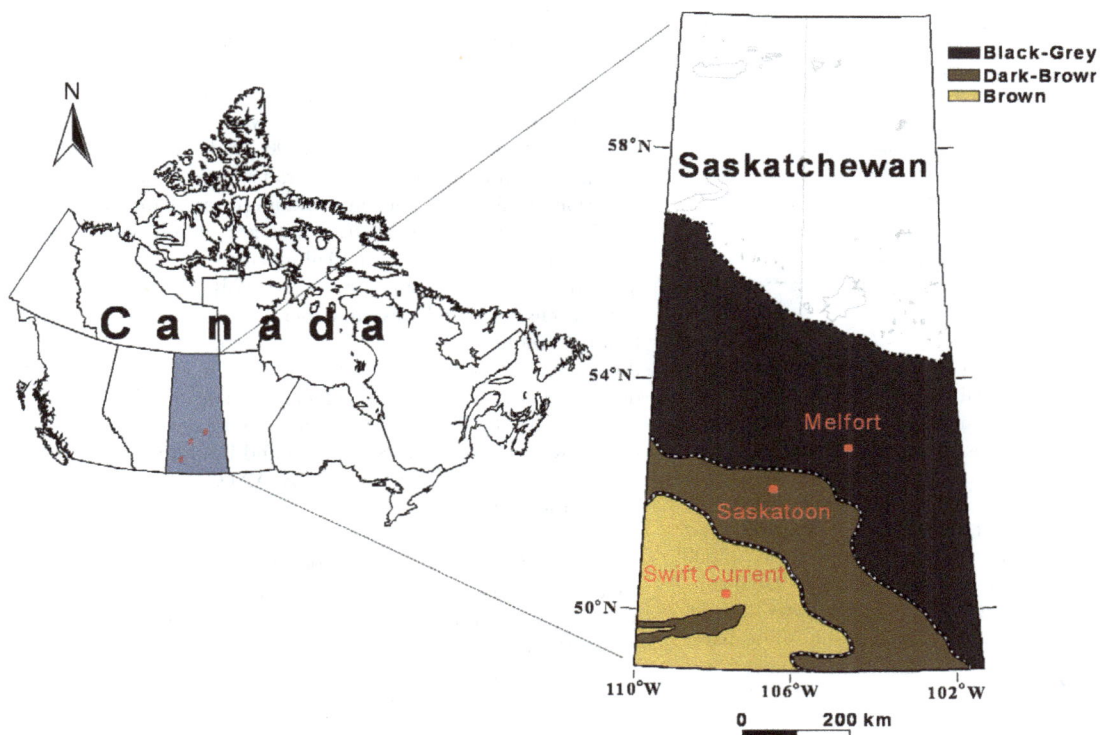

Figure 1. Map of Saskatchewan in Canada, showing major soil zones, and locations of the sites used in this study.

model. Since historic soil water content at each site was unknown, the model was initialized to start five years prior to the period of analysis. A spin-up period of five years was found to be adequate to provide the model with sufficient time for the soil water values to stabilize and not be affected by the initial input values. Through

Table 1. Soil physical properties in the experimental field.

	Depth (cm)	Particle size distribution (%)			Bulk density g cm^{-3}	SLLL[a]	SDUL[b]	SSKS[c] cm d^{-1}	SSAT[d]	Organic C (%)	pH
		Clay	Silt	Sand							
	0–15	18	50	31	1.15	0.09	0.28	1.42	0.30	1.8	6.5
	15–30	24	49	26	1.25	0.09	0.28	1.43	0.30	1.5	6.6
	30–45	23	52	24	1.35	0.09	0.28	2.76	0.32	1.3	7.0
Swift Current	45–60	25	33	41	1.4.0	0.10	0.27	4.32	0.34	0.9	7.7
	60–90	32	28	40	1.45	0.10	0.27	5.61	0.34	0.5	7.8
	90–120	33	26	41	1.55	0.10	0.25	3.04	0.34	0.3	8.1
	120–150	33	28	40	1.65	0.10	0.25	3.04	0.34	0.1	8.5
	0–15	60	29	11	1.30	0.11	0.37	7.40	0.59	2.9	7.0
	15–45	60	28	12	1.40	0.11	0.37	7.68	0.57	1.0	7.3
Saskatoon	45–80	60	28	12	1.50	0.11	0.36	5.08	0.53	0.3	7.5
	80–100	60	37	3	1.50	0.11	0.36	1.38	0.53	0.3	8.0
	0–19	50	40	10	1.20	0.20	0.45	0.34	0.55	4.5	6.8
Melfort	19–58	49	44	7	1.30	0.20	0.46	0.29	0.51	1.3	7.0
	58–100	52	45	3	1.30	0.20	0.46	0.25	0.51	0.5	8.0

[a]Lower limit of plant extractable soil water (cm^3 cm^{-3}).
[b]Drained upper limit (cm^3 cm^{-3}).
[c]Saturated hydraulic conductivity cm d^{-1}.
[d]Saturated upper limit (cm^3 cm^{-3}).

Table 2. Criteria used in estimating seeding dates of spring wheat.

Variable	Criterion
Daily maximum air temperature	>10°C
Daily precipitation	<2 mm
Snow on ground	<10 mm
Soil moisture	<0.9 soil water capacity

repeated wetting and drying of soils over time, the water content becomes less dependent on the initial estimates [9].

To get the initial conditions (such as soil moisture and snow cover) that are not available from observations, an initial (first) run of the DSSAT model was conducted. It is possible to run only a water balance sub-model with no crop, which is often the case when there is a wheat-fallow rotation system. However, the modelling result with a crop will produce more vivid initial conditions that are closer to the true condition. For this initial run, 15, 20 and 25 March were chosen as a fixed seeding dates in Swift Current, Saskatoon and Melfort, respectively. While those fixed days are a rough approximation of seeding dates, those dates are not expected to have a significant effect on soil moisture during the first few weeks of seedling emergence and early crop growth. During this time, a crop in its very early stages will not influence the water use amounts to any great extent since the main source of water removal would be through evaporation from the soil surface rather than transpiration from the crop. Following the initial model run, the daily spring soil moisture contents were applied to the seeding date criteria.

Before prediction of seeding date, the initial modelling day should be set in advance of expected seeding since the criteria may be met at a time (such as the beginning of March) where the spring wheat can be killed by frost in the following days. To avoid such a circumstance, Bootsma and DeJong [3] selected 15 April as initial day of modelling. This may have been suitable for their study sites. However, the initial day may vary from place to place and time to time due to spatial and temporal variability of weather conditions. A reasonable determination of the initial modelling data should be set according to long term weather records, so the crop has a lower probability of suffering a killing frost in spring. In our study, we set −8°C as a standard of last killing frost in the spring based on previous studies on spring wheat [19]. The lower and upper deciles (10% and 90% probability) of last killing frost were calculated according to the baseline and projected climate. According to the qualitative description provided by the Intergovernmental Panel on Climate Change (IPCC) [20], an event with a probability <10% is very unlikely to occur. In addition, 10% risk is also used as a criterion for crop insurance companies in western Canada [21]. Therefore, this characteristic of probability distribution can effectively describe the probability of killing frost, which is a key characteristic to evaluate for avoidance of agroclimatic risk. However, the temperature of killing frost may be higher or lower depending on the hardening process. In the case of hardening, the prior sequence of temperature events is important when determining the killing frost temperature, a complicated process beyond the scope of this paper.

Baseline and Climate Change Scenarios

Thirty years of current climate data are normally used in developing a baseline climate scenario [22]. A 30 year period is

considered adequate to include a good representation of wet, dry, warm, or cool periods. Selecting a recent 30 year period is preferred because it not only represents the current climate, but also, in most cases, has the most accurate data. In this study, the climate database for these three sites were derived from historic databases between 1961 and 1990 and consisted of daily maximum and minimum air temperature, precipitation and solar radiation. Daily maximum and minimum air temperatures and precipitation were obtained from weather stations located on or nearby the research sites [11]. Daily solar radiation was calculated using the Mountain Climate Simulator [23].

Four climate databases that included a baseline (treated as historic weather climate during the period of 1961–1990) and three climate change scenarios in 2050s (2040–2069) were developed by a Canadian global climate model (CGCM3) ([24,25] with the forcing of three greenhouse gas (GHG) emission scenarios (i.e., IPCC SRES A2, A1B and B1) [26]. Synthetic 300-yr weather data were generated by AAFC Stochastic Weather Generator (AAFC-WG) for the baseline period and for each scenario [27]. The three commonly used emission scenarios A2, A1B and B1 were used in order of greatest greenhouse gas emissions to least emissions by the end of the century [20]. Qian et al. [28] compared extremes between observed weather and synthetic data. They found the AAFC-WG was capable of reproducing extremes. Therefore, we used the synthetic baseline data instead of real historical weather recording to represent current climate because it only mimics observed weather data on concerned statistical properties and the synthetic data may be more matchable when compared with climate change scenarios [28]. These generated data were used to predict the climate effect on wheat production using the DSSAT-CSM model. Qian et al. [28] found that crop model simulations with 30-yr observed and 300-yr synthetic weather data generated by AAFC-WG with parameters calibrated from the same 30-yr observed data, in general and without considering the effect of seeding dates, do not show significant differences, with regard to timing of biomass accumulation, crop maturity date, as well as final biomass and grain yield at maturity.

Data Analysis and DSSAT model test

Statistical analyses were performed using SAS [29]. Means and standard deviation of synthetic air temperature and precipitation were calculated and compared among baseline and climate change scenarios by PROC MEANS. Mean difference of predicted and calculated variables were compared between scenarios with PROC MIXED.

Two statistical procedures were used to assess the level of agreement between the predicted value and observed data:

i.   Root mean square error (RMSE):

$$RMSE = \sqrt{\sum_{i=1}^{n} \frac{(P_i - O_i)^2}{n}} \qquad (1)$$

where P_i is the predicted value corresponding to the observed value O_i.

ii. Index of agreement (d) [30]:

$$d = 1 - \frac{\sum_{i=1}^{n}(O_i - P_i)^2}{\sum_{i=1}^{n}(|O'_i| - |P'_i|)^2} \qquad (2)$$

where $O'_i = O_i - O$ and $P'_i = P_i - P$, O is observed mean and P is predicted mean.

The closer the root mean square error (RMSE) is to 0, the more accurate the model is. Index of agreement (*d*) is a measure of the degree of deviation between observed and predicated. The value of *d* is unity when there is a perfect agreement [31].

Results and Discussion

Model test

Before application of the DSSAT model, the performance of the DSSAT model was evaluated. A ten year data set (1981–1990) in Swift Current with seeding dates, maturity dates and grain yield were used to test the DSSAT model. The variables tested included maturity date and the final grain yield, as recommended by Hunt et al. [32]. In general, the model gave good predictions of maturity dates and the final grain yield except for the maturity dates in years of 1983, 1986 and 1990 and grain yield in years of 1982 and 1987 (Table 3). Overall, the simulation of maturity dates and grain yield were acceptable with an overall root mean squared error (RMSE) value of 7.3 d and 0.86 t ha^{-1}, respectively. Bannayan et al. [33] simulated wheat growth with DSSAT and the model achieved root mean squared errors (RMSEs) of 10.0 d, for maturity dates. Jamieson et al. [34] compared 5 different wheat models in Australia, including CERES-Wheat, and found an RMSE of 0.9 t ha^{-1}. Considering that the model doesn't account for pest and disease incidence, the goodness of fit statistics of our results compared favourably to the above reported studies.

Baseline and Climate Change Scenarios

Air temperature.   The distribution of air temperature varied among sites under baseline climate. That is, the maximum temperature rose from about 20°C at the beginning of June to about 26°C by the end of July. Compared to the baseline climate, air temperature in all climate scenarios would increase (Fig. 2). The greatest differences in temperature occurred in the winter, followed by summer, and relatively small differences occurred in the spring and fall. Scenario A2 had the greatest increase in annual mean temperature (3.9, 3.6 and 3.5°C for Swift Current, Saskatoon and Melfort, respectively) (Table 4). Annual mean temperature increased from south-west (Swift Current) to north-

east (Melfort). The change in pattern and difference between scenarios in daily maximum and minimum air temperatures were similar to that in daily mean temperature (data not shown).

Precipitation.   Under baseline climate, the pattern of precipitation might be fortuitous for spring wheat, with more than 40% falling during the growing season (May to August). The climate change scenarios showed an increase in annual precipitation, compared to the baseline period (Fig. 3 and Table 4) for all sites. The increase in annual precipitation was greatest for A1B and followed by A2 and B1. Total precipitation is predicted to increase 52% above the baseline climate while the precipitation in July and August would decrease for all scenarios. Because precipitation was projected to be less in July and Aug, negative impacts on spring wheat. When precipitation declines, temperatures will increase and negatively impact spring wheat [35]. Precipitation from the A1B scenario was higher than the baseline in every month except from June to August when less rain was projected than the baseline for this period. Scenario A2 was similar to A1B in terms of precipitation distribution except that it was markedly less than A1B in June. Precipitation for scenario B1 was slightly less than that of A1B, except in July and August when B1 had more rain than A1B. Scenario A2 had much more precipitation in Melfort than Swift Current or Saskatoon.

Seeding dates

The initial dates used in the seeding date modelling (initial modelling day) for Swift Current, Saskatoon and Melfort are shown in Fig. 4A. The initial modelling day was estimated with a less than 10% probability of experiencing a killing frost. Using 100 years of historic weather recording may provide more reliable statistical results. However, due to the "smoothing" effect of long term data, 100 years of data may not reflect the variety of current weather conditions. Therefore, selection of recent 30 year historic weather data may be more reasonable [36].

As shown in Fig. 4A, under a baseline climate, the initial day for modelled seeding dates at Swift Current and Saskatoon were close to each other due to similar annual mean temperature. The initial modelling day for Melfort under the baseline scenario was 11 days later because of its much lower temperature than either Swift Current or Saskatoon. Among the three scenarios, Scenario A1B was closer to the baseline than B1 and A2, which coincides with the temperature distribution pattern.

Simulated baseline seeding dates for spring wheat were averaged for 30 years (Fig. 4B). The earliest averaged seeding date occurred on 16 April (day 106) in Swift Current. Seeding dates were delayed from Swift Current to Melfort, coinciding with lower temperatures influenced by latitude (south-north). The last average seeding occurred in Melfort after 3 May (day 123). The seeding dates under baseline conditions were closer for Swift Current and Saskatoon since their annual mean temperatures were similar.

Due to the temperature increase in the climate change scenarios, the seeding dates were advanced for all sites. For Scenario A2, the earliest seeding date occurred on 5 April (day 95) in Swift Current. For Scenario A1B, the seeding dates were also earliest in Swift Current, and occurred on 9 April (day 99). Again, because of having similar increases in annual mean temperature, there was little difference in seeding dates between Swift Current and Saskatoon in scenarios A1B and B1(Fig. 2 and Fig. 4B).

The results showed that the seeding dates for all sites advance significantly in comparison with the future scenarios and the baseline (Fig. 4B). The change in seeding date between the scenarios and baseline exhibited a south-north trend with the southern areas experiencing a larger increase. In Swift Current,

Table 3. Index of agreement (*d*) of maturity dates and grain yield.

| Year | d-index | |
	Maturity	Yield
1981	−1.37	0.674
1982	0.945	0.260
1983	0.276	0.662
1984	0.593	0.972
1985	0.960	0.796
1986	0.442	0.521
1987	0.748	0.001
1988	0.992	1.000
1989	0.950	0.959
1990	0.111	0.674

Note: both maturity and yield were measured only once each year.

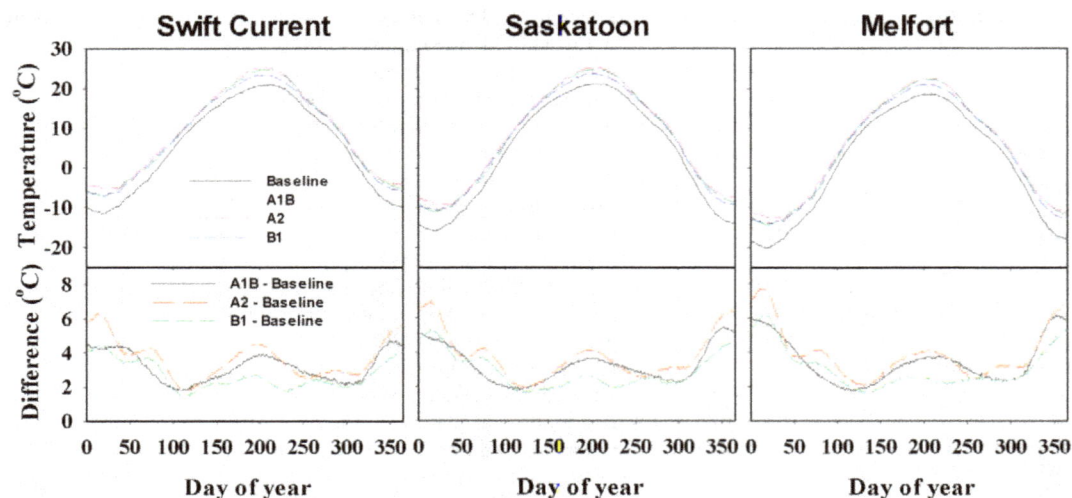

Figure 2. Daily mean temperature under baseline and three scenarios for 2050s, relative to the baseline period 1961–1990, and difference between the baseline and scenarios.

the seeding date was advanced by at least 7 days. Such advancement, however, is possible because i) the seeding date was predicted after the initial modelling day, which has <10% probability of being killed by spring frost; ii) the earlier seeding as recommended by Smith et al. [37] not only ensures a longer growing season, it also reduces the incidence of drought and heat stress during sensitive crop development stages and iii) an acceptable early seeding date should be set based on its corresponding yield, which will be discussed below.

In Melfort, the seeding dates under climate change were close to the baseline seeding dates. Although the increase in annual mean temperature at Melfort was higher, it still wasn't large enough to meet the temperature criterion (10°C), which may explain the minor advance of seeding date at Melfort. Our results are similar to McGinn et al. [8], who also modeled seeding date under climate change scenarios on the Canadian Prairies.

Phasic Development and precipitation received during critical period

Understanding phasic development of spring wheat is important in matching management decisions and crop inputs with plant development. Although the growth stages under current climate can be recorded, the phenological development and growth in response to environmental factors (soils, weather and manage-

ment) information is not available in future or projected climate. This problem highlights the importance of using a crop model.

The predicted seeding dates under all climate change scenarios were earlier than the prediction under the baseline climate for all sites. Because of the earlier seeding and higher temperatures, predicted dates of anthesis and early dough were 7 days earlier than simulations based on baseline climate for all sites (Fig. 5A, B). Crop maturity was advanced by more than 10 days (Fig. 5C). It is noticeable that the RMSE for maturity dates of model evaluation could be due to differences between sites or scenarios. This is understandable since the seeding date, climate and other input parameters are highly variable and thus affect the model output similarly. However, caution is still required since using such process-based crop models is not an error free technique.

The influence of water stress at various growth stages on spring wheat was investigated by Sionit et al. [38] and Mogensen et al. [39]. They found that the stage of anthesis was most sensitive to water stress, followed by the early dough stage. As shown in Table 4 and Fig. 3, although the total annual precipitation will be increased under a projected climate, monthly precipitation in June, July and August will decrease. Historically, average seeding dates are May 9, May 14 and May 15, for Swift Current, Saskatoon and Melfort, respectively. These dates are quite late when compared with our predicted seeding dates. If spring wheat

Table 4. Mean annual temperature and precipitation total of baseline and three scenarios for 2050s.

Items	Sites	Baseline	A2	A1B	B1
Annual mean temperature (°C)	Swift Current	6.9	10.4 (3.5)	10.0 (3.1)	9.5 (2.7)
	Saskatoon	5.6	9.2 (3.6)	8.7 (3.1)	8.4 (2.8)
	Melfort	1.1	5.0 (3.9)	4.5 (3.4)	4.1 (3.0)
Annual precipitation total (mm)	Swift Current	332	371 (38.8)	387 (54.7)	369 (37.0)
	Saskatoon	349	394 (44.3)	402 (53.2)	391 (42.0)
	Melfort	431	482 (51.0)	483 (52.6)	469 (38.0)

Values in parentheses indicate the change from baseline values.

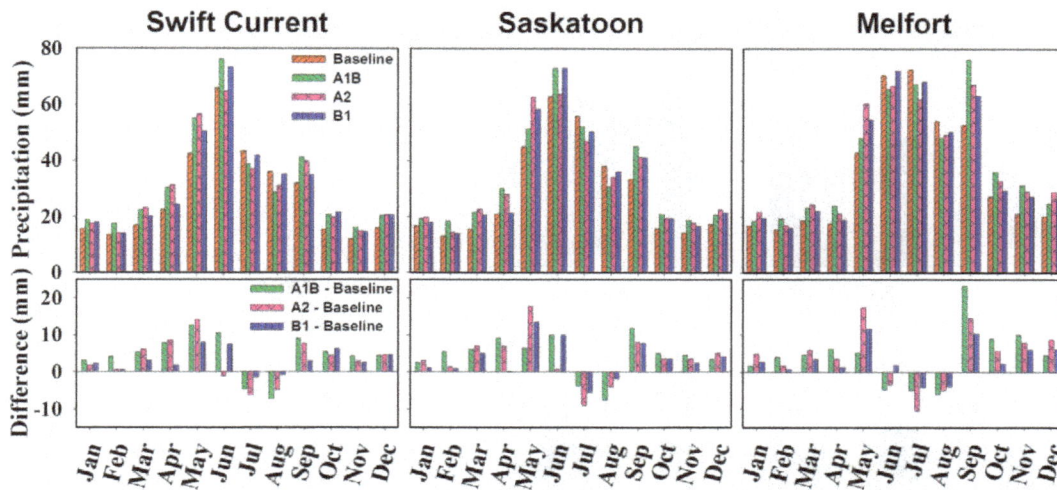

Figure 3. Monthly precipitation under baseline and three scenarios for 2050s, relative to period 1961–1990, and difference between baseline and scenarios.

is not seeded earlier than it is currently, it will undergo a severe water stress period under our climate change scenarios.

In order to compare the water stress experienced by spring wheat under baseline climate and projected climate change, we calculated the accumulated precipitation received between seeding and the early dough stage. As shown in Table 5, all scenarios will receive more precipitation during this period when compared with baseline climate due to advanced seeding dates. The increase in precipitation varied among scenarios, of which A1B has highest precipitation in Swift Current and Saskatoon. Compared with Melfort, Swift Current and Saskatoon will receive more precip-

itation both for baseline and projected climate, which means these two sites have higher yield potential in yield for increase.

The growth period can be obtained by calculating the difference between seeding and maturity date. The average simulated growth period for baseline, A1B, A2 and B1 in Swift Current is 102, 103, 104 and 106 days, respectively. This is 5 days longer than our observed (97 days) in Swift Current (1981–1990). Based on a three year cultivar×environment test in a rainfed area, Anderson et al. [40] reported that early seeded long-season cultivars tend to outyielded mid-season cultivars with earlier or late seeding treatment. In a short season environment, Kerr et al. [41] also

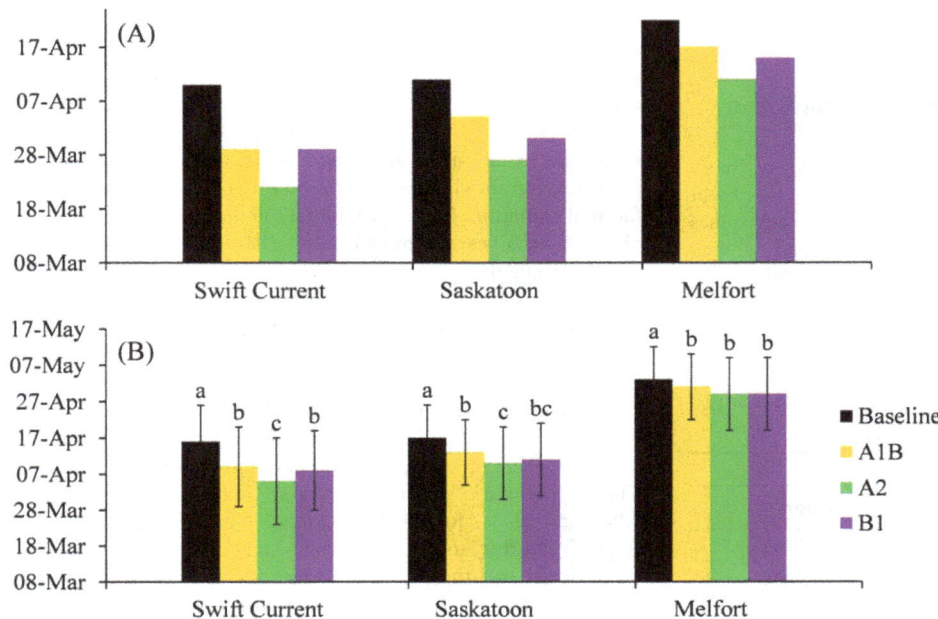

Figure 4. Initial modelling day of baseline and three scenarios for 2050s (A) and averaged seeding dates of baseline and three scenarios for 2050s (B). Within each site, values followed by the same letter are not significantly different at the 0.05 level of probability. Vertical bar indicate standard deviation (days).

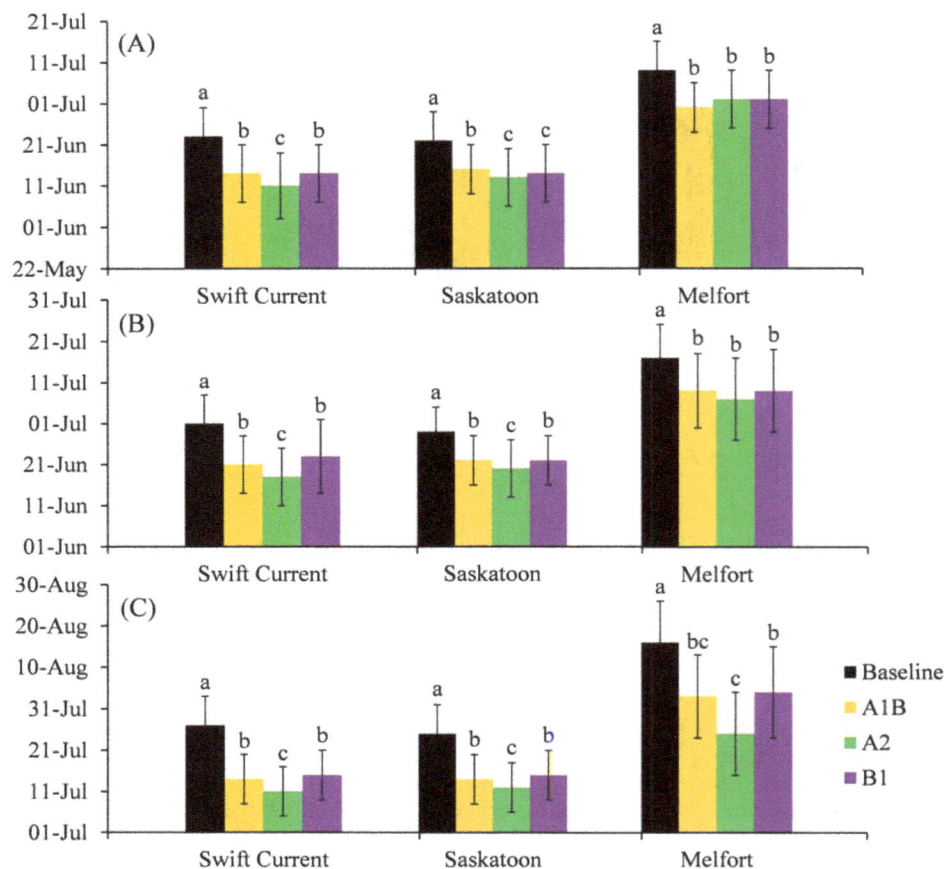

Figure 5. Effects of climate change on phasic development of wheat in (A) anthesis, (B) early dough and (C) maturity. Within each site, values followed by the same letter are not significantly different at the 0.05 level of probability. Vertical bar indicate standard deviation (days).

found that cultivars with longer growth period can achieve higher yield if seeded early. Our results agree with these reports.

Grain Yield response to simulated seeding dates and climate change

Compared with long-term observations (1961–1990) of maximum grain yield, the grain yield under baseline conditions at Swift Current and Saskatoon were very close to observed values (Table 6). This means the advancement of predicted seeding dates were reasonable under current weather conditions. In Melfort, the simulated grain yield under baseline conditions was much higher than the observed value. The higher grain yield

Table 5. Precipitation received during growth stage from seeding to early dough.

Scenarios	Swift Current	Saskatoon	Melfort
Baseline	128 (45.3)d†	134 (48.5)c	142 (51.4)c
A1B	164 (57.5)a	163 (57.6)a	147 (52.3)b
A2	155 (51.8)b	162 (56.3)a	158 (54.8)a
B1	150 (49.7)c	151 (56.4)b	158 (54.7)a

†Within columns, values followed by the same letter are not significantly different at the 0.05 level of probability.

simulated in Melfort was likely due to model insensitivity to disease and pest damage. This site may experiences losses due to insects such as grasshoppers, wheat stem maggots, leaf spots and wireworms [42,15].

Chipanshi et al. [15] modeled spring wheat grain yield at the same sites as our study and found the most sensitive parameters influencing grain yield were initial soil moisture, seeding date and thermal condition. In our case initial soil moisture conditions were obtained by 5 years of model training and the thermal condition was determined by weather conditions. Therefore, the grain yield might have been affected by combination of seeding dates and weather condition. The effect of seeding dates in the short-term may not obviously be due to its combination with weather conditions. Through repeated wetting and drying or warm and cold years, the long term effect of seeding date becomes obvious.

The model predicted that all three climate change scenarios had a positive effect on grain yield compared to the baseline (Table 6). The A1B scenario showed the largest yield increases, with 40%, 32% and 25% for Swift Current, Saskatoon and Melfort, respectively due to the projected highest precipitation, while Scenario B1 had the smallest increase for all sites. McGinn et al. [8] and Barklacich and Stewart [43] modeled spring wheat grain yield under climate change scenarios on the Canadian Prairies with similar increased annual mean temperature and total precipitation. They also reported a similar increasing trend in grain yield. Therefore, early seeding for farmers in these areas can

Table 6. Effects of climate change on grain production of wheat.

	Grain Yield t ha^{-1}		
Scenario	Swift Current	Saskatoon	Melfort
Baseline	2.16 (0.141)c[1]	2.16 (0.223)b	3.42 (0.334)b
A1B	3.02 (0.239)a	2.85 (0.176)a	4.29 (0.267)a
A2	2.73 (0.180)b	2.83 (0.132)a	4.32 (0.150)a
B1	2.58 (0.223)b	2.79 (0.189)a	4.04 (0.346)a
Historic maximum recording (1960–1990)[2]	2.21	2.24	2.53

[1]Within columns, values followed by the same letter are not significantly different at the 0.05 level of probability.
[2]From reference [15].
Values in parentheses are the standard deviation.

benefit from increasing yield but also an employing an important strategy to avoid drought stress in future climate.

Conclusions

The climatic scenarios in 2040–2069, projected by CGCM3, were used to identify the potential of early seeding dates in agricultural regions on the Canadian Prairies as an adaptation strategy to projected climate change. We predicted earlier seeding dates in a projected future climate for three representative sites located at different latitudes with different soil and weather conditions on the Canadian Prairies. These estimates seemed appropriate as the anticipated yields with earlier seeding dates as predicted by the crop model were similar to or higher than those with actual seeding dates. The average advancement is dependent on the site and the climate scenario. The Swift Current (south-west) site has highest potential for earlier seeding whereas such advancement was small in Melfort (north-east). Our results demonstrated that advancing seeding dates could be helpful in avoiding moisture stress during the sensitive wheat growth stages and to extend the growth period of wheat.

This study was not intended to provide optimal seeding dates but the potential of early ones. The optimum seeding date involves not only reducing unfavorable risk factors (like frost) but also favorable factors like optimum radiation, temperature and moisture. Optimum seeding dates will be investigated in future studies. Caution is required when using process-based crop models to check the simulated seeding dates, since such modelling is not an error free technique and the ability of the model to predict the complicated trait×environment interactions depends on the assumptions made in the model. Nevertheless, such an approach allows us to investigate interactions among seeding date and other environment factors in a way that is not easy in a field experiment. One limitation of the seeding date model in our study is the assumption that the future sowing techniques will be similar to the current era, whereas the seeding date based on "real techniques" in the future may be different. Caution also must be exercised when interpreting the model-simulated results as the effect of heat stress on wheat growth is not well described by the model. Heat stress occurs often in wheat on the Canadian Prairies especially during reproductive growth, which has markedly negative impacts on yield [44]. Under climate change scenarios, the occurrence of heat stress or heat shock may become more frequent. Adaptation measures must be taken with regard to the projected high temperature under climate change. One possible strategy is early seeding. This would allow wheat to mature earlier, avoiding heat shock, which will mostly occur in July (Fig. 2). In addition, the impact of seeding dates in our study depends upon the GHG scenarios used to generate the results. New revisions may produce different impacts of the agroclimate on Prairie agriculture.

Acknowledgments

We are grateful to Stewart Brandt (Retired Agronomy Scientist- AAFC-Scott, SK), who gave us constructive suggestions on selection criteria for seeding dates. We also thank Kelsey Brandt and Chantal Hamel (AAFC-Swift Current, SK) for their assistance on this paper.

Author Contributions

Conceived and designed the experiments: HW YH. Analyzed the data: HW YH BQ. Contributed reagents/materials/analysis tools: HW YH BQ. Wrote the paper: YH HW BQ RD. Gave advice and suggestions on restrictions of seeding date of wheat in Saskatchewan, Canada: BM.

References

1. Khan M, Donald WW, Prato T (1996) Spring Wheat (Triticum aestivum) Management Can Substitute for Diclofop for Foxtail (Setaria spp.) Control. Weed 44: 362–372.
2. Hassan AE, Broughton RS (1975) Soil moisture criteria for tractability. Canadian Agricultural Engineering 17(2): 124–129.
3. Bootsma A, De Jong R (1988) Estimates of seeding dates of spring wheat on the Canadian Prairies from climate data. Canadian Journal of Plant Science 68: 513–517.
4. Gan YT, Zentner B, McConkey B (2000) Wise seeding date decisions-discover the hidden value. Research Letter (Semiarid Prairie Agricultural Research Centre). April 14.
5. Major DJ, Hill BD, Toure A (1996) Prediction of seeding date in southern Alberta. Canadian Journal of Plant Science 76: 59–75.
6. Bootsma A, Suzuki M (1986) Zonation of optimum seeding period of winter wheat based on autumn temperatures. Canadian Journal of Plant Science 66: 789–793.
7. Bootsma A, Andrews CJ, Hoekstra Gl, Seaman WL, Smid AE (1992) Estimated optimum seeding dates for winter wheat in Ontario. Canadian Journal of Plant Science 73: 389–396.
8. McGinn SM, Touré A, Akinremi OO, Major DJ, Barr AG (1999) Agroclimatic and crop response to climate change in Alberta. Canada. Outlook on Agriculture 28: 19–28
9. Wang H, Cutforth H, McCaig T, McLeod G, Brandt K, et al. (2009) Predicting the time to 50% seedling emergence in wheat using a Beta model. Wageningen Journal of Life Sciences 57: 65–71.
10. Wang H, Flerchinger GN, Lemke R, Brandt K, Goddard T, et al. (2010) Improving SHAW long–term soil moisture prediction for continuous wheat rotations, Alberta, Canada. Canadian Journal of Plant Science 90: 37–53.
11. Environment Canada (1993) Canadian Climate Normals. Canadian Climate Program, Ottawa, Canada.

12. Quiring SM, Papakryiakou TN (2003) An evaluation of agricultural drought indices for the Canadian Prairies. Agricultural and Forest Meteorology 118: 49–62.

13. Jones JW, Hoogenboom G, Porter CH, Boote KJ, Batchelor WD, et al. (2003) The DSSAT cropping system model. European Journal of Agronomy 18: 235–265.

14. Moulin AP, Beckie HJ (1993) Evaluation of the CERES and EPIC models for predicting spring wheat grain yield. Canadian Journal of Plant Science 73: 713–719.

15. Chipanshi AC, Ripley EA, Lawford RG (1999) Large–scale simulation of wheat yields in a semi–arid environment using a crop–growth model. Agricultural Systems 59: 57–66.

16. Jame YW, Cutforth HW (2004) Simulating the effects of temperature and seeding depth on germination and emergence of spring wheat. Agricultural and Forest Meteorology 124: 207–218.

17. Zentner RP, Campbell CA, Selles F, McConkey BG, Jefferson PG, et al. (2003) Cropping frequency, wheat classes and flexible rotations: Effects on production, nitrogen economy, and water use in a Brown Chernozem. Canadian Journal of Plant Science 83: 667–680.

18. De Jong R, Shields JA, Sly WK (1984) Estimated soil water reserves applicable to a wheat-fallow rotation for general soil areas mapped in southern Saskatchewan. Canadian Journal of Soil Science 64: 667–680.

19. Macdowa FDH (1974) Growth kinetics of Marquis wheat. VI. Genetic dependence and winter hardening. Canadian Journal of Botany 52: 151–157.

20. IPCC (2007) Climate Change 2007: The Physical Science Basis. Contribution of Working Group I to the Fourth Assessment Report of the Intergovernmental Panel on Climate Change [Solomon, S., D. Qin, M. Manning, Z. Chen, M. Marquis, K.B. Averyt, M. Tignor and H.L. Miller (eds.)]. Cambridge University Press, Cambridge, United Kingdom and New York, NY, USA.

21. Manitoba Agriculture, Food and Rural Initiatives. Available: http://www.gov.mb.ca/agriculture/climate/waa50s00.html. Accessed 2012 July 1.

22. Alexandrov VA, Hoogenboom G (2000) Vulnerability and adaptation assessments of agricultural crops under climate change in the Southeastern USA. Theoretical and Applied Climatology 67: 45–63.

23. Thornton PE, Hasenauer H, White MA (2000) Simultaneous estimation of daily solar radiation and humidity from observed temperature and precipitation: an application over complex terrain in Austria. Agricultural and Forest Meteorology 104: 255–271.

24. Kim SJ, Flato GM, Boer GJ, McFarlane NA (2002) A coupled climate model simulation of the Last Glacial Maximum. I. Transient multi-decadal response. Climate Dynamics 19: 515–537.

25. Kim SJ, Flato GM, Boer GJ (2003) A coupled climate model simulation on the Last Glacial Maximum. II. Approach to equilibrium. Climate Dynamics 20: 636–661.

26. Nakicenovic N, Swart R (2000) Emissions Scenarios IPCC Special Report, 2000. Nebojsa Nakicenovic and Rob Swart (Eds.) – Cambridge University Press, UK, pp 570.

27. Qian BD, Gameda S, Hayhoe H, De Jong R, Bootsma A (2004) Comparison of LARS–WG and AAFC–WG stochastic weather generators for diverse Canadian climates. Climate Research 26: 175–191.

28. Qian B, De Jong R, Yang J, Wang H, Gameda S (2011) Comparing simulated crop yields with observed and synthetic weather data. Agricultural and Forest Meteorology 151: 1781–1791.

29. SAS Institute, Inc (1999) SAS procedures guide. Version 8. SAS Institute, Inc., Cary, NC, USA.

30. Willmott CJ (1981) On the validation of models. Physical Geography 2: 184–194.

31. Nangia V, Gowda PH, Mulla DJ, Sands GR (2008) Water quality modeling of fertilizer management impacts on nitrate losses in tile drains at the field scale. Journal of Environmental Quality 37: 296–307.

32. Hunt LA, Pararajasingham S, Jones JW, Hoogenboom G, Imamura DT, et al. (1993) GENCALC-software to facilitate the use of crop models for analyzing field experiments. Agronomy Journal 85: 1090–1094.

33. Bannayan M, Crout NMJ, Hoogenboom G (2003) Application of the CERES-Wheat model for within-season prediction of winter wheat yield in the United Kingdom. Agronomy Journal 95: 114–125.

34. Jamieson PD, Porter JR, Goudriaan J, Ritchie JT, van Keulen H, et al. (1998) A comparison of the models AFRCWHEAT2, CERESWheat, Sirius, SUCROS2 and SWHEAT with measurements from wheat grown under drought. Field Crops Research 55: 23–44.

35. Prasad PVV, Pisipati SR, Momcčilović I, Ristic Z (2011) Independent and combined effects of high temperature and drought stress during grain filling on plant yield and chloroplast EFTu expression in spring wheat. Journal of Agronomy and Crop Science 197: 430–441.

36. Bootsma A (1994) Long term (100 yr) climatic trends for agriculture at selected in Canada. Climatic Change 26: 65–88.

37. Smith CW (1995) Crop production: evolution, history, and technology. JohnWiley and Sons, New York.

38. Sionit N, Teare ID, Kramer PJ (1980) Effects of repeated application of water stress on water status and growth of wheat. Physiologia Plantarum 50: 11–15.

39. Mogensen VO (1985) Growth rate of grains and grain yield of wheat in relation to drought. Acta Agriculturae Scandinavica 35: 353–360.

40. Anderson WK, Heinrich A, Abbotts R (1996) Long-season wheats extend sowing opportunities in the central wheat belt of Western Australia. Australian Journal of Experimental Agriculture 36: 203–208.

41. Kerr NJ, Siddique KHM, Delane RJ (1992) Early sowing with wheat cultivars of suitable maturity increases grain yield of spring wheat in a short season environment. Australian Journal of Experimental Agriculture 32: 717–733.

42. Richards JH, Fung KI (1969) Atlas of Saskatchewan, Saskatoon: Modern Press, Saskatoon, Canada.

43. Barklacich M, Stewart RB (1995) Impacts of climate change on wheat yields in the Canadian Prairies. In: Rosenzwieg C (ed) Climate change and agriculture: analysis of potential international impacts, special publication No. 59, American Society of Agronomy, Madison, WI, pp 147–162.

44. McCaig TN (1997) Temperature and precipitation effects on durum wheat grown in southern Saskatchewan for fifty years. Canadian Journal of Plant Science 77: 215–223.

20

Effect of Different Fertilizer Application on the Soil Fertility of Paddy Soils in Red Soil Region of Southern China

Wenyi Dong[1], Xinyu Zhang[1]*, Huimin Wang[1], Xiaoqin Dai[1], Xiaomin Sun[1], Weiwen Qiu[2], Fengting Yang[1]

1 Key Laboratory of Ecosystem Network Observation and Modeling, Institute of Geographic Sciences and Natural Resources Research, Chinese Academy of Sciences, Beijing, People's Republic of China, **2** The New Zealand Institute for Plant and Food Research Limited, Christchurch, New Zealand

Abstract

Appropriate fertilizer application is an important management practice to improve soil fertility and quality in the red soil regions of China. In the present study, we examined the effects of five fertilization treatments [these were: no fertilizer (CK), rice straw return (SR), chemical fertilizer (NPK), organic manure (OM) and green manure (GM)] on soil pH, soil organic carbon (SOC), total nitrogen (TN), C/N ratio and available nutrients (AN, AP and AK) contents in the plowed layer (0–20 cm) of paddy soil from 1998 to 2009 in Jiangxi Province, southern China. Results showed that the soil pH was the lowest with an average of 5.33 units in CK and was significantly higher in NPK (5.89 units) and OM (5.63 units) treatments ($P<0.05$). The application of fertilizers have remarkably improved SOC and TN values compared with the CK, Specifically, the OM treatment resulted in the highest SOC and TN concentrations (72.5% and 51.2% higher than CK) and NPK treatment increased the SOC and TN contents by 22.0% and 17.8% compared with CK. The average amounts of C/N ratio ranged from 9.66 to 10.98 in different treatments, and reached the highest in OM treatment ($P<0.05$). During the experimental period, the average AN and AP contents were highest in OM treatment (about 1.6 and 29.6 times of that in the CK, respectively) and second highest in NPK treatment (about 1.2 and 20.3 times of that in the CK). Unlike AN and AP, the highest value of AK content was observed in NPK treatments with 38.10 mg·kg^{-1}. Thus, these indicated that organic manure should be recommended to improve soil fertility in this region and K fertilizer should be simultaneously applied considering the soil K contents. Considering the long-term fertilizer efficiency, our results also suggest that annual straw returning application could improve soil fertility in this trial region.

Editor: Peter Shaw, Roehampton University, United Kingdom

Funding: This work was supported by the Knowledge Innovation Program of the Chinese Academy of Sciences (KZCX2-EW-310) and National Natural Science Foundation of China (No. 41171153, 41001179). The authors declare that no additional external funding was received for this study. The funders had no role in study design, data collection and analysis, decision to publish, or preparation of the manuscript.

Competing Interests: The authors have declared that no competing interests exist.

* E-mail: zhangxy@igsnrr.ac.cn

Introduction

Red soils, which can be classified as Ultisols in the Soil Taxonomy System of the USA and Acrisols and Ferralsols in the FAO legend [1], occupy approximately 2.04 million km^2 in tropical and subtropical regions of China [1–3]. In these red soil regions, Rice (*Oryza sativa* L.) is the main cereal crop, contributing 19% and 29% of the world rice area and rice production, respectively [4]. Paddy soil, formed under interchange between drying and wetting rice field conditions, is considered to be the most important soil resource for the food security of China [5]. In recent years, due to the rapid population growth and a continuous decline in the amount of cultivated land area, the rate of fertilizer application keeps on rising in these regions in order to obtain high crop production in agriculture [6]. Nevertheless, instead of improving the soil structure and fertility, the long-term inappropriate fertilization has caused severe degradation of red soils, characterized by high acidity, low nutrients and a disturbed, unbalanced ecosystem [7]. Therefore, how to ameliorate degraded paddy soils in the red soil region and maintain the region's sustainable development of agricultural production has become an urgent problem.

Recently, soil quality has gained attention as a result of environmental issues related to soil degradation and production sustainability under different farming systems [8]. It has been considered by previous researches that the concentrations of soil nutrients (e.g., organic C, N, P, and K) are good indicators of soil quality and productivity because of their favorable effects on the physical, chemical, and biological properties of soil [9]. Soil pH affects the chemical reactions in soil [10]. Extremes of pH in soils, for example, will lead to a rapid increase in net negative surface charge and thus increases the soil's affinity for metal ions [11,12]. Soil organic components, such as soil organic carbon (SOC) or total N (TN) are the most critical indices of paddy soil fertility [13]. Dynamics of SOC and TN storage in agricultural soils drives microbial activity and nutrient cycles, promotes soil physical properties and water retention capacity, and reduces erosion [14]. Moreover, it has been recognized that soil available nutrients (including N, P and K), coming from mineralization and available components of fertilizer, can be directly absorbed by plants, contributing greatly to the soil fertility [15].

With the development of agricultural production, fertilization has been widely used as a common management practice to maintain soil fertility and crop yields [16]. Long-term field experiments (LTFEs) using different agronomic management can provide direct observations of changes in soil quality and fertility and can be predictions of future soil productivity and soil environment interactions [17,18]. Over past decades, a great number of long-term experiments were initiated to examine the effects of fertilization on soil fertility in the world [19–22]. Some studies have documented that the use of fertilizers was necessary, and that continuous fertilizer application increased the concentrations of SOC, TN and other nutrients in plough layers compared with the initial value at the beginning of the experiment [23–25]. Manure amendments markedly increased the contents of SOC, TN, and other available nutrients, and reduced soil acidification [13,26]. However, other studies have shown that the continued use of fertilizers may result in the decline of soil quality and productivity [27,28]. Long-term application of fertilizer was inadequate to maintain levels of nutrients, the SOC and TN significantly decreasing under the fertilizer treatment and the available N (AN), available P (AP) and available K (AK) did not show clear changes with time or between treatments despite some variation [29,30]. Thus, there's still some debate over the effect of different fertilization treatments on soil fertility.

Qianyanzhou experimental station, which was founded in 1982 by Chinese Academy of Sciences (CAS), is noted for its studies on the integrated development and management of natural resources in red earth hilly regions. It has been identified as the red soil hilly system of international experimental demonstration research station in UNESCO's programme on Man and the Biosphere (MAB), and now has become an essential station of the Chinese Ecosystem Research Network (CERN) [31]. In paddy soil of southern China, some LTFEs have been initiated [32,33]. These experiments have provided basic data for research into paddy soil. However, most of them dealt with different chemical fertilizer rates in long-term experiments but few studies focused on different types of fertilizers (such as green manure and rice straw return) affecting the soil fertility. A thorough understanding of how these fertilizers and varying management practices affect the long-term soil fertility of conventional cropping systems is still lacking in Qianyanzhou region, one of the most important typical red earth hill regions in China. In this study, five fertilization treatments (no fertilizer, rice straw return, chemical fertilizer, organic manure and green manure) were applied. Our objectives were to (i) assess the changes of soil fertility parameters in Qianyanzhou from 1998 to 2009 (ii) evaluate the effects of different fertilization treatments on soil fertility parameters, and (iii) put forward suggestions to improve soil fertility in agricultural regions of southern China.

Materials and Methods

Experimental Site

Qianyanzhou Experimental Station (115°03′29.2″E, 26°44′29.1″N) of CAS is situated on the typical red earth hill region in the mid-subtropical monsoon landscape zone of Taihe County, Jiangxi Province, China. The average elevation is approximately 100 m, and relative relief is 20–50 m. Qianyanzhou Experimental Station has a subtropical monsoon climate. According to the statistics of the meteorological data, the mean annual temperature at this site is 17.8°C, and the annual active accumulated temperatures (above 0 and 10 degrees Celsius) are 6543.8 and 5948.2 degrees Celsius respectively. The annual precipitation and evaporation are 1471.2 mm and 259.9 mm

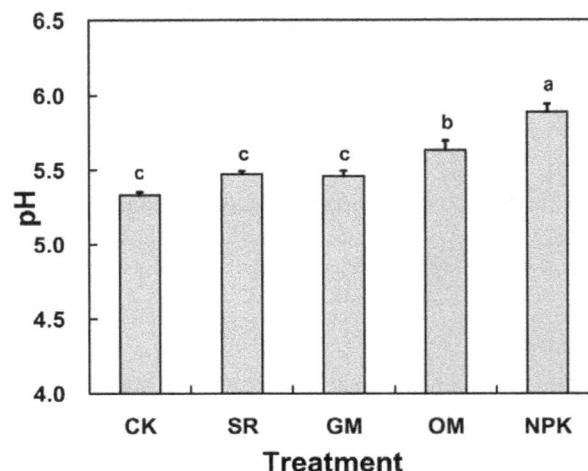

Figure 1. Average soil pH in the different fertilizer treatments. CK: no fertilizer; SR: straw returning rice field; NPK: chemical fertilizer; OM: organic manure and GM: green manure.

respectively, with the mean relative humidity of 83%. The frost-free period is 290 d and global radiation is 4223 MJ·m^{-2}. Our experimental field is located in the flat floodplain where the soil-forming parent material consists of red sandstone and sandy conglomerate. Based on the investigation and analysis before our experiment, it can be concluded that paddy soil is the main soil type with bulk density of 1.50 g·cm^{-3} (0–20 cm), pH of 5.97, soil organic carbon of 9.71 g·kg^{-1}, total N content of 1.02 g·kg^{-1}, available P content of 1.56 mg·kg^{-1} and available K content of 17.61 mg·kg^{-1}.

Experimental Design

A long-term fertilization experiment was conducted initially in 1998 under a double rice cropping system (rice-rice-winter fallow) which is one of the most common cropping systems in the region. Summer rice was sown at the end of April and harvested in July. Winter rice was sown at the end of July and harvested in November. During the growing season, hand weeding was done to control weeds.

There were five treatments in total: no fertilizer (CK), straw returning rice field (SR), chemical fertilizer (NPK), organic manure (OM) and green manure (GM). All treatments were arranged in a randomized block design with three replications [25,34,35], totalling 15 plots. Each plot was 15 m^2 (3 m×5 m) and was isolated by concrete walls (50 cm depth and 15 cm above the soil surface). These fertilization systems were chosen based on several common fertilization experiences from local farmers. In SR treatment, all the aboveground rice residues were returned to the soil after harvest, about 4500 kg·hm^{-2} on dry weight. In NPK treatments, inorganic fertilizers were applied at the rates of N-P$_2$O$_5$-K$_2$O at 225–135–225 kg·hm^{-2} by using urea, calcium-magnesium phosphate and potassium chloride. Before sowing, 60% of N, P and K fertilizer were applied as base fertilizers and the remaining fertilizer was applied as top-dressing. In OM treatment, organic manure which came from the faeces of pigs was applied at the rate of 4100 kg·hm^{-2} fresh weight. All pig manure used in our experiment came from a pig farm in Taihe County, where the composting-process was conducted at high temperatures and a good organic fertilizer was obtained after a few months of fermentation by sterilization, deodorization, and so on [36,37]. In GM treatment, fresh Chinese milk vetch (*Astragalus sinicus* L.) was

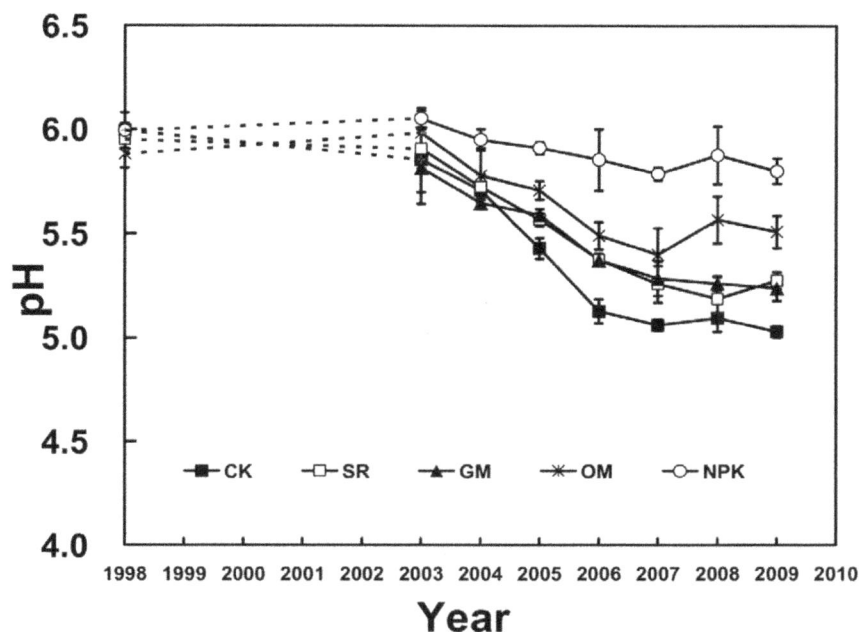

Figure 2. Dynamics of soil pH in the different fertilizer treatments during 1998–2009. Soil samples for 1999–2002 were not analyzed (dashed lines).

applied at the rate of 22500 kg·hm^{-2} fresh weight according to the local conventional green manure application rate.

Soil Sampling and Analysis

Soil samples in the 15 plots were collected annually during 2003–2009 at 7–10 days after the harvest of the late rice. To reflect the real effect of long-term fertilization on the soil fertility, the data in the first 5 years of this experiment were not obtained to analyze in our paper. In each plot, soils were sampled with an auger with 5 cm internal diameter in the plough layer (0–20 cm) at five randomly selected locations and then mixed as one sample [35,38,39]. All fresh soil samples were air-dried and sieved through a 2.0 mm and 0.25 mm sieve and stored for nutrient analysis [38,39].

Soil physical and chemical properties were measured using the methods described by Bao [40]. Soil pH was measured with glass electrode in a 1:2.5 soil/water suspension. SOC was measured by a K_2CrO_7-H_2SO_4 oxidation procedure and TN by the Kjeldahl method. Soil C/N values were calculated as the ratio SOC to TN. AN was determined by using a micro-diffusion technique after alkaline hydrolysis. AP was determined by the Olsen method. AK was measured by flame photometry after NH_4OAc neutral extraction.

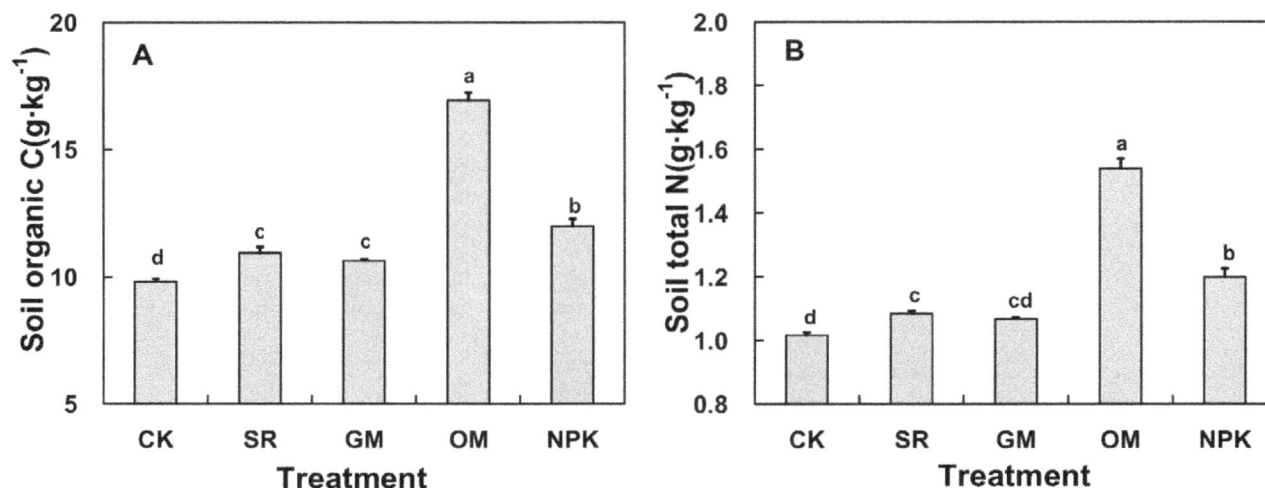

Figure 3. Average SOC and TN in the different fertilizer treatments. CK: no fertilizer; SR: straw returning rice field; NPK: chemical fertilizer; OM: organic manure and GM: green manure.

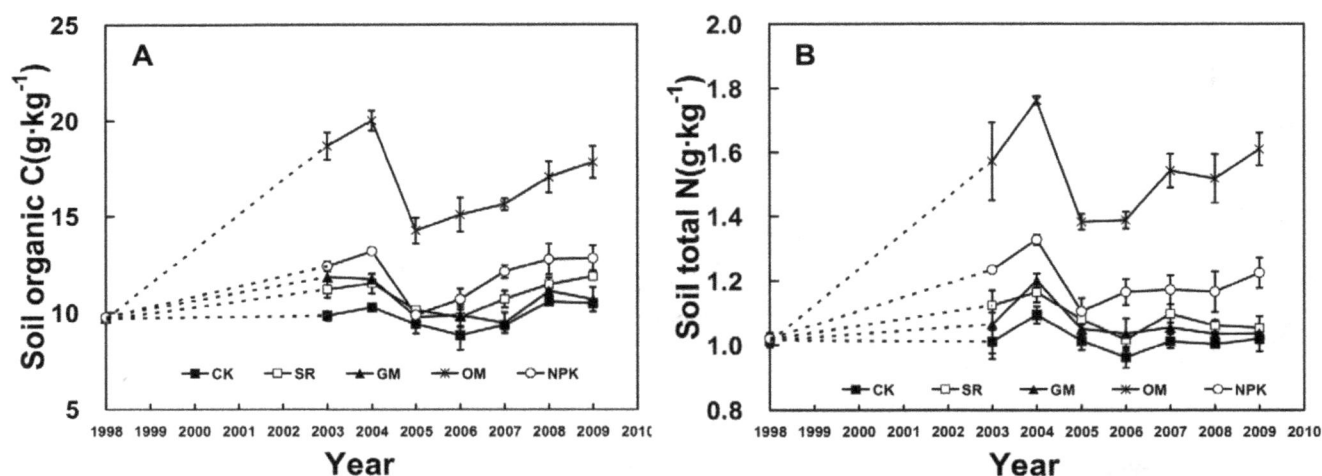

Figure 4. Dynamics of SOC and TN in the different fertilizer treatments during 1998–2009. Soil samples for 1999–2002 were not analyzed (dashed lines).

Data Analyses

All results were reported as means ± standard error (SE) for three replicates. One-way of variance (ANOVA) and Duncan's multiple comparisons were performed to determine the differences among the fertilizer treatments in terms of the long-term soil nutrient contents means (during 2003–2009). Then the annual values under different treatments were used to investigate the dynamics of nutrient contents during the whole period. All statistical analyses were performed using the SPSS software package (version 15.0) (Statistical Graphics Crop, Princeton, USA). A difference at $P<0.05$ level was considered as statistically significant.

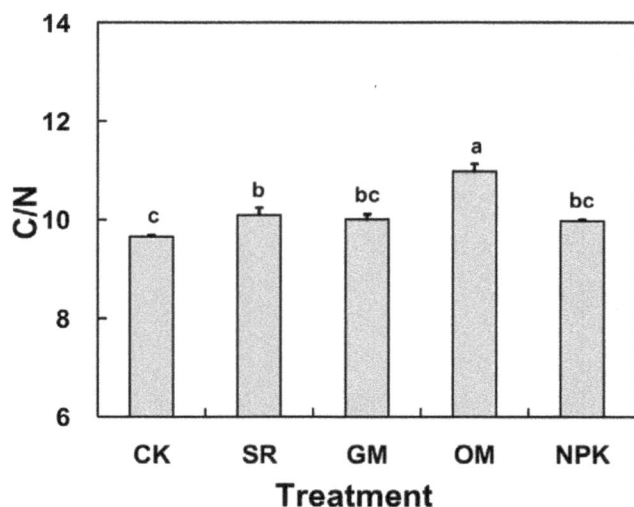

Figure 5. Average C/N ratios in the different fertilizer treatments. CK: no fertilizer; SR: straw returning rice field; NPK: chemical fertilizer; OM: organic manure and GM: green manure.

Results

Effects of Different Fertilizer Treatments on Soil pH

The average soil pH was shown in Fig.1. Statistical analysis revealed that fertilization treatments led to a significant increase in soil pH compared with the CK treatment ($P<0.05$). The soil pH was the lowest in CK with an average of 5.33 units and highest in NPK treatment with 5.89 units. In OM. treatment, the soil pH was relatively higher than CK (reaching 5.63 units).

During 1998 to 2009, the soil pH in NPK treatment appeared relatively stable despite some slight drop with the time (Fig.2). However, the values in other treatments showed a clear decline trend with time despite some variations. In the beginning years after fertilization, there was no evident difference in pH among all treatments, but eventually the soil pH in CK reduced dramatically and declined sharply from 5.71 to 5.03 (0.68 units lower). In SR, GM and OM treatments, the soil pH values declined by 0.57, 0.57 and 0.27 units respectively during the experimental period.

Effects of Different Fertilizer Treatments on Soil Organic C and Total N

The SOC and TN contents showed statistically significant differences among the five treatments (Fig.3). We observed that the application of fertilizers (especially OM and NPK fertilizers) had remarkably improved SOC and TN values compared with the CK. Specifically, the OM treatment resulted in the highest SOC and TN concentrations (16.93 and 1.54 $g \cdot kg^{-1}$, respectively), which was 72.5%. and 51.2%[1] higher than that of CK. The SOC and TN in NPK treatment were significantly higher than CK, reaching 11.97 and 1.20 $g \cdot kg^{-1}$, respectively. While in SR and GM treatment, the SOC was remarkably higher than CK, reaching 10.94 and 10.64 $g \cdot kg^{-1}$ respectively, but significant differences in TN contents between GM and CK were not observed.

The SOC in different treatments had a similar trend over time (Fig.4A). From 1998 to 2004, the SOC showed a clear increase with time due to fertilization, rising from initial 9.65–9.78 $g \cdot kg^{-1}$ to 11.51–20.00 $g \cdot kg^{-1}$ in 2004, respectively. Then SOC content dropped sharply but quickly reached at stable level. It was also obtained that the SOC content in OM was obviously higher than

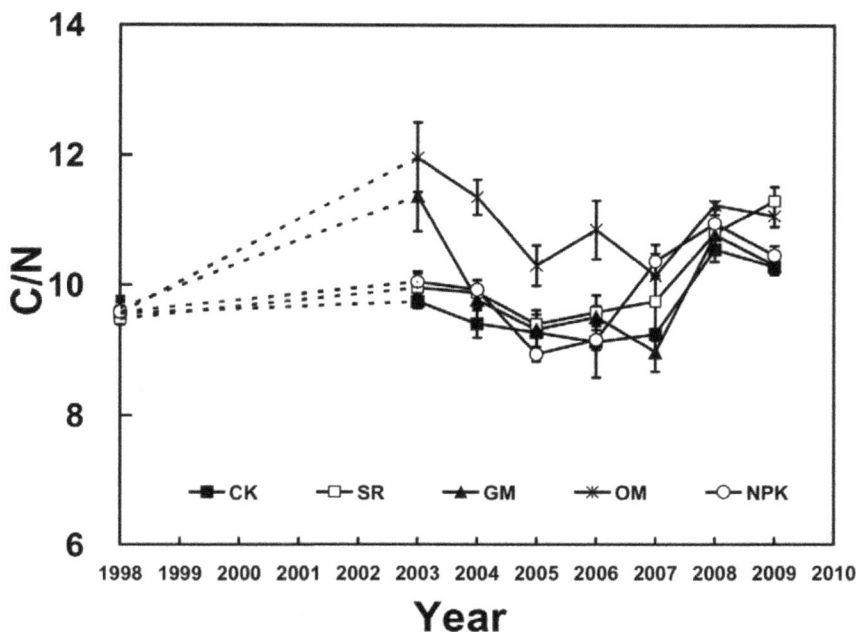

Figure 6. Dynamics of soil C/N ratios in the different fertilizer treatments during 1998–2009. Soil samples for 1999–2002 were not analyzed (dashed lines).

the other treatments during the experiment period, whereas that in CK remained relative stable (about 10 g·kg⁻¹).

The dynamics of TN content in the five treatments followed similar patterns with SOC during 1998–2009 (Fig.4B). In the first few years, TN content tended to increase rapidly in the OM treatment (from 1.01 to 1.76 g·kg⁻¹), followed by NPK treatment (from 1.02 to 1.33 g·kg⁻¹). Thereafter, both of them declined and then maintained a certain level. Meanwhile, the soil TN contents in SR, GM and CK treatments were relatively steady, at approximately 1.05 g·kg⁻¹.

Effects of Different Fertilizer Treatments on Soil C/N Ratio

There were marked differences in soil C/N ratio among different treatments due to fertilizer application (Figure5). The average C/N ratio in the OM treatment (10.98) was obviously higher than the other treatments ($P<0.05$, Fig.5). Similarly, in SR treatment, the C/N ratio was significantly higher than CK. Nevertheless, there were no significant differences of C/N ratios in the CK, SR, GM and NPK treatments, ranging from 9.66 to 10.00.

Dynamics of soil C/N ratios during 1998–2009 are shown in Fig.6. From 1998 to 2003, the C/N ratios increased sharply in OM and GM treatments (25.0%and 17.9% higher than the initial amount), and then the values declined slowly and constantly till 2007 (reaching 10.14 and 8.98 respectively). However, the other treatments including CK, SR and NPK fluctuated at a stable level (approximately 10.0) from 1998 to 2007. In the last two years of the experiment period, all the five treatments displayed similar trends without a significant difference varying slightly between 8.98 and 11.29.

Effects of Different Fertilizer Treatments on Soil Available Nutrients

A comparison of available nutrients among the treatments indicated that the fertilizer had a notable influence on soil AN, AP

and AK ($P<0.05$, Fig.7A–C). During the experiment period, the average AN and AP contents in OM were highest (about 1.6 and 29.6 times of the CK, respectively) and second was NPK treatment (about 1.2 and 20.3 times of the CK). However, there were no obvious differences of AN and AP between SR, GM and CK treatments. Unlike AN and AP, the highest value of AK content was found in NPK treatment, which was 38.10 mg·kg⁻¹ (about 2.2 times of the CK), and there were no obvious differences among the other four treatments (Fig.7C).

AN in the OM treatment obviously increased with time due to fertilization at the beginning, and then the value tended to rise with a slight fluctuation before remaining at the highest level in comparison to the other treatments. We also found a similar trend in the NPK treatment but lower than OM and did not find the significant differences between SR, GM and CK (Fig.8A).

During the entire experiment period, AP concentrations in CK, SR and GM treatments remained at an extremely low level (approximately 1.80 mg·kg⁻¹) and were almost the same as the initial values. On the contrary, AP in both OM and NPK treatments displayed similar changes over time. Specifically, the value rose sharply in the first few years of fertilization (63.59 and 45.54 mg·kg⁻¹ respectively in 2003) then progressively reduced till 2007 (25.37 and 16.68 mg·kg⁻¹ respectively) and later increased slightly (Fig.8B).

The NPK treatment significantly increased AK content, especially from 2007 to 2009. However, the significant changes of AK in other treatments with time were not observed, maintaining at a stable and low level. (Fig.8C).

Discussion

Many experiments have been conducted on the relationship between fertilization and soil pH [35,41]. Some studies demonstrated that the soil pH was decreased to a certain extent with different fertilizer treatments [6]. In our study, the soil pH tended to drop in different treatments with time (Fig.2), and the decline

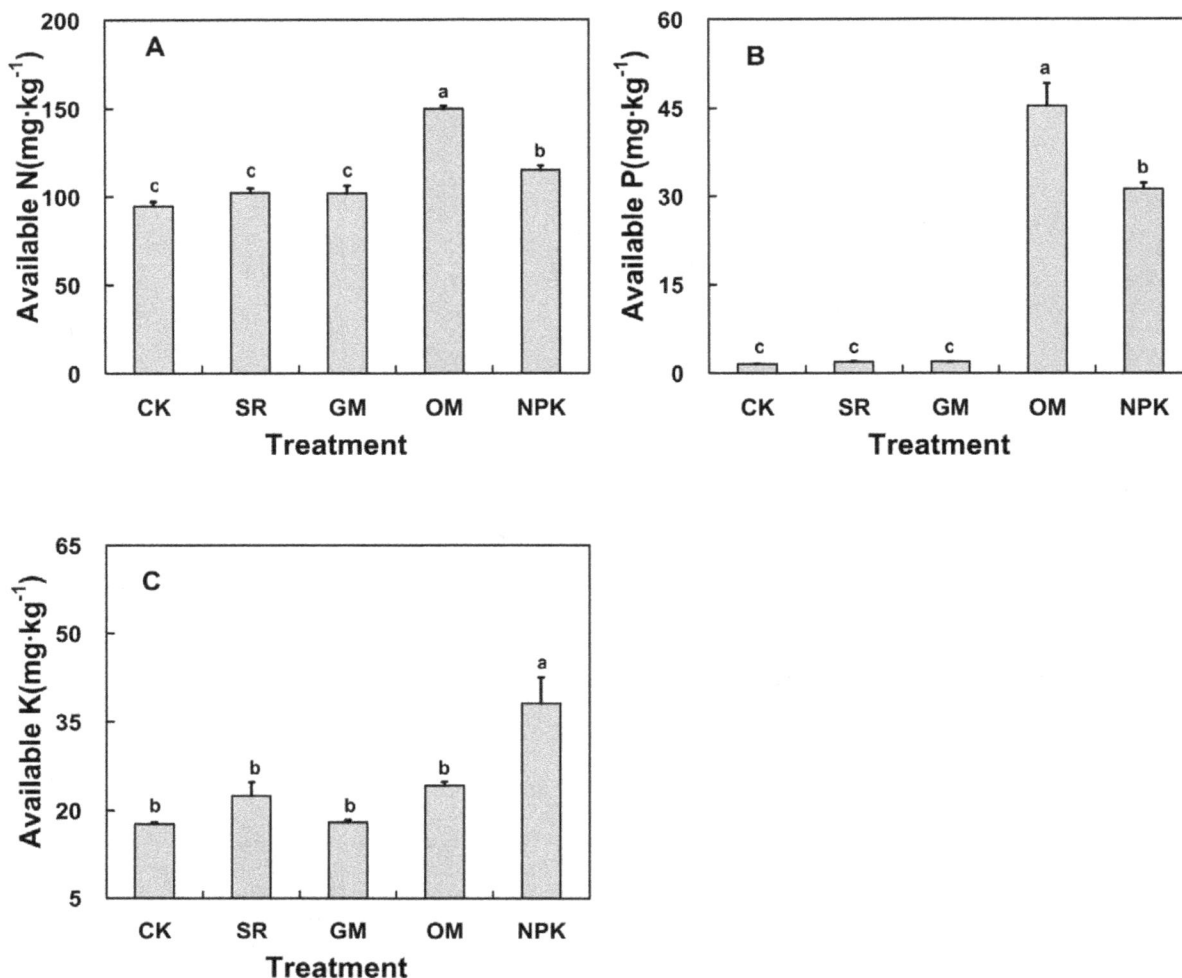

Figure 7. Average soil AN, AP and AK in the different fertilizer treatments. CK: no fertilizer; SR: straw returning rice field; NPK: chemical fertilizer; OM: organic manure and GM: green manure.

rate in the NPK and OM treatments were relatively lower than CK, but producing higher pH values in the NPK and OM treatments than in the CK (Fig.1), which suggested that chemical fertilizer and organic manure could alleviate soil acidification to some extent. It has been reported that the application of alkaline fertilizer (e.g. calcium magnesium phosphate fertilizer) would return some alkaline substance to soils and thus increase the soil pH [42]. In addition, the application of organic manure could improve soil acidity by increasing the soil organic matter, promoting the soil maturation, improving the soil structure, and enhancing the soil base saturation percentage, which is in line with Zhang [35] and Li [43]. Moreover, studies showed that the soil pH in CK was lower than the initial value, which indicated that the acid deposition could have a great influence on the soil acidification in this trial region [44]. Since a too high or too low pH is harmful to the crop growth, it might be a practicable measure to establish the proper range of soil pH through fertilizer use.

Soil organic matter is a key contributor to soil due to its capacity to affect plant growth indirectly and directly [45]. As SOC and TN constitute heterogeneous mixtures of organic substances, they are widely used as the main parameters for evaluating soil fertility [46]. Meanwhile, human activities such as fertilizer practices and

cropping systems play a key role in the regulation of C and N contents in agricultural soils [47]. In our experiments, SOC and TN contents increased considerably in the fertilization treatments compared with CK, especially in OM and NPK treatments (Fig.3), suggesting that organic and chemical fertilizer are beneficial to the accumulation of soil organic matter and thus improves soil fertility. This may be because both the application of organic manure and chemical fertilizer can improve soil aggregation, soil water retention, and reduce bulk density of the soil in the plough layer, promoting crop growth and the return of more root residues to the soil [48]. Under the SR application, SOC content was increased gradually with time, suggesting that the continuous SR supply had a positive effect in sustaining SOC, in accordance with the finding of Nie [49]. In farmland ecosystems, pig manure is easy to accumulate in soil and has lower ammonia losses than other fertilizers [33], so the TN concentration was significantly increased by continual annual OM applications compared to the other treatments (Fig.4B).

Previous studies have pointed out that the soil C/N ratio plays a key role in mineralization and accumulation of SOC and TN contents and that a high C/N ratio generally slows the mineralization process [47,50].As found by Zhang [51], the C/N ratio was obviously higher in OM applications than the other

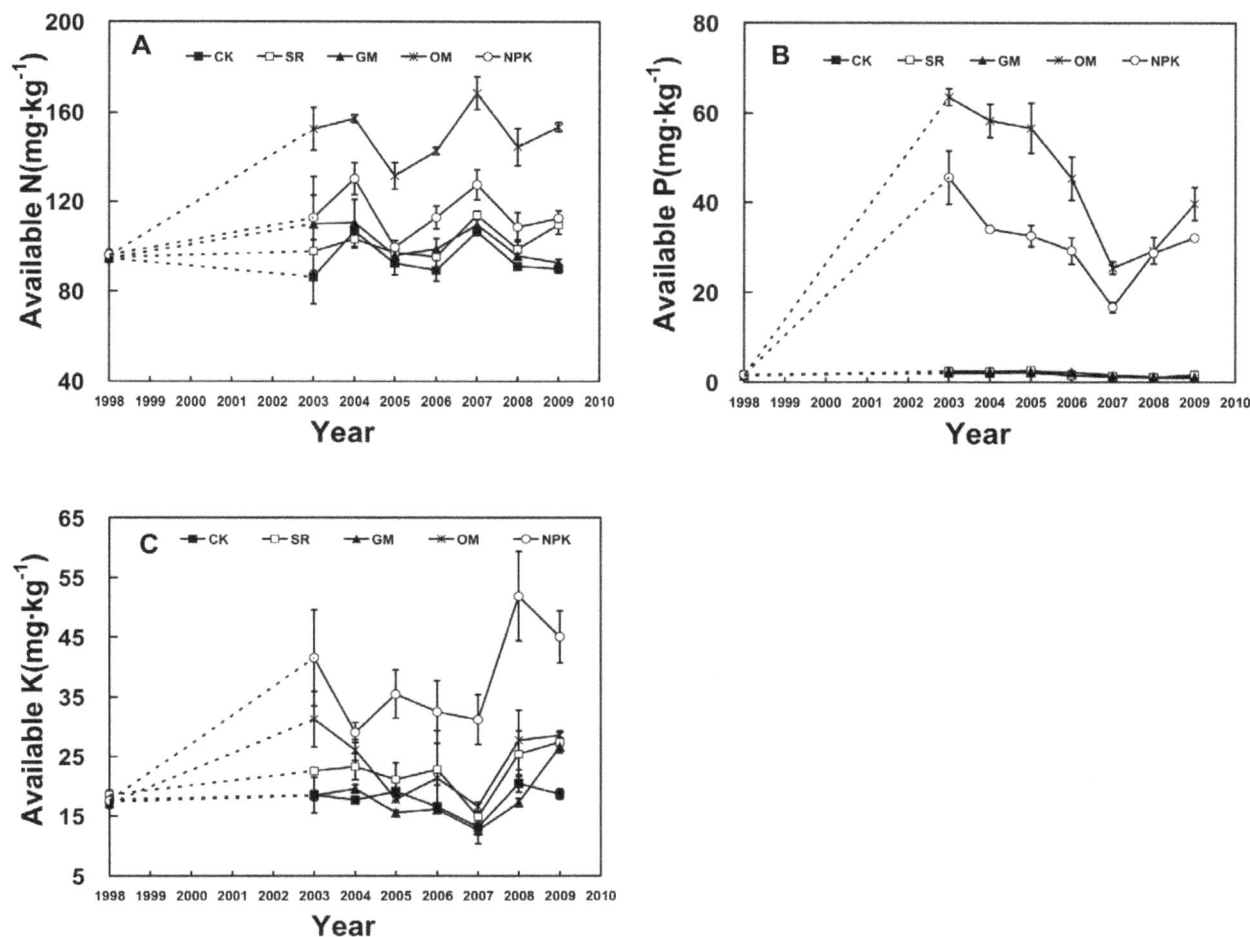

Figure 8. Dynamics of soil AN, AP and AK in the different fertilizer treatments during 1998–2009. soil samples for 1999–2002 were not analyzed (dashed lines).

treatments attributed to the SOC was more markedly enriched relative to TN. Moreover, we observed that there was a slight rise in soil C/N ratios with time in SR and NPK treatments (Fig.6), suggesting both SOC and TN accumulated gradually with time and the increase in SOC buildup was more rapid than TN over the past years in SR and NPK treatments [52].

It is known that fertilization is crucial for maintaining soil available nutrient levels, because fertilization ensures the largely constant presence of active microorganisms and the regular dynamic of biomass carbon [53]. Our research also showed that soil available nutrient contents were significantly affected by different fertilization treatments. Long-term OM application led to significantly higher values of soil AN and AP, compared to the other fertilization treatments (Fig.8A and 8B). It has also been reported by Huang [25] that significant AN and AP increases were observed in the manure-applied treatments. In addition, the AN and AP contents were maintained at a very higher level in the NPK treatments (Fig.8C) as a result of the long-term high inputs of N and P fertilizers [54], whereas AN and AP did not show statistically significant differences between CK, SR and GM treatments. Unlike AN and AP, AK was the highest in NPK treatments in our results, followed by the OM treatments, demonstrating that the K supplement from chemical and manure fertilizers are important and this may be the most advantageous

way to solve the problem in China, where K resources are quite limited. Previous studies have shown that application of rice straw significantly increased available K while increasing organic matter contents [54,55]. Similarly, the AK in SR treatment was higher than CK in our study. It was noted that the SOC, TN, C/N ratios and available nutrients had no significant difference under GM treatment in our study. Previous study had described that the decomposition process of green manure (as a fresh organic matter) was very slow and complicated, affected by soil temperature, moisture, plough back time and so on [56]. Hence, the real mechanism of the nutrient release of green manure deserves further study in order to make better use of green manure and increase its fertilizer efficiency in our study region.

Considering the dynamic changes of soil available nutrient content, we found that AN significantly increased with time from 1998 to 2009 in OM and NPK treatments, suggesting that the long-term soil organic matter played a major role in releasing soil AN. The evident increase of AK in NPK treatment over time in our study, suggests that continuously applying K fertilizer would dramatically improve the soil AK supply, In addition, AP values in both OM and NPK treatments were increased greatly compared to the other three treatments during the whole experiment period. This is consistent with many previous studies showing the accumulation of P is the most obvious and long-term manure or inorganic fertilizer application can significantly alter the amounts and proportion of labile and stable soil P pools [57–59]. That the OM treatment resulted in even higher AP value than NPK, also support the view that organic fertilizer is much more conducive to soil P availability rather than commercial P fertilizers in cropping systems that receive predominantly organic P amendments [59].

In conclusion, significant differences in soil fertility of paddy soils in the red soil region of southern China among different fertilization treatments were found in our study. Application of OM and NPK resulted in a substantial increase of SOC, TN, C/N ratios, AN and AP contents relative to the other fertilization treatments. Thus, it is likely that the OM and NPK application improves soil fertility. Meanwhile, the application of NPK would increase soil AK, leading to the highest AK contents. Continuous application of SR also had a positive effect in sustaining SOC, TN and C/N ratios. However, the effect of GM application on soil fertility was not remarkable compared to CK. Hence, organic manure should be recommended to improve soil fertility in this region and K fertilizer should be simultaneously applied considering the soil K contents. Moreover, in terms of the long-term

fertilizer efficiency, annual straw returning application year by year could be adopted in this trail region.

Supporting Information

Table S1 Data statistic and analysis for samples in this study. The detailed data of soil nutrient contents in different fertilization treatments for each samples in this study is shown.

Acknowledgments

We thank Jingdong Zou in the Qianyanzhou experimental Station of Chinese Ecosystem Research Network (CERN) for many sampling and analysis work.

Author Contributions

Conceived and designed the experiments: WD XZ XS. Performed the experiments: WD XZ XS. Analyzed the data: WD XZ. Contributed reagents/materials/analysis tools: HW XD FY. Wrote the paper: WD XZ XS WQ.

References

1. FAO-Unesco(1974)Soil Map of the World 1: 5000000. Legend, Volume 1. Unesco, Paris.
2. Li CK (1983) Red Soils of China. Science Press, Beijing.
3. Zhang MK, Xu JM (2005) Restoration of surface soil fertility of an eroded red soil in southern China. Soil and Tillage Research 80: 13–21.
4. Sun W, Huang Y (2011) Global warming over the period 1961–2008 did not increase high-temperature stress but did reduce low-temperature stress in irrigated rice across China. Agricultural and Forest Meteorology 151: 1193–1201.
5. Ma L, Yang LZ, Ci E, Cheng YQ, Wang Y, et al. (2009) Effects of long-term fertilization on distribution and mineralization of organic carbon in paddy soil. Acta Pedologica Sinica 46: 1050–1058.
6. Wang BR, Cai ZJ, Li DC (2010) Effect of different long-term fertilization on the fertility of red upland soil. Journal of Soil and Water Conservation 24: 85–88.
7. Chen WC, Wang KR, Xie XL (2009) Effects on distributions of carbon and nitrogen in a reddish paddy soil under long-term different fertilization treatments. Chinese Journal of Soil Science 40: 523–528.
8. Galantini J, Rosell R (2006) Long-term fertilization effects on soil organic matter quality and dynamics under different production systems in semiarid Pampean soils. Soil and Tillage Research 87: 72–79.
9. Cao C, Jiang S, Ying Z, Zhang F, Han X (2011) Spatial variability of soil nutrients and microbiological properties after the establishment of leguminous shrub Caragana microphylla Lam. plantation on sand dune in the Horqin Sandy Land of Northeast China. Ecological Engineering 37: 1467–1475.
10. Zhao J, Dong Y, Xie X, Li X, Zhang X, et al. (2011) Effect of annual variation in soil pH on available soil nutrients in pear orchards. Acta Ecologica Sinica 31: 212–216.
11. Wu Z, Gu Z, Wang X, Evans L, Guo H (2003) Effects of organic acids on adsorption of lead onto montmorillonite, goethite and humic acid. Environmental Pollution 121: 469–475.
12. Yang JY, Yang XE, He ZL, Li TQ, Shentu JL, et al. (2006) Effects of pH, organic acids, and inorganic ions on lead desorption from soils. Environmental Pollution 143: 9–15.
13. Liu M, Li ZP, Zhang TL, Jiang CY, Che YP (2011) Discrepancy in response of rice yield and soil fertility to long-term chemical fertilization and organic amendments in paddy soils cultivated from Infertile upland in subtropical China. Agricultural Sciences in China 10: 259–266.
14. Manna M, Swarup A, Wanjari R, Mishra B, Shahi D (2007) Long-term fertilization, manure and liming effects on soil organic matter and crop yields. Soil and Tillage Research 94: 397–409.
15. Vogeler I, Rogasik J, Funder U, Panten K, Schnug E (2009) Effect of tillage systems and P-fertilization on soil physical and chemical properties, crop yield and nutrient uptake. Soil and Tillage Research 103: 137–143.
16. Shen JP, Zhang LM, Guo JF, Ray JL, He JZ (2010) Impact of long-term fertilization practices on the abundance and composition of soil bacterial communities in Northeast China. Applied Soil Ecology 46: 119–124.
17. Blair N, Faulkner R, Till A, Poulton P (2006) Long-term management impacts on soil C, N and physical fertility: Part I: Broadbalk experiment. Soil and Tillage Research 91: 30–38.
18. Li BY, Huang SM, Wei MB, Zhang HL, Shen AL, et al. (2010) Dynamics of soil and grain micronutrients as affected by long-term fertilization in an aquic Inceptisol. Pedosphere 20: 725–735.
19. Mitchell CC, Westerman RL, Brown JR, Peck TR (1991) Overview of long-term agronomic research. Agronomy Journal 83: 24–25.
20. Bhandari A, Sood A, Sharma K, Rana D (1992) Integrated nutrient management in a rice-wheat system. Journal of the Indian Society of Soil Science 40: 742–747.
21. Dawe D, Dobermann A, Moya P, Abdulrachman S, Singh B, et al. (2000) How widespread are yield declines in long-term rice experiments in Asia? Field Crops Research 66: 175–193.
22. Ladha J, Dawe D, Pathak H, Padre A, Yadav R, et al. (2003) How extensive are yield declines in long-term rice-wheat experiments in Asia? Field Crops Research 81: 159–180.
23. Whitbread A, Blair G, Konboon Y, Lefroy R, Naklang K (2003) Managing crop residues, fertilizers and leaf litters to improve soil C, nutrient balances, and the grain yield of rice and wheat cropping systems in Thailand and Australia. Agriculture, Ecosystems & Environment 100: 251–263.
24. Bi L, Zhang B, Liu G, Li Z, Liu Y, et al. (2009) Long-term effects of organic amendments on the rice yields for double rice cropping systems in subtropical China. Agriculture, Ecosystems & Environment 129: 534–541.
25. Huang S, Zhang W, Yu X, Huang Q (2010) Effects of long-term fertilization on corn productivity and its sustainability in an Ultisol of southern China. Agriculture, Ecosystems & Environment 138: 44–50.
26. Gu YF, Zhang XP, Tu SH, Lindström K (2009) Soil microbial biomass, crop yields, and bacterial community structure as affected by long-term fertilizer treatments under wheat-rice cropping. European Journal of Soil Biology 45: 239–246.
27. Kumar A, Yadav DS (2001) Long-term effects of fertilizers on the soil fertility and productivity of a rice–wheat System. Journal of Agronomy and Crop science 186: 47–54.
28. Yang S (2006) Effect of long-term fertilization on soil productivity and nitrate accumulation in Gansu oasis. Agricultural Sciences in China 5: 57–67.
29. Shen J, Li R, Zhang F, Fan J, Tang C, et al. (2004) Crop yields, soil fertility and phosphorus fractions in response to long-term fertilization under the rice monoculture system on a calcareous soil. Field Crops Research 86: 225–238.
30. Su YZ, Wang F, Suo DR, Zhang ZH, Du MW (2006) Long-term effect of fertilizer and manure application on soil-carbon sequestration and soil fertility under the wheat–wheat–maize cropping system in northwest China. Nutrient Cycling in Agroecosystems 75: 285–295.
31. Li J, Liu Y, Yang X (2006) Studies on water-vapor flux characteristic and the relationship with environmental factors over a planted coniferous forest in Qianyanzhou Station. Acta Ecologica Sinica 26: 2449–2456.
32. Yuan YH, Li HX, Huang QR, Hu F, Pan GX, et al. (2008) Effects of long-term fertilization on dynamics of soil organic carbon in red paddy soil. Soils 40: 237–242.
33. Chen AL, Xie XL, Wen WY, Wang W, Tong CL (2010) Effect of long term fertilization on soil profile nitrogen storage in a reddish paddy soil. Acta Ecologica Sinica 30: 5059–5065.
34. Sikka R, Kansal BD (1995) Effect of fly-ash application on yield and nutrient composition of rice, wheat on pH and available nutrient status of soil. Bioresource Technology 51: 199–203.
35. Zhang HM, Wang BR, Xu MG, Fan TL (2009) Crop yield and soil responses to long-term fertilization on a red soil in southern China. Pedosphere 19: 199–207.
36. Bhamidimarri SMR, Pandey SP (1996) Aerobic thermophilic composting of piggery solid wastes. Water Science and Technology 33: 89–94.
37. Imbeah M (1998) Composting piggery waste: a review. Bioresource Technology 63: 197–203.

38. Wang YC, Wang EL, Wang DL, Huang SM, Ma YB, et al. (2010) Crop productivity and nutrient use efficiency as affected by long-term fertilization in North China Plain. Nutrient Cycling in Agroecosystems 86: 105–119.

39. Kapkiyai JJ, Karanja NK, Qureshi JN, Smithson PC, Woomer PL (1999) Soil organic matter and nutrient dynamics in a Kenyan nitisol under long-term fertilizer and organic input management. Soil Biology and Biochemistry 31: 1773–1782.

40. Bao SD (2005) Soil and Agricultural Chemistry Analysis. Agriculture Press, Beijing. (In Chinese).

41. Daugelene N, Butkute R (2008) Changes in phosphorus and potassium contents in soddy-podzolic soil under pasture at the long-term surface application of mineral fertilizers. Eurasian Soil Science 41: 638–647.

42. Wu XC, Li ZP, Zhang TL (2008) Long-term effect of fertilization on organic carbon and nutrients content of paddy soils in red soil region. Ecology and Environment 17: 2019–2023. (In Chinese).

43. Li BY, Huang SM, Wei MB, Zhang HL, Shen AL, et al. (2010) Dynamics of soil and grain micronutrients as affected by long-term fertilization in an aquic Inceptisol. Pedosphere 20: 725–735.

44. Liu KH, Fang YT, Yu FM, Liu Q, Li FR, et al. (2010) Soil acidification in response to acid deposition in three subtropical forests of subtropical China. Pedosphere 20: 399–408.

45. Lee SB, Lee CH, Jung KY, Park KD, Lee D, et al. (2009) Changes of soil organic carbon and its fractions in relation to soil physical properties in a long-term fertilized paddy. Soil and Tillage Research 104: 227–232.

46. Huang QR, Hu F, Huang S, Li HX, Yuan YH, et al. (2009) Effect of long-term fertilization on organic carbon and nitrogen in a subtropical paddy soil. Pedosphere 19: 727–734.

47. Tong C, Xiao H, Tang G, Wang H, Huang T, et al. (2009) Long-term fertilizer effects on organic carbon and total nitrogen and coupling relationships of C and N in paddy soils in subtropical China. Soil and Tillage Research 106: 8–14.

48. Hyvönen R, Persson T, Andersson S, Olsson B, Ågren GI, et al. (2008) Impact of long-term nitrogen addition on carbon stocks in trees and soils in northern Europe. Biogeochemistry 89: 121–137.

49. Nie J, Zhou J, Wang H, Chen X, Du C (2007) Effect of long-term rice straw return on soil glomalin, carbon and nitrogen. Pedosphere 17: 295–302.

50. Khalil M, Hossain M, Schmidhalter U (2005) Carbon and nitrogen mineralization in different upland soils of the subtropics treated with organic materials. Soil Biology and Biochemistry 37: 1507–1518.

51. Zhang M, He Z (2004) Long-term changes in organic carbon and nutrients of an Ultisol under rice cropping in southeast China. Geoderma 118: 167–179.

52. Darilek JL, Huang B, Wang Z, Qi Y, Zhao Y, et al. (2009) Changes in soil fertility parameters and the environmental effects in a rapidly developing region of China. Agriculture, Ecosystems & Environment 129: 286–292.

53. Nardi S, Morari F, Berti A, Tosoni M, Giardini L (2004) Soil organic matter properties after 40 years of different use of organic and mineral fertilisers. European Journal of Agronomy 21: 357–367.

54. Li Z, Zhang T, Chen B (2006) Changes in organic carbon and nutrient contents of highly productive paddy soils in Yujiang county of Jiangxi province, China and their environmental application. Agricultural Sciences in China 5: 522–529.

55. Chen XW, Li BL (2003) Change in soil carbon and nutrient storage after human disturbance of a primary Korean pine forest in Northeast China. Forest Ecology and Management 186: 197–206.

56. Tejada M, Gonzalez JL, García-Martínez AM, Parrado J (2008) Application of a green manure and green manure composted with beet vinasse on soil restoration: Effects on soil properties. Bioresource technology 99: 4949–4957.

57. Lwkin LY, Kosilova AN, Dubanina GV (1994) The effect of long-term application of fertilizers on soil fertility and winter hardiness and productivity of winter wheat on typical chernozen. Agrokhimiya, 1: 38–43.

58. Qu JF, Dai JJ, Xu MG, Li JM (2009) Advances on effects of long-term fertilization on soil phosphorus. Chinese Journal of Tropical Agriculture, 29: 75–80.

59. Motavalli PP, Miles RJ (2002) Soil phosphorus fractions after 111 years of animal manure and fertilizer application. Biology and Fertility of Soils, 36: 35–42.

21

Simultaneous Simulations of Uptake in Plants and Leaching to Groundwater of Cadmium and Lead for Arable Land Amended with Compost or Farmyard Manure

Charlotte N. Legind[1]*, Arno Rein[1], Jeanne Serre[2], Violaine Brochier[2], Claire-Sophie Haudin[3], Philippe Cambier[3], Sabine Houot[3], Stefan Trapp[1]

1 Department of Environmental Engineering, Technical University of Denmark, Lyngby, Denmark, 2 Veolia Environnement – Research and Innovation, Rueil-Malmaison, France, 3 INRA, UMR 1091 Environment and Arable Crop Research Unit, Thiverval-Grignon, France

Abstract

The water budget of soil, the uptake in plants and the leaching to groundwater of cadmium (Cd) and lead (Pb) were simulated simultaneously using a physiological plant uptake model and a tipping buckets water and solute transport model for soil. Simulations were compared to results from a ten-year experimental field study, where four organic amendments were applied every second year. Predicted concentrations slightly decreased (Cd) or stagnated (Pb) in control soils, but increased in amended soils by about 10% (Cd) and 6% to 18% (Pb). Estimated plant uptake was lower in amended plots, due to an increase of K_d (dry soil to water partition coefficient). Predicted concentrations in plants were close to measured levels in plant residues (straw), but higher than measured concentrations in grains. Initially, Pb was mainly predicted to deposit from air into plants (82% in 1998); the next years, uptake from soil became dominating (30% from air in 2006), because of decreasing levels in air. For Cd, predicted uptake from air into plants was negligible (1–5%).

Editor: Jack Anthony Gilbert, Argonne National Laboratory, United States of America

Funding: The work received funding from Veolia Environnement R&I and INRA (French National Institute for Agricultural Research) de Grignon. The funders had a role in study design, data collection and analysis, decision to publish and preparation of the manuscript.

Competing Interests: Violaine Brochier and Jeanne Serre are hired as research engineers in 2006 by Veolia Environnement - Research and Development. They are mainly working with environmental impacts of agronomic valorization.

* E-mail: chanl@env.dtu.dk

Introduction

Amending soils with compost or sewage sludge is beneficial to the soil fertility due to the high content of organic matter and positive effects on the release of nutrients [1]. On the other hand, amendments may contain various metals and organic micro pollutants that could induce some potential adverse effects to terrestrial ecosystems and human health. A recent review [2] that compared municipal solid waste composts (MSW) to sewage sludge in terms of heavy metal availability in amended soils concluded that the application to soil of both types of amendments in the long run increase the total concentration of several metals in soils. However, the metal availability in compost amended soils tends to be decreased and of less risk to humans concerning exposure through the food chain, whereas amending soils with digested sludge can increase the metal availability.

The QualiAgro long-term field experiment on agronomic effects and environmental impacts of amending various composts on soil and crops has been started in September 1998 at Feucherolles, France (about 30 km west of Paris). Amendments included urban composts (biowaste compost, BIOW; municipal solid waste compost, MSW; co-compost of green waste and sewage sludge,

GWS) as well as farm yard manure (FYM) and applications were compared to controls without amendment (CTR) [3].

Factors affecting uptake of heavy metals into vegetation are type of metal, plant species and cultivar, plant-related parameters such as transpiration and growth, and soil parameters like pH, organic matter, soil texture and redox status [4]. Metals that are available to the plant in the soil solution can be taken up and this fraction is often assessed from mild extractions of soil. However, robust tools for predicting the transfers of metals from soil and air to plants are scarce and often error prone due to the large variability of metal uptake in plants [5]. For Cd and Pb, most regressions for predicting plant uptake from soil correlate the concentration in the plant with soil parameters like pH, organic matter content and total metal concentration in soil e.g. [4,6]. These are the same parameters that are applied for estimating the solubility of these metals in soil water [7]. This indicates that the water soluble fraction in soil is important for plant uptake and that dissolved metal species are transported together with the water into plants.

Plants also change the water balance of the soil: about 2/3rd of the precipitated water is transpired in most ecosystems [8]. In summer, evapotranspiration is typically higher than precipitation, and the soil dries out. Hereby, also leaching of water and solute to groundwater is reduced or stopped. On the other hand, water and solute uptake into vegetation also depends on the distribution and availability of both in soil. Consequently, water balance, solute transport, leaching to groundwater and plant uptake of solute and water are coupled processes. Recently, a coupled plant and groundwater transport model for NaCl could simulate the transpiration-induced changes in groundwater salinity [9]. However, for metals, no models that simultaneously predict plant uptake as well as leaching to groundwater were found.

The objective of this work is to present and test a model framework for the simulation of the coupled transport of water and dissolved heavy metals, the uptake of both into crops, and leaching of solute and water to groundwater. It is hypothesized that uptake of Cd and Pb from soil can be simplified as a passive uptake with soil water only. The model is dynamic and iterative and can be run for a variable number of periods (n). The same superposition principle as for the dynamic plant uptake model for organic compounds [10,11] was applied, where changes in emission and input between periods were considered by superposition of the results of n periods. This model for uptake into plants was coupled with a tipping buckets soil water model [12], which calculates the water budget, solute transport and root uptake in the vadose zone. The model is parameterized with data derived from the ten-year field study and tested versus measured concentrations of lead (Pb) and cadmium (Cd) in soil and plants [3]. The accuracy of the model predictions can thus be evaluated. Furthermore, the simulation results will also be used to interpret the measured data.

Results

Measured K_d's versus Regression K_d's

The K_d estimates based on the regression equations of Sauvé et al. [7] were compared to measured K_d values based on $CaCl_2$ extractions of the soil surface horizons from 2002 to 2007 (Figure 1). The median ratios between predicted and measured K_d's are 1.9 (1.1; 4.4) for Cd and 0.68 (0.31; 1.3) for Pb (values in brackets are the 5th and 95th percentiles). Only predicted K_d-values were applied for the modeling of metal adsorption in all horizons and for the whole period. The predicted K_d-values for Cd in the control soil surface layer decreased over the ten-years period from 609 to 423 L kg^{-1}. For the GWS plot, they were first decreasing, but the final K_d was the same as the initial (588 L kg^{-1}). On the contrary, the predicted K_d-values for Cd in the FYM plot were increasing (from 538 to 858 L kg^{-1}), as for BIOW (785 to 1437 L kg^{-1}) and MSW (507 to 965 L kg^{-1}). The same tendency – decreasing predicted K_d-values for control and GWS plots and increasing K_d for the FYM, BIOW and MSW plots, was observed for Pb. These variations are mainly related to variations of soil pH, and of organic carbon in the case of Cd (equations 23–24; Table S1).

Results for Top Soil

Simulated and measured concentrations of Cd and Pb in top soil are shown in Figure 2. The differences between the five treatments are generally rather small. Measured values of Cd range between 0.21 and 0.27 mg kg dw^{-1} (median of four replicates), with seeming random variations versus time for the amended plots, and a decreasing trend for the control (Figure 2). Predicted concentrations of the control plot decline from 0.24 to 0.234 mg kg dw^{-1} (-2.6%), showing the slightly negative balance between the estimated air input and outputs by leaching and plant uptake.

Predicted concentrations for amended plots display non monotonous curves, related to the successive inputs from amendments and seasons dominated by outputs; however they

increase overall after 10 years, GWS by 9.8%, FYM, BIOW and MSW by 10.1%, 12.1% and 10.8%, respectively. Therefore, deviations between predicted and measured concentrations of Cd occur towards the end of the simulation period, the predicted values becoming about 10% higher than the measured ones.

The predicted concentrations of Pb in the top soil of the control plot are almost constant (+0.2%) over the ten years (Figure 2). Predicted concentrations from the other plots all increase after 10 years, between 6.7% (GWS) and 18% (FYM). The medians of measured data follow this trend with the highest value being found in 2006 for all treatments and the control. However, the relative dispersion of data and unexplained drops of Pb contents toward the last year weaken the possible links between measured and simulated variations.

The modeled fluxes from top soil are presented for one simulation event with growth of maize in Table 1 for the control and the treatment with the highest input of metal by amendment (BIOW, Table 2). Similar deposition values from air were measured by Azimi et al. [13] in 2002 at Versailles, about 20 km from the study site (0.05 mg Cd m^{-2} year^{-1} and 2.20 mg Pb m^{-2} year^{-1} compared to our estimates of 0.03 and 1.97, respectively). Table 1 also shows that the predicted plant uptake is 22% (Cd) and 10% (Pb) lower for the amended soil compared to the control soil.

Model Results for Plants

Simulated and measured concentrations of Cd and Pb in plants are shown in Figure 3. The predicted results for Cd are near the measured concentrations for harvest residues (leaves and stems). The measured concentrations of Cd in grains are lower than simulated and not shown in the figure, since most values were below the limit of quantification. The measured concentrations of Cd in stem and leaves are particularly low for the year 2007, which may be related to the exceptional crop barley and to a temporary change of method this year (see Chemical analyses). Before 2007, the simulated values are lower every second year, because concentrations for maize (1999, 2001, 2003 and 2005) are predicted to be lower than concentrations for wheat. The reason is that the model takes in account that maize is a C4-plant and has a lower transpiration coefficient (Table 3), i.e. less uptake of water per produced biomass [8]. The measured values for plant residues do not show this trend, the opposite is the case: the highest concentrations were measured 2001, for maize. Assuming no difference between the years, the measured concentration of Cd and Pb is significantly ($p<0.01$) higher in maize residues. It is known that Cd concentrations are lower in maize grain compared to wheat grain [14], but only a few studies allow comparison between wheat and maize residues. Lavado et al. [15] found for both residues and grains higher Cd concentrations in wheat compared to maize and the opposite for Pb.

Initially, the FYM and MSW plot (the curves overlap) have the highest simulated Cd concentration in plants, and the BIOW plot the lowest. The simulated concentrations of the MSW and FYM scenario decrease with time, those for the control (CTR) and GWS scenario increase. This shows that the total concentration in soil is less important for the predicted concentration in plants than the K_d. According to the simulations, deposition of Cd from air is of minor relevance. Only 1 to 5% of the simulated Cd in plants stems from atmospheric deposition. Also, the harvested plant mass has only little influence on the predicted concentration. Overall, the predicted concentrations of Cd in plants are rather similar for the five plots, and do depend only marginally on the Cd applied with amendment. This is confirmed by the measured concentrations of Cd in plants: Only in three instances, maize FYM 1999 (higher),

K_d (L/kg dw): Cd

K_d (L/kg dw): Pb

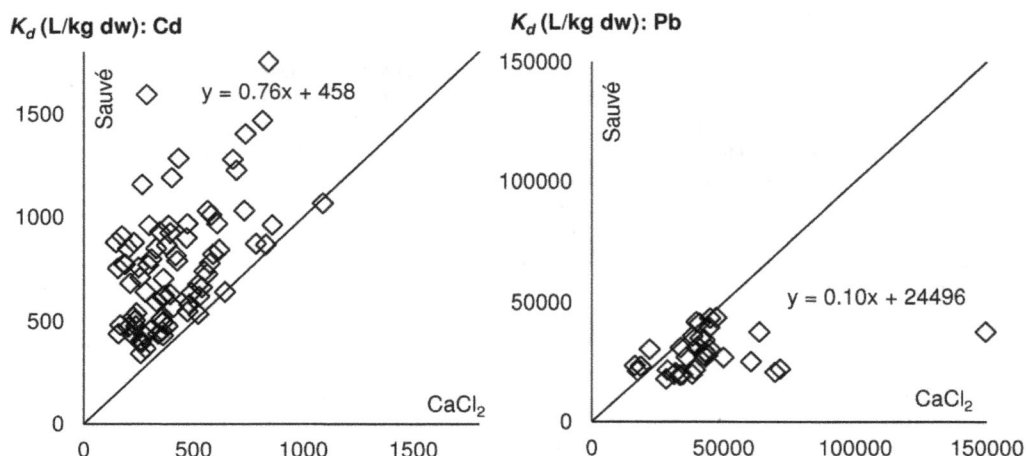

Figure 1. Estimated soil-water partition coefficient K_d (Sauvé regression) vs. measured K_d (determined from $CaCl_2$ extractions). The dotted line indicates a ratio of one.

wheat GWS 2000 (lower) and wheat BIOW 2006 (lower) was the measured concentration of Cd in leaf and stem significantly ($p < 0.05$) different for treatment and control.

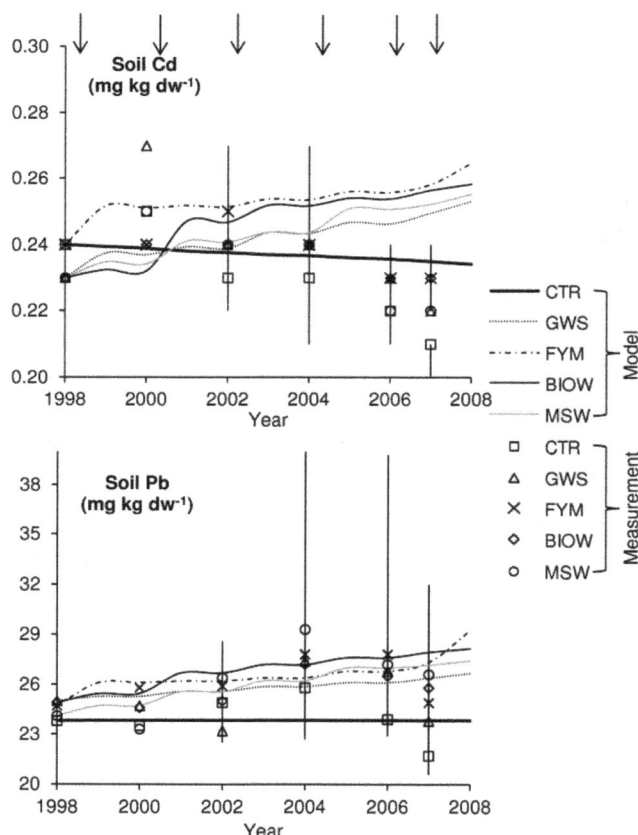

Figure 2. Comparison of predicted and measured concentrations in top soil for the five treatments. September 1998 to July 2007. Model predictions are connected by lines for clearer comparison to measured values. Vertical lines denote the range of measured values and arrows the time of amendment application. Measured concentrations represent median of four replicates.

Both the simulations and the measured results show a clear decreasing trend for Pb in plants in the period between 1998 and 2007 (Figure 3). According to the model simulations, the reason for the decline is the declining deposition of Pb from air. Measured concentrations of Pb in air in Paris (which served as input data for the simulation) decrease from 0.21 μg m^{-3} in 1998 to 0.01 μg m^{-3} in 2007 [16]. Subsequently, deposition from air declines, too. The fraction of Pb uptake from air into plants falls from 82% in 1998/9 (year 1) to about 30% from 2003/4 (year 6). For most of the years, the simulated Pb concentration in plants lies within the range of measured concentrations for stems or leaves. Again measured concentrations in grains are not shown, since most were below the limit of quantification. As before for Cd, the model predicts higher concentrations in wheat as in maize. The measured data are only for the first years (as long as aerial deposition dominates) higher for wheat. For the first years, as long as plants take up Pb mainly from air, there is little or no difference between the five plots. This is confirmed by the statistical analysis of the measured concentrations: only in three events (wheat BIOW 2000, higher, barley GWS and MSW 2007, higher) was the concentration of Pb in leaf and stem significantly ($p < 0.05$) different between treatment and control. Thus, for most events, no significant difference in concentration of plants from amended and control treatments could be found. Towards the end, the control scenario (CTR) and the GWS have slightly higher simulated

Table 1. Modeled fluxes for top soil, CTR and BIOW treatments (August 2000–October 2001).

1st soil layer (mg m^{-2})	CTR		BIOW	
	Cd	Pb	Cd	Pb
$m_{Soil, initial}$	91.4	9121	88.8	9741
Amendment	0	0	+6	+470
Air	+0.03	+1.97	+0.03	+1.97
Leaching	−0.25	−0.57	−0.21	−0.53
Plant uptake	−0.09	−0.20	−0.07	−0.18
ΔSoil	−0.31	+1.20	+5.75	+471.3

Table 2. Amendment application (second half of September in each given year) and input of Cd and Pb with amendment for the different treatments.

Year	Amendment application (kg dw m^{-2})				Cd Input (mg m^{-2})				Pb Input (mg m^{-2})			
	GWS	FYM	BIOW	MSW	GWS	FYM	BIOW	MSW	GWS	FYM	BIOW	MSW
1998	1.07	1.31	1.62	1.00	3.1	4.7	1.1	2.1	90	527	198	224
2000	1.98	1.10	2.45	1.92	1.3	0.6	6.0	2.9	117	36	470	324
2002	1.85	1.56	2.58	0.95	2.1	1.1	2.1	1.3	110	69	190	250
2004	1.73	1.37	1.97	1.46	1.5	1.1	1.0	2.9	91	151	160	295
2006	1.77	1.49	1.94	1.00	1.5	1.7	1.1	0.7	104	210	125	65
2007	1.58	1.33	1.62	1.05	1.6	2.7	0.8	1.3	113	730	84	101

GWS: Co-compost of green waste and sewage sludge, BIOW: Biowaste compost, FYM: Farmyard manure and MSW: Municipal solid waste compost.

concentrations of Pb. Again, this is due to the decreasing K_d of these plots. The input of amendment, which was highest for the FYM and the BIOW treatment (Table 2), did not lead to elevated concentrations in plants, but to a predicted decrease, because pH and thus the K_d increased.

Leaching to Groundwater

An annual water balance for the control scenario and the period from August 1998 to October 1999 is shown in Figure 4a. The precipitation is rather equally distributed over the whole season. Transpiration of plants (maize) occurs only during the vegetation period (from May to October). Evaporation from soil is relevant from March to June, then it stops, due to the drying of the upper soil, and continues when the plants are ripening and do not take up water anymore, after September. Leaching from the lowest soil layer to groundwater takes place in winter (December to March), and in periods with elevated precipitation (April, May). The simulation of the water content of the five soil layers (Figure 4b) starts with empty soil, i.e., the initial water content is set to the permanent wilting point (which differs for the five soil layers, see Table 4). In autumn, the layers are filled up again with water, due to precipitation and low or no transpiration of the vegetation, beginning with the top layer and then downwards. The water content remains at field capacity until the vegetation starts to draw larger amounts of water for transpiration. From May, the water storage of the soil is depleted, again starting with the top layer and then downwards. In July, all five soil layers have reached the permanent wilting point. From end of August, when the plants reduce their transpiration and ripen, the water content of the soil layers increases again, starting with the top layers (Figure 4b).

The leaching of Cd and Pb is closely coupled to the leaching of water. In fact, the pattern of leaching is identical for both compounds, only the level is different. Like water, leaching of compounds occurs in the winter time, and in periods with heavy precipitation and thus water surplus. Table 5 shows the leaching of water (L m^{-2}), Cd (mg m^{-2}) and Pb (mg m^{-2}) over the ten-year simulation for the control scenario. The annual leaching of water is very variable; the range is from 0 to 457 L m^{-2}. To consider is that the lengths of the simulation periods are not equal, due to different vegetation periods of maize and wheat. The average leaching of water is 157 L water m^{-2} per year, which is 23% of the average precipitation (Table S2). The average leaching of metals is 0.07 mg Cd m^{-2} year^{-1} and 0.16 mg Pb m^{-2} year^{-1} (Table 5).

According to the simulations, heavy metals applied in top soil via the various amendments do not affect the leaching of these metals, because neither Cd nor Pb are transported from top soil to bottom soil within the considered ten years. Thus, the amounts of Cd and Pb that leach to groundwater do not depend on the type of amendment. Some differences are seen because of the different initial concentrations of the five plots.

Calculated leaching of water from the second to the third soil layer was compared to water collected in situ at 40 cm depth with lysimeters, in 3 plots of the field (CTR, MSW and GSW) for the period from January 2005 to December 2007 [17] (Figure 5). Estimated leaching of 805, 815 and 819 L for the entire period for the GWS, MSW and CTR treatments are higher than the measured values (GWS: 474–535 L, MSW: 488–539 L, CTR: 648–741 L).

Discussion

Concept

The concept to couple the flux-based plant uptake model to a simulation model for water and solutes in discrete soil layers seems promising to us and allows the simultaneous simulation of water budget and plant uptake. The model can simulate various scenarios with different crops, soil and water conditions. However, the model has only one plant compartment (i.e. internal distribution is not accounted for) and the concept is limited to non-essential heavy metals, because the plant uptake of essential heavy metals from soil is regulated [4]. The full potential of the model concept could not be realized, because most soil properties (including concentrations) were determined only for the top soil. Thus, the simulations should rather be considered as illustrative.

Accuracy of Predictions

The simulated concentrations of Cd and Pb in soil and plants can be compared to the measured ones. The predicted increasing trend for Cd in top soil is not seen in the measured data. The samples taken last (July 2007) show for all soil variants the lowest concentration (Figure 2). There is a significant correlation ($p < 0.05$) between some of the measured concentrations from the five treatments (CTR-GWS, CTR-MSW, GWS-MSW and BIOW-MSW), indicating that the sampling or analysis method has some influence on the results. On the other hand, the measured concentrations are rather consistent (all median values range between 0.21 and 0.27 mg kg dw^{-1}), which shows that the analytical method is precise. But not precise enough to show the small changes predicted by the model. For Pb, too, the measured soil concentrations from the last samples are comparatively low. All other measured data confirm the upward trend of top soil concentrations for amended soils. Measured Pb concentrations in

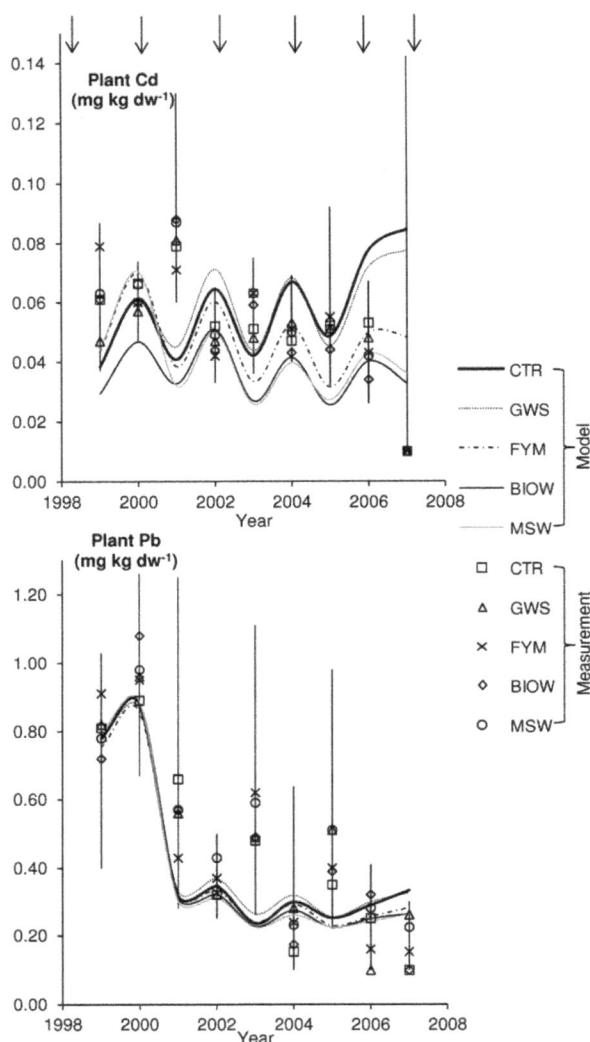

Figure 3. Comparison of predicted and measured concentrations in plants (mg kg^{-1}) for the five treatments. October 1999 to July 2007. Model predictions are connected by lines for clearer comparison to measured values. Vertical lines denote the range of measured values and symbols the medians of the four replicates (values below QL were set equal to ½ QL (note that QLs from 1999–2005 were applied for all years). Top arrows recall the time of amendment application.

control soil are the lowest, and have a constant trend. Predicted and measured concentrations are in this regard in good agreement.

The predicted concentrations of Cd in plants range between 0.025 and 0.085 mg kg dw^{-1}, the measured ones in stems and leaves between 0.02 and 0.087 mg kg dw^{-1}. For Pb, the predicted concentrations in plants are between 0.22 and 0.89 mg kg dw^{-1}, those measured in stems and leaves between 0.2 and 1.08 mg kg dw^{-1}. This is a rather good agreement, given the fact that the model is purely based on the calculation of passive transport with the water flux and deposition from air. In some cases, other factors than passive uptake with soil water may play a significant role, for instance the presence of competing ions like Ca^{2+} [18] and a high Cl$^-$ content of soils [4]. However, these effects do not seem to be relevant in our study. Furthermore, the model allows an interpretation of the relevant processes: for Cd, uptake from soil is dominating, while for Pb, deposition from air is the most relevant uptake process for the first three years.

In some details, the model has limitations. Measured concentrations in grains, which are more relevant for human consumption than stems and leaves, are lower than the simulated concentrations in plants. One reason for this could be that the water within plants flows mainly to leaves, from where it evaporates, while grains receive less water (about 1–2% of the xylem flow) and additionally are supplied with phloem sap. Therefore, the relation between growth and water uptake, which was used to calculate the transpiration stream and passive uptake from soil, does not hold for grains. The model has only one plant compartment and is not intended to simulate the internal distribution of metals within the plant, such as decreased concentration with distance from the roots due to sorption [19]. For the transport of metals to the grain via phloem, enzymatic processes could be involved, because the phloem sieve tubes are living cells [20] and some studies have suggested that ions like Cu^{2+} and Zn^{2+} are competing with Cd in the transport to grains [21,22]. In this study, measured concentrations of Cd and Pb in grains are mostly below the quantification limit and much lower than concentrations in stems and leaves. Mench et al. [21] also reported Cd content in wheat grains as being lower than that in the shoot, whereas Lavado et al. [15] found similar concentrations of Cd in both shoot and grain. It is uncertain, whether the modeling approach used here, i.e. physiologically based simulation of passive transport processes, can be modified so that it will successfully predict concentrations of non-essential metals in grains.

Another detail where the model does not meet the data is that measured concentrations of Cd are significantly ($p < 0.05$) lower for wheat straw than for maize straw, about one third. The same is seen for Pb, but less pronounced the first years, when deposition from air plays a major role [23]. The predictions are opposite, because maize, as C4-plant [20] needs less water per produced biomass, and thus the passive transport of solutes into the plant is, relatively seen, less. The model offers no explanation for this deviation, but it is known that genetic factors influence uptake [14]. Also, Lavado et al. [15] measured the same trend as the

Table 3. Estimated plant parameters (initial plant mass normalized to an area of 1 m^2).

Parameter	Symbol	Unit	Wheat	Maize	Barley	Source
Fraction of attached soil	SA	g ww g fw^{-1}	0.001	0.001	0.001	[37]
Transpiration coefficient	T_C	L kg fw^{-1}	100	60	as wheat	[8]
Overall growth rate	$k_{G,0}$	d^{-1}	0.094	0.081	as wheat	Estimated [11]
Initial plant mass	$M_{Initial}$	kg fw^{-1}	0.031	as wheat	as wheat	[11]

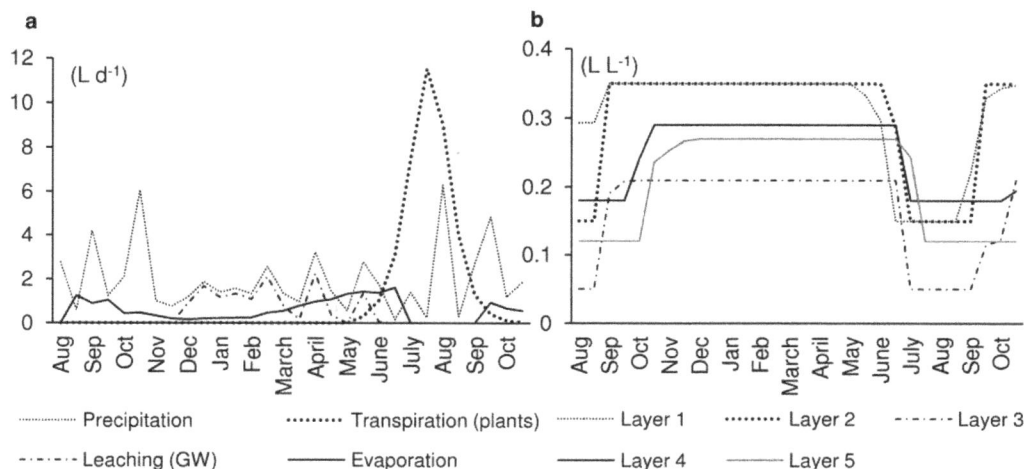

Figure 4. Simulated water balance and content of soil. (a) Simulated annual water balance, control scenario, August 1998 to October 1999; (b) simulated water content of the five soil layers, same simulation event.

model predictions with Cd concentrations being higher in wheat straw compared to maize straw. Measured concentrations in grains are in this study similar for maize and wheat, since most data are below or slightly above the quantification limit.

Concentrations of Cd and Pb in deeper soil layers and groundwater were in this study set equal to the concentration in top soil, since no measured data were available. However, this is probably not exact and we would expect lower concentrations in deeper layers. This would decrease the predicted concentrations in the soil solution of these layers, in leaching water and in plant tissues, since deep layers contribute to transpiration part of the year. E.g., by assuming the concentration in groundwater equal to half of the concentration in the lowest soil layer; Cd concentrations in maize are decreased by 16% for 1999.

The simulation of the water balance including leaching mimicked the timing and amount of soil water, as can be seen from the comparison of predicted and measured leaching of water (Figure 5). The model results were not fitted, and the measured results were only available after all simulations had been done. Two important simplifications were made with respect to the role of plants in the water balance. First, transpiration was calculated from measured plant growth data (Table 6), using a constant factor, the transpiration coefficient (Table 3). Second, and that is a novelty in the model approach, we skipped the calculation of root distribution, and assumed instead that roots grow to the soil layer

where they find water [24]. Both assumptions avoid parameter-intensive calculations, and were a prerequisite for simulations with the available data set.

Effect of Soil Amendments

The predicted increase in top soil concentrations of Cd and Pb was solely due to the application of Cd and Pb contained in the amendments (Table 2), whereas concentrations in control soils were predicted to decrease slightly or stay constant. For Cd, the BIOW, GWS and MSW soils in 2002 and all amended soils in 2007 had a statistically higher concentration of Cd than the control soil ($p < 0.05$). For Pb, the BIOW, GWS and FYM soils in 2006 and the BIOW, FYM and MSW soils in 2007 had a statistically higher concentration of Pb than the control soil ($p < 0.05$). But despite this increase for soil, both measured and simulated concentrations in harvest products from the amended plots did not increase. The opposite was observed: simulated concentrations of Cd in plants increased for the control soil (and GWS amendment), but it decreased for the BIOW, MSW and the

Table 4. Measured soil parameters (depth, dry density $\rho_{S,dry}$, field capacity, *FC*, and permanent wilting point, *PWP*) of the soil layers.

Soil layer	Depth	$\rho_{S,dry}$	FC	PWP
	(cm)	(kg L^{-1})	(L L^{-1})	(L L^{-1})
1	0–29	1.32	0.35	0.15
2	29–35	1.53	0.35	0.15
3	35–50	1.46	0.21	0.05
4	50–90	1.50	0.29	0.18
5	90–150	1.45	0.27	0.12

Table 5. Leaching of water (L m^{-2}), Cd (mg m^{-2}) and Pb (mg m^{-2}) for the ten simulation events in the control scenario.

Period	(L m^{-2})	Cd (mg m^{-2})	Pb (mg m^{-2})
Aug 98–Oct 99	199	0.078	0.181
Nov 99–Jul 00	113	0.044	0.103
Aug 00–Oct 01	457	0.192	0.431
Nov 01–Jul 02	124	0.052	0.117
Aug 02–Oct 03	229	0.102	0.216
Nov 03–Jul 04	0.0	0.000	0.000
Aug 04–Oct 05	125	0.065	0.130
Nov 05–Jul 06	0.0	0.000	0.000
Aug 06–Jul 07	86	0.049	0.093
Aug 07–Jul 08	236	0.134	0.286
Total	**1568**	**0.715**	**1.557**
Average	157	0.072	0.156

Leaching 40 cm (L d⁻¹)

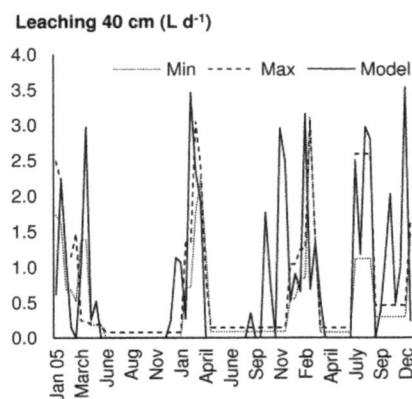

Figure 5. Leaching of water from soil layer 2. Model compared to measurement for CTR, MSW and GSW treatments. Model is average of all predictions, min and max is minimum and maximum lysimeter measurements, MSW(I) and CTR(II), respectively.

Table 6. Calculated total mass of harvested plant parts (from results in dry weight, please see Text S2, Table S4).

Date of harvest	Crops	Total plant mass (kg fw m⁻²)				
		GWS	BIOW	FYM	MSW	Control
20 Oct 1999	Maize	9.92	9.73	10.64	9.86	9.88
25 July 2000	Wheat	6.64	6.42	7.02	6.69	6.52
20 Oct 2001	Maize	9.26	9.18	9.18	9.55	8.91
20 July 2002	Wheat	5.22	5.75	5.68	6.05	5.05
15 Oct 2003	Maize	8.32	9.22	9.10	8.50	9.11
25 July 2004	Wheat	7.80	7.78	7.87	7.68	7.21
15 Oct 2005	Maize	8.75	8.63	8.71	8.15	7.81
19 July 2006	Wheat	7.00	6.70	7.26	6.89	6.42
19 July 2007	Barley	7.33	5.79	6.80	6.18	5.04

GWS: Co-compost of greenwaste and sewage sludge, BIOW: Biowaste compost, FYM: Farmyard manure and MSW: Municipal solid waste compost.

FYM amendment. Measured and simulated concentrations of Pb fell in harvest products from all plots (Figure 3). The reason is that deposition from air, which was responsible for higher concentrations in plants the first years decreased dramatically (concentrations of Pb in air were factor 21 higher in 1998 than in 2007). For Cd, deposition from air played only a minor role, and predicted concentrations in harvest products fell due to increasing K_d. All amendments increased the organic carbon content of the soil, while it fell slightly in the control soil (Table S1). Three of the amendments (FYM, BIOW, MSW) furthermore increased pH. Consequently, the calculated K_d of Cd and Pb increased in these three soils. For the GWS soil, only the K_d of Cd at the end was the same as initial (but with falling trend in between), the K_d of Pb fell. The calculated K_d of both metals decreased in control soil over the ten-years period (Table S1), leading to a predicted increase in plant uptake. This means, within the considered period, the application of FYM, BIOW and MSW amendments led to a reduction of the simulated heavy metal content in harvested crops. However, the organic carbon may be degraded again. In control soils, the average organic carbon fell from 1.072 g g⁻¹ in 1998 to 1.045 g g⁻¹ in 2007.

Comparison to Other Findings

The reduction of bioavailability of heavy metals by soil amendments was also mentioned in the review of Smith [2]. Accordingly, compost typically increases pH. A study comparing MSW amended soil to soil receiving mineral salts found a slight increase of soil concentrations but reduced transfer of Pb and Cd to field grown fodder crops from the MSW amended soil after 4 years of application, compared to mineral salt fertilized soils [25]. Similar, Gondek et al. [26] found no difference in aboveground concentrations of Pb in maize for maize grown in sewage sludge amended soils, compared to maize grown in soils fertilized with minerals only.

Comparison to Other Model Approaches

A variety of approaches is used to predict the uptake of heavy metals from soil into crops [4]. Commonly used are empirical bioconcentration factors (BCFs). These BCFs often have the form of a regression between soil concentration, soil properties and concentration in plants and are easy to apply. The disadvantage of such regressions is that they are typically limited to their regression range, and often only hold for a certain type of plant species, and within a limited range of soil properties. In a recent study, we could not confirm that multi-parameter regressions are superior to simple empirical, crop-specific transfer factors [5]. The model applied in the present study belongs to the so-called physiological models [4]. Their advantage is that conditions at site (such as plant growth and water budget) can be considered, and may explain uptake differences between the years. Peijnenburg et al. [27] used soil pore water concentrations and the water use efficiency (which is the inverse of the transpiration coefficient used in our study) multiplied with the weight change of plants to predict successfully the uptake of Cd and Zn into lettuce. Also, Ingwersen and Streck [28] estimated the concentration of Cd in wheat, sugar beet and potato using transpiration, concentration in soil solution and a plant specific empirical uptake efficiency parameter. For wheat, passive uptake was assumed and the uptake efficiency set to one. These approaches are thus very similar to the one used here. A difference is that we additionally considered deposition from air, but this process was only relevant for Pb. A more complex approach is the *Barber-Cushman* model which simulates advection and diffusion into roots using root geometry and soil properties by solving the underlying partial differential equation [29–31]. The approach may be useful to explain uptake processes of nutrients and heavy metals, but it is troublesome to derive the required input parameters on a field scale [4].

The coupling of physiological plant uptake models with water and solute transport models for soil is rare. Bauer-Gottwein et al. [9] combined a groundwater transport model with a physiological model for salt uptake and simulated the formation of salt islands in the Okawango delta. No publication about an approach to couple heavy metal transport in soil and groundwater to physiological plant uptake models is known to us. Therefore, our approach is probably unique. A common problem, namely the description of root distribution, root growth and root water uptake, was solved by the following assumption: Roots grow to where the water is; roots take up water from the highest soil layer where water is available; if this layer is depleted, roots continue to take up water from the next (deeper) layer. This description may be oversimplified in some cases, e.g., when the water content of the soil with depth changes rapidly. This may, for example, happen when precipitation events with high intensity appear after longer periods of drought. On the other hand, this algorithm allows an easy and efficient description of otherwise quite complex and largely unknown processes.

Conclusions

The long-term simultaneous simulation of the water budget of soil, the uptake into plants and the leaching to groundwater of two heavy metals on field scale succeeded by coupling a physiological plant uptake model to a buckets soil model. Concentration in soil of Cd and Pb, plant uptake, leaching, and deposition from air were simulated for a ten-years field experiment where biowaste compost, municipal solid waste compost, co-compost of green waste and sewage sludge and farm yard manure were applied to soil. In top soils from the control plot, calculated concentrations of Cd were slowly declining (2.6% in 10 years), mainly due to leaching. The calculated concentration of Pb in the control top soil was practically constant (+0.2%).

When soils were amended, calculated Cd and Pb concentrations in top soil were in all cases increasing, about 10% for Cd, and between 6% and 18% for Pb. Most organic soil amendments led to a reduction in the simulated plant uptake, because soil pH and organic carbon and thus the calculated K_d was increasing. Deposition from air was the dominating process for Pb before 2001, but hereafter was less relevant, due to steeply declining concentrations of Pb in air [16]. The comparison between simulated and measured concentrations in soils and plants showed overall good agreement, but also deviations in details. The uptake into plants using water flux and heavy metal concentration in soil pore water yielded concentrations which are comparable to those measured in leaves and stems, but the approach does not seem applicable for concentrations in grains.

The model can predict other scenarios and future trends for plant uptake and leaching of Cd and Pb. Also, future work should focus on variation of concentration in soil with depth and the possibility of adding an extra plant compartment to the plant model, so that the concentration in grains can be predicted.

Materials and Methods

Field Study

The "QualiAgro" long term field experiment has been initiated in 1998 by INRA de Grignon and Veolia Environnement R&I in order to study the benefits and environmental impacts of repeated urban compost applications on soil, water and plant qualities. The field is located at Feucherolles, Ile de France, 35 km west of Paris and is equipped with a meteorological station nearby that records climatic parameters [3]. Mean annual temperature is 11°C. Mean annual rainfall amounted to 582 mm yr^{-1} (average data between 1989 and 2009 measured in the nearby weather station). The soil is a silt loam Luvisol and contains on average 15% clay, 78% silt and 7% sand in the ploughed layer. Amendments included three urban composts (bio waste compost, BIOW; municipal solid waste compost, MSW; co-compost of green waste and sewage sludge, GWS) as well as farm yard manure (FYM). Applications were compared to controls without amendment (CTR). The experimental field was divided into 20 plots of 450 m^2 with 4 replicates for each treatment (the four amendments and the control). Amendments were applied and incorporated to soil in September 1998, 2000, 2002, 2004, 2006 and 2007 after wheat harvest. Composts and manure were applied at doses equivalent to 4 t carbon ha^{-1}, corresponding to 15 to 20 t dry weight (DW) ha^{-1} depending on the organic products (Table 2). In May the following year (1999, 2001, 2003 and 2005) maize was seeded and it was harvested in October, giving it a growth period of 5.5 months (169 days). After harvest of maize, wheat was seeded in November

(1999, 2001, 2003 and 2005) and harvested in July of the following year, assuming starting point of growth in March and thereby a growth period of 5 month (150 days). Barley was only seeded in October 2006, 1 month after application of amendment and harvested in July 2007. It was seeded in replacement of maize, because of a pest (*Diabrotica virgifera*) alert. Additional mineral N fertilizer was added in each treatment to reach optimum crop yield. Figure 6 gives an overview on amendment application and succession of crop cultivation. Plants (harvested grains and plant residues) were analyzed for metals and other characteristics. Wheat residues were always exported and maize residues incorporated into soils after harvest. Sampling of soil plough layers and organic amendments was done prior to each amendment application, in early September or late August. For each plot a representative soil sample was obtained from 10 sampling points. Amendments were sampled in triplicates. All samples were conditioned and analyzed for metal content and other parameters according to normalized methods (see Chemical analyses section). The median of the replicates (three for amendment and 4 for soil and plant samples) is reported here. Three plots of the field, corresponding to the 3 treatments CTR, MSW and GWS, were equipped by mid 2004 with lysimeters at a depth of 40 cm in order to collect soil water during the drainage periods [17].

All necessary permits were obtained for the described field studies. An agreement was made between the land owner, Mr Bignon, and INRA.

Chemical Analyses

All analyses of soils and amendments were performed at the INRA Laboratoire d'Analyses des Sols (Arras, France). All analyses of plants were performed at the USRAVE laboratory (INRA Bordeaux, France). Both laboratories are accredited according to NF ISO/CEI 17025 for the soil and plant analyses reported here. Concerning the analysis of amendments, INRA Arras applies the same quality controls and the same validation methods (norm NF V3-110 and T90-210) as for soil analysis.

Soil samples were dried at 40°C and passed through a 2 mm sieve. Representative aliquots were ground and sieved at 250 μm before C and total metal analyses. Organic C was determined by catalysed combustion-oxidation (norm ISO 10694). Metal analyses were performed by ICP-MS after heating at 450°C and complete digestion in HF-HClO$_4$ (norm NF X 31–147). pH was measured on a 10 g aliquot of the sample <2 mm dispersed in pure water (soil:water 1:5; norm ISO 10390). Exchangeable metals were determined from extraction in 0.01 M CaCl$_2$ solution (ratio 1:10, shaking 2 h, centrifugation and filtration; NL norm NEN 5704).

Organic amendment samples were freeze-dried, ground and sieved at 5 mm. Total metals and C were analysed on aliquots like for soils. The carbonate content was also analysed and, when significant, inorganic C was subtracted from total C to get the organic C. Total metal contents were obtained by the same digestion method as for soil followed by ICP-AES (NF ISO 2203).

Plant samples were ground and homogenized with a rotary homogenizer. 1 gram of dry plant powder was weighed in a silicon capsule and incinerated in a muffle furnace at 480°C for 5 h. Afterwards, it was digested with concentrated nitric acid in several steps. The remaining powder collected on ash-free paper was incinerated at 550°C for 2 h. Ashes were dissolved in a Teflon capsule by 5 mL of concentrated hydrofluoric acid, evaporated and dissolved again in two steps with concentrated nitric acid. All obtained solutions were collected in a volumetric flask and completed to 100 mL with distilled water. An ICP-AES Iris Intrepid (Thermo Fischer Scientific Inc., Waltham, USA) and an

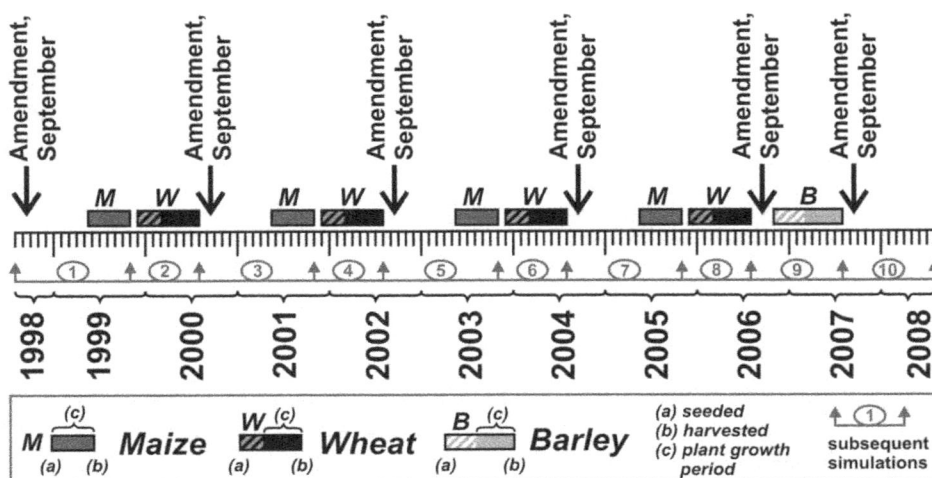

Figure 6. Overview of the field and simulation study. For wheat and barley the starting point of growth takes place after seeding.

ICP-AES Liberty Serie 2 (Varian, Mulgrave, Australia) equipped with an ultrasonic nebulizer U-5000AT+ (CETAC, Omaha, Nebraska, USA) was used for Cd and Pb analysis. Spectrometer operating conditions are fully described elsewhere [32,33]. Analysis quality was controlled using an in-house laboratory reference sample V463 (entire maize plant) and blanks which have undergone the entire analysis process. Concentration values measured in blanks were subtracted from concentration values measured in the samples.

All chemical contents were expressed per dry weight (dw) 105°C according to the norms NF ISO 11465 and NF U44-171 for soils and amendments, respectively. The weighing and humidity correction for plants were done using a meteorologically controlled scale (Mettler Toledo S.A., Viroflay, France) and a meteorologically controlled drying cupboard at the temperature of 103±5°C.

From 1999 to 2005, the quantification limits (QLs) in plants were ≤0.03 mg kg dw^{-1} for Cd and ≤0.2 mg kg dw^{-1} for Pb. In 2006 and 2007, probably due to a temporary change of method, QLs were much higher, i.e. up to 0.3 and 1 mg kg dw^{-1} for Cd and Pb, respectively. In 2008 and 2009 (data not shown), QLs were again down to previous levels (0.03 mg kg dw^{-1} for Cd and 0.2 mg kg dw^{-1} for Pb). Only the QLs from 1999–2005 and 2008–2009 were applied in this manuscript. QLs for the total metal contents of soil are 0.02 mg Cd and 0.2 mg Pb per kg dw (down to 0.1 mg Pb per kg dw after 2006). For organic amendments QLs are higher, 0.5 mg Cd and 2 mg Pb per kg dw. For aqueous samples, QLs are 0.05 μg L^{-1} of Cd and 0.2 μg L^{-1} of Pb.

Modeling Approach

Modeling of metal transport in the soil-air-plant system was done by coupling a model for water and solute transport in soil including a discrete cascade approach for the water balance (tipping buckets model) [12,34] to a dynamic plant uptake model similar to the multi-cascade approach [10,11] (Figure 7). The tipping buckets soil water and substance transport model was chosen because its step-wise and periodic simulation mode makes it easily compatible to the step-wise solution method of the analytical multi-cascade plant model. A second reason was that the time period for simulation was ten years, and the buckets approach needs only a reasonable number of input data. In each time period, the water and substance balance in the five soil layers is solved iteratively considering precipitation, infiltration, leaching and transpiration (i.e., water uptake from soil by growing plants). Uptake of heavy metals into plants is with the water taken up by the roots at various depths. The coupled soil water and solute transport and plant uptake model was realized as Microsoft Excel^{TM} spreadsheet.

Tipping Buckets Model for Transport of Water in Soil

The discrete tipping buckets water balance model [12,34] considers m soil layers located above the groundwater table, for which the water balance is calculated. Five soil layers ($m = 5$, Table 4) were specified in the applied model. The soil layers are considered to be a series of "tipping buckets", which have an upper and lower limit for water storage capacity: the water content at the upper limit is the field capacity, FC (L), and that at the lower limit is the permanent wilting point, PWP (L). Flow is discontinuous, i.e. the soil layers are considered as buckets that can be filled up to field capacity, after which they tip, and by putting the soil layers in series, tipping buckets arise that transport water and solutes. The model considers downwards (leaching) as well as upwards (transpiration and evaporation) movement of water and solutes. Transpiration, i.e. water extraction by plants, is calculated from plant growth (see later section). It is assumed that plant roots always extract water from the highest possible soil layer, and until the PWP is reached [8]. Capillary rise from the groundwater table to the plant roots was not included in the model, except as part of the transpiration in the growing season of the plants. Also, groundwater elevation due to leaching was neglected. Precipitation, evaporation and transpiration were considered and each calculation was done in eight steps as detailed in the following. All calculations were done for an area of 1 m^2.

Step 1. Initial (absolute) water content in the top soil layer (soil layer 1), W_{Ini1} (L), is obtained from initial volumetric water content, $\theta_{W,Ini1}$ (L L^{-1}), and the volume of soil layer 1, V_{S1} (L), as

$$W_{Ini1} = \theta_{W,Ini1} \times V_{S1} \qquad (1)$$

Step 2. Infiltration, Inf (L d^{-1}), is calculated from precipitation, P (L d^{-1}), and evaporation, E (L d^{-1}), (soil layer 1):

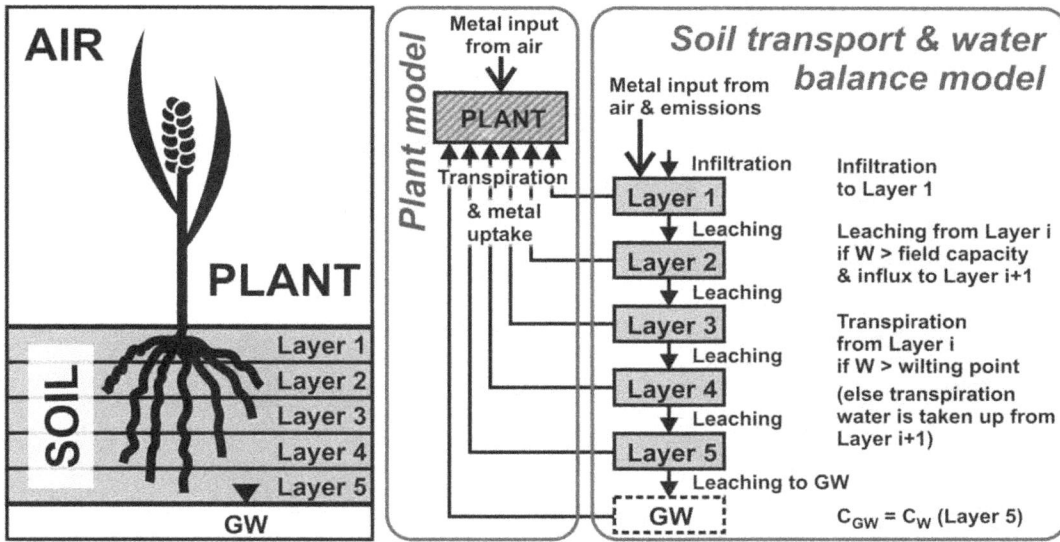

Figure 7. Processes and compartments in the coupled soil solute transport, water balance and plant uptake model. W: water content, GW: groundwater, C_{GW}: groundwater concentration, C_W: soil pore water concentration.

$$Inf = \begin{cases} P - E & if \ P > E \\ 0 & if \ P \leq E \end{cases} \qquad (2)$$

Step 3. After infiltration, a new water content, W_{Inf1} (L), is established in soil layer 1:

$$W_{Inf1} = W_{Ini1} + Inf \times \Delta t \qquad (3)$$

where Δt (d) is the length of the time period.

Step 4. Leaching from soil layer 1, $Leach_1$ (L), occurs if the water content is now above field capacity FC (L):

$$Leach_1 = \begin{cases} W_{Inf1} - FC_1 & if \ W_{Inf1} > FC_1 \\ 0 & if \ W_{Inf1} \leq FC_1 \end{cases} \qquad (4)$$

Step 5. After leaching, the water content of soil layer 1 changes to W_{Leach1} (L):

$$W_{Leach1} = W_{Inf1} - Leach_1 \qquad (5)$$

Step 6. Transpiration, i.e. water flux to plants from soil layer 1, q_1 (L), takes place if the water content is now above the permanent wilting point PWP_1 (L):

$$q_1 = \begin{cases} 0 & if \ W_{Leach1} \leq PWP_1 \\ W_{Leach1} - PWP_1 & if \ W_{Leach1} > PWP_1 \ and \ W_{Leach1} - PWP_1 < Q \times \Delta t \\ Q \times \Delta t & if \ W_{Leach1} > PWP_1 \ and \ W_{Leach1} - PWP_1 \geq Q \times \Delta t \end{cases} \qquad (6)$$

where Q (L d^{-1}) is the total transpiration of the plant in this period (see later section).

Step 7. After transpiration, again a new water content, W_{q1} (L), is established in soil layer 1:

$$W_{q1} = W_{Leach} - q_1 \qquad (7)$$

Step 8. Finally, remaining transpiration $q_{Total-1}$ (L), i.e. transpiration water that needs to be taken from deeper soil layers, is obtained by:

$$q_{Total-1} = Q \times \Delta t \ q_1 \qquad (8)$$

For the next soil layers (soil layer i, with $i > 1$), steps 3 to 8 are repeated. However, Step 3 (Eq. 3) is the new water content of layer i due to leaching from above:

$$W_{Inf,i} = W_{Ini,i} + Leach_{i-1} \qquad (9)$$

Step 6 (Eq. 6) changes to

$$q_i = \begin{cases} 0 & if \ W_{Leach,i} \leq PWP_i \\ W_{Leach,i} - PWP_i & if \ W_{Leach,i} > PWP_i \ and \ W_{Leach,i} - PWP_i < q_{Total-(i-1)} \\ q_{Total-(i-1)} & if \ W_{Leach,i} > PWP_i \ and \ W_{Leach,i} - PWP_i \geq q_{Total-(i-1)} \end{cases} \qquad (10)$$

and Step 8 (Eq. 8) changes to

$$q_{Total-i} = q_{Total-(i-1)} \ q_i \qquad (11)$$

The water balance was established iteratively for all soil layers i in each time period p. The calculated water content after transpiration from one time period, $W_{q,i,p}$, was entered as initial water content for the following time period, $W_{Ini,i,p+1}$.

If the plant does not find sufficient water in the five soil layers (i.e., $Q > \sum q_i$), it is assumed that the remaining water required for transpiration is drawn from groundwater. This does not affect water or substance content of the five soil layers. In the present model formulation we assume that the groundwater has the same substance concentration as the water in the lowest soil layer.

Solute Transport in Soil

Solutes passively follow the water movement. The change of solute concentration in soil is given by input from air and pulse emissions (amendment application) to soil layer 1 minus loss of solute by leaching and plant uptake via transpiration. As heavy metals are considered in this study, loss by degradation and by volatilization from the top layer is not of relevance. In discrete form, the concentration in soil layer 1, $C^*_{S,1}$ (mg L^{-1}) (referred to the volume of bulk soil, V_S), at time t is:

$$C^*_{S,1}(t) = C^*_{S,1}(t-1) + \frac{A_S \, v_{dep}}{V_{S1}} \Delta t \, C_{A,p} + \frac{I}{V_{S,1}}$$
$$- \frac{Leach_1(t) + q_1(t)}{V_{S,1}} K_{WS1} \times C^*_{S,1}(t-1) \qquad (12)$$

where $C^*_{S,1}(t-1)$ is metal concentration in soil layer 1 at time $t-1$ (preceding time period), A_S (1 m^2) is the surface area of the soil, v_{dep} (m d^{-1}) is the deposition velocity of particles, $C_{A,p}$ (mg m^{-3}) is the total concentration (usually at particles) in air and I (mg) is the pulse input of metal (from amendment application). The water to dry soil partition coefficient K_{WS1} (−) in soil layer 1 was calculated as

$$K_{WS1} = \frac{1}{K_d \times \rho_{S1,dry}} \qquad (13)$$

where Kd (L kg dw−1) is the dry soil to water partition coefficient and ρS,dry (kg dw L−1) is the density of dry soil.

The change of metal concentration in the second and following soil layers (index i, with $i>1$) is given by influx of solute from the upper soil layer via leachate minus loss by leaching to deeper soil layers and transpiration. Soil concentration $C^*_{S,i}$ at time t is accordingly:

$$C^*_{S,i}(t) = C^*_{S,i}(t-1) + \frac{Leach_{i-1}(t)}{V_{S,i}} K_{WS,i-1} \, C^*_{S,i-1}(t-1)$$
$$- \frac{Leach_i(t) + q_i(t)}{V_{S,i}} K_{WS,i} \times C^*_{S,i}(t-1) \qquad (14)$$

The volume-based concentrations in bulk soil, C^*_S (mg L^{-1}), can be converted to soil dry weight, C_S (mg kg dw^{-1}), by dividing by the dry soil density, $\rho_{S,dry}$ (kg dw L^{-1}). For solutes, the Courant criterion [12] needs to be fulfilled, which says that in one step not more compound can flow out of a layer than is in it. This limits thickness of the layers and length of time steps.

Plant Growth and Transpiration

Transpiration was coupled to plant growth and implemented in the model according to Rein et al. [11]. Logistic plant growth was assumed (following e.g. Richards [35]), where the change of plant mass M (kg fw) with time t (d) can be expressed as

$$\frac{dM}{dt} = k_{G,O} \times M \left(1 - \frac{M}{M_{Harvst}}\right) \qquad (15)$$

where $k_{G,O}$ (d^{-1}) is the overall first-order growth rate constant and $M_{Harvest}$ (kg fw) is harvested (assumed maximum) plant mass. With

initial plant mass $M_{Initial}$ (kg fw) (plant mass at time $t=0$), the analytical solution is:

$$M(t) = \frac{M_{Harvest}}{1 + \left(\frac{M_{Harvest}}{M_{Initial}} - 1\right) e^{-k_{G,O} \, t}} \qquad (16)$$

Transpiration, Q (L d^{-1}), is coupled to plant mass growth via the transpiration coefficient, T_C (L kg fw^{-1}) [8]. In discrete form, transpiration Q at time t induced by changing (growing) plant mass is accordingly given by

$$Q(t) = T_C \frac{M(t) - M(t-1)}{\Delta t} \qquad (17)$$

where $M(t)$ and $M(t-1)$ are plant mass (kg) at time t and $t-1$ (preceding time period) and Δt (d) is the length of the time period. First-order growth rate constants, $k_G(t)$ (d^{-1}), specific to each time period were obtained by

$$k_G(t) = \frac{\ln [M(t)/M(t-1)]}{\Delta t} \qquad (18)$$

These were used for step-wise (i.e. time-period-wise) approximation of logistic growth and applied as first-order loss rate constants for growth dilution (see Eq. 19; please refer to Rein et al. [11] for more details). The application of these formulae requires only four input data ($M_{Initial}$, $M_{Harvest}$, k_G and T_C) for the whole simulation, instead of plant mass and transpiration data for each period.

Plant Uptake Model for Non-essential Metals

Non-essential heavy metals show plant uptake linearly related to their concentration in soil solution [4]. We assumed passive uptake of heavy metals with soil water into the plant. The transpiration of the plant depends on its transpiration coefficient and growth (see above). The model contains only one plant compartment and the change of concentration in the plant compartment was calculated from input via wet and dry particle deposition plus input via uptake from soil minus loss by growth dilution:

$$\frac{dC_P}{dt} = \frac{A_S \left(f_{wet} \Lambda_{part} \, Rain + f_{dry} \, v_{dep}\right)}{M_P} C_{A,p}$$
$$+ \frac{I_P}{M_P} - k_G \, C_P \qquad (19)$$

where C_P (mg kg fw^{-1}) is the concentration of metal in plant tissue, Λ_{part} (m^3 air m^{-3} rain) is the rainfall scavenging ratio for particles, $Rain$ (m d^{-1}) is precipitation, M_P (kg fw) is plant mass, I_P (mg d^{-1}) is the uptake of metal from soil and k_G (d^{-1}) is the first-order growth rate constant of plants (for consideration of growth dilution). This equation can be used to predict the overall concentration in plants.

The fractions of metal in rainfall and at particles that are intercepted by and transferred to the plant, f_{wet} and f_{dry} (−), were calculated from plant mass and absorption coefficients, μ_{wet} and μ_{dry} (m^2 kg dw^{-1}) [36]:

$$f_{wet} = 1 - \exp (- \mu_{wet} \, M_P \, DW_P) \qquad (20a)$$

$$f_{dry} = 1 - \exp(- \mu_{dry} M_P DW_P) \quad (20b)$$

where DW_P (kg dw kg fw^{-1}) is the dry matter content of the plant, which is equal to one minus the water content of the plant $(1 - W_P)$ and M_P is plant mass in units of kg fw m^{-2} (for an area of 1 m^2). The uptake of metal from soil into plant, I_P (mg d^{-1}), in each period was calculated as the sum of the uptake from all m soil layers and from groundwater via transpiration, q (see above):

$$I_P = \frac{\sum_{i=1}^{m} q_i \, C_{S,i}^* \, K_{WS,i} + q_{Total-n} \, C_{GW}}{\Delta t} \quad (21)$$

where CGW (mg L−1) is groundwater concentration, which is assumed equal to pore water concentration in the deepest soil layer (CGW = C*S,n x KWS,n), and qTotal-n (L) is the remaining transpiration of the plant that cannot be satisfied from soil water. Attachment of soil particles was considered subsequently as an additional process, assuming that a fraction of soil particles (default 0.1% for cereals) is attached to plant surfaces [37]:

$$C_{P,total} = C_P + 0.001 \times C_{S,1} \quad (22)$$

For the dynamic calculation of the concentration in plant, the principle of superposition was applied, i.e. the simulation was divided into n periods during which all parameters are kept constant (each period was then further subdivided into 30 time intervals, at which intermediate results were calculated). This procedure allowed the application of analytical solutions of the differential equation (Eq. 19) for each period, i.e. the result from one period is entered as initial value for the following period. This also allowed varying all rates and constants from period to period, and thus, to model time-varying contaminant input as well as to approach non-linear input (such as logistic growth of plants, or changing weather conditions).

Simulation Study and Model Parameterization

The total simulation was ten years (August 1998 to July 2008) and was subdivided into ten consecutive simulation events, each ending with harvest (Figure 6). The simulation events were further subdivided into periods of two weeks (first and second half of each month). Five simulations were carried out, one for the control plot (the only source of pollutants for crops can be the background level or the aerial deposition) and one for each type of amendment (bio waste compost, BIOW; municipal solid waste compost, MSW; co-compost of green waste and sewage sludge, GWS and farm yard manure, FYM).

Soil data. Density, porosity, thickness of soil layers, field capacity and permanent wilting point of the soils were considered equal for all soils (Table 4). The simulation in August 1998 started with "empty" soil, i.e. the water content of all soil layers was set to the permanent wilting point, corresponding to the typical situation towards the end of the growing season. For all other years, the initial water content of the soil layers was the calculated final water content of the year before.

Total metal contents of soils measured in 1998, before the first amendments, slightly differed among the five treatments; the median values from the 4 field replicates were input for the simulation (Table S1). For the following years, the calculated concentrations of the year before served as starting concentration. Organic carbon content and pH varied slightly with plot and year (Table S1). The soil to water distribution coefficients, K_d (L kg dw^{-1}), for Cd and Pb, was estimated by the following regressions [7]:

$$\log K_d(\text{Cd}) = 0.48 \, pH + 0.82 \log(OC) - 0.65 \quad (23)$$

$$\log K_d(\text{Pb}) = 0.37 \, pH + 0.44 \log(C_{S,t}) + 1.19 \quad (24)$$

where pH is the pH of soil water, OC (% (dw dw^{-1})) is the percentage of organic carbon in soil and $C_{S,t}$ (mg kg dw^{-1}) is the measured total concentration of metal in soil (Table S1). Measured data for the top soil were applied to estimate soil concentrations and K_d for all five soil layers (Table 4).

Water balance. Recorded daily precipitation rates representative for the QualiAgro site were averaged to give one precipitation estimate per half month (Table S2). Evaporation from soil, E (L m^{-2} d^{-1}), was estimated from reference evapotranspiration, ET_0 (L m^{-2} d^{-1}), as

$$E = \begin{cases} K_{C,Ini} \times ET_0 & \text{if } W_{S1} > PWP_1 \\ 0 & \text{if } W_{S1} = PWP_1 \end{cases} \quad (25)$$

where $K_{C,Ini}$ (−) is the crop coefficient from the initial growth stage of the crop (a value of 0.3 used as best estimate for cereal crops [38]). Reference evapotranspiration, averaged for 15 days, was calculated using the Penman-Monteith equation [39] (Method see Text S1, results Table S3). Transpiration by plants was calculated as described above. Surface run-off of water was neglected.

Plants. Crop-specific parameters are transpiration coefficient, growth rates and initial and final plant mass (Table 3, Text S2, Table S4). The harvested amounts of grains and residues (consisting of leaves and stems) were measured in the field experiment on a dry weight basis (Table 6).

Air. Concentrations of Cd (0.68 ng m^{-3} in 1999 to 0.28 ng m^{-3} in 2008) and Pb (0.21 µg m^{-3} in 1998 to 0.01 µg m^{-3} in 2007) measured in air in Paris were taken as input parameters [16]. For the calculation of particle deposition, a rainfall scavenging ratio was applied (Λ_{part} in Eq. 1); literature values range from 1000 to 200 000 [12], and a value of 20 000 m^3 m^{-3} was chosen. The default particle deposition velocity (v_{dep}) is 0.001 m s^{-1} (fine particles [12]).

Input via amendment. The input (mg m^{-2}) of Cd and Pb to soil via amendment is simulated as pulse input by the model. Data in Table 2 were derived from measured amendment concentrations by multiplying with applied amendment mass.

Supporting Information

Table S1 Estimation of soil-water partition coefficient K_d. Measured concentration of Cd and Pb in soils, organic carbon content and pH of soils together with estimated K_d's from Sauvé et al's equations.

Table S2 Precipitation. Precipitation rates (m d^{-1}), average half monthly values from August 1998 to December 2002 (1998 to August 2002: records from Parc Meteo Grignon, Grignon, France, about 5 km southwest of the test site; November 2002 to 2008: records from Feucherolles, France, adjacent to the test site).

Table S3 Evaporation. Calculated evaporation rates $(0.3 \times ET_0)$ (L m^{-2} d^{-1}), average half monthly values from August 1998 to July 2008.

Table S4 Plant mass. Estimated grain, leaf, stem and root mass (mg kg fw m^{-2}).

Text S1 Evaporation. Method for calculating reference evapotranspiration.

Text S2 Plant mass. Method for estimating plant mass.

Acknowledgments

We thank V. Mercier, J.N. Rampon and M. Jolly from INRA de Grignon and Veolia Environnement R&I for their help to manage the site of Qualiagro and A. Guérin, M. Barbaste and other contributors from INRA Arras and Bordeaux for the chemical analyses.

Author Contributions

Conceived and designed the experiments: SH. Performed the experiments: VB PC SH CNL ST AR. Analyzed the data: CNL ST AR PC SH. Contributed reagents/materials/analysis tools: CNL ST AR JS VB CH PC SH. Wrote the paper: CNL ST AR PC VB. Provided intellectual input: CNL ST AR JS VB CSH PC SH.

References

1. Diacono M, Montemurro F (2010) Long-term effects of organic amendments on soil fertility. A review. Agron Sustain Dev 30: 401–422.
2. Smith SR (2009) A critical review of the bioavailability and impacts of heavy metals in municipal solid waste composts compared to sewage sludge. Environ Int 35: 142–156.
3. Houot S, Bodineau G, Rampon JN, Annabi M, Francou C, et al. (2005) Agricultural use of different residual waste composts - current situation and experiences in France. The Future of Residual Waste Management in Europe. (http://www.compost.it/biblio/2005_luxembourg/vortraege/houot-doc.pdf).
4. McLaughlin MJ, Smolders E, Degryse F, Rietra R (2011) Uptake of metals from soil into vegetables. In: Swartjes FA, editor. Dealing with contaminated sites. Springer, Dordrecht. 325–367.
5. Legind CN, Trapp S (2010). Comparison of prediction methods for the uptake of As, Cd and Pb in carrot and lettuce. SAR QSAR Environ Res 21: 513–525.
6. Hough RL, Breward N, Young SD, Crout NMJ, Tye AM, et al. (2004). Assessing potential risk of heavy metal exposure from consumption of home-produced vegetables by urban populations. Environ Health Persp 112: 215–221.
7. Sauve S, Hendershot W, Allen HE (2000) Solid-solution partitioning of metals in contaminated soils: Dependence on pH, total metal burden, and organic matter. Environ Sci Technol 34: 1125–1131.
8. Larcher W (1995). Physiological Plant Ecology. Springer, Berlin, Germany.
9. Bauer-Gottwein P, Rasmussen NF, Feificova D, Trapp S (2008) Phytotoxicity of salt and plant salt uptake: Modeling ecohydrological feedback mechanisms. Water Resour Res 44.
10. Legind CN, Kennedy CM, Rein A, Snyder N, Trapp S (2011) Dynamic plant uptake model applied for drip irrigation of an insecticide to pepper fruit plants. Pest Manag Sci 67: 521–527.
11. Rein A, Legind CN, Trapp S (2011) New concepts for dynamic plant uptake models. SAR QSAR Environ Res 22: 191–215.
12. Trapp S, Matthies M (1998) Chemodynamics and environmental modeling: An introduction. Springer, Berlin-Heidelberg.
13. Azimi S, Cambier P, Lecuyer I, Thevenot D (2004) Heavy metal determination in atmospheric deposition and other fluxes in northern France agrosystems. Water Air Soil Poll 157: 295–313.
14. Chaney RF (2010) Cadmium and Zinc. In: Hooda PS, editor. Trace Elements in Soils, John Wiley & Sons, Ltd, Chichester, UK. 409–439.
15. Lavado RS, Rodríguez M, Alvarez R, Taboada MA, et al. (2007) Transfer of potentially toxic elements from biosolid-treated soils to maize and wheat crops. Agr Ecosyst Environ 118: 312–318.
16. AIRPARIF (2008) Rapport d'activité 2008. http://www.airparif.asso.fr/airparif/pdf/2008.pdf.
17. Cambier P, Benoit P, Bodineau G, Jaulin A, Trouve A, et al. (2007) Lixiviation des ETM et des produits phytosanitaires après épandages répétés de produits résiduaires organiques. Journée technique "Retour au sol des produits résiduaires organiques", Ademe, Colmar, 27/11/2007, 108–122.
18. Sterckeman T, Redjala T, Morel JL (2011) Influence of exposure solution composition and of plant cadmium content on root cadmium short-term uptake. Environ Exp Bot 74: 131–139.
19. Peralta-Videa JR, Lopez ML, Narayan M, Saupe G, Gardea-Torresdey J (2009) The biochemistry of environmental heavy metal uptake by plants: Implications for the food chain. Int J Biochem Cell B 41: 1665–1677.
20. Sitte P, Ziegler H, Ehrendorfer F, Bresinsky A (1991) Lehrbuch der Botanik für Hochschulen. Stuttgart: Gustav Fischer, 33rd ed.
21. Mench M, Baize D, Mocquot B (1997) Cadmium availability to wheat in five soil series from the Yonne district, Burgundy, France. Environ Pollut 95: 93–103.
22. Herren T, Feller S (1997) Transport of cadmium via xylem and phloem in maturing wheat shoots: Comparison with the translocation of zinc, strontium and rubidium. Ann Bot-London 80: 623–628.
23. De Temmerman L, Hoenig M (2004) Vegetable crops for biomonitoring lead and cadmium deposition. J Atmos Chem 49: 121–135.
24. Walter H (1986) Allgemeine Geobotanik. Ulmer, Stuttgart.
25. Montemurro F, Charfeddine M, Maiorana M, Convertini G (2010) Compost use in agriculture: The fate of heavy metals in soil and fodder crop plants. Compost Sci Util 18: 47–54.
26. Gondek K, Filipek-Mazur B, Koncewicz-Baran M (2010) Content of heavy metals in maize cultivated in soil amended with sewage sludge and its mixtures with peat. Int Agrophys 24: 35–42.
27. Peijnenburg W, Baerselman R, de Groot A, Jager T, Leenders Det al. (2000) Quantification of metal bioavailability for lettuce (Lactuca sativa L.) in field soils. Arch Environ Con Tox 39: 420–430.
28. Ingerwersen J, Streck T (2005) A regional-scale study on the crop uptake of cadmium from sandy soils: Measurement and modeling. J Environ Qual 34: 1026–1035.
29. Barber SA, Cushman JH (1981) Nitrogen uptake model for agronomic crops. In: Iskandar IK, editor. Modeling Waste Water Renovation- Land Treatment Wiley Interscience, New York. 382–409.
30. Barber SA (1995) Soil nutrient bioavailability: a mechanistic approach. John Wiley and Sons, New York.
31. Penn State University Laboratory (PSU): Barber-Cushman Model. http://roots.psu.edu/node/934, last accessed 11 July 2011.
32. Masson P, Orignac D, Vives A, Prunet T (1999) Matrix effects during trace elements analysis in plant samples by inductively coupled plasma atomic emission spectrometry with axial view configuration and ultrasonic nebulizer. Analusis 27: 813–820.
33. Masson P, Vives A, Orignac D, Prunet T (2000) Influence of aerosol desolvation from the ultrasonic nebulizer on the matrix effect in axial view inductively coupled plasma atomic emission spectrometry. J Anal At Spectrom 15: 543–547.
34. Da Silva CC, De Jong E (1986) Comparison of two computer models for predicting soil water in a tropical monsoon climate. Agric For Meteorol 36: 249–262.
35. Richards FJ (1959) A flexible growth function for empirical use. J Exp Bot 10: 290–300.
36. Proehl G (2009) Interception of dry and wet deposited radionuclides by vegetation. J Environ Radioact 100: 675–682.
37. Legind CN, Trapp S (2009) Modeling the exposure of children and adults via diet to chemicals in the environment with crop-specific models. Environ Pollut 157: 778–785.
38. Allen RG, Pereira LS, Raes D, Smith M (1998) Crop Evapotranspiration. Guidelines for computing crop water requirements. FAO Irrigation and Drainage Paper, Rome, Italy.
39. Kay AL, Davies HN (2008) Calculating potential evaporation from climate model data: A source of uncertainty for hydrological climate change impacts. J Hydrol 358: 221–239.

22

Modeling Impacts of Alternative Practices on Net Global Warming Potential and Greenhouse Gas Intensity from Rice–Wheat Annual Rotation in China

Jinyang Wang[1], Xiaolin Zhang[1], Yinglie Liu[1], Xiaojian Pan[1], Pingli Liu[1,2], Zhaozhi Chen[1], Taiqing Huang[1], Zhengqin Xiong[1]*

1 Jiangsu Key Laboratory of Low Carbon Agriculture and GHGs Mitigation, College of Resources and Environmental Sciences, Nanjing Agricultural University, Nanjing, China, **2** Hebi Academy of Agricultural Sciences, Hebi, Henan, China

Abstract

Background: Evaluating the net exchange of greenhouse gas (GHG) emissions in conjunction with soil carbon sequestration may give a comprehensive insight on the role of agricultural production in global warming.

Materials and Methods: Measured data of methane (CH_4) and nitrous oxide (N_2O) were utilized to test the applicability of the Denitrification and Decomposition (DNDC) model to a winter wheat – single rice rotation system in southern China. Six alternative scenarios were simulated against the baseline scenario to evaluate their long-term (45-year) impacts on net global warming potential (GWP) and greenhouse gas intensity (GHGI).

Principal Results: The simulated cumulative CH_4 emissions fell within the statistical deviation ranges of the field data, with the exception of N_2O emissions during rice-growing season and both gases from the control treatment. Sensitivity tests showed that both CH_4 and N_2O emissions were significantly affected by changes in both environmental factors and management practices. Compared with the baseline scenario, the long-term simulation had the following results: (1) high straw return and manure amendment scenarios greatly increased CH_4 emissions, while other scenarios had similar CH_4 emissions, (2) high inorganic N fertilizer increased N_2O emissions while manure amendment and reduced inorganic N fertilizer scenarios decreased N_2O emissions, (3) the mean annual soil organic carbon sequestration rates (SOCSR) under manure amendment, high straw return, and no-tillage scenarios averaged 0.20 t C ha^{-1} yr^{-1}, being greater than other scenarios, and (4) the reduced inorganic N fertilizer scenario produced the least N loss from the system, while all the scenarios produced comparable grain yields.

Conclusions: In terms of net GWP and GHGI for the comprehensive assessment of climate change and crop production, reduced inorganic N fertilizer scenario followed by no-tillage scenario would be advocated for this specified cropping system.

Editor: Ben Bond-Lamberty, DOE Pacific Northwest National Laboratory, United States of America

Funding: This research was jointly supported by the National Science Foundation of China (41171238 and 40971139), the National Basic Research Program of China (2009CB118603), the Nonprofit Research Foundation for Agriculture (200903003), the Program for New Century Excellent Talent in Universities (NCET-10-0475), the Doctoral Program of Higher Education of China (20110097110001), the Fundamental Research Funds for the Central Universities (KYZ201110) and the PAPD (Priority Academic Program Development of Jiangsu Higher Education Institutions). The funders had no role in study design, data collection and analysis, decision to publish, or preparation of the manuscript.

Competing Interests: The authors have declared that no competing interests exist.

* E-mail: zqxiong@njau.edu.cn

Introduction

Agricultural activities are responsible for approximately 50% of global atmospheric methane (CH_4) emissions, and agricultural soils account for 75% of global nitrous oxide (N_2O) emissions [1]. Rice paddies have been identified as one of the major sources of atmospheric CH_4 and N_2O emissions [2–4]. China is one of the most important rice producing countries, rice planting area accounts for 20% of the world total and occurs on 23% of all cultivated land in China [5]. Flooded rice and upland crop, such as winter wheat and rice annual rotation system dominates in Chinese rice paddies [6]. The total CH_4 emissions from Chinese

rice paddies were estimated to be 6–10 Tg yr^{-1} in the 1990s [7,8], while N_2O emissions accounted for 25–35% of the total N_2O emissions from Chinese croplands [3,9]. These facts indicate that there is great potential for greenhouse gas (GHG) mitigations from Chinese rice agriculture [10].

Over the past decades, management practices affecting CH_4 and N_2O emissions from rice paddies have been well documented [4,11–14]. Shifting water regimes from continuous flooding to midseason drainage can significantly reduce CH_4 emissions, however, during the same period, N_2O emissions can increase due to trade-off among the emissions of CH_4 and N_2O [2,12,15]. Pronounced differences in CH_4 emissions from the rice season between straw returning time (i.e., the time of straw incorporation into soil after harvest) of on rice season (i.e., before soil flooded for rice transplanting) and off-rice season (i.e., after rice harvest and before wheat sowing), have also been demonstrated [13]. Altering the applications of inorganic fertilizer could either increase or decrease N_2O emissions, depending on the amounts and the timing applied. Conversion of conventional tillage to reduced or no-tillage may benefit soil carbon (C) stocks, but this conversion can also lead to anaerobic zones and thereby stimulate N_2O emissions [16,17].

A systematic approach for comprehensive assessment of GHG mitigation potential in agriculture is urgently needed [18]. Integrating the net exchanges of GHG with changes in the surface layer of soil organic carbon (SOC) has been proposed to analyze the effect of management practices on the net GWP of ecosystems [18,19]. The status of soil C pool plays an important role in regulating terrestrial ecosystem processes through the dynamic equilibrium of C gains and losses and is strongly dependent on current anthropogenic activities [20]. Although soil C sequestration is a separate issue from increasing crop productivity and protecting environmental health, the great potential of increasing SOC, to offset fossil fuel emissions and thus retard global warming, should be highlighted.

A process-based biogeochemical model – the Denitrification and Decomposition (DNDC) model– was originally developed to simulate N_2O emissions and SOC levels in US crop systems [21–23]. It has been widely used to simulate N_2O, N_2, nitric oxide (NO), CH_4, and carbon dioxide (CO_2) emissions for a wide range of ecosystems, such as cropland, grassland, and forests around the world [24–30]. Recently, the DNDC model was employed to estimate GHG emissions from uplands [27,29] and rice paddies [31–34] after validation with field measurement data. Via integrations of remote maps of soil and climate information with changing the alternative practice scenarios, or scaling up site-specific results to regional scale, DNDC simulations provided better understanding of the effect of site-specific management on global warming potential (GWP) at regional or large scales. However, large uncertainties still existed when estimating GHG emissions under certain managements on regional scale due to the spatial heterogeneity of soil properties such as texture, SOC content and pH [35]. Moreover, available evidence suggested that certain calibrations in the DNDC default parameters were essential for site-specific systems or scaling up to regional or large scales [36,37]. Due to the highly temporal and spatial variability of GHG emissions and their complex relationship to climatic and soil conditions, short-term field measurements may not capture the long-term effects of management practices on GHG emissions [38,39]. Although a number of field measurements have been conducted on GHG emissions and SOC change [2,40–42], the long-term impacts of the alternative management practices are poorly understood for the winter wheat – single rice cropping system.

The objectives of this study were to assess the applicability of the DNDC model tested against the field measurement data for the emissions of CH_4 and N_2O, and to utilize the validated model to evaluate the long-term (45-year) effects of alternative management practices on net GWP and GHGI for this specific rotation system of winter wheat – single rice in southern China.

Materials and Methods

Field Experiment

A field trial was carried out in Nanjing (31°52′N, 118°50′E), southern China for a winter wheat – single rice rotation system since 2008. The field studies did not involve endangered or protected species and the location is not protected in any way, No specific permits were required for the described field studies due to the local typical cropping and ambient air sampling.

The experimental soil was classified as *Stagnic Anthrosols* [43]. The texture of this studied soil was silt loam, consisting of 14% clay, 6% sand, and 80% silt with an initial pH of 5.7. Total organic C and N in the surface cultivated layer (0–20 cm) were 14.7 and 1.32 g kg^{-1}, and soil bulk density was 1.28 g cm^{-3} [44]. The annual mean temperature and total precipitation were 16.9°C and 136.5 cm, respectively in 2009, and 16.8°C and 130.4 cm, respectively in 2010, which were listed as baseline in Table 1.

During the period of 2009–2010, field measurements of the emissions of CH_4 and N_2O from this rotation system were conducted and adopted for the model test. Three treatments, each with three replicates, including a control treatment without N fertilization or straw return (CK), and the N fertilized treatments without straw (N) or with straw return (NS), were utilized to test the DNDC model (Table 2). For the straw return treatment, air-dried rice straw at the amount of 3 t ha^{-1} (C:N = 52:1) was applied to the surface before rice seedling transplantation, whereas no straw was returned during the winter wheat season. After crops were harvested, no above-ground residues were left *in situ*. The area of each plot was 20 m^2 (4 m×5 m), and cement bulkheads were placed between plots. The periods of crop planting and harvesting were November 13, 2009 and June 6, 2010, respectively, for winter wheat, and June 20, 2010 and October 10, 2010, respectively, for rice. The soil was conventionally tilled twice with a plough at a depth of approximately 10 cm, on November 8, 2009 and June 18, 2010, before crops were planted during the rotation cycle. For the fertilized treatment, urea was used as N fertilizer at the rate of 250 kg N ha^{-1} per crop with a split ratio of 4:3:3 for both crops. For winter wheat, the basal fertilization date was on November 13, 2009, and the two top dressings occurred on February 21, 2010 and March 18, 2010. The corresponding dates for single rice were June 18, July 6, and August 11, 2010. For each treatment, calcium superphosphate, used as phosphorus fertilizer, was applied at the local rate of 120 kg P$_2$O$_5$ ha^{-1}, and potassium chloride was applied at the local rate of 60 kg K$_2$O ha^{-1} as a basal fertilizer for each crop season. The common water management strategy of flooding – midseason drainage (June 29, 2010– August 7, 2010) – reflooding – final drainage (starting October 3, 2010) was employed for the rice season in this study, and no additional irrigation was used, with plots receiving only precipitation during the winter wheat season (Figure 1).

The fluxes of CH_4 and N_2O were measured using the static opaque chamber method [45]. The chamber was made of PVC, and consisted of two parts, one sized 45 cm×45 cm×50 cm and the other with an extended height of 60 cm to accommodate plant growth. The chamber was equipped with a circulating fan to ensure complete gas mixing and was wrapped with sponge and aluminum foil to minimize temperature changes inside the chamber during gas sampling. During gas sampling, these chambers were placed on permanently installed PVC collars of the same size and fitted into the groove of the collar sealed by water. The frequency of gas sampling was approximately once a week from November 13, 2009 to October 10, 2010, intensive gas sampling once every two days was performed after N fertilization and during drainage. For each plot, four gas samples were withdrawn from the chamber through a three-way stopcock using a 25-ml airtight syringe at 10 min intervals (0, 10, 20, and 30 min after the chamber closure). Gas samples were taken from 8:00 am through 11:00 am since the soil temperature during this period was close to the mean daily soil temperature.

Table 1. Input values utilized to the validated DNDC model for baseline scenario and the sensitivity tests.

Parameter	Baseline	Range tested
Environmental factors		
Annual mean temperature (°C)	16.9 (2009)/16.8 (2010)	Decrease by 2°C and 4°C and increase by 2°C and 4°C
Total annual precipitation (cm)	136.5 (2009)/130.4 (2010)	Decrease by 20% and increase by 20%
Soil texture	Silt loam	Loamy sand, sandy clay loam and sandy clay
SOC content (0–5 cm)	0.125%	0.05%, 0.1%, 0.15% and 0.2%
Soil pH	5.7	4.7, 6.7 and 7.7
Management alternatives		
Tillage	Conventional tillage (ploughed about 10 cm)	No-tillage and reduced tillage
Total annual N input (kg N ha^{-1} yr^{-1})	500 (250 for each crop season)	300 (150, 150) and 700 (350, 350)
Straw return (rice straw) (t ha^{-1})	3	1.5 and 6
Manure amendment (kg N ha^{-1} yr^{-1})	No amendment	250 and 500 applied as basal fertilizer instead of the equivalent annual rate of inorganic N

The gas samples were analyzed using a gas chromatograph (Agilent 7890A, USA) that was equipped with two detectors [45]. Methane was detected using a hydrogen flame ionization detector (FID), and N_2O was detected using an electron capture detector (ECD). Argon-CH_4 (5%) and N_2 were used as the carrier gas at a flow rate of 40 ml min^{-1} for N_2O and CH_4 analysis, respectively. The temperatures of the column and the ECD were maintained at 40°C and 300°C, respectively. The oven and the FID were operated at temperatures of 50°C and 300°C, respectively. The concentrations of CH_4 and N_2O were quantified by comparing their peak areas with those of reference gases (Nanjing Special Gas Factory). The fluxes were determined from the change in the slope of the mixing ratio of the collected samples after the chamber was closed. The seasonal or cumulative amounts of CH_4 and N_2O emissions were sequentially accumulated from the emissions between every two adjacent intervals of the measurements.

DNDC Model and the Sensitivity Test

The DNDC model is adopted to simulate the daily flux rates of CH_4 and N_2O from a winter wheat – single rice rotation system in this study. The DNDC model is available online at http://www.dndc.sr.unh.edu/, and consists of six sub-models for simulating soil climate, plant growth, decomposition, nitrification, denitrification and fermentation. Briefly, the soil climate sub-model calculates hourly and daily soil temperature and moisture fluxes in one dimension, the plant growth sub-model simulates plant biomass accumulation and partitioning, the decomposition sub-model simulates soil organic matter decay, N mineralization, CO_2 and dissolved organic carbon production, the nitrification and denitrification sub-models track the sequential biochemical reaction from ammonium to nitrate production and consumption, net NO and N_2O production and N_2 production, the fermentation sub-model simulates CH_4 production, consumption, transport and net flux. Numerous studies suggested that the DNDC model generally produced good performances for modeling SOC dynamics from the paddy cropping system across China [31,39,46]. Nonetheless, due to the lack of the long-term monitoring at the experimental site, the model validation and sensitivity test for SOC change cannot be conducted in this study.

To better understand the effects of both environmental factors and management practices on GHG emissions, a sensitivity test was conducted to isolate the most sensitive factors. A baseline scenario was chosen based on the local climatic and soil conditions and typical management for a winter wheat – single rice rotation system (Table 1). The sensitivity test was conducted by varying a single input parameter in a predefined range while keeping all other input parameters constant as those in the baseline scenario (Table 1). The DNDC model was run with each of the predefined scenarios for one rotation to produce annual emissions of CH_4 and N_2O and thereafter to calculate the total GWP of these gases on a 100-year time horizon.

Table 2. Measured and simulated data of cumulative emissions of CH_4 and N_2O from a winter wheat – single rice rotation system from November 13, 2009 to October 10, 2010.

Treatment [a]	CH_4 (kg C ha^{-1})				N_2O (kg N ha^{-1})			
	Observed		Simulated		Observed		Simulated	
	Wheat	Rice	Wheat	Rice	Wheat	Rice	Wheat	Rice
CK	0.78(3.05) [b]	83.9(29.9)	−0.28	5.3	0.18(0.07)	0.08(0.03)	0.22	0.01
N	1.41(2.02)	66.2(34.5)	−0.28	93.6	3.19(0.63)	0.22(0.18)	4.43	1.75
NS	1.67(1.87)	156.4(22.3)	−0.28	172.4	2.79(0.88)	0.18(0.04)	2.94	0.69

[a]CK, without both N fertilization and straw incorporation; N, with N fertilization; NS, with both N fertilization and straw incorporation;
[b]Data in the parenthesis indicate the standard deviation of three replicated experiments.

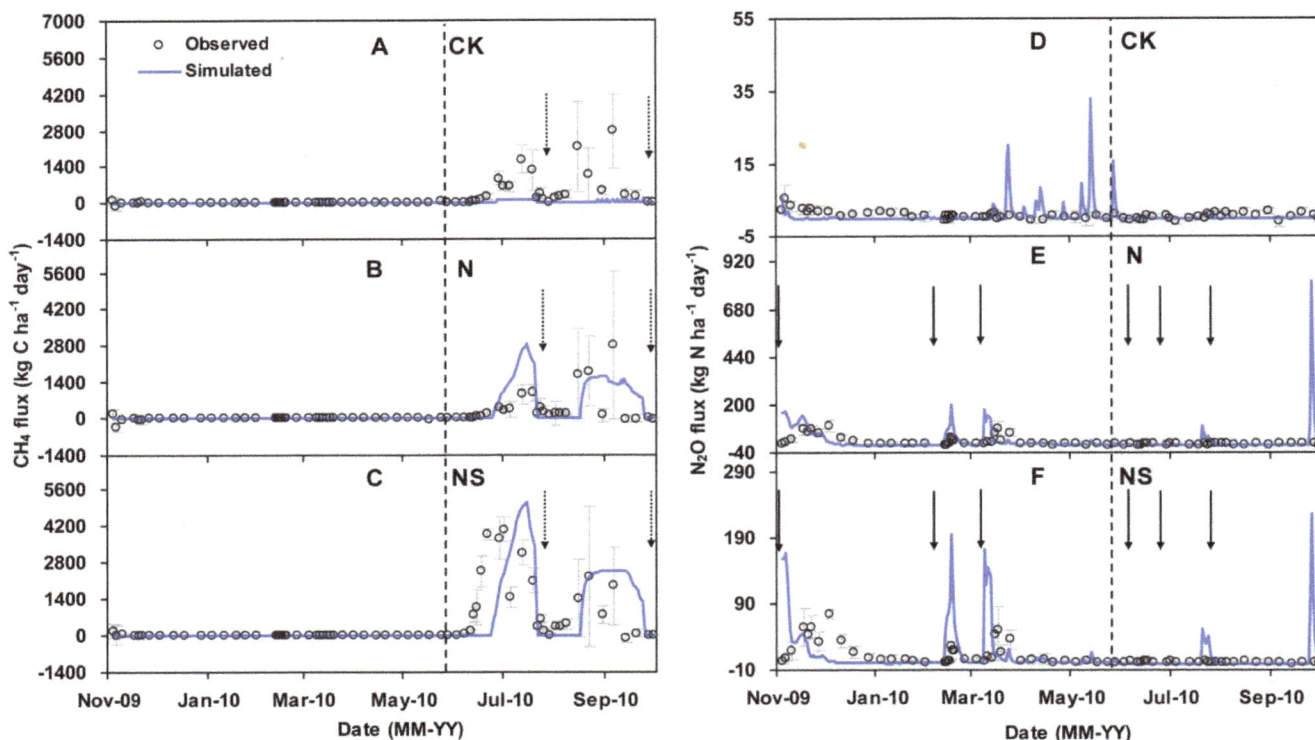

Figure 1. Dynamics variations of field observed (in dot) and simulated (in line) emissions of CH₄ (A, B, C) and N₂O (D, E, F) for the treatments of CK, N, and NS, respectively. CK, without both N fertilization and straw incorporation, N, with N fertilization, NS, with both N fertilization and straw incorporation. The vertical line in each panel divided the whole rotation into wheat (left) and rice (right) seasons. The vertical bars indicate the standard deviation of three replicates for each treatment. The dotted and solid arrows represent midseason or final drainage and N fertilization, respectively.

Design of Alternative Management Practices

Large uncertainties commonly existed in the evaluation of soil C and N processes when evaluating short-term anthropogenic perturbations. To accurately identify the consequences of major management practices, a predictive process-based model was employed to evaluate the long-term impacts for a 45-year period. Six alternative management practice scenarios were designed (Tables 1, 3). As compared to the baseline scenario, only the targeted parameters were changed by (1) reducing the annual inorganic N fertilizer rate to 300 kg N ha^{-1} (N300), (2) increasing the annual inorganic N fertilizer rate to 700 kg N ha^{-1} (N700), (3) reducing the amount of straw return by 50% (S1.5), (4) doubling the amount of straw return to 6 t ha^{-1} (S6), (5) replacing half of the annual inorganic N fertilizer (250 kg N ha^{-1}) with an equivalent N amount of bean cake (C:N = 6.8:1) incorporated as basal manure fertilizer (125 kg N ha^{-1}) for each crop season (OM250), and (6) changing the conventional tillage practice to no-tillage (No-tillage). The climate data used in these simulations were the present data from our field measurements during the corresponding wheat and rice seasons. Soil properties were obtained from the experimental measurements as those in the baseline scenario.

Analysis of Net GWP and GHGI

To quantitatively identify the impacts of alternative scenarios on net GWP over the 45-year period for this system, the IPCC factors [47] were adopted for calculating the combined GWPs on a time horizon of 100-year. The equation used was as follows:

Net GWP (kg CO$_2$-equiv. ha^{-1} yr^{-1}) = kg CH$_4$ ha^{-1} yr^{-1} ×25+ kg N$_2$O ha^{-1} yr^{-1} ×298– SOCSR ×44/12.

In addition, to associate the net GWP with crop production, greenhouse gas intensity (GHGI) was introduced and calculated by the following equation [38,48,49]:

GHGI (kg CO$_2$-equiv. kg^{-1} yield C) = Net GWP/grain yield.

Statistical Analysis

A linear regression analysis was performed to determine the variance between the cumulative emissions of CH$_4$ and N$_2$O from the three treatments, and then to reflect "the goodness of fit" of applying the DNDC model to test against our field measurement data. This statistical analysis was carried out using SigmaPlot 12.0 (Systat, San Jose, CA, USA).

Results and Discussion

Modeling Validation

Based on the baseline scenario as listed in Table 1 and also the same as the field managements, we used the DNDC model to simulate the daily flux rates of CH$_4$ and N$_2$O from a winter wheat – single rice rotation system. The simulated results were then compared with the field measurement data in daily flux dynamics (Figure 1) and in cumulative emissions (Figure 2).

During the winter wheat season, the rates of CH$_4$ were usually negligible in all the treatments, occasionally acted as a small source of atmospheric CH$_4$, and the simulated CH$_4$ emissions were in good agreement with the observed data (Figure 1A–C). The simulated seasonal CH$_4$ emissions fell within the statistical

Table 3. Impacts of alternative scenarios on averaged annual GHG emissions, soil organic carbon sequestration rate (SOCSR), N loss, grain yield, net GWP, and GHGI over the 45-year simulation.

Scenario [a]	CH$_4$ (kg C ha^{-1} yr^{-1})	N$_2$O (kg N ha^{-1} yr^{-1})	SOCSR (t C ha^{-1} yr^{-1})	N loss (kg N ha^{-1} yr^{-1})	Grain yield (kg C ha^{-1} yr^{-1})	Net GWP (kg CO$_2$-equiv. ha^{-1} yr^{-1})	GHGI (kg CO$_2$-equiv. kg^{-1} yield C)
Baseline	115.1	3.6	0.10	61.8	5682	5148	0.91
N300	114.0	2.2	0.10	30.9	5666	4439	0.78
N700	115.1	5.1	0.10	96.5	5683	5853	1.03
S1.5	113.9	3.6	0.06	60.9	5684	5291	0.93
S6	143.7	3.6	0.19	64.9	5689	5773	1.01
OM250	149.3	2.2	0.22	69.2	5714	5203	0.91
No-tillage	117.0	3.8	0.18	62.3	5686	5023	0.88

[a]The baseline scenario see Table 1; N300, total N fertilizer rate of 300 kg N ha^{-1} yr^{-1}; N700, total N fertilizer rate of 700 kg N ha^{-1} yr^{-1}; S1.5, straw return rate of 1.5 t ha^{-1} yr^{-1}; S6, straw return rate of 6 t ha^{-1} yr^{-1}; OM250, replacing half of the annual inorganic N fertilizer (250 kg N ha^{-1}) with an equivalent N amount of bean cake (C:N = 6.8:1) used as manure and incorporated as basal fertilizer for each season; No-tillage, zero-tillage.

standard deviation ranges of the observed data for all the treatments (Table 2). The simulated CH$_4$ emission rates were similar among treatments, suggesting that N fertilization did not affect CH$_4$ emissions during the winter wheat season (Table 2). During the rice season, the simulated and observed fluxes of CH$_4$ showed similar seasonal patterns in both the N and NS treatments (Figure 1B, C).

The simulated cumulative CH$_4$ rates of 93.6 kg C ha^{-1} and 172.4 kg C ha^{-1} for N and NS treatments, respectively, were close to the observed results of 66.2±34.5 kg C ha^{-1} and 156.4±22.3 kg C ha^{-1}, respectively (Table 2). In agreement with previous studies from rice paddies in Asia [15,24,25,34], CH$_4$ emissions were well simulated by the DNDC model with substrate input from straw return and crop growth (Figure 1B,C, 2B,C, Table 2). For the control treatment the DNDC model failed to simulate the peaks of CH$_4$ emission and resulted in a huge difference between the cumulative emissions in the simulation and measurement (Figure 1A, 2A, Table 2).

During the winter wheat season, the numbers of N$_2$O emission peaks were the same between the simulations and the measurements, the simulated peaks generally occurred earlier and greater than the observed ones when receiving N fertilizer (Figure 1D–F). Small discrepancies, with a mean value of 22%, existed between the observed and simulated seasonal rates of N$_2$O emission from the winter wheat season (Table 2). The simulated results of N$_2$O emissions coincide with previous studies from upland soils [29]. However, the differences between the observed and simulated cumulative emissions of N$_2$O during the rice season were big, ranging from −1.53 kg N ha^{-1} to 0.07 kg N ha^{-1} (Table 2). A previous study by Cai et al. [24] reported such discrepancies of up to eight-fold between the observed and simulated values for the rice field treated with urea of 300 kg N ha^{-1} at the same region. Great relative deviations, as high as -238 to 29%, were also reported by Babu et al. [25] in a study of several rice paddies in India. As compared to the automatic measurement, the manual chamber method might have missed the episodic N$_2$O emission peaks, particularly after midseason drainage and final drainage (Figure 1E, F) [2,24,44].

The regression slopes around 1 between the observed and simulated cumulative emissions of CH$_4$ and N$_2$O demonstrated the good performance of the DNDC model while their determination coefficients showed the covariance between the observed and simulated cumulative emissions (Figure 2). Thus, except for

the control treatment, the regression slopes were close to 1 (0.87–1.34) and the determination coefficients between the observed and simulated annual total emissions of CH$_4$ and N$_2$O were high ($r^2 = 0.84$–0.96) in the present study (Figure 2A–F), indicating that the applicability of the DNDC model is conservatively feasible for this site-specific rotation system of winter wheat – single rice in southern China.

Sensitivity Test

With one year rotation of winter wheat – single rice, sensitivity tests were conducted to identify the most sensitive factors that affect total GWP of CH$_4$ and N$_2$O by varying one single factor as listed in Table 1. An increase or reduction in CH$_4$ and N$_2$O emissions was thus converted into CO$_2$-equivalent in terms of their GWPs on 100-year time horizon as shown in Figure 3. The dominant source of the total GWP in this study was CH$_4$ emissions during the rice season and N$_2$O emissions during the wheat season, and their corresponding contributions were also shown in Figure 3.

Among all the selected environmental factors, soil texture and initial SOC content were the most influential factors associated with the total GWP. The GWP of this rotation system substantially decreased with increased clay fraction, the increased clay fraction had negative effect on CH$_4$ emissions although significantly promoted N$_2$O emissions (Figure 3A). The total GWP increased substantially in response to elevated SOC content due to the stimulatory effect on CH$_4$ emissions (Figure 3B). −20% low precipitation significantly increased the GWP mainly through the N$_2$O emissions during the wheat season while +20% precipitation had no obvious effect (Figure 3C). N$_2$O emissions during the wheat season increased in response to reduced annual mean temperature and remained similar in response to increased temperature, CH$_4$ emissions during the rice season increased with increasing temperature (Figure 3D). Due to the fact that the ratio of N$_2$O to N$_2$ decreases with increasing soil pH, soil pH was associated with lower N$_2$O emissions and with an insignificant impact on CH$_4$ emissions (Figure 3E). The above simulated results were in good agreement with previous field or model studies [29,34,35].

Among all selected management alternatives, N fertilization, straw return, and manure amendment were the most influential factors affecting the GWP of this rotation system (Figure 3F, G, H). For example, the GWPs rose from 6652 to 8191 kg CO$_2$-equiv.

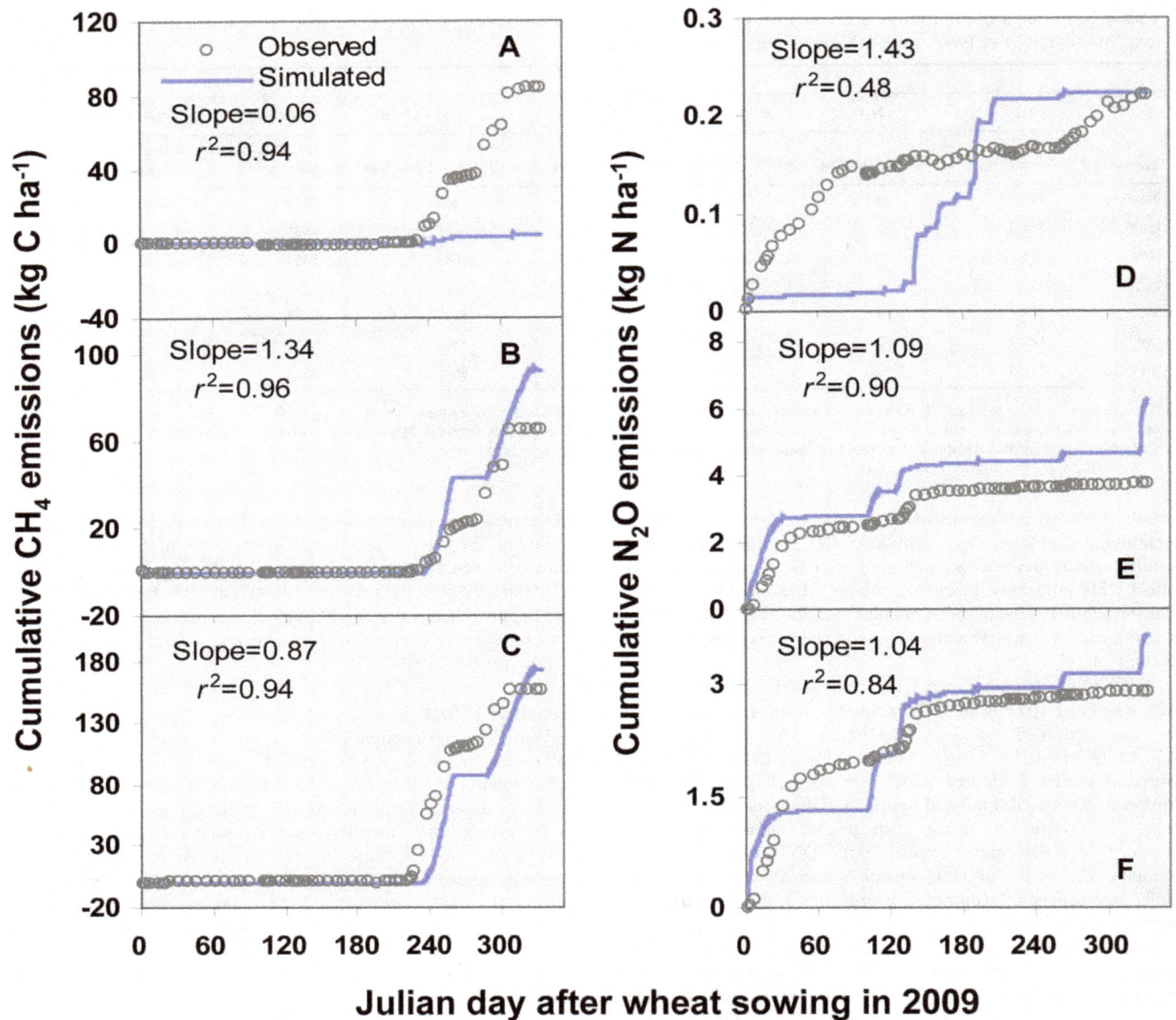

Figure 2. Cumulative emissions of CH₄ (A, B, C) and N₂O (D, E, F) for the treatments of CK, N, and NS, respectively. CK, without both N fertilization and straw incorporation, N, with N fertilization, NS, with both N fertilization and straw incorporation. The slopes and determination coefficients were calculated by the linear regression of observed and simulated cumulative emissions of CH₄ and N₂O.

ha^{-1} yr^{-1} with increasing the rates of N fertilizer from 300 to 700 kg N ha^{-1} yr^{-1}. And the increase in GWPs was mainly caused by increased N₂O emissions during the wheat season (Figure 3F). The increased GWPs were resulted from the stimulated CH₄ emissions when straw return and manure amendment increased (Figure 3G, H). As for the changes in tillage and the split ratio of N fertilization, there was no significant difference between the baseline scenario and the other alternative practices (Figure 3I, J).

Modeling the Long-term Effects of Alternative Management Practices on C and N Cycles and GHG Emissions

The long-term impacts of these practices on C and N cycles and GHG emissions were emphasized by repeatedly running the

DNDC model for 45-year period (Figure 4A, B). Compared with the baseline scenario, high straw return and manure amendment scenarios significantly increased CH₄ emissions by 25% and 30%, respectively, whereas other scenarios had negligible effects on CH₄ emissions (Table 3). Over the 45-year time course, CH₄ emissions under all management practices gradually decreased, which was probably due to the enhanced SOC stock that increased the capacity of soil to oxidize CH₄ [50]. The increasing trend of CH₄ emissions was in accordance with the short-term effect of organic matter incorporation [2,13,14,38]. No change in CH₄ emissions was observed due to varying inorganic N fertilizer rate (Table 3). Nitrogen fertilizer generally had statistically insignificant effect on CH₄ emissions (Figure 4B) [51], although some previous studies reported decreased or increased CH₄ emissions [2,38].

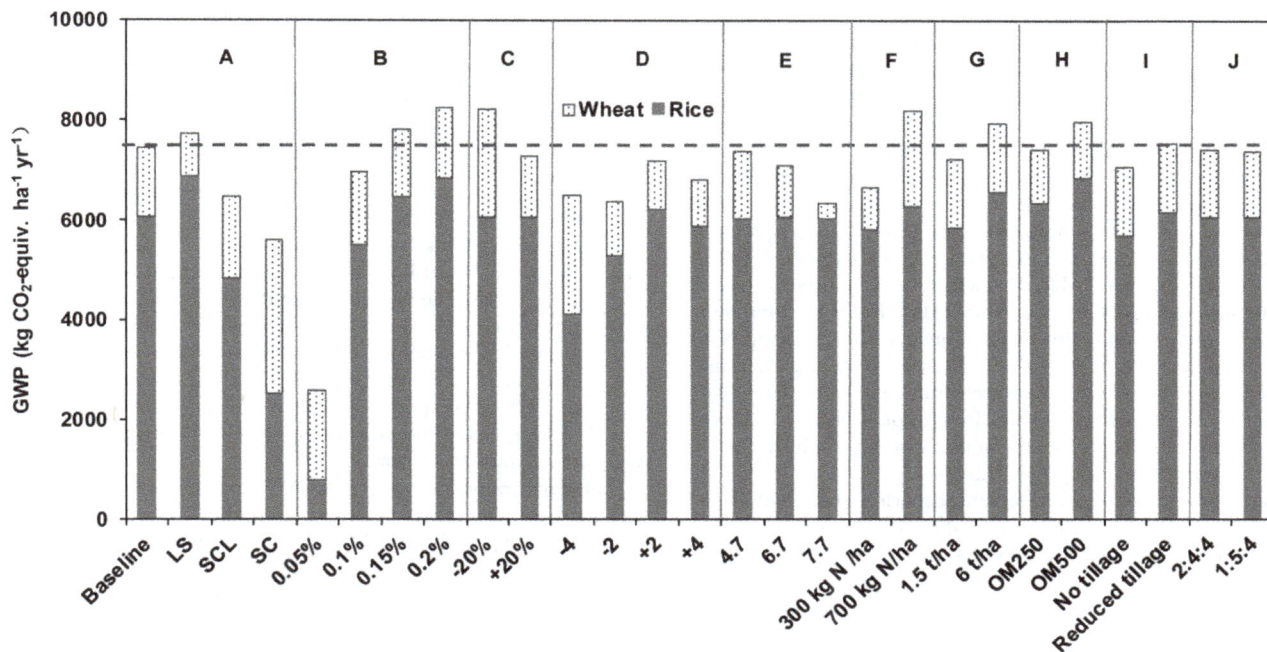

Figure 3. Sensitivity tests of GWP of CH_4 and N_2O emissions to environmental factors and alternative management practices. Starting from the baseline management conditions, change in (A) soil texture, (B) initial SOC content, (C) total precipitation, (D) annual mean temperature, (E) soil pH, (F) N fertilizer input, (G) straw return, (H) manure amendment, (I) tillage, and (J) split application ratio altered the GWP for the rotation system. Abbreviations for soil texture are as follows: LS, loamy sand; SCL, sandy clay loam; SC, sandy clay.

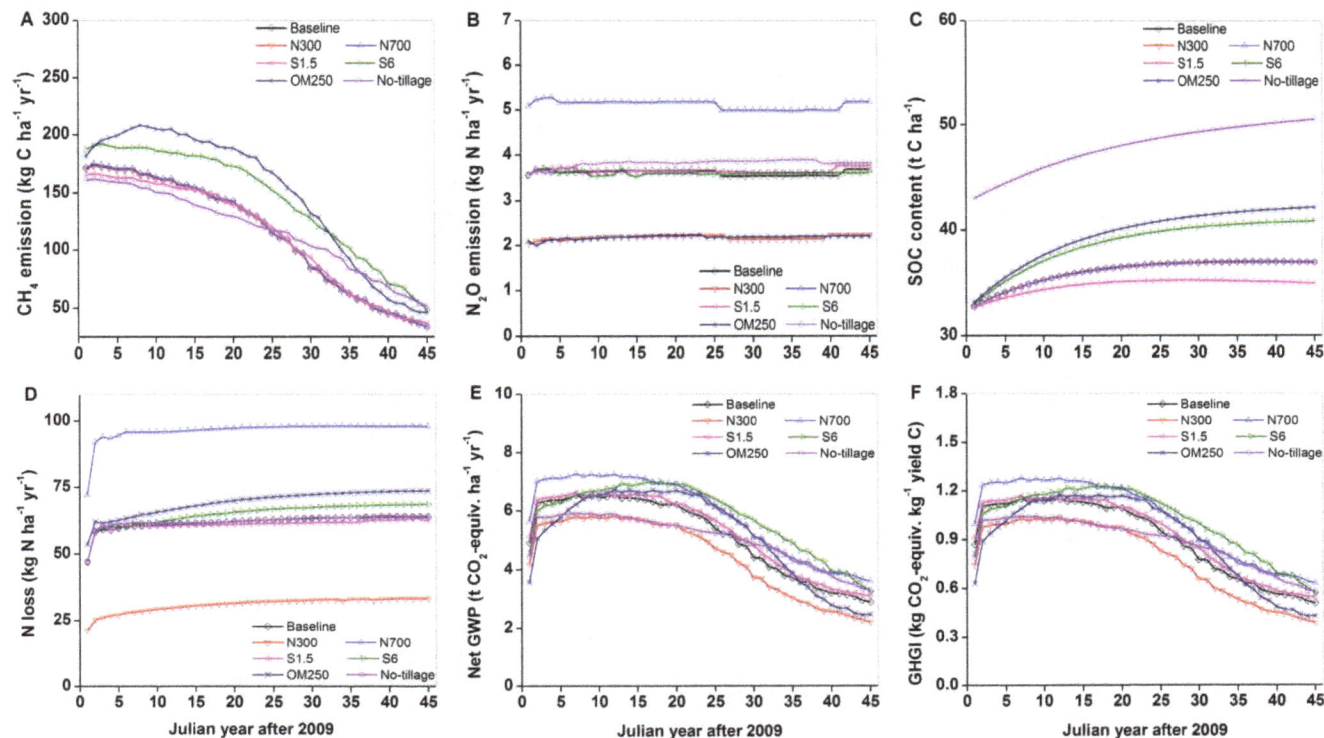

Figure 4. Long-term (45-year) impacts of alternative management scenarios on (A) CH_4, (B) N_2O, (C) SOC content, (D) N loss, (E) net GWP, and (F) GHGI. See Table 1 for the baseline scenario, N300 and N700, total N fertilizer rate of 300 and 700 kg N ha^{-1} yr^{-1}, respectively, S1.5 and S6, straw return rate of 1.5 and 6.0 t ha^{-1} yr^{-1}, respectively, OM250, replacing half of the annual inorganic N fertilizer (250 kg N ha^{-1}) with an equivalent N amount of bean cake (C:N = 6.8:1) used as manure and incorporated as basal fertilizer for each season, No-tillage, zero-tillage.

Among the alternative management scenarios, N_2O emissions had similar dynamics curves with small increasing trend over the 45-year time course, the increasing trend was more obvious under no-tillage scenario. N_2O emissions were significantly different under different levels of inorganic N fertilizer scenarios (Figure 4B). A significant linear relationship between total N input (inorganic plus organic N) and N_2O emissions was found for the six alternative scenarios (Table 3, $r^2 = 0.69$, $P = 0.02$), which was in good agreement with numerous previous studies [2,4,44]. Increasing straw return did not reduce N_2O emissions in this long-term simulation (Table 3), which was contrary to our previous short-term field measurement [44]. From the long term simulation, manure amendment was beneficial for reducing N_2O emissions as a result of reducing inorganic N fertilization rate (Table 3), although organic manure return would increase N_2O emissions in the long-term double rice cropping system [38]. No-tillage did not reduce N_2O emissions from soils, which was in support of previous findings [17,37,50,52].

Among the alternative management scenarios, SOC content (0–30 cm) gradually increased over the 45-year time course (Figure 4C), indicating that the soil has not reached its maximum capacity for C sequestration. The increasing trend of SOC supported previous model study that organic matter amendment on SOC sequestration would last across the 45-year period for paddy fields in the Yangtze Delta [41]. There may be great potential for C sequestration in this region due to low SOC density, and thus for mitigating the increasing atmospheric CO_2. Among the six management alternatives, the simulated SOC sequestration rates ranged between 0.06 and 0.22 t C ha^{-1} yr^{-1} (Table 3), which were comparable to previous field and modeling studies [39,40]. The amounts of SOC stock did not vary with inorganic N fertilizer rates, being around 0.10 t C ha^{-1} yr^{-1} of SOCSR (Figure 4C, Table 3). Scenarios with high straw return, manure amendment, and no-tillage produced greater SOCSR around 0.20 t C ha^{-1} yr^{-1} than other scenarios (Table 3). A positive correlation between soil C sequestration and the amount of incorporated C was found in this study ($r^2 = 0.76$, $P = 0.01$), suggesting that large amounts of crop residue inputs are necessary for enhancing SOC levels, especially in paddy soils [20,39–42,53]. Enhanced SOC helped to not only mitigate climate change but also enhance soil fertility and hence sustain crop productivity [20,40], and thus there was closely relationship between SOC and crop yield (Table 3). Thus straw return and manure amendment are of vital importance. No-tillage practices significantly elevated SOC content (Table 3, Figure 3C), suggesting that no-tillage as alternative practice should be deserved more attention as proposed by previous studies [49,50,53].

Among the management alternatives, N loss was simulated over the 45-year time course (Figure 4D). The elevated N loss was due to not only the increased N input but organic manure amendment [29] (Table 3). In terms of the GWPs of CH_4 and N_2O, the scenario of reduced inorganic N fertilizer was advocated for

attenuating global warming under this rotation system over the long-term period.

Modeling the Long-term effects of Alternative Management Practices on Grain Yield, Net GWP and GHGI

Due to the already-high N rate in the baseline scenario, increased N input level did not further increase grain yield under current managements without improving N use efficiency (Table 3). Moreover, it is not surprising that annual N fertilization reduced by 40% of the baseline value did not reduce crop yield in this study, which was confirmed by the previous study that 36% reduction in N fertilizer for the rice-wheat rotation in the same region did not reduce crop yield [54]. The similar trend for crop yield and SOC also reported by Wang et al. [39] through running the DNDC model at the same site for rice-wheat cropping system with similar alternative practices.

The temporal variations of net GWP and GHGI were different among these alternative scenarios across the 45-year time course (Figure 4E, F), indicating that different strategies should be employed at different time scales [29,32]. For example, during the first 25-year period, the scenarios of reduced inorganic N fertilizer and no-tillage obviously reduced net GWP, whereas during the following 20-year period, the reducing effect became complex for the no-tillage scenario (Figure 4E, F).

Scenarios of reduced inorganic N fertilizer and no-tillage reduced both the net GWP and GHGI, while the remaining scenarios tended to increase the net GWP and GHGI as compared with the baseline scenario (Table 3).

Overall, among all the alternative scenarios the scenarios of reduced inorganic N fertilizer and no-tillage could therefore contribute to mitigate global warming potential while sustain crop production, particularly for reduced inorganic N scenario with obviously the least N loss.

Conclusions

The applicability of the DNDC model and the long-term assessments on various management alternatives were tested for the rice-wheat rotation system in this study. The validation, sensitivity tests, and long-term prediction provided a sound basis for comprehensive understanding of the alternative management practices on soil C and N cycles involved in global warming. Therefore, reduced inorganic N fertilizer scenario followed by no-tillage scenario would be advocated for mitigating global warming without decreasing crop yield.

Author Contributions

Conceived and designed the experiments: ZQX JYW. Performed the experiments: XLZ YLL XJP PLL ZZC TQH. Analyzed the data: JYW XLZ. Contributed reagents/materials/analysis tools: PLL XLZ TQH. Wrote the paper: JYW ZQX.

References

1. USEPA (2006) Global anthropogenic non-CO_2 greenhouse gas emissions: 1990–2020 (June 2006 Revised), available at: http://www.epa.gov/climatechange/economics/downloads/GlobalAnthroEmissionsReport.pdf, Office of Atmospheric Programs, USEPA, Washington, DC.

2. Cai Z, Xing G, Yan X, Xu H, Tsuruta H, et al. (1997) Methane and nitrous oxide emissions from rice paddy fields as affected by nitrogen fertilisers and water management. Plant Soil 196 (1): 7–14.

3. Zheng X, Han S, Huang Y, Wang Y, Wang M (2004) Re-quantifying the emission factors based on field measurements and estimating the direct N_2O emission from Chinese croplands. Glob Biogeochem Cycles 18 (1): 1–19.

4. Zou J, Huang Y, Qin Y, Liu S, Shen Q, et al. (2009) Changes in fertilizer-induced direct N_2O emissions from paddy fields during rice-growing season in China between 1950s and 1990s. Glob Chang Biol 15 (1): 229–242.

5. Frolking S, Qiu J, Boles S, Xiao XM, Liu JY, et al. (2002) Combining remote sensing and ground census data to develop new maps of the distribution of rice agriculture in China. Glob Biogeochem Cycles 16 (4). Doi:10.1029/2001gb001425.

6. Xu Z, Lu Y (1992) Ecological environment of paddy soil. In: Li QK, editor. Paddy Soils of China. Science Press, Beijing, China. 108–125.

7. Huang Y, Zhang W, Zheng X, Li J, Yu Y (2004) Modeling methane emission from rice paddies with various agricultural practices. J Geophys Res 109 (D8): D08113.

8. Yan X, Akiyama H, Yagi K, Akimoto H (2009) Global estimations of the inventory and mitigation potential of methane emissions from rice cultivation conducted using the 2006 Intergovernmental Panel on Climate Change Guidelines. Glob Biogeochem Cycles 23 (2): GB2002.

9. Liu S, Qin Y, Zou J, Liu Q (2010) Effects of water regime during rice-growing season on annual direct N_2O emission in a paddy rice-winter wheat rotation system in southeast China. Sci Total environ 408 (4): 906–913.

10. Smith P, Martino D, Cai Z, Gwary H, Janzen H, et al. (2007) Agriculture. In: Metz B, Davidson OR, Bosch PR, et al (eds) Climate Change 2007: Mitigation. Contribution of Working Group III to the Fourth Assessment Report of the Intergovernmental Panel on Climate Change. Cambridge University Press, Cambridge, 497–540.

11. Yan X, Ohara T, Akimoto H (2003) Development of region-specific emission factors and estimation of methane emission from rice fields in the East, Southeast and South Asian countries. Glob Chang Biol 9 (2): 237–254.

12. Cai Z, Tsuruta H, Gao M, Xu H, Wei C (2003) Options for mitigating methane emission from a permanently flooded rice field. Glob Chang Biol 9 (1): 37–45.

13. Yan X, Yagi K, Akiyama H, Akimoto H (2005) Statistical analysis of the major variables controlling methane emission from rice fields. Glob Chang Biol 11 (7): 1131–1141.

14. Wang J, Zhang X, Xiong Z, Khalil MAK, Zhao X, et al. (2012) Methane emissions from a rice agroecosystem in South China: Effects of water regime, straw incorporation and nitrogen fertilizer. Nutr Cycl Agroecosyst 93: 103–112.

15. Li C, Frolking S, Xiao XM, Moore B III, Boles S, et al. (2005) Modeling impacts of farming management alternatives on CO_2, CH_4, and N_2O emissions: A case study for water management of rice agriculture of China. Glob Biogeochem Cycles 19 (3). Doi:10.1029/2004gb002341.

16. Smith P, Goulding KW, Smith KA, Powlson DS, Smith JU, et al. (2001) Enhancing the carbon sink in European agricultural soils: including trace gas fluxes in estimates of carbon mitigation potential. Nutr Cycl Agroecosyst 60 (1): 237–252.

17. Six J, Ogle SM, Breidt FJ, Conant R, Mosier AR, et al. (2004) The potential to mitigate global warming with no-tillage management is only realized when practised in the long term. Glob Chang Biol 10 (2): 155–160.

18. Robertson GP, Field C, Raupach M (2004) Abatement of nitrous oxide, methane, and the other non-CO_2 greenhouse gases: the need for a systems approach. The global carbon cycle: integrating humans, climate and the natural world. 493–506.

19. Fornara D, Steinbeiss S, McNamara N, Gleixner G, Oakley S, et al. (2011) Increases in soil organic carbon sequestration can reduce the global warming potential of long-term liming to permanent grassland. Glob Chang Biol 17(5): 1925–1934.

20. Lal R (2004) Soil carbon sequestration to mitigate climate change. Geoderma 123 (1–2): 1–22.

21. Li C, Frolking S, Frolking TA (1992) A model of nitrous oxide evolution from soil driven by rainfall events: 1. Model structure and sensitivity. J Geophy Res 97 (D9): 9759–9776.

22. Li C, Frolking S, Harriss R (1994) Modeling carbon biogeochemistry in agricultural soils. Glob Biogeochem Cycles 8 (3): 237–254.

23. Li C, Narayanan V, Harriss RC (1996) Model estimates of nitrous oxide emissions from agricultural lands in the United States. Glob Biogeochem Cycles 10 (2): 297–306.

24. Cai Z, Sawamoto T, Li C, Kang G, Boonjawat J, et al. (2003) Field validation of the DNDC model for greenhouse gas emissions in East Asian cropping systems. Glob Biogeochem Cycles 17(4): 1107.

25. Babu YJ, Li C, Frolking S, Nayak DR, Adhya TK (2006) Field validation of DNDC model for methane and nitrous oxide emissions from rice-based production systems of India. Nutr Cycl Agroecosyst 74 (2): 157–174.

26. Beheydt D, Boeckx P, Sleutel S, Li C, Vancleemput O (2007) Validation of DNDC for 22 long-term N_2O field emission measurements. Atmos Environ 41 (29): 6196–6211.

27. Abdalla M, Wattenbach M, Smith P, Ambus P, Jones M, et al. (2009) Application of the DNDC model to predict emissions of N_2O from Irish agriculture. Geoderma 151 (3–4): 327–337.

28. Li D, Lanigan G, Humphreys J (2011) Measured and simulated nitrous oxide emissions from ryegrass- and ryegrass/white clover-based grasslands in a moist temperate climate. PLoS ONE 6(10):e26176. Doi:10.1371/journal.pone.0026176.

29. Li H, Qiu J, Wang L, Tang H, Li C, et al. (2010) Modelling impacts of alternative farming management practices on greenhouse gas emissions from a winter wheat–maize rotation system in China. Agric Ecosyst Environ 135 (1–2): 24–33.

30. Kang X, Hao Y, Li C, Cui X, Wang J, et al. (2011) Modeling impacts of climate change on carbon dynamics in a steppe ecosystem in Inner Mongolia, China. J Soils Sediments 11: 562–576.

31. Pathak H, Li C, Wassmann R (2005) Greenhouse gas emissions from Indian rice fields : calibration and upscaling using the DNDC model. Climat Res: 113–123.

32. Li C, Salas W, DeAngelo B, Rose S (2006) Assessing alternatives for mitigating net greenhouse gas emissions and increasing yields from rice production in China over the next twenty years. J Environ Qual 35 (4): 1554–1565.

33. Zhang L, Yu D, Shi X, Weindorf DC, Zhao L, et al. (2009) Simulation of global warming potential (GWP) from rice fields in the Tai-Lake region, China by coupling 1: 50,000 soil database with DNDC model. Atmos Environ 43 (17): 2737–2746.

34. Zhang Y, Wang Y, Su S, Li C (2011) Quantifying methane emissions from rice paddies in Northeast China by integrating remote sensing mapping with a biogeochemical model. Biogeosciences 8 (5): 1225–1235.

35. Li C, Mosier A, Wassmann R, Cai Z, Zheng X, et al. (2004) Modeling greenhouse gas emissions from rice-based production systems: Sensitivity and upscaling. Glob Biogeochem Cycles 18 (1): 1–19.

36. Fumoto T, Kobayashi K, Li C, Yagi K, Hasegawa T (2008) Revising a process-based biogeochemistry model (DNDC) to simulate methane emission from rice paddy fields under various residue management and fertilizer regimes. Glob Chang Biol 14 (2): 382–402.

37. Ludwig B, Bergstermann A, Priesack E, Flessa H (2011) Modelling of crop yields and N_2O emissions from silty arable soils with differing tillage in two long-term experiments. Soil Till Res 112 (2): 114–121.

38. Shang Q, Yang X, Gao C, Wu P, Liu J, et al. (2011) Net annual global warming potential and greenhouse gas intensity in Chinese double rice-cropping systems: a 3-year field measurement in long-term fertilizer experiments. Glob Chang Biol 17 (6): 2196–2210.

39. Wang L, Qiu J, Tang H, Li H, Li C, et al. (2008) Modelling soil organic carbon dynamics in the major agricultural regions of China. Geoderma 147 (1–2): 47–55.

40. Pan G, Li L, Wu L, Zhang X (2004) Storage and sequestration potential of topsoil organic carbon in China's paddy soils. Glob Chang Biol 10: 79–92.

41. Rui W, Zhang W (2010) Effect size and duration of recommended management practices on carbon sequestration in paddy field in Yangtze Delta Plain of China: A meta-analysis. Agric Ecosyst Environ 135(3): 199–205.

42. Zhang W, Xu M, Wang X, Huang Q, Nie J, et al. (2012) Effects of organic amendments on soil carbon sequestration in paddy fields of subtropical China. J Soils Sediments 12: 457–470.

43. RGCST (Research Group on Chinese Soil Taxonomy (Institute of Soil Science, Chinese Academy of Sciences), Cooperative Research Group on Chinese Soil Taxonomy) (2001) Chinese Soil Taxonomy. Science Press, Beijing, New York. 1–203.

44. Wang J, Jia J, Xiong Z, Khalil MAK, Xing G (2011) Water regime–nitrogen fertilizer–straw incorporation interaction: Field study on nitrous oxide emissions from a rice agroecosystem in Nanjing, China. Agric Ecosyst Environ 141: 437–446.

45. Wang J, Pan X, Liu Y, Zhang X, Xiong Z (2012) Effects of biochar amendment in two soils on greenhouse gas emissions and crop production. Plant Soil. Doi:10.1007/s11104–012–12503.

46. Li C, Zhuang Y, Frolking S, Galloway J, Harriss R, et al. (2003) Modeling soil organic carbon change in croplands of China. Ecol Appl 13 (2): 327–336.

47. Intergovernmental Panel on Climate Change (IPCC) (2007) Changes in atmospheric constituents and in radiative forcing. In: Solomon S, Qin D, Manning M, et al (editors) Climate Change 2007: The Physical Science Basis, Contribution of Working Group I to the FourthAssessment Report of the Intergovernmental Panel on Climate Change. Cambridge, United Kingdom and New York, NY, USA, Cambridge University Press.

48. Jia J, Ma Y, Xiong Z (2012) Net ecosystem carbon budget, net global warming potential and greenhouse gas intensity in intensive vegetable ecosystems in China. Agric Ecosyst Environ 150(15): 27–37.

49. Mosier A, Halvorson A, Reule C, Liu X (2006) Net global warming potential and greenhouse gas intensity in irrigated cropping systems in northeastern Colorado. J Environ Qual 35 (4): 1584–1598.

50. Six J, Feller C, Denef K, Ogle SM, de Moraes Sa JC, et al. (2002) Soil organic matter, biota and aggregation in temperate and tropical soils - Effects of no-tillage. Agronomie 22: 755–775.

51. Cai Z, Shan Y, Xu H (2007) Effects of nitrogen fertilizer on CH_4 emissions from rice paddies. Soil Sci Plant Nutr 53: 353–361.

52. Rochette P (2008) No-till only increases N_2O emissions in poorly-aerated soils. Soil Till Res 101: 97–100.

53. Tang H, Qiu J, Van Ranst E, Li C (2006) Estimations of soil organic carbon storage in cropland of China based on DNDC model. Geoderma 134: 200–206.

54. Ju X, Xing G, Chen X, Zhang S, Zhang L, et al. (2009) Reducing environmental risk by improving N management in intensive Chinese agricultural system. Proc Nat Acad Sci 106: 3041–3046.

23

Sodic Soil Properties and Sunflower Growth as Affected by Byproducts of Flue Gas Desulfurization

Jinman Wang[1,2]*, **Zhongke Bai[1,2]**, **Peiling Yang[3]**

1 College of Land Science and Technology of China University of Geosciences, Handian District, Beijing, People's Republic of China, **2** Key Laboratory of Land Consolidation and Land Rehabilitation Ministry of Land and Resources, Beijing, People's Republic of China, **3** College of Hydraulic and Civil Engineering, China Agricultural University, Handian District, Beijing, People's Republic of China

Abstract

The main component of the byproducts of flue gas desulfurization (BFGD) is $CaSO_4$, which can be used to improve sodic soils. The effects of BFGD on sodic soil properties and sunflower growth were studied in a pot experiment. The experiment consisted of eight treatments, at four BFGD rates (0, 7.5, 15 and 22.5 t ha^{-1}) and two leaching levels (750 and 1200 m^3 ha^{-1}). The germination rate and yield of the sunflower increased, and the exchangeable sodium percentage (ESP), pH and total dissolved salts (TDS) in the soils decreased after the byproducts were applied. Excessive BFGD also affected sunflower germination and growth, and leaching improved reclamation efficiency. The physical and chemical properties of the reclaimed soils were best when the byproducts were applied at 7.5 t ha^{-1} and water was supplied at 1200 m^3·ha^{-1}. Under these conditions, the soil pH, ESP, and TDS decreased from 9.2, 63.5 and 0.65% to 7.8, 2.8 and 0.06%, and the germination rate and yield per sunflower reached 90% and 36.4 g, respectively. Salinity should be controlled by leaching when sodic soils are reclaimed with BFGD as sunflower growth is very sensitive to salinity during its seedling stage.

Editor: James C. Nelson, Kansas State University, United States of America

Funding: This research was supported by the Fundamental Research Funds for the Central Universities of China, and the National Natural Science Foundation of China (50749032). The funders had no role in study design, data collection and analysis, decision to publish, or preparation of the manuscript.

Competing Interests: The authors have declared that no competing interests exist.

* E-mail: wangjinman2002@163.com

Introduction

Desulfurization technologies reduce SO_2 emissions in the flue gas of coal combustion, but they also produce large amounts of byproducts from flue gas desulfurization (BFGD) [1]. Substantial evidence exists that these BFGD can, with proper use, be valuable for soil reclamation, but improper use can be detrimental to the soil quality and to the environment when they enter the water and soil system via rainfall and surface runoff [2–4].

Typical sodic soils contain an excess of exchangeable sodium (ES) in the soil colloids, and the soluble carbonates are in the form of Na_2CO_3 and $NaHCO_3$. The pH value, sodium adsorption ratio (SAR) and exchangeable sodium percentage (ESP) are greater than 8.5, 13 and 15 respectively, and the electrical conductivity of the saturated paste (EC_{sat}) is less than 4.0 dS m^{-1}. The key to reclamation is to remove the ES and replace it with more favorable calcium ions in the topsoil [5]. Historically, sodic soils have been reclaimed with gypsum ($CaSO_4·2H_2O$), but gypsum has become unpopular because of its high price. The main character and components of BFGD are similar to those of gypsum, and they also contain sufficient amounts of the minerals that are necessary to crops, such as Ca, S and so on. At present, most BFGD are discarded, primarily into landfills, using up land resources and with increasing disposal costs [6]. Interest has been growing in using BFGD to reclaim sodic soils. Recent research has shown that BFGD can appreciably decrease soil pH and ESP, with no environmental impacts from trace metals, and significantly increase germination rates and crop outputs [2,4,7–10]. However, research on reclaiming sodic soils with BFGD is still in an elementary stage, and in China such reclamation is used only on a small scale (e.g., in Kangping county of Liaoning Province, Tumochuan county and Wulateqian county of Inner Mongolia, and on the Hetao plain of Ningxia Province).

Some studies have indicated that different crops have different degrees of sensitivity to salinity at different growth stages [11–12]. The tolerance of salinity was relatively high for the crops that were sensitive to salinity in the seedling stage, but relatively low for the crops that were not sensitive to salinity [11]. When BFGD are used in the reclamation of sodic soils, their high content of soluble salts may adversely affect plant growth, as they reduce the osmotic potential of plant roots, even though they are a good source of $CaSO_4$ [12].

Thus, it is necessary to determine a suitable application rate for BFGD that allows crops to tolerate not only sodicity, but also salinity. The objective of this research was to analyze the responses of sodic soil properties and sunflower growth to BFGD and to select an application rate and mode of application that would simultaneously control sodicity and salinity.

Materials and Methods

Physical and Chemical Properties of Soils and BFGD

The soil tested was sampled from the Changsheng Experimental Station of Baoyannur League Institute of Water Resources in the northwest of China (N 40°20′, E 108°31′). The soil had typical characteristics of sodic soil, that is, a high pH and exchangeable sodium percentage, and low hydraulic conductivity. The soil texture is clay, and its physical and chemical properties are listed in Table 1. The soil was air-dried, crushed and passed through a 2-mm sieve before the pot experiments. The BFGD tested was from Huaneng Power International, Inc. The particle size of the BFGD is between 20 μm and 80 μm. The $CaSO_4$ content in BFGD was 89.8%, the amount of moisture in BFGD was almost 10.1% (Table 2), and the concentrations of pollution elements in the BFGD were far below the tolerance limits regulated by the Control Standards for Pollutants in Fly Ash for Agricultural Use (GB8173-87) and the Control Standards for Pollutants in Sludge for Agricultural Use (GB4284-84) [13–14].

BFGD Application Rate and Experimental Design

The application rate of the BFGD was determined using a modified method based on gypsum application rates [15–16]. It was calculated according to the relative content of calcium in BFGD and gypsum as follows.

$$R_{BFGD} = 1.11R_G = 0.095H\gamma(ES)(E_{Nai} - E_{Naf}) \quad (1)$$

where R_{BFGD} is BFGD application rate in t ha^{-1}; R_G is the gypsum application rate in t ha^{-1}; 1.11 is the modified coefficient, which was determined by comparing the content of $CaSO_4$ in gypsum and BFGD; H is the soil depth to be reclaimed in cm, 20 cm was used in the research; γ is the soil bulk density in g cm^{-3}, which is 1.48 g cm^{-3} according to Table 1; ES is the exchangeable sodium in cmol kg^{-1}, which is 3.85 cmol kg^{-1} according to Table 1; E_{Nai} is the initial exchangeable sodium fraction, which is 1; and E_{Naf} is the desired final exchangeable sodium fraction, 0.1 was used in the research. The BFGD application rate was 9.75 t ha^{-1}, based on the initial properties of the sodic soil tested.

The pot experiments were carried out in the experimental field of China Agricultural University (N 39°48′, E116°28′) using sunflower G101. The pots were 23 cm in height, 19 cm in bottom diameter and 25 cm in top diameter, and were filled with 12.31 kg of dried soil, 3 g urea and a measured amount of BFGD. The soil bulk density in the pots was 1.48 g cm^{-3}, and the BFGD was mixed with the soil of 0–10 cm depth prior to packing. The irrigation cycle is 12 days, and aluminum containers were placed in the bottom of each pot to collect leachate. On June 8, 2004, 20 sunflower seeds were sown in each pot and the plants grew under natural conditions. The sunflowers were harvested on September 21, after 106 days of growth. The leachate were collected four time, the dates were May 21, June 5, July 4 and July 10

respectively; Three soils were sampled with an auger at soil depths of 0 to 10 cm and 10 to 20 cm on June 30, August and September 9 respectively.

The experiment consisted of eight treatments (Table 3), at four BFGD rates (0, 7.5, 15 and 22.5 t ha^{-1}; or 0, 28.5, 57.0 and 83.6 g per pot) and two leaching levels (750 and 1200 m^3 ha^{-1}; or 3 l per pot and 4.5 l per pot). There were eight replicates for each treatment.

Analytical Methods and Statistical Analyses

The samples were air-dried after the removal of the plant roots and passed through a 1-mm sieve. The EC, pH, soluble anions, and soluble cations were measured using 1:5 water extracts. The soluble cations were measured using an atomic absorption spectrophotometer; soluble anions were determined by anion chromatography; exchangeable cations were determined in 1 M ammonium acetate (pH = 7) extract, and cation exchange capacity was determined by the removal of ammonium ions by distillation following this extraction and washing with 96% alcohol [17]. Na and K were determined by flame emission spectroscopy in the extract, and Ca and Mg were determined by atomic absorption spectrophotometer. Soil pH was determined with the glass electrode method. The salt content was measured using a 1 cm conductivity cell, dip-type probe. Saturated hydraulic conductivities (HC) were determined by using cutting ring and calculated by using Darcy's law. Particle size distribution was determined with the hydrometer method.

Measurements of physical and chemical items were duplicated, and there were three replicates for the chemical analysis. Standard errors of the means of the three samples from each treatment were calculated. The variation among the treatments was analyzed with STATISTICA software (StatSoft, Inc; USA).

Results

The Effects of BFGD and Leaching on Sunflower Germination

All of the sunflowers germinated between the 6th and 14th day after sowing (Fig. 1). The germination of the sunflower plants occurred in two stages. The quantity of germination was greater in the first stage, and these time slots were 6–10 d, 6–8 d, 6–9 d and 6–9 d with 0 t·ha^{-1}, 7.5 t ha^{-1}, 15 t ha^{-1} and 22.5 t ha^{-1} BFGD rates, respectively. The quantity of germination was relatively less during the second stage, and these time slots were 10–14 d, 8–14 d, 9–14 d and 9–14 d, respectively. The different BFGD rates and

Table 1. Physical and chemical properties of soil.

Property	Parameter	Value	Property	Parameter	Value
Particle size distribution	2.0–0.02 mm (%)	23	Chemical	ES(cmol kg^{-1})	3.85
	0.02–0.002 mm (%)	35		CEC(cmol kg^{-1})	8.98
	<0.002 mm (%)	42		ESP(%)	42.85
Physical	Bulk density(g cm^{-3})	1.48		EC(dS m^{-1})	2.15
	HC (cm min^{-1})	2.51×10^{-5}		pH	9.15
Soluble cations	Na$^+$(cmol L^{-1})	0.81	Soluble anions	HCO$_3^-$ (cmol L^{-1})	0.16
	K$^+$(cmol L^{-1})	0.01		CO$_3^{2-}$ (cmol L^{-1})	0.04
	Ca^{2+}(cmol L^{-1})	0.05		SO$_4^{2-}$ (cmol L^{-1})	0.36
	Mg^{2+}(cmol L^{-1})	0.03		Cl$^-$(cmol L^{-1})	0.34

Note: Value of the mean of three replicates; HC is hydraulic conductivity.

Table 2. Component and properties of the BFGD.

pH	Density (g·cm^{-3})	Free water (%)	CaSO$_4$·2H$_2$O (%)	CaSO$_4$·1/2H$_2$O (%)	CaCO$_3$ (%)
5.90	1.02	10.10	89.8	0.20	5.55

Cd (mg·kg^{-1})	Cr (mg·kg^{-1})	As (mg·kg^{-1})	Se (mg·kg^{-1})	Ni (mg·kg^{-1})	Cu (mg·kg^{-1})	Hg (mg·kg^{-1})	Pb (mg·kg^{-1})
0.01	83.4	5.04	4.24	21.35	83.26	0.32	99.38
5[a]	250[a]	75[a]	15[a]	200[a]	250[a]	5.0[b]	250[a]

Note: [a]Control Standards for Pollutants in Fly Ash for Agricultural Use (GB8173–87);
[b]Control Standards for Pollutants in Sludge for Agricultural Use (GB4284–84).

leaching levels had different effects on the germination rate ($P<0.05$; Fig. 2). The germination rate with low BFGD rates was higher than that of plants with high BFGD rates, and the germination rate with high leaching levels was higher than that of plants with low leaching levels. It can be concluded that BFGD both increased and accelerated the germination, but an excessive BFGD rate also suppressed the germination and delayed the time of the germination.

The variation in leachate concentration and the buildup of the salt content of the leachate are shown in Figs. 3 and 4, respectively. The leachate concentration of the high leaching level was lower than that of low leaching level, and the salt content was also different because of variation in the volume of leachate among different treatments. The salt content of the high leaching level gradually decreased, whereas the salt content of the low leaching water decreased after an initial increase. The accumulated salt content of the leachate was basically the same in all treatments. Although the total salt content of the leachate was basically equal, the effects were not equal as the time that the salt was held in the soil varied and was higher in some treatments during the seedling stage of the sunflowers. Therefore, the growth of the sunflowers varied. The sunflower heights with different treatments in the seedling stage are shown in Fig. 2. The height of plants treated with high leaching water was higher than plants treated with low leaching water, and the height of sunflower plants with a low BFGD rate was higher than that of plants with a high BFGD rate.

The Effects of BFGD and Leaching on Sodic Soil Properties

The ESP, pH and TDS for different treatments were compared with each other on June 30, August 13, and September 9 and the results are depicted in Fig. 5. The ESP, pH and TDS significantly decreased at soil depths of 0 to 10 cm and 10 to 20 cm after applying BFGD and leaching. At the 0–10 cm soil depth, the pH, ESP, and TDS of soils treated with a low BFGD rate and a high leaching level were lower than in the 10–20 cm depth of soils treated with a high BFGD rate and a low leaching level. The physical and chemical properties of the reclaimed soil were best when the by-products were applied at 7.5 t·ha^{-1} and the water

was supplied at 1200 m^3 ha^{-1}; under these conditions the soil pH, ESP, and TDS decreased from 9.2, 63.5 and 0.65% to 7.8, 2.8 and 0.06% respectively.

The Effects of BFGD and Leaching on Sunflower Growth

The variations in sunflower height and dry matter weight are shown in Fig. 6. The sunflower height and dry matter weight were markedly increased by the application of the byproducts and leaching. Height and dry matter weight were greatest when the byproducts were applied at 7.5 t ha^{-1} and water was supplied at 1200 m^3 ha^{-1}. So, BFGD application made good initial sunflower stands possible and enhanced sunflower growth, presumably increasing leaching through root channels and the exchange of sodium to a deeper soil depth.

The Response of Sunflower Yield to BFGD and Leaching

Fig. 7 compares the effects of BFGD and leaching on the yield per sunflower and on the one thousand seeds weight. The output of the sunflower had no positive correlation with BFGD rate. BFGD could increase the output of the sunflowers when they were applied in the right amount, but the output of the sunflowers decreased if the applied amount of BFGD increased beyond the sunflowers peak value output. The output of sunflowers under the high leaching condition was higher than under the low leaching condition. The sunflower output was best when the byproducts were applied at 7.5 t ha^{-1} and water was supplied at 1200 m^3 ha^{-1}; under these conditions the output of one sunflower could reach 36.4 g. The variation in the one thousand seeds weight was not the same as the variation in the output of one sunflower; the one thousand seeds weight at the high leaching level was lower than that at the low leaching level.

The effects of BFGD and leaching on the yield per sunflower were analyzed with progressive regression analysis using STATISTICA software. The regression equation was as follows.

$$y = -12.2852 + 3.6912x_1 + 0.0184x_2 - 0.1418x_1^2$$
$$\left(R^2 = 0.89, \alpha = 0.02 < 0.05\right) \tag{2}$$

Table 3. Experimental design.

Experimental treatments	T1	T2	T3	T4	T5	T6	T7	T8
BFGD rate (t ha^{-1})	0	0	7.5	7.5	15	15	22.5	22.5
Leaching water (m^3 ha^{-1})	750	1,200	750	1,200	750	1,200	750	1,200

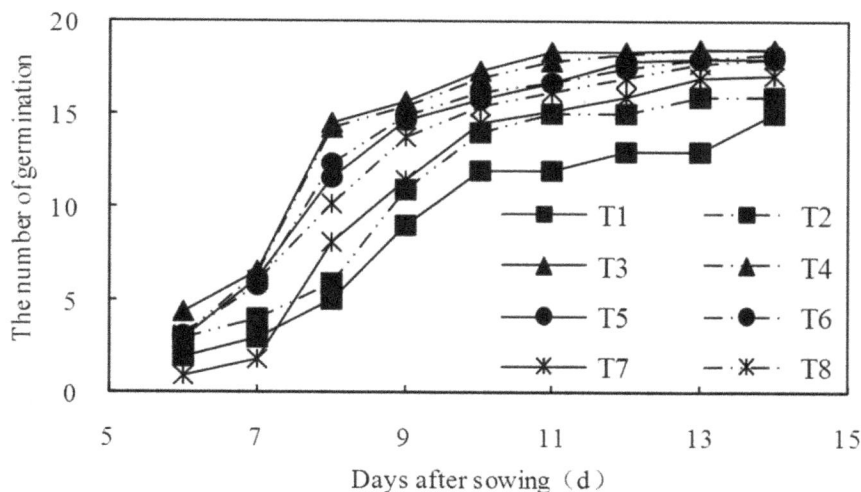

Figure 1. Variation of sunflower germination with days after sowing.

where y is the yield per sunflower in g; x_1 is the application rate of BFGD in t ha^{-1}; and x_2 is the amount of leaching water in m^3 ha^{-1}.

Discussion

Mechanism of the Effect of BFGD on Sodic Soil and Sunflower Growth

The dispersion of clay in sodic soils can be reduced three-fold when soil pH decreases from 9 to 7, and soil pH is an important factor in managing dispersive soils [18]. The by-products clearly reduced the ESP, pH, TDS and clay dispersion and thus improved the physical properties of the soil. Through the intervention of calcium, BFGD can reduce the electric potential produced by the mutual exclusion of negative charges in the soil colloids surface, promote the formation of a soil aggregate structure, raise soil water holding capacity, lower soil bulk density, increase soil porosity, improve the growth environment of root systems, increase the germination rate of sunflowers, and promote the growth of sunflowers [5,19]. So, BFGD application in poorly structured soils helped crop establishment and increased crop growth [20–21].

Soil Salinity Control in the Process of Reclaiming Sodic Soil with BFGD

With the application of BFGD the salt content of the soil will increase. As crops are very sensitive to salinity, especially in the seedling stage, the salt in the soil must be promptly removed [11–12]. In this study, although the total salt content of the leachate was basically equal in all treatments, the residence time of salinity in the soil varied, which led to different effects. The improvement effects had no positive correlation with BFGD rate. The effect of a high leaching level was better than that of a low leaching level. Initially, the increase of BFGD application rate resulted in soil improvement, but beyond a certain application rate threshold, the improvement effects began to decrease. Thus, when reclaiming sodic soil with BFGD, the soil salinity needs to be reasonably controlled. Under field conditions the salts should be immediately removed by constructing good drainage conditions. On the other hand, microbial activities are very important to the reclamation of sodic soils and can accelerate the improvement of soil structure [22–23]. So, the regulation of microbial activity is also our future research focus.

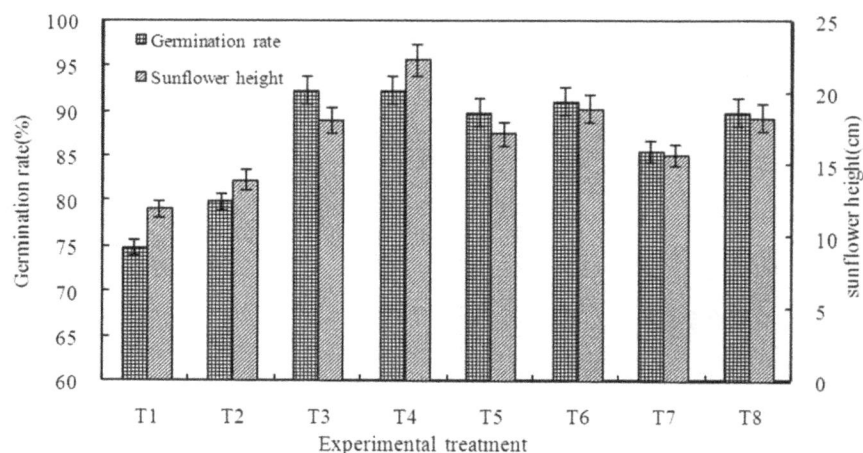

Figure 2. Germination rate and sunflower height in the seedling stage.

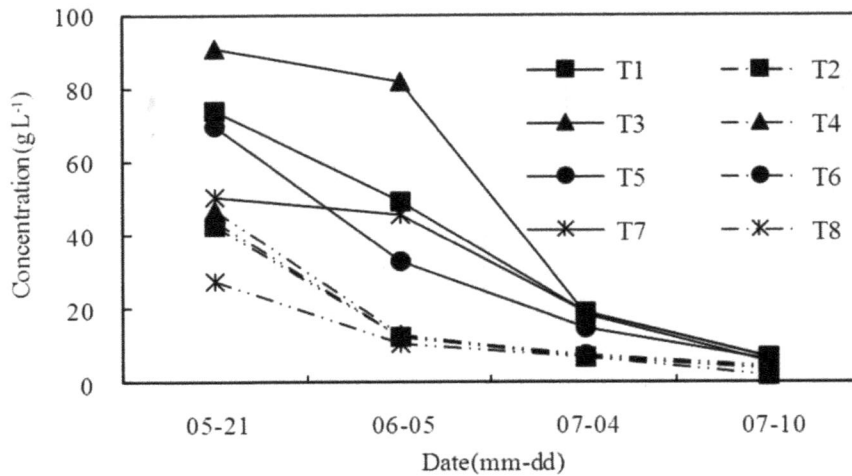

Figure 3. Variation of leachate concentration.

The Best BFGD Application Rate

The relationship between the application rate of BFGD and sunflower yield per plant fits a parabola shape. The relationship is:

$$y = -0.1418x^2 + 3.6912x + 5.644, \qquad (3)$$

where y is the yield per sunflower in g and x is the application rate of BFGD in t ha^{-1}.

The above equation was differentiated and the following formula was obtained.

$$-0.2836x + 3.6912 = 0 \qquad (4)$$

It was calculated that the yield per sunflower was the highest when the application rate of BFGD was 13 t·ha^{-1}. This application rate is 1.34 times the amount of the theoretical calculations. The best application rate was obtained in the pot experiment conditions, so the higher application rates might be used in field conditions due to the limitations of various factors [2,8].

Conclusions

The following conclusions can be drawn from our findings.

(1) Application of the by-products from flue gas desulfurization is an efficient method to reclaim sodic soil. The germination rate and output of sunflower plants significantly increased, and the exchangeable sodium percentage (ESP), the pH, and the total dissolved salts (TDS) decreased. However, excessive amounts of BFGD also negatively affected germination and growth of sunflowers.

(2) Leaching improves the efficiency of the reclamation. Salinity should be controlled by leaching when sodic soils are reclaimed with BFGD, as the sunflower growth is very sensitive to salinity in the seedling stage.

(3) The physical and chemical properties of the soil with a value of pH = 9.2 and ESP = 63.5 were best when the by-product

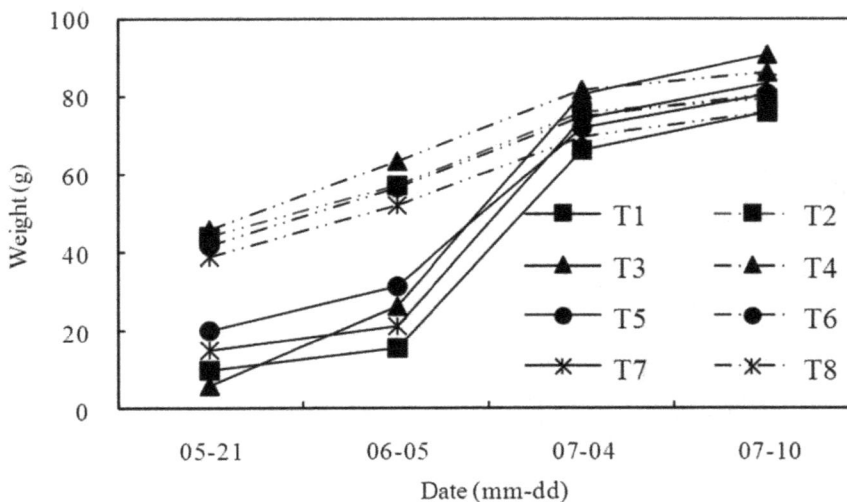

Figure 4. Buildup of salt content of leachate.

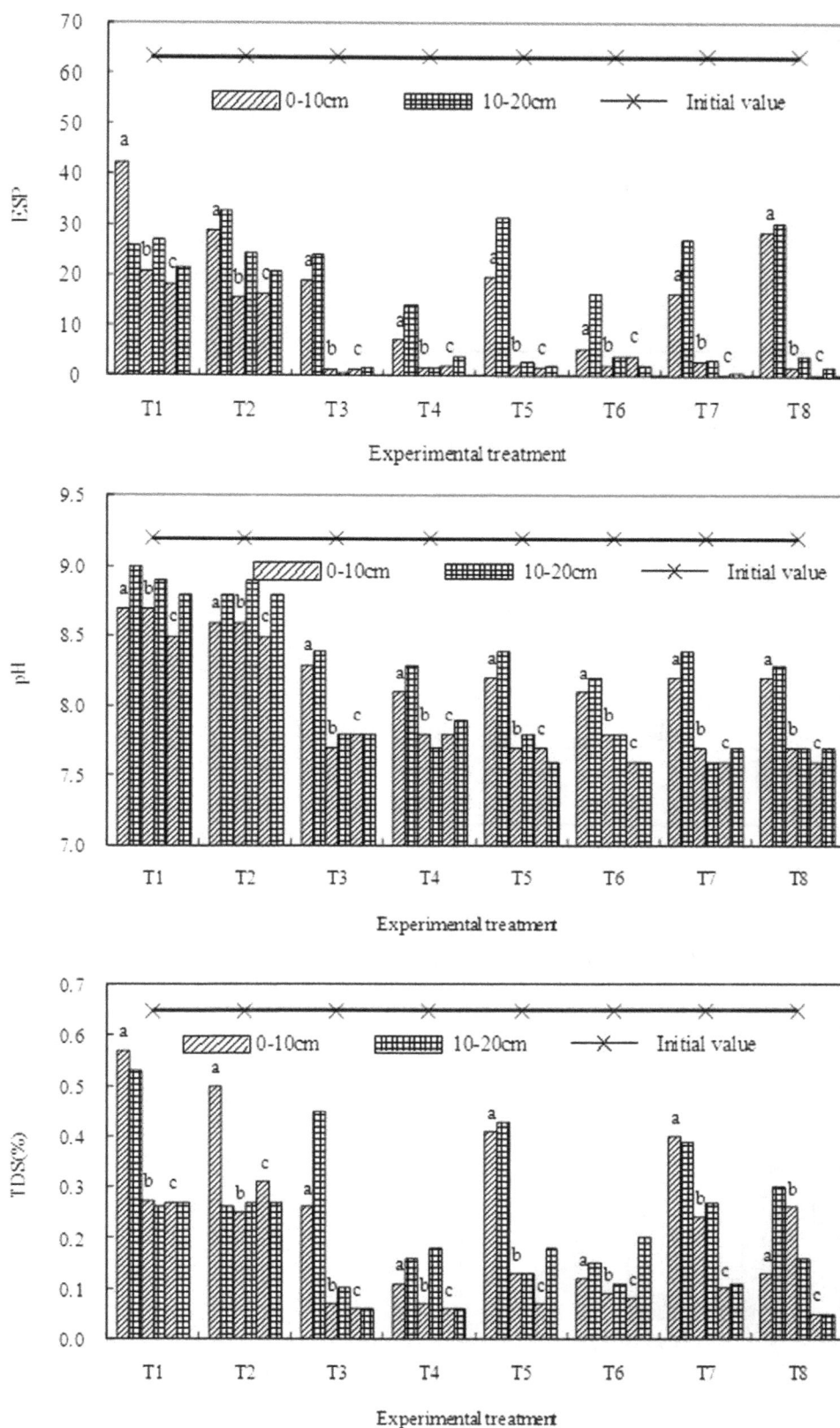

Figure 5. The ESP, pH and TDS of the soil as affected by the BFGD and leaching (a) June 30; b) August 13; c) September 9).

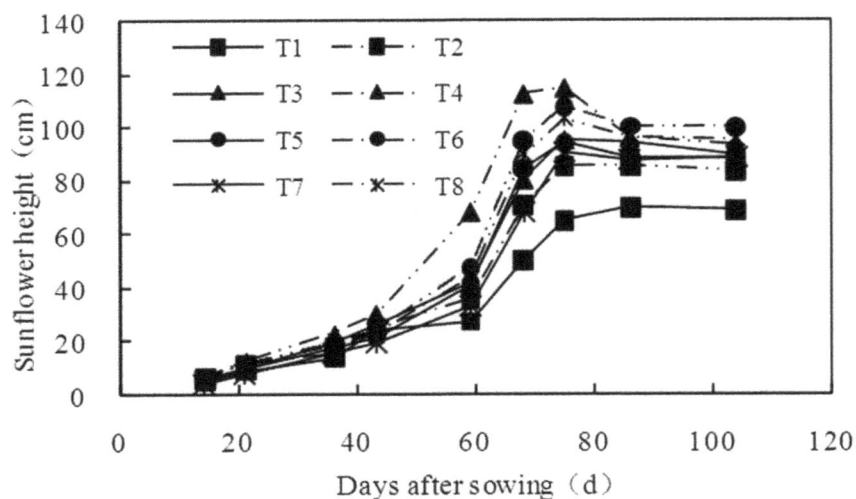

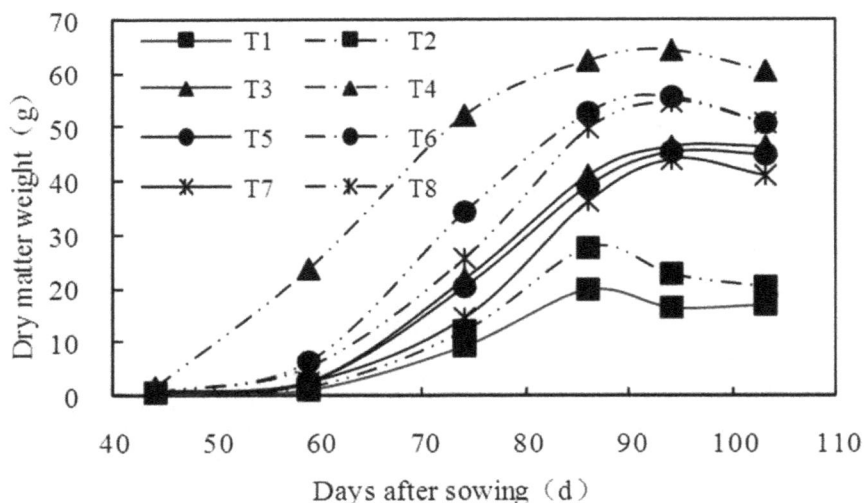

Figure 6. Variation of sunflower height and dry matter weigh with days after sowing.

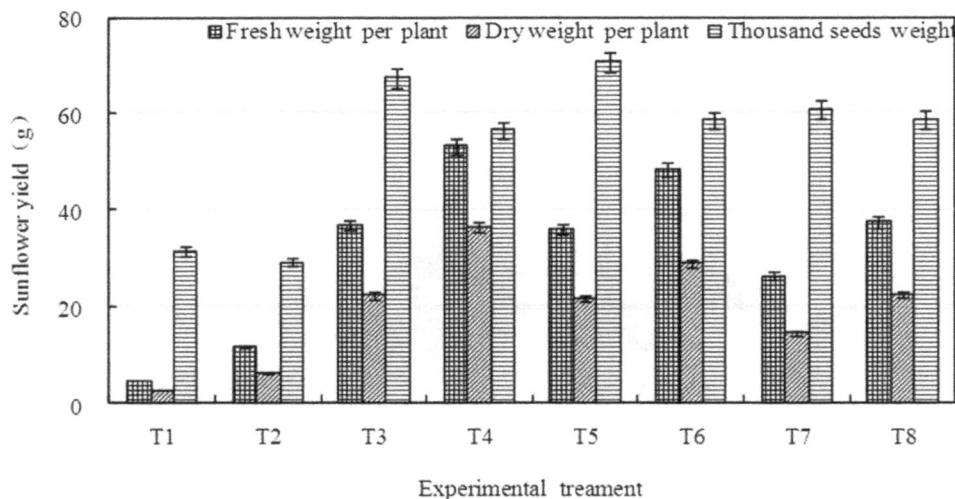

Figure 7. Variation of sunflower yield.

was applied at 7.5 t ha^{-1} and water was supplied at 1200 m^3 ha^{-1}. Under these conditions the germination rate and yield per sunflower could reach 90% and 36.4 g, respectively. Therefore, a reasonable application rate should be selected when BFGD are used to reclaim sodic soils.

References

1. Carlson CL, Adriano DC (1993) Environmental impacts of coal combustion residues. Journal of Environmental Quality 22: 227–247.
2. Chun S, Nishiyama M, Matsumoto S (2001) Sodic soils reclaimed with by-product from flue gas desulphurization: corn production and soil quality. Environmental Pollution 114: 453–459.
3. Stuczynski TL, McCarty GW, Wright RJ (1998) Impact of coal combustion product amendments on soil quality: Mobilization of soil organic nitrogen. Soil Science 163: 953–959.
4. Wang SJ, Chen CH, Xu XC, Li YJ (2008) Amelioration of alkali soil using flue gas desulfurization byproducts: Productivity and environmental quality. Environmental Pollution 151: 200–204.
5. Frenke H, Gerstle Z, Alperovitch N (1989) Exchange-induced dissolution of gypsum and the reclamation of sodic soils. Soil Science 40: 599–611.
6. Clark RB, Ritchey KD, Baligar VC (2001) Benefits and constraints for use of FGD products on agricultural land. Fuel 80: 821–828.
7. Sloan JJ, Dowdy RH, Dolan MS, Rehm GW (1999) Plant and soil responses to field-applied flue gas desulphurization residue. Fuel 78: 169–174.
8. Wang J M, Yang PL, Ren SM, Xiang GM (2005) Variation of chemical indices of alkaline soil ameliorated with desulphurization byproducts. Acta Pedologica Sinica (in Chinese) 42(1): 98–105.
9. Sakai Y, Matsumoto S, Sadakata M (2004) Alkali soil reclamation with flue gas desulphurization gypsum in China and assessment of metal content in corn grains. Soil & Sediment Contamination 13(1): 65–80.
10. Chen L, Ramsier C, Bigham J, Slater B, Kost D, et al. (2009) Oxidation of FGD-CaSO$_3$ and effect on soil chemical properties when applied to the soil surface. Fuel 88: 1167–1172.
11. Katerji N, Van Hoorn JW, Hamdy A, Mastrorilli M, Mou Karzel E (1997) Osmotic adjustment of sugar beets in response to soil salinity and its influence on stomata conductance, growth and yield. Agricultural Water Management 34: 57–69.
12. Shannon MC (1997) Adaptation of plants to salinity. In Advances in Soil Science. Lal, R. and Steward, B. A. (eds). Springer-Verlag, 75–120.
13. Ministry of Agriculture and fisheries (1987) Control Standards for Pollutants in Fly Ash for Agricultural Use (GB8173–87). Beijing: China Standards Press (in Chinese).
14. Ministry of Agriculture and fisheries (1984) Control Standards for Pollutants in Sludge for Agricultural Use (GB4284–84). Beijing: China Standards Press (in Chinese).
15. Wang JM, Yang PL (2004) The effect on physical and chemical properties of saline and sodic soils reclaimed with byproduct from flue gas desulphurization, In Land and Water Management Decision Tools and Practices Vol.II, Huang GH. and S.Pereria L (eds). China Agriculture Press, Beijing: 1015–1021.
16. Oster JD, Frenkel H (1980) The chemistry of the reclamation of sodic soils with gypsum and lime. Soil Science Society of America Journal 44: 41–45.
17. Rhodes DR (1982) Cation exchange capacity. In Methods of soil analysis Part 2, Chemical methods. Page AL, Miller RH and Keeney DR. (eds). American Society of Agronomy: Madison, Wisconsin, 149–165.
18. Chorom M, Rengasamy P, Murry RS (1994) Clay dispersion as influenced by pH and net particle of sodic soils. Australian Journal of Soil Research 32: 1152–1163.
19. Truman CC, Nuti RC, Truman LR, Dean JD (2010) Feasibility of using FGD gypsum to conserve water and reduce erosion from an agricultural soil in Georgia. Catena 81: 234–239.
20. Carter MR, Pearen JR (1989) Amelioration of a saline-sodic soil with low application of calcium and nitrogen amendments. Arid Soil Research 3: 1–9.
21. Ilyas M, Qureshi RH, Qadir MA (1997) Chemical changes in saline-sodic soils after gypsum application and cropping. Soil Technologies 10: 247–260.
22. Pandey VC, Singh K, Singh B, Singh RP (2011) New approaches to enhance eco- restoration efficiency of degraded sodic lands: critical research needs and future prospects. Restoration Ecology 29: 322–325.
23. Singh K, Singh B, Singh RR (2012) Changes in physico-chemical, microbial and enzymatic activities during restoration of degraded sodic lands: Ecological suitability of mixed forest over plantation. Catena 96: 57–67.

Author Contributions

Conceived and designed the experiments: JW PY. Performed the experiments: JW. Analyzed the data: JW ZB. Contributed reagents/materials/analysis tools: JW ZB. Wrote the paper: JW.

p-Coumaric Acid Influenced Cucumber Rhizosphere Soil Microbial Communities and the Growth of Fusarium oxysporum f.sp. cucumerinum Owen

Xingang Zhou, Fengzhi Wu*

Department of Horticulture, Northeast Agricultural University, Xiangfang, Harbin, People's Republic of China

Abstract

Background: Autotoxicity of cucumber root exudates or decaying residues may be the cause of the soil sickness of cucumber. However, how autotoxins affect soil microbial communities is not yet fully understood.

Methodology/Principal Findings: The aims of this study were to study the effects of an artificially applied autotoxin of cucumber, p-coumaric acid, on cucumber seedling growth, rhizosphere soil microbial communities, and Fusarium oxysporum f.sp. cucumerinum Owen (a soil-borne pathogen of cucumber) growth. Abundance, structure and composition of rhizosphere bacterial and fungal communities were analyzed with real-time PCR, PCR-denaturing gradient gel electrophoresis (DGGE) and clone library methods. Soil dehydrogenase activity and microbial biomass C (MBC) were determined to indicate the activity and size of the soil microflora. Results showed that p-coumaric acid (0.1–1.0 µmol/g soil) decreased cucumber leaf area, and increased soil dehydrogenase activity, MBC and rhizosphere bacterial and fungal community abundances. p-Coumaric acid also changed the structure and composition of rhizosphere bacterial and fungal communities, with increases in the relative abundances of bacterial taxa Firmicutes, Betaproteobacteria, Gammaproteobacteria and fungal taxa Sordariomycete, Zygomycota, and decreases in the relative abundances of bacterial taxa Bacteroidetes, Deltaproteobacteria, Planctomycetes, Verrucomicrobia and fungal taxon Pezizomycete. In addition, p-coumaric acid increased Fusarium oxysporum population densities in soil.

Conclusions/Significance: These results indicate that p-coumaric acid may play a role in the autotoxicity of cucumber via influencing soil microbial communities.

Editor: Jack Anthony Gilbert, Argonne National Laboratory, United States of America

Funding: This work was supported by the National Basic Research Program of China (2009CB119004-05) and National Staple Vegetable Industrial Technology Systems of China (CARS-25-08). The funders had no role in study design, data collection and analysis, decision to publish, or preparation of the manuscript.

Competing Interests: The authors have declared that no competing interests exist.

* E-mail: fzwu2006@yahoo.com.cn

Introduction

Soil sickness is a reduction in both crop yield and quality caused by continuous mono-cropping in the same land. It is one of the major problems in agricultural production, especially for greenhouse crops [1]. Cucumber (Cucumis sativus L.), a crop of high economic importance in many countries, is vulnerable to soil sickness [1]. Recently, cultivation of cucumber under greenhouse conditions has greatly expanded in China, but significant agricultural loss is observed each year because continuous mono-cropping practice is becoming more and more popular.

The accumulation of autotoxins is probably responsible for the soil sickness of cucumber [1]. Autotoxicity is an intraspecific allelopathy, where a plant species inhibits the growth of plants of the same species through releasing toxic chemicals into the environment [2]. Cucumber root exudates and plant debris were shown to have autotoxicity potential [1]. Autotoxins, including some phenols, have been identified in cucumber root exudates [3]. Some phenols from living and decomposing plant tissues can be active allelochemicals and they can accumulate in soil and have detrimental effects on the growth of associated and next-season plants [4].

Soil microorganisms may influence the persistence, availability and biological activities of allelochemicals in soil [5,6] and root exudates or allelochemicals can affect soil microbial communities [7]. Thus, it is suggested that allelopathy can be better understood in terms of soil microbial ecology [5,8]. Recently, a great research effort has addressed the effects of plant root exudates, such as low molecular carbohydrate, organic acids, amino acids, and plant secondary metabolites (e.g. flavonoids and glucosinolate) on soil microbial communities [9–13]. Phenols were shown to affect the growth of microorganisms in vitro [14], but little information is available on how soil microbial communities respond to putative allelochemicals in natural soils [15,16].

Our knowledge about effects of autotoxins, such as phenols, on soil microorganism populations has been mainly achieved by traditional cultivation-dependent methods, which is limited in that only a small fraction of the microorganisms are accessible to study [17–19]. Analysis of soil microbial communities with molecular techniques, such as real-time PCR, PCR-denaturing gradient gel

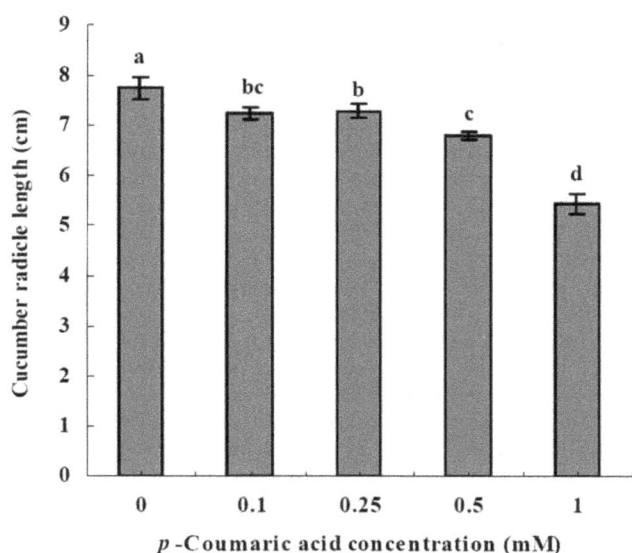

Figure 1. Effects of *p*-coumaric acid on cucumber radicle elongation. Data are represented as the means of three independent replicates with standard error bars. *Different letters* indicate significant difference between treatments (P<0.05, Tukey's HSD test).

electrophoresis (DGGE) and phylogenetic analysis, can improve our understanding of how autotoxins affect abundance, structure and composition of soil microbial communities [20,21].

Several studies have reported that soil may become suppressive in the case of long-term monoculture, for example, *Gaeumannomyces graminis* var. *tritici* (take-all disease) of wheat [22], *Rhizoctonia solani* of sugar beet [23], *Thielaviopsis basicola* of tobacco [24] and Fusarium wilt of watermelon [25]. However, it is also suggested that the accumulation of soil-borne pathogens is responsible for the soil sickness [26–28]. The *Fusarium* (*Ascomycota, Fungi*) community size was found to be linked to the soil sickness associated with cucumber cultivation [28]. Autotoxins would accumulate under continuous mono-cropping conditions; therefore, how autotoxins, such as phenols, affect soil-borne pathogens in soil needs to be further clarified.

p-Coumaric acid (*p*-hydroxycinnamic acid) has been identified in plant root exudates or residues [14] and in soils under many plant species, including cucumber [29,30]. We hypothesized that *p*-coumaric acid could influence cucumber growth and soil

microbial communities. The primary aims of this research were to study: 1) the phytotoxic effects of *p*-coumaric acid on cucumber radicle elongation and seedling growth; 2) the effects of *p*-coumaric acid on structure, composition, abundance, activity and size of cucumber rhizosphere microbial communities; 3) the effects of *p*-coumaric acid on the growth of *F. oxysporum* f.sp. *cucumerinum* Owen (a soil-borne pathogen of cucumber) both *in vitro* and in soil. Abundance, structure and composition of bacterial and fungal communities were analyzed by real-time PCR, PCR-DGGE and clone library methods. Soil dehydrogenase activity and microbial biomass C (MBC) were determined to indicate the activity and size of soil microflora.

Materials and Methods

Cucumber Radicle Elongation Experiment

Ten germinated cucumber seeds (cv. 'Jinlv 3') with radicles of 1 mm length were separately placed in a Petri dish (9 cm diameter), which contained two layers of sterilized filter papers. Five milliliter of different concentrations of *p*-coumaric acid solutions (0.1, 0.25, 0.5 or 1.0 mM) with pH adjusted to 7.0 with 0.1 M NaOH solution were added in the Petri dish. Cucumber seeds treated with distilled water (pH 7.0) were used as the control. There were five treatments (four concentrations of *p*-coumaric acid and one control) in total. We incubated three replicates for each treatment, and had five Petri dishes per replicate. The radicle length was measured 10 days after incubation at 25°C in the dark.

Cucumber Seedling Experiment

Cucumber seedlings with two cotyledons were transplanted into cups (5.5 cm bottom diameter, 8 cm top diameter, 8.5 cm height) containing 250 g soil. The soil was sampled from the upper soil layer (0–15 cm) of an open field in the experimental station of Northeast Agricultural University, Harbin, China (45°41′N, 126°37′E), which was covered with grass and undisturbed for more than 15 years. The soil was a black soil (Mollisol) with sandy loam texture: organic matter, 3.67%; available N, 89.02 mg/kg; available P, 63.36 mg/kg; available K, 119.15 mg/kg; EC (1:2.5, w:v), 0.33 mS cm^{-1}; and pH (1:2.5, w:v), 7.78. No fertilizer was added to soil. There was one cucumber seedling per cup. The seedlings were incubated in the greenhouse (32°C day/22°C night, relative humidity of 60–80%, 16 h light/8 h dark).

Phenols can be rapidly depleted after their addition to soil due to microbial degradation [17]. In this study, *p*-coumaric acid was added into soil periodically to maintain the desired concentrations. Our previous study found that *p*-coumaric acid in soils repeatedly

Table 1. Effects of *p*-coumaric acid on cucumber seedling growth.

Concentration (μmol/g soil)	Leaf area (cm²/plant)	Plant height (mm)	Dry weight (mg/plant DW)
0	209±6 a	75±1 a	638±13 a
0.1	179±11 ab	70±3 ab	593±53 ab
0.25	164±9 b	67±3 ab	554±21 ab
0.5	160±7 b	64±4 ab	537±9 ab
1.0	145±8 b	63±3 b	498±9 b
ANOVA			
F$_{4,12}$	8.48	3.69	4.01
P-value	0.003	0.043	0.034

Values (mean ± standard error) in the same column followed by different letters are significantly different (P<0.05, Tukey's HSD test).

Table 2. Effects of p-coumaric acid on soil pH, available N, p-coumaric acid in soil, dehydrogenase activity, MBC and bacterial and fungal abundance in cucumber rhizosphere.

Concentration (μmol/g soil)	pH (1:2.5, w/v)	Available N (mg/kg)	p-Coumaric acid retained in soil (μmol/g soil)	Dehydrogenase activity (μg TPF/g soil/24 h)	MBC (mg C/kg soil)	Bacterial abundance (10^{11} copies/g soil)	Fungal abundance (10^8 copies/g soil)	Bacteria-to-fungi ratio (10^3)
0	7.72±0.15 a	60.83±4.01 ab	0.14±0.02 d	1.21±0.04 b	152±1 d	4.29±0.10 d	1.30±0.02 d	3.31±0.13 a
0.1	7.94±0.10 a	66.2±1.93 a	0.18±0.02 cd	2.34±0.18 a	199±4 c	7.15±0.42 c	3.72±0.04 c	1.92±0.12 c
0.25	7.77±0.08 a	56.92±2.29 bc	0.28±0.04 c	2.63±0.21 a	220±6 bc	11.25±0.13 a	4.09±0.06 b	2.75±0.06 b
0.5	7.93±0.12 a	58.32±2.42 bc	0.48±0.05 b	2.76±0.13 a	265±5 a	8.87±0.21 b	4.58±0.03 a	1.94±0.04 c
1.0	7.98±0.08 a	52.78±2.47 c	0.87±0.06 a	2.84±0.10 a	235±5 b	8.26±0.11 b	3.74±0.03 c	2.21±0.01 c
ANOVA								
$F_{4,12}$	1.09	9.99	173.57	20.25	88.89	123.31	1057.35	48.81
P-value	0.411	0.002	<0.0001	<0.0001	<0.0001	<0.0001	<0.0001	<0.0001

Values (mean ± standard error) in the same column followed by different letters are significantly different (P<0.05, Tukey's HSD test).

cultivated with cucumber ranged from 0.51 to 0.62 μmol/g soil (data not shown). Therefore, cucumber seedlings at the one-leaf stage were treated with different concentrations of p-coumaric acid every 48 h to achieve final concentrations of 0.1, 0.25, 0.5, 1.0 μmol/g soil DW. The pH of p-coumaric acid solution was adjusted to 7.0 with 0.1 M NaOH solution because soil pH is a dominant factor that regulates soil microbial communities [31]. The soil treated with distilled water (pH 7.0) was used as the control. p-Coumaric acid in the soil before cucumber planting (the field soil) was 0.06 μmol/g soil DW, which was considered when adding p-coumaric acid: the final concentration of p-coumaric acid was the sum of the concentration in the field soil and the concentration added. Soil water content was adjusted every two days with distilled water to maintain a constant weight of cups. There were five treatments (four concentrations of p-coumaric acid and one control) in total. We had three replicates for each treatment, and had five seedlings per replicate.

Ten days later, cucumber seedlings were destructively sampled and separated into leaves, stems and roots. Five plants per replicate were destructively sampled, and different tissues of the five plants were pooled to make three composite replicates per treatment. Cucumber leaves were scanned with a Microtek ScanMaker i800 plus system (WSeen, China) and leaf area was calculated with a LA-S Leaf Area Analysis software (WSeen, China). Plant dry weight was measured after oven drying at 70°C to constant weight.

Rhizosphere soil

Soil sampling. Ten days later, rhizosphere soil samples were collected by shaking off from the roots in the air [28]. Soils from five plants per replicate were pooled to make a composite sample. After sieving (2 mm), subsamples of rhizosphere soils were monitored for soil pH, available N content, dehydrogenase activity and MBC estimation, whereas the other subsamples were stored at −70°C prior DNA extraction.

Soil phenol concentration. Soil phenols were extracted as previously described [32]. Briefly, 15 g soil was shaken in 100 ml of 2 M NaOH for 24 h (dark, 25°C). After centrifuged at 6000 g for 15 min, the supernatant was acidified to 2.5 with 5 M HCl, and extracted with ethyl acetate. The resulting extracts were evaporated to dryness at 40°C. The residue was dissolved in 5 ml of 80% methanol and filtered (0.22 μm). The methanol solution of soil extracts was analyzed with a Waters HPLC system (Waters, Milford, MA). The mobile phase was a mixture of 80% methanol and 20% water (0.8 ml/min). Detection was performed at 280 nm. Identification and quantification of p-coumaric acid were done by comparing the retention time and area with pure standard.

Soil pH, available N, dehydrogenase activity and MBC. Soil pH was determined in water suspensions at a soil/water ratio of 1:2.5 with a glass electrode. Available N (nitrate and ammonium) was extracted with 1 M KCl solution (1:10, w/v) and analyzed with an autoanalyzer. Dehydrogenase activity was determined by the reduction of 2,3,5-triphenyltetrazolium chloride (TTC) [33]. Soil MBC was determined by the chloroform-fumigation-extraction method. An extractability factor of 0.38 was used to calculate MBC [34].

Real-time PCR. Bacterial and fungal community abundances were estimated by measuring bacterial 16S rRNA gene and fungal internal transcribed spacer (ITS) rRNA gene densities. Total soil DNA was extracted with an E.Z.N.A. Soil DNA Kit (Omega Bio-Tek, Inc., GA, USA). SYBR Green real-time PCR assays were conducted on an IQ5 real-time PCR system (Bio-Rad Lab, LA, USA) with 338F/518R [20] and ITS1F/ITS4 [35,36] as

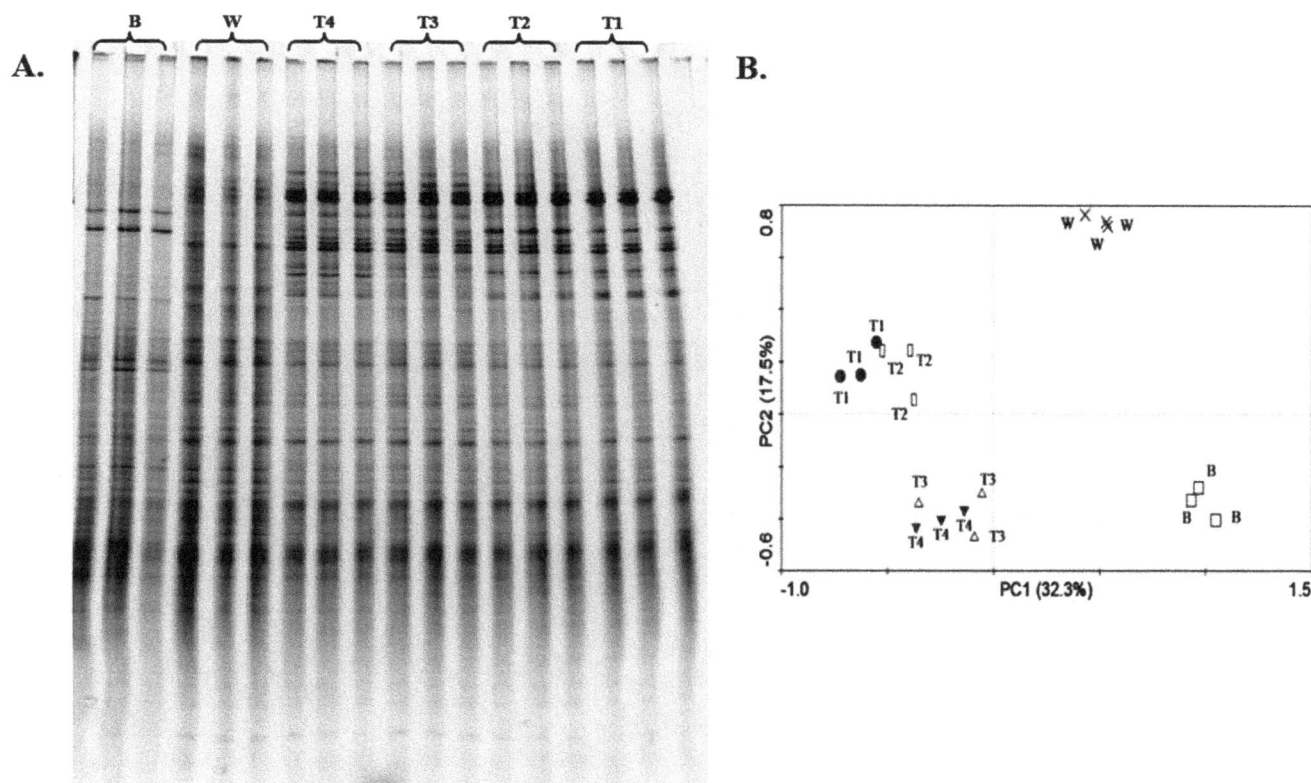

Figure 2. Effects of _p_-coumaric acid on bacterial community structure. (A) DGGE profiles of partial bacterial 16S rRNA gene sequences. (B) Principal component analysis (PCA) of bacterial community based on DGGE profiles. _B_ and _W_ represent soil sample before cucumber planting and control, respectively. _T1, T2, T3_ and _T4_ represent soil treated with 0.1, 0.25, 0.5, 1.0 μmol _p_-coumaric acid/g soil, respectively.

primers, respectively. The PCR conditions were: 94°C for 5 min; 94°C for 45 s, 56°C for 45 s for bacterial 16S rRNA gene (57.5°C for 45 s for fungal ITS region), 72°C for 90 s, 30 cycles in total; a final elongation at 72°C for 10 min. Standard curves were created with a 10-fold dilution series of plasmids containing the 16S rRNA gene or ITS region. Soil DNA extracts were tested for

inhibitory effect by diluting soil DNA extracts and by mixing a known amount of plasmid DNA with the DNA extracted from soil. In all cases, no inhibition was detected. Sterile water was used as a negative control to replace templates. All amplifications were replicated three times. The specificity of the products was confirmed by melting curve analysis and agarose gel electropho-

Table 3. Number of visible bands (_S_), Shannon index (_H_) and evenness index (_E_) based on DGGE analysis of bacterial and fungal communities on DGGE profiles.

_p_-Coumaric acid concentration (μmol/g soil)	Bacterial community			Fungal community		
	_S_	_H_	_E_	_S_	_H_	_E_
B	47±1 ab	3.70±0.04 ab	0.88±0.01 ab	29±3 b	2.96±0.07 a	0.80±0.02 a
0	51±3 a	3.81±0.04 a	0.91±0.01 a	34±1 a	3.09±0.04 a	0.83±0.01 a
0.1	47±2 ab	3.68±0.05 b	0.88±0.01 b	29±1 b	2.98±0.06 a	0.80±0.02 a
0.25	45±1 b	3.66±0.01 b	0.87±0.00 b	31±2 ab	3.00±0.07 a	0.81±0.02 a
0.5	39±2 c	3.47±0.05 c	0.83±0.01 c	33±0 ab	3.09±0.04 a	0.83±0.01 a
1.0	38±2 c	3.45±0.05 c	0.82±0.01 c	32±1 ab	3.04±0.03 a	0.82±0.01 a
ANOVA						
$F_{5,12}$	18.89	29.51	29.51	4.88	3.45	3.45
P-value	<0.0001	<0.0001	<0.0001	0.012	0.036	0.036

Values (mean ± standard error) in the same column followed by different letters are significantly different (P<0.05, Tukey's HSD test).
_B_ soil sample before cucumber planting.

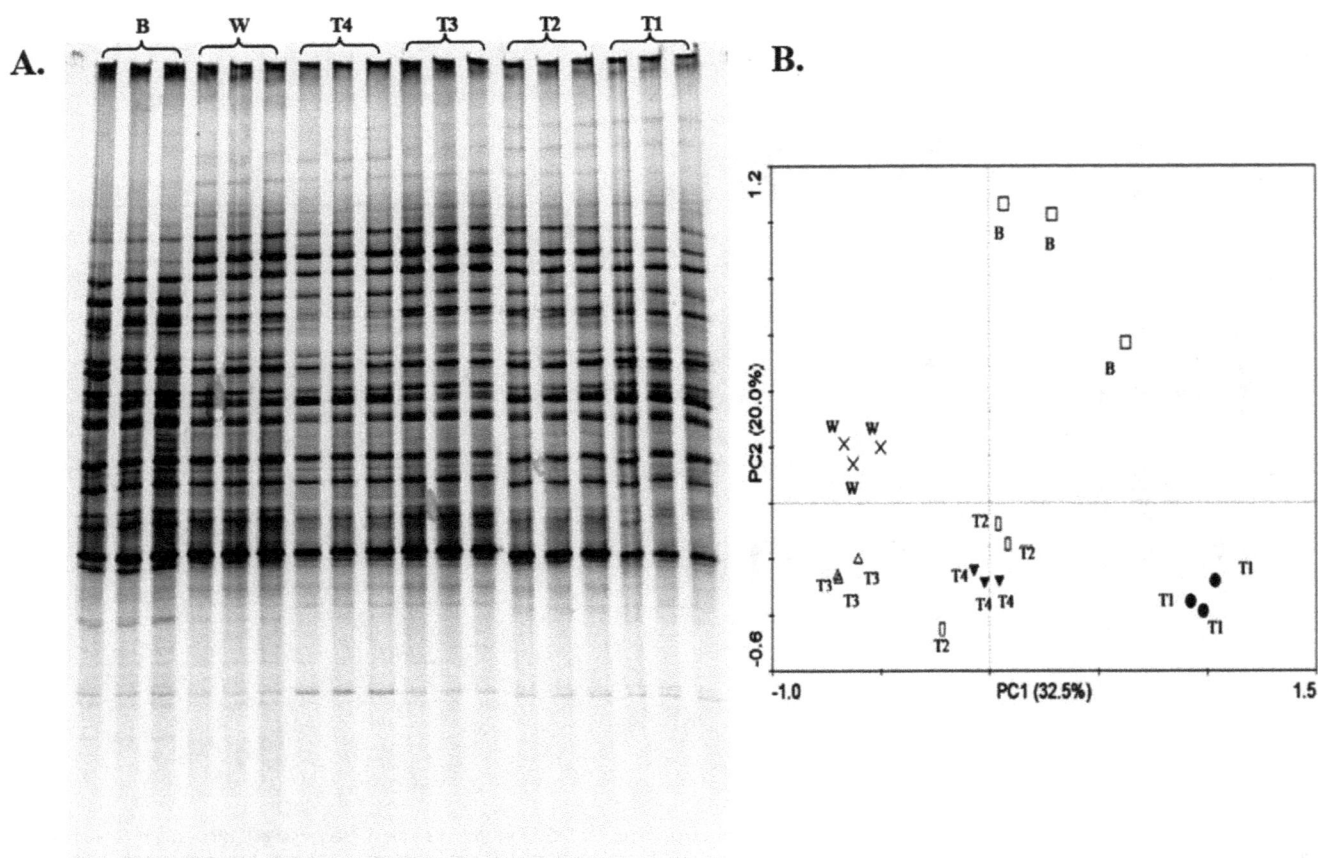

Figure 3. Effects of *p*-coumaric acid on fungal community structure. (A) DGGE profiles of partial fungal ITS regions. (B) Principal component analysis (PCA) of fungal community based on DGGE profiles. *B* and *W* represent soil sample before cucumber planting and control, respectively. *T1, T2, T3* and *T4* represent soil treated with 0.1, 0.25, 0.5, 1.0 μmol *p*-coumaric acid/g soil, respectively.

resis. The threshold cycle (*Ct*) values obtained for each sample were compared with the standard curve to determine the initial copy number of the target gene.

PCR-DGGE. Bacterial and fungal community structures were estimated by the PCR-DGGE method. PCR amplification of partial bacterial 16S rRNA gene was performed with the GC-338F/518R primer [20]. A nested PCR protocol was used to amplify fungal ITS regions of the rRNA gene [35] with TS1F/ITS4 [35,36] and GC-ITS1F/ITS2 [35] as primers for the first and second round of PCR amplifications, respectively. DGGE was performed using an 8% (w/v) acrylamide gel with 30–70% (bacteria) and 20–60% (fungi) denaturant gradient and run in a 1×TAE (Tris-acetate-EDTA) buffer for 14 h at 60°C and 80 V with a DCode universal mutation detection system (Bio-Rad Lab, LA, USA). After the electrophoresis, the gel was stained in 1:3300 (v/v) GelRed (Biotium, CA, USA) nucleic acid staining solution for 20 min. DGGE profiles were photographed with an AlphaImager HP imaging system (Alpha Innotech Crop., CA, USA) under UV light.

Phylogenetic analysis. Bacterial and fungal community compositions were estimated by the clone library method. The bacterial 16S rRNA genes were amplified from soil DNA extracts with primers 27F/1492R and fungal ITS regions with primers ITS1F/ITS4 under previously described conditions [37,38]. PCR products were cleaned using a TIANgel Midi Purification Kit (TIANGEN Biotech, Beijing, China) and then were inserted into

pMD19-T vectors (Takara Biotech, Dalian, China) and transformed into JM109 competent cells (Takara Biotech, Dalian, China). Colonies were screened by *Eco*RI restriction endonuclease digestions [37] and were commercially sequenced using universal M13 forward and reverse primers.

F. oxysporum Experiment

In vitro. A strain of *F. oxysporum* f.sp. *cucumerinum* Owen was isolated and identified from a *Fusarium* wilted cucumber grown in a greenhouse. The potato-dextrose-agar (PDA) medium was autoclaved at 120°C for 15 min [14]. When the medium was at 53°C, 2 ml of *p*-coumaric acid solution (filtered through 0.22 μm filter membranes) at pH 7.0 were added into 18 ml of PDA medium, and then thoroughly mixed and poured into Petri dishes. The final concentrations of *p*-coumaric acid were 0.1, 0.25, 0.5 or 1.0 mM, respectively. The medium added with the same amount of distilled water (pH 7.0) was used as the control. Afterwards, an agar plug (0.5 cm diameter) of *F. oxysporum* taken from a 7-d old PDA culture was inoculated on the center of these Petri dishes. There were five treatments (four concentrations of *p*-coumaric acid and one control) in total. We had three replicates for each treatment, and had five Petri dishes per replicate. After four days of incubation at 28°C in the dark, colony diameter was measured.

Microcosm. The soil used in the cucumber seedling experiment was used here. After sieving (2 mm), 50 g soil were inoculated with 10 ml of 10^4 conidia/ml of *F. oxysporum* conidia

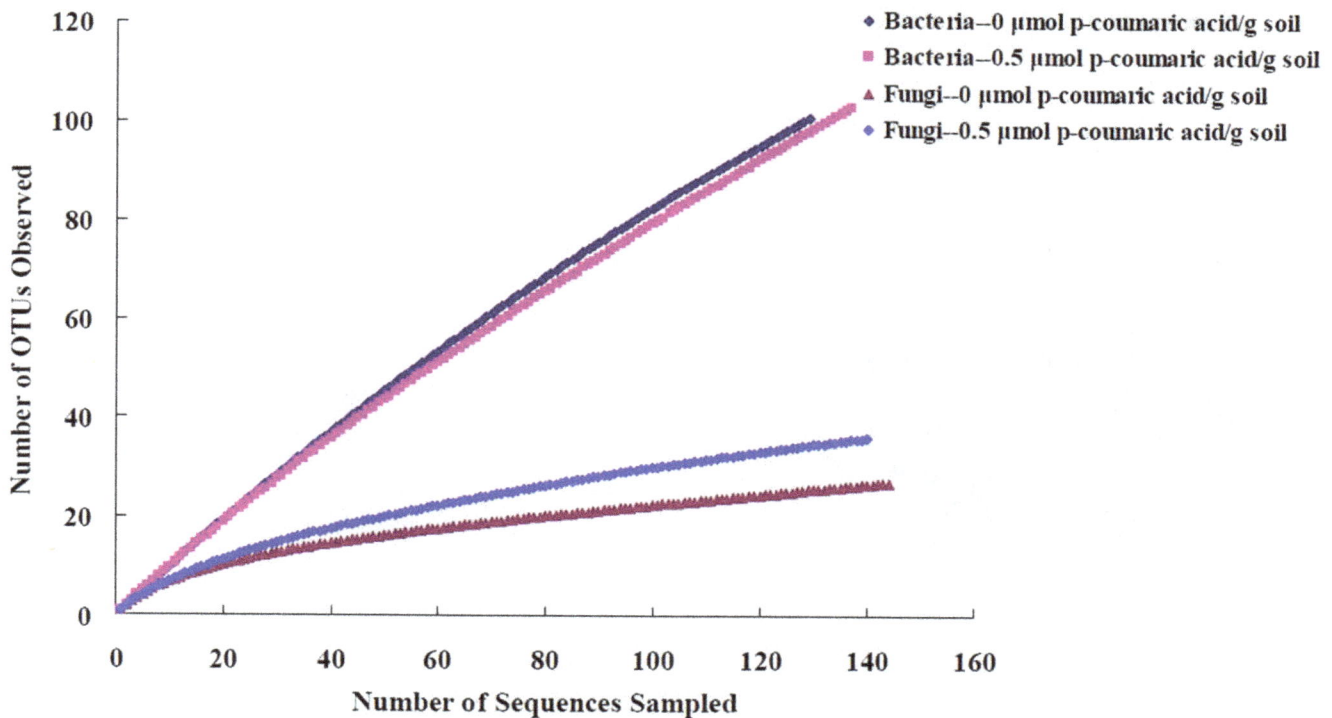

Figure 4. Rarefaction analysis of bacterial 16S rRNA and fungal ITS gene clone libraries. The total number of sequences per library is plotted against the number of unique OTUs encountered within the same library. OTUs were defined at the 97% similarity level.

suspension [27]. Then, different concentrations of p-coumaric acid (pH 7.0) were added to these *F. oxysporum*-inoculated soils every 48 h to achieve final concentrations of 0.1, 0.25, 0.5, 1.0 μmol/g soil DW. The soil treated with the same amount of distilled water (pH 7.0) was used as the control. Then, jars containing these treated soils were incubated at 28°C in the dark. The final soil water content was maintained at 30% of its water holding capacity. There were five treatments (four concentrations of p-coumaric acid and one control) in total. We had three replicates for each treatment, and had five jars per replicate. The colony forming unit (CFU) of culturable *F. oxysporum* population in the soil was measured 10 days after incubation with the plate counting method using the Komada medium [27].

Statistical Analyses

Data were analyzed by analysis of variance (ANOVA) and mean comparison between treatments was performed based on the Tukey's honestly significant difference (HSD) test at the 0.05 probability level with SAS 8.0 software. Banding patterns of the DGGE profiles and principal component analysis (PCA) were analyzed by the Quantity One software (version 4.5) and Canoco for Windows 4.5 software, respectively [39]. The DGGE evenness (E) and Shannon diversity (H) indices were calculated as described before [37].

Sequences from the clone libraries were edited manually to correct falsely identified bases and trimmed at both the 5′ and 3′ ends using the Chromas software. The presence of chimeras were screened using Bellerophon [40], and potential chimeras were excluded from further analysis. Mothur was used to assign sequences to operational taxonomic units (OTUs, 97% similarity)

Table 4. Diversity indices for the soil bacterial and fungal communities as represented by clone libraries.

	p-Coumaric acid concentration (μmol/g soil)	No of sequences	No of OTUs	Coverage (%)	Chao 1	Shannon (H)	Evenness (E)
Bacterial community							
	0	129	101	40	259	4.54	0.98
	0.5	137	103	38	400	4.48	0.97
Fungal community							
	0	144	27	90	57	2.62	0.79
	0.5	140	36	86	50	2.78	0.78

Calculations were based on OUTs formed at the 97% similarity level.

Control (0 µmol *p*-coumaric acid/g soil)   **0.5 µmol *p*-coumaric acid/g soil**

A.

B.

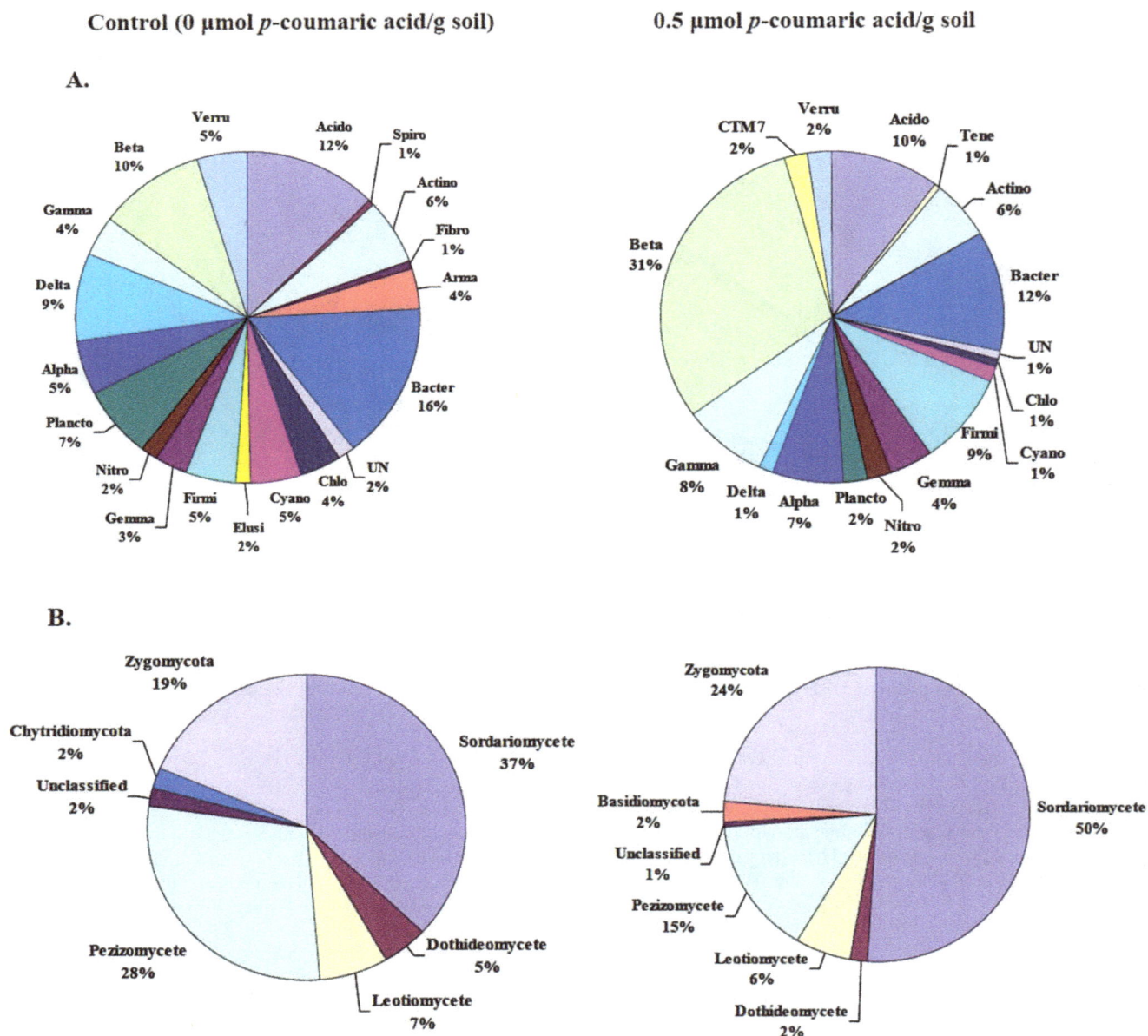

Figure 5. Effects of *p*-coumaric acid on bacterial and fungal community composition. (A) relative abundances of bacterial 16S rRNA gene phylotypes. (B) relative abundances of fungal ITS gene phylotypes. Bacterial phylum *Proteobacteria* and fungal phylum *Ascomycota* are subdivided into class. Shown fractions indicate the relative percentage to the total number of clones. For bacteria: Acido, *Acidobacteria*; Spiro, *Spirochaetes*; Tene, *Tenericutes*; Actino, *Actinobacteria*; Fibro, *Fibrobacteres*; Arma, *Armatimonadetes*; Bacter, *Bacteroidetes*; UN, Unclassified; Chlo, *Chloroflexi*; Cyano, *Cyanobacteria*; Elusi, *Elusimicrobia*; Firmi, *Firmicutes*; Gemma, *Gemmatimonadetes*; Nitro, *Nitrospirae*; Plancto, *Planctomycetes*; Alpha, *Alphaproteobacteria*; Delta, *Deltaproteobacteria*; Gamma, *Gammaproteobacteria*; Beta, *Betaproteobacteria*; CTM7, *Candidate division*; TM7 Verru, *Verrucomicrobia*.

Table 5. LIBSHUFF statistical analysis for differences between bacterial 16S rRNA and fungal ITS gene sequences in clone libraries.

p-Coumaric acid concentration (µmol/g soil)	Bacterial community		Fungal community	
	0	0.5	0	0.5
0		0.0114, P<0.0001		0.033343, P<0.0001
0.5	0.0037, P = 0.0002		0.0367, P<0.0001	

Coverage (C) values of libraries are given in ΔC_{AB} (top diagonal) and ΔC_{BA} (lower diagonal) scores and significances are given in P values.

A.

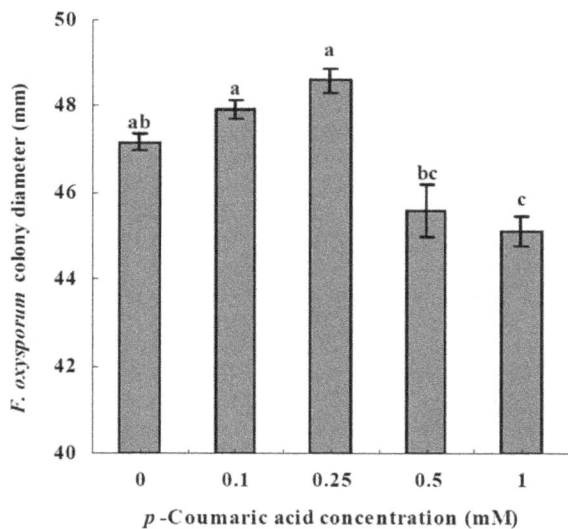

B.

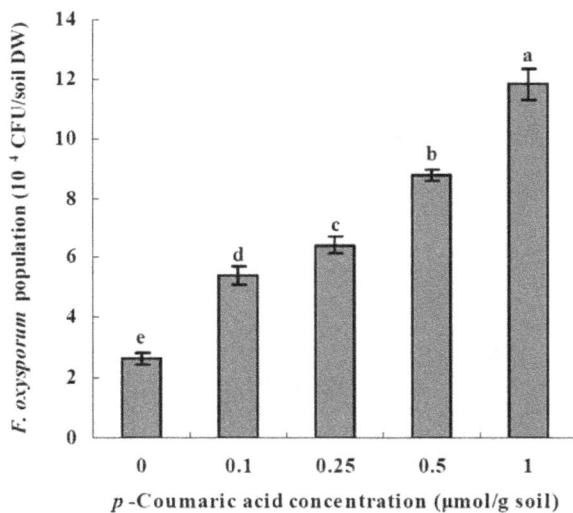

Figure 6. Effects of *p*-coumaric acid on *F. oxysporum* growth. (A) *F. oxysporum* colony diameter. (B) *F. oxysporum* population in the soil. Data are represented as the means of three independent replicates with standard error bars. *Different letters* indicate significant difference between treatments (P<0.05, Tukey's HSD test). *CFU* colony forming unit.

and calculate rarefaction curves, Chao 1 richness and Shannon diversity (*H*) indices [41]. LIBSHUFF in Mothur was used to test whether there were significant differences in bacterial or fungal community compositions between clone libraries. The Classifier function of the Ribosomal Database Project (http://rdp.cme.msu.edu/classifier/classifier.jsp) was used for broad taxonomic characterization of these sequences. The closest relatives of these sequences were determined at GenBank by the BLASTn algorithm. Phylogenetic analyses were conducted using MEGA (version 5.0) and the neighbour-joining tree was constructed using Kimura 2-parameter distance with 1000 replicates to produce Bootstrap values [42].

Results

Cucumber Radicle Elongation

p-Coumaric acid significantly inhibited cucumber radicle elongation ($F_{treatment} = 92.40$; df = 4,10; P<0.0001) and the inhibitory effects increased by increasing *p*-coumaric acid concentration (Fig. 1).

Cucumber Seedling Growth

Generally, *p*-coumaric acid inhibited cucumber seedling growth and the magnitude of inhibition increased by increasing *p*-coumaric acid concentration (Table 1); *p*-coumaric acid significantly inhibited cucumber leaf area at concentrations equal or higher than 0.25 µmol/g soil, and significantly inhibited plant height and seedling dry weigh at 1.0 µmol/g soil (P<0.05).

p-Coumaric Acid Retained in Soil

At the end of the cucumber seedling growth experiment, *p*-coumaric acid concentrations in cucumber-cultivated soils were significantly higher than in the soil before cucumber planting (P<0.05). Addition of *p*-coumaric acids increased *p*-coumaric acid concentration in soil, and the stimulatory effects increased by increasing concentration (Table 2).

Soil pH and Available N

p-Coumaric acid had not significant effects on soil pH and available N content at concentrations equal or lower than 0.5 µmol/g soil, but significantly decreased soil available N content at 1.0 µmol/g soil (P<0.05) (Table 2).

Soil Dehydrogenase Activity and MBC

p-Coumaric acid significantly increased soil dehydrogenase activity and MBC at all concentrations (P<0.05) (Table 2). Compared to the control soil, soil dehydrogenase activity and MBC were increased by 93.4–134.7% and 31.1–54.7%, respectively.

Bacterial and Fungal Community Abundance

p-Coumaric acid significantly increased rhizosphere bacterial and fungal community abundances and decreased bacteria-to-fungi ratio (P<0.05) (Table 2). The soil treated with 0.25 µmol *p*-coumaric acid/g soil had the highest bacterial community abundance, while the soil treated with 0.5 µmol *p*-coumaric acid/g soil had the highest fungal community abundance.

Bacterial and Fungal Community Structure

PCR-DGGE analyses showed that *p*-coumaric acid changed structures of rhizosphere bacterial (Fig. 2A) and fungal (Fig. 3A) communities. The DGGE banding patterns were similar in triplicate samples of each treatment for both bacterial (Fig. 2A) and fungal (Fig. 3A) communities, while were different among the soil before cucumber planting, the control soil (soil treated with water) and *p*-coumaric acid-treated soils. Number of visible bands, Shannon (*H*) and Evenness (*E*) indices of the bacterial community were smaller in *p*-coumaric acid-treated soils than in the control soil and the soil before cucumber planting (Table 3). These diversity indices decreased by increasing *p*-coumaric acid concentration. No definitive trend was observed in the indices of fungal community.

PCA analyses of bacterial (Fig. 2B) and fungal (Fig. 3B) DGGE banding patterns showed that *p*-coumaric acid-treated soils were different from the soil before cucumber planting and the control soil. For bacterial community, soils treated with 0.5 and 1.0 µmol/

g soil *p*-coumaric acid grouped together, while soils treated with 0.1 and 0.25 μmol/g soil *p*-coumaric acid grouped together, indicating that high concentrations (0.5 and 1.0 μmol/g soil) and low concentrations (0.1 and 0.25 μmol/g soil) of *p*-coumaric acid had different influences on bacterial community structure.

Bacterial and Fungal Community Composition

In total, we recovered 266 partial sequences of bacterial 16S rRNA and 284 fungal ITS gene sequences from the control soil and soil treated with 0.5 μmol *p*-coumaric acid/g soil (Table 4). The rarefaction analysis showed that bacterial 16S rRNA gene clone libraries were far from saturation with low coverages of 40% and 38% at 3% cut-off (Fig. 4 and Table 4); however, declines in the rate of OTUs were observed at 10% cut-off (data not shown), indicating that only the most dominant bacterial phyla were detected. For fungal ITS clone libraries, relatively higher coverages (90% and 86%) were observed and the rarefaction curves appeared to be closer to leveling off at 3% cut-off (Fig. 4 and Table 4). The Chao 1 richness index of the bacterial community was higher and the Shannon diversity index was smaller in *p*-coumaric acid-treated soils than in the control soil (Table 4).

Phylum-level characterization of the bacterial 16S rRNA gene clones showed that the bacterial clone libraries were largely composed of individuals representing the *Proteobacteria*, *Bacteroidetes* and *Acidobacteria* (Fig. 5A and Fig. S1). The relative abundances of *Bacteroidetes*, *Chloroflexi*, *Cyanobacteria*, *Planctomycetes* and *Verrucomicrobia* decreased while that of *Proteobacteria* and *Firmicutes* increased in the soil treated with 0.5 μmol *p*-coumaric acid/g soil. Within the *Proteobacteria* phylum, the relative abundances of *Betaproteobacteria* and *Gammaproteobacteria* classes increased while that of *Deltaproteobacteria* class decreased in the soil treated with 0.5 μmol *p*-coumaric acid/g soil. In the control soil, the *Deltaproteobacteria* class was dominated by *Myxococcales* (data not shown).

At the phylum-level, the fungal libraries were dominated by members of the *Ascomycota* and *Zygomycota* (Fig. 5B and Fig. S2). The relative abundance of *Zygomycota* increased in the soil treated with 0.5 μmol *p*-coumaric acid/g soil. With in the *Ascomycota* phylum, the relative abundance of *Sordariomycete* class increased and that of the *Pezizomycete* class decreased in the soil treated with 0.5 μmol *p*-coumaric acid/g soil.

The LIBSHUFF test revealed statistically significant differences in the bacterial and fungal community compositions between the libraries derived from the control soil and soil treated with 0.5 μmol *p*-coumaric acid/g soil (Table 5).

F. oxysporum Mycelial Growth and Population

In the Petri dish experiment, *p*-coumaric acid slightly increased *F. oxysporum* colony diameter at 0.1 and 0.25 mM, but significantly decreased *F. oxysporum* colony diameter at 1.0 mM *in vitro* ($F_{treatment} = 18.31$; df = 4,10; P<0.0001) (Fig. 6A). In the microcosm experiment, all concentrations of *p*-coumaric acid significantly increased *F. oxysporum* population density in soil ($F_{treatment} = 341.12$; df = 4,10; P<0.0001) (Fig. 6B).

Discussion

p-Coumaric acid concentrations in *p*-coumaric acid-treated soils were lower than the sum of the concentration in the control soil and the concentration applied, suggesting that part of this compound was consumed by microorganisms [43]. The physiological alterations caused by allelochemicals are concentration dependent, and for many phenols the range of bioactivity is between 0.1 and 1 mM [30]. Consistent with previous findings,

detrimental effects of *p*-coumaric acid on cucumber radicle elongation and seedling growth were observed in this study. However, the concentration of phenols available to microorganisms may be different at different plant growth stages because phenols secreted by cucumber varied with plant growth stages [3]. Therefore, cucumber at different growth stages may respond differently to the same concentration of artificially applied phenols.

The inhibition of plant growth by plant litter or other types of organic matter may be due to microbial N immobilization or phytotoxicity [5,44]. In this study, soil available N was only significantly decreased at the highest concentration (1.0 μmol *p*-coumaric acid/g soil). A previous study also showed that, in some cases, addition of N solution could not eliminate the phytotoxicity [45]. Therefore, N immobilization may not be responsible for the inhibition of cucumber seedling growth, at least at concentrations lower than 1.0 μmol *p*-coumaric acid/g soil. In fact, it was shown that phenols did not inhibit nitrification or ammonifiers in soil [46] and not retard ammonia oxidation by autotrophic microorganisms *in vitro* [47]. Special attention should be given to the long-term effects of phenols on soil N dynamics and microorganisms involved in N transformation.

In this study, soil dehydrogenase activity, an indicator of microbial activity, was increased by *p*-coumaric acid, which was consistent with the findings that phenol(s) can increase soil microbial activity [19,48]. Soil MBC, and bacterial and fungal community abundance were also increased by *p*-coumaric acid, thus confirming what already observed that phenols can increase MBC [19] and the number of culturable microorganisms [17]. These increases were not surprising, considering that, phenols can be used as carbon sources by soil microorganisms [43].

Based on cultivation-dependent methods, previous studies showed that phenols affected the number of culturable soil microorganisms [17,19]. With cultivation-independent methods (real-time PCR, PCR-DGGE and clone library analysis), we found that *p*-coumaric acid changed the abundance, structure and composition of soil microbial communities. Notably, *p*-coumaric acid stimulated and inhibited certain species of rhizosphere bacteria and fungi. The addition of phenols was shown to stimulate the growth of culturable soil microorganisms that are capable of degrading these compounds [17]. Members of the *Firmicutes* (e.g. *Bacillus licheniformis* [49]), *Gammaproteobacteria* (e.g. *Pseudomonas putida* and *Pseudomonas nitroreducens* [50,51]), *Basidiomycota* (e.g. *Rhodotorula glutinis* [50]), *Betaproteobacteria* (e.g. *Alcaligenes faecalis* [52] and *Comamonas* [51]) and *Sordariomycetes* (e.g. *Phomopsis liquidambari* [53] and *Trichoderma harzianum* [29]) could degrade phenols. Thus, these increased bacteria and fungi may be responsible for degrading *p*-coumaric acid. The *Myxococcales* order contains a group of bacteria that predominantly feed on insoluble organic matter [54]. The *Verrucomicrobia* phylum is saccharolytic, oligotrophic and ubiquitous in soil [55]. The decreases in the relative abundances of *Myxococcales* and *Verrucomicrobia* implied that *p*-coumaric acid may decrease the soil microbial community's ability to hydrolyze insoluble macromolecules.

PCR-DGGE analysis showed that the response of soil bacterial community structure to *p*-coumaric acid was concentration dependent, and PCR-DGGE and clone library analyses showed that *p*-coumaric acid decreased the diversity indices of bacterial community, however, these were not observed in fungal community. These indicated that rhizosphere soil bacterial and fungal communities responded differently to *p*-coumaric acid, confirming what already reported that soil bacterial and fungal communities responded differently to labile soil carbon (sugars, organic acids, and amino acids) inputs [10]. The different responses to these organic C additions to soil probably depend on the fact that bacteria use readily available organic compounds more quickly than fungi, whereas fungi are able to synthesize enzyme activities catalyzing the decomposition of more complex organic compounds [10].

The Petri dish experiment showed that p-coumaric acid inhibited *F. oxysporum* mycelial growth at 0.5 and 1.0 mM. *In vitro* inhibition of p-coumaric acid on *F. oxysporum* f.sp. *niveum* has been already demonstrated [14]. However, the microcosm experiment showed that p-coumaric acid stimulated *F. oxysporum* population in soil. These inconsistencies may be due to the different nutrient status of the environment in which *F. oxysporum* grew: nutrients are limited in soil while are in sufficient concentration in the artificial medium. Ye et al. [27] showed that phenols promoted the incidence of cucumber *Fusarium* wilt caused by *F. oxysporum*. Therefore, phenols may increase the population of *F. oxysporum* and the severity of *Fusarium* wilt under field conditions.

Phenols can affect plant root membrane permeability [56]; therefore, it is possible that p-coumaric acid could lead to the root leakage of organic compounds that can affect soil microbial communities. Effects of phenols (phenol 2,4-di-tert-butylphenol and vanillic acid) on bulk soil microorganism communities, such as stimulation of soil *Basidiomycota*, have been already observed [16], suggesting that changes in microbial communities observed in this study may be due to the added p-coumaric acid and the changed root exudation pattern.

Overall, our results validated the hypothesis that p-coumaric acid inhibited cucumber growth and changed rhizosphere soil microbial communities. Soil microorganisms are involved in many processes crucial to plant survival and performance [57]. Changes in soil microbial communities may affect functions performed by the microbial communities, which can influence plant growth [58,59]. Thus, the bad performance of cucumber seedlings may be attributed to the changes in structure, composition, abundance, activity and size of soil microbial communities as affected by p-coumaric acids. Cucumber plant residues, which contain autotoxins, can accumulate in soil under continuously mono-cropped conditions. These autotoxins may affect soil microbial communities over time in the field. We have shown that, in the continuously mono-cropped cucumber system, the structure of rhizosphere fungal and *Fusarium* communities changed, whereas the abundances increased in the season that cucumber showed retarded growth [28]. A decline in the bacteria/fungi ratio was observed in 'sick' soils from continuous mono-cropping systems [26]. p-

Coumaric acid increased the relative abundances of *Firmicutes* and *Proteobacteria* and decreased the relative abundance of *Cyanobacteria*, which was also found in soils under mono-cropped soybean [60]. The increase in the population of soil-borne pathogens (e.g. *F. oxysporum*) is thought to be responsible for the soil sickness [27]. Therefore, as an autotoxin, p-coumaric acid may account for the soil sickness of cucumber through changing soil microbial communities in natural ecosystems. Further researches should focus on gaining direct evidences that changes in soil microbial communities caused by accumulations of phenols are responsible for the bad performance of cucumber in the continuous mono-cropping system.

Supporting Information

Figure S1 Neighbour-joining tree constructed from sequences from bacterial 16S rRNA gene clone libraries. (A) the control. (B) soil treated with p-coumaric acid (0.5 μmol/g soil). Trees display one representative clone per OTU (defined at the 97% similarity level) and their respective best match sequence with GenBank accession number. The phylogenetic distances of each sequence were calculated using the Kimura 2-parameter model and the tree was constructed using the neighbor-joining algorithm. Scale bars represent 5% changes. Bootstrap values (1 000 repetitions) above 50% are represented.

Figure S2 Neighbour-joining tree constructed from sequences from fungal ITS gene clone libraries. (A) the control. (B) soil treated with p-coumaric acid (0.5 μmol/g soil). Trees display one representative clone per OTU (defined at the 97% similarity level) and their respective best match sequence with GenBank accession number. The phylogenetic distances of each sequence were calculated using the Kimura 2-parameter model and the tree was constructed using the neighbor-joining algorithm. Scale bars represent 5% changes. Bootstrap values (1 000 repetitions) above 50% are represented.

Author Contributions

Conceived and designed the experiments: XZ FW. Performed the experiments: XZ. Analyzed the data: XZ. Contributed reagents/materials/analysis tools: XZ FW. Wrote the paper: XZ FW.

References

1. Yu JQ, Shou SY, Qian YR, Zhu ZZ, Hu WH (2000) Autotoxic potential of cucurbit crops. Plant Soil 223: 147–151.
2. Singh HP, Batish DR, Kohli RK (1999) Autotoxicity: Concept, organisms and ecological significance. Crit Rev Plant Sci 18: 757–772.
3. Yu JQ, Matsui Y (1994) Phytotoxic substances in the root exudates of *Cucumis sativus* L. J Chem Ecol 20: 21–31.
4. Inderjit, Duke SO (2003) Ecophysiological aspects of allelopathy. Planta 217: 529–539.
5. Inderjit (2005) Soil microorganisms: An important determinant of allelopathic activity. Plant Soil 274: 227–236.
6. Ehlers BK (2011) Soil microorganisms alleviate the allelochemical effects of a thyme monoterpene on the performance of an associated grass species. PloS one 6: e26321.
7. Bais HP, Park SW, Weir TL, Callaway RM, Vivanco JM (2004) How plants communicate using the underground information superhighway. Trends Plant Sci 9: 26–32.
8. Kaur R, Kaur R, Kaur S, Baldwin IT, Inderjit (2009) Taking ecological function seriously: soil microbial communities can obviate allelopathic effects of released metabolites. PloS one 4: e4700.
9. Bressan M, Roncato MA, Bellvert F, Comte G, el Zahar Haichar F, et al. (2009) Exogenous glucosinolate produced by *Arabidopsis thaliana* has an impact on microbes in the rhizosphere and plant roots. ISME J 3: 1243–1257.
10. de Graaff MA, Classen AT, Castro HF, Schadt CW (2010) Labile soil carbon inputs mediate the soil microbial community composition and plant residue decomposition rates. New Phytol 188: 1055–1064.
11. Eilers KG, Lauber CL, Knight R, Fierer N (2010) Shifts in bacterial community structure associated with inputs of low molecular weight carbon compounds to soil. Soil Biol Biochem 42: 896–903.
12. Héry M, Herrera A, Vogel TM, Normand P, Navarro E (2005) Effect of carbon and nitrogen input on the bacterial community structure of Neocaledonian nickel mine spoils. FEMS Microbiol Ecol 51: 333–340.
13. Shi S, Richardson AE, O'Callaghan M, Deangelis KM, Jones EE, et al. (2011) Effects of selected root exudate components on soil bacterial communities. FEMS Microbiol Ecol 77: 600–610.
14. Hao W, Ren L, Ran W, Shen Q (2010) Allelopathic effects of root exudates from watermelon and rice plants on *Fusarium oxysporum* f.sp. *niveum*. Plant Soil 336: 485–497.
15. Kong CH, Wang P, Zhao H, Xu XH, Zhu YD (2008) Impact of allelochemical exuded from allelopathic rice on soil microbial community. Soil Biol Biochem 40: 1862–1869.
16. Qu XH, Wang JG (2008) Effect of amendments with different phenolic acids on soil microbial biomass, activity, and community diversity. Appl Soil Ecol 39: 172–179.
17. Shafer SR, Blum U (1991) Influence of phenolic acids on microbial populations in the rhizosphere of cucumber. J Chem Ecol 17: 369–389.

18. Sparling GP, Ord BG, Vaughan D (1981) Changes in microbial biomass, activity in soils amended with phenolic acids. Soil Biol Biochem 13: 455–460.

19. Wang ML, Gu Y, Kong CH (2008) Effects of rice phenolic acids on microorganisms and enzyme activities of non-flooded and flooded paddy soils. Allelopathy J 22: 311–319.

20. Muyzer G, de Waal EC, Uitterlinden AG (1993) Profiling of complex microbial populations by denaturing gradient gel electrophoresis analysis of polymerase chain reaction-amplified genes encoding for 16S rRNA. Appl Environ Microbiol 59: 695–700.

21. Fierer N, Jackson JA, Vilgalys R, Jackson RB (2005) Assessment of soil microbial community structure by use of taxon-specific quantitative PCR assays. Appl Environ Microbiol 71: 4117–4120.

22. Sanguin H, Sarniguet A, Gazengel K, Moënne-Loccoz Y, Grundmann G (2009) Rhizosphere bacterial communities associated with disease suppressiveness stages of take-all decline in wheat monoculture. New Phytol 184: 694–707.

23. Mendes R, Kruijt M, De Bruijn I, Dekkers E, Van Der Voort M, et al. (2011) Deciphering the rhizosphere microbiome for disease-suppressive bacteria. Science 332: 1097–1100.

24. Kyselková M, Kopecký J, Frapolli M, Défago G, Ságová-Marečková M, et al. (2009) Comparison of rhizobacterial community composition in soil suppressive or conducive to tobacco black root rot disease. ISME J 3: 1127–1138.

25. Larkin R, Hopkins D, Martin F (1993) Ecology of *Fusarium oxysporum* f.sp. *niveum* in soils suppressive and conducive to *Fusarium* wilt of watermelon. Phytopathology 83: 1105–1116.

26. Li C, Li X, Kong W, Wu Y, Wang J (2010) Effect of monoculture soybean on soil microbial community in the Northeast China. Plant Soil 330: 423–433.

27. Ye S, Yu J, Peng Y, Zheng J, Zou L (2004) Incidence of Fusarium wilt in *Cucumis sativus* L. is promoted by cinnamic acid, an autotoxin in root exudates. Plant Soil 263: 143–150.

28. Zhou X, Wu F (2012) Dynamics of the diversity of fungal and *Fusarium* communities during continuous cropping of cucumber in the greenhouse. FEMS Microbiol Ecol 80: 469–478.

29. Chen L, Yang X, Raza W, Li J, Liu Y, et al. (2011) *Trichoderma harzianum* SQR-T037 rapidly degrades allelochemicals in rhizospheres of continuously cropped cucumbers. Appl Microbiol Biotechnol 89: 1653–1663.

30. Muscolo A, Sidari M (2006) Seasonal fluctuations in soil phenolics of a coniferous forest: effects on seed germination of different coniferous species. Plant Soil 284: 305–318.

31. Fierer N, Jackson RB (2006) The diversity and biogeography of soil bacterial communities. Proc Natl Acad Sci USA 103: 626.

32. Dalton BR, Weed SB, Blum U (1987) Plant phenolic acids in soils: a comparison of extraction procedures. Soil Sci Soc Am J 51: 1515–1521.

33. Tabatabai MA (1994) Soil enzymes. In: Weaver RW, Angle JR, Bottomley PS, editors. Methods of Soil Analysis. Madison: Soil Society of America. 775–833.

34. Vance ED, Brookes PC, Jenkinson DS (1987) An extraction method for measuring soil microbial biomass C. Soil Biol Biochem 19: 703–707.

35. Gardes M, Bruns TD (1993) ITS primers with enhanced specificity for basidiomycetes: application to the identification of mycorrhiza and rusts. Mol Ecol 2: 113–118.

36. White TJ, Buns TD, Lee S, Taylor J (1990) Analysis of phylogenetic relationships by amplification and direct sequencing of ribosomal RNA genes. In: Innis MA, Gefland DH, Sninsky JJ, White TJ, editors. PCR protocols: A guide to methods and applications. New York: Academic. 315–322.

37. Ibekwe AM, Poss JA, Grattan SR, Grieve CM, Suarez D (2010) Bacterial diversity in cucumber (*Cucumis sativus*) rhizosphere in response to salinity, soil pH, and boron. Soil Biol Biochem 42: 567–575.

38. Yarwood SA, Myrold DD, Högberg MN (2009) Termination of belowground C allocation by trees alters soil fungal and bacterial communities in a boreal forest. FEMS Microbiol Ecol 70: 151–162.

39. Zhou X, Yu G, Wu F (2011) Effects of intercropping cucumber with onion or garlic on soil enzyme activities, microbial communities and cucumber yield. Eur J Soil Biol 47: 279–287.

40. Huber T, Faulkner G, Hugenholtz P (2004) Bellerophon: a program to detect chimeric sequences in multiple sequence alignments. Bioinformatics 20: 2317–2319.

41. Schloss PD, Westcott SL, Ryabin T, Hall JR, Hartmann M, et al. (2009) Introducing mothur: open-source, platform-independent, community-supported software for describing and comparing microbial communities. Appl Environ Microbiol 75: 7537–7541.

42. Tamura K, Peterson D, Peterson N, Stecher G, Nei M, et al. (2011) MEGA5: molecular evolutionary genetics analysis using maximum likelihood, evolutionary distance, and maximum parsimony methods. Molecul Biol Evol 28: 2731–2739.

43. Souto XC, Chiapusio G, Pellissier F (2000) Relationships between phenolics and soil microorganisms in spruce forests: Significance for natural regeneration. J Chem Ecol 26: 2025–2034.

44. Bonanomi G, Incerti G, Barile E, Capodilupo M, Antignani V, et al. (2011) Phytotoxicity, not nitrogen immobilization, explains plant litter inhibitory effects: evidence from solid-state ^{13}C NMR spectroscopy. New Phytol 191: 1018–1030.

45. Meier CL, Keyserling K, Bowman WD (2009) Fine root inputs to soil reduce growth of a neighbouring plant via distinct mechanisms dependent on root carbon chemistry. J Ecol 97: 941–949.

46. McCarty GW, Bremner JM (1986) Effects of phenolic compounds on nitrification in soil. Soil Sci Soc Am J 50: 920–923.

47. McCarty GW, Bremner J, Schmidt EL (1991) Effects of phenolic acids on ammonia oxidation by terrestrial autotrophic nitrifying microorganisms. FEMS Microbiol Lett 85: 345–349.

48. Wu F, Wang X, Xue C (2009) Effect of cinnamic acid on soil microbial characteristics in the cucumber rhizosphere. Eur J Soil Biol 45: 356–362.

49. Koschorreck K, Richter SM, Ene AB, Roduner E, Schmid RD, et al. (2008) Cloning and characterization of a new laccase from *Bacillus licheniformis* catalyzing dimerization of phenolic acids. Appl Microbiol Biotechnol 79: 217–224.

50. Zhang ZY, Pan LP, Li HH (2010) Isolation, identification and characterization of soil microbes which degrade phenolic allelochemicals. J Appl Microbiol 108: 1839–1849.

51. Watanabe K, Yamamoto S, Hino S, Harayama S (1998) Population dynamics of phenol-degrading bacteria in activated sludge determined by *gyr*B-targeted quantitative PCR. Appl Environ Microbiol 64: 1203–1209.

52. Ribeiro Bastos AE, Moon DH, Rossi A, Trevors JT, Tsai SM (2000) Salt-tolerant phenol-degrading microorganisms isolated from Amazonian soil samples. Arch Microbiol 174: 346–352.

53. Chen Y, Peng Y, Dai CC, Ju Q (2011) Biodegradation of 4-hydroxybenzoic acid by *Phomopsis liquidambari*. Appl Soil Ecol 51: 102–110.

54. Dworkin M (1966) Biology of the myxobacteria. Annu Rev Microbiol 20: 75–106.

55. Bergmann GT, Bates ST, Eilers KG, Lauber CL, Caporaso JG, et al. (2011) The under-recognized dominance of *Verrucomicrobia* in soil bacterial communities. Soil Biol Biochem 43: 1450–1455.

56. Baziramakenga R, Leroux G, Simard R (1995) Effects of benzoic and cinnamic acids on membrane permeability of soybean roots. J Chem Ecol 21: 1271–1285.

57. Garbeva P, van Veen JA, van Elsas JD (2004) Microbial diversity in soil: selection of microbial populations by plant and soil type and implication on disease suppressiveness. Annu Rev Phytopathol 42: 243–270.

58. Acosta-Martínez V, Burow G, Zobeck TM, Allen VG (2010) Soil microbial communities and function in alternative systems to continuous cotton. Soil Sci Soc Am J 74: 1181–1192.

59. Allison SD, Martiny JBH (2008) Resistance, resilience, and redundancy in microbial communities. Proc Natl Acad Sci USA 105: 11512–11519.

60. Tang H, Xiao C, Ma J, Yu M, Li Y, et al. (2009) Prokaryotic diversity in continuous cropping and rotational cropping soybean soil. FEMS Microbiol Lett 298: 267–273.

Permissions

List of Contributors

Yuzhou Luo and Minghua Zhang
Wenzhou Medical College, Wenzhou, China
Department of Land, Air, and Water Resources, University of California Davis, Davis, California, United States of America

Jorge Barriuso, José R. Valverde and Rafael P. Mellado
Centro Nacional de Biotecnologı´a (CSIC), Campus de la Universidad Autónoma, Cantoblanco, Madrid, Spain

Zhen-Fang Li, Yan-Qiu Yang, Dong-Feng Xie, Lan-Fang Zhu, Zi-Guan Zhang and Wen-Xiong Lin*
Agroecological Institute, Fujian Agriculture and Forestry University, Fuzhou, China

Guirui Yu
Key Laboratory of Ecosystem Network Observation and Modeling, Institute of Geographic Sciences and Natural Resources Research, Chinese Academy of Sciences, Beijing, China

Chunyan Luo
Institute of Agricultural Resources and Regional Planning, CAAS, Beijing, China

Pei Zhou
School of Agriculture and Biology, Shanghai Jiaotong University, Shanghai, China

Yang Gao
School of Agriculture and Biology, Shanghai Jiaotong University, Shanghai, China

Li Cheng-Fang, Kou Zhi-Kui, Zhang Zhi-Sheng, Wang Jin-Ping, Cai Ming-Li and Cao Cou-Gui
College of Plant Science and Technology, Huazhong Agricultural University, Wuhan, Hubei, China

Zhou Dan-Na
Institute of Animal Husbandry and Veterinary Science, Hubei Academy of Aguicultural Sciences, Wuhan, Hubei, China

A. Mark Ibekwe
United States Salinity Laboratory, Agriculture Research Service, United States Department of Agriculture, Riverside, California, United States of America

David E. Crowley
Department of Environmental Sciences, University of California Riverside, Riverside, California, United States of America

Jincai Ma
United States Salinity Laboratory, Agriculture Research Service, United States Department of Agriculture, Riverside, California, United States of America
Department of Environmental Sciences, University of California Riverside, Riverside, California, United States of America

Xuan Yi, Akihiro Yamazaki and Ching-Hong Yang
Department of Biological Sciences, University of Wisconsin, Milwaukee, Wisconsin, United States of America

Haizhen Wang
United States Salinity Laboratory, Agriculture Research Service, United States Department of Agriculture, Riverside, California, United States of America
Department of Environmental Sciences, University of California Riverside, Riverside, California, United States of America
Institute of Soil and Water Resources and Environmental Science, Zhejiang University, Hangzhou, China
Zhejiang Provincial Key Laboratory of Subtropical Soil and Plant Nutrition, Zhejiang University, Hangzhou, China

Cheng-Jie Wang, Shi-Ming Tang, Yuan-Yuan Jiang and Guo-Dong Han
College of Ecology and Environmental Science, Inner Mongolia Agricultural University, Huhhot, China

Andreas Wilkes
World Agroforestry Centre, 12 Zhongguancun, Beijing, China

Ding Huang
Institute of Grassland Science, China Agricultural University, Beijing, China

Yun-Liang Yang
Department of Biological Science and Technology, National Chiao Tung University, Hsinchu, Taiwan
Institute of Molecular Medicine and Bioengineering, National Chiao Tung University, Hsinchu, Taiwan

Chih-Chao Lin, Tsai-Ling Lauderdale, Pei-Chen Chen and Te-Pin Chang
National Institute of Infectious Diseases and Vaccinology, National Health Research Institutes, Miaoli, Taiwan,

Hui-Ting Chen
Institute of Molecular Medicine and Bioengineering, National Chiao Tung University, Hsinchu, Taiwan
National Institute of Infectious Diseases and Vaccinology, National Health Research Institutes, Miaoli, Taiwan

Ching-Fu Lee and Chih-Wen Hsieh
Department of Applied Science, National Hsinchu University of Education, Hsinchu, Taiwan

Hsiu-Jung Lo
National Institute of Infectious Diseases and Vaccinology, National Health Research Institutes, Miaoli, Taiwan
School of Dentistry, China of Medical University, Taichung, Taiwan

Maged Saad and Xavier Perret
Department of Botany and Plant Biology, University of Geneva, Geneva, Switzerland

Anna Mariotti and Mauro Tonolla
Department of Botany and Plant Biology, University of Geneva, Geneva, Switzerland
Institute of Microbiology, Bellinzona, Switzerland

Valentin Pflüger and Guido Vogel
Mabritec AG, Riehen, Switzerland

Dominik Ziegler
Department of Botany and Plant Biology, University of Geneva, Geneva, Switzerland
Mabritec AG, Riehen, Switzerland

Jan Dirk van Elsas and Cristiane C. P. Hardoim
Department of Microbial Ecology, University of Groningen, Centre for Ecological and Evolutionary Studies, Groningen, The Netherlands,

Leonard S. van Overbeek
Plant Research International, Wageningen, The Netherlands

Pablo R. Hardoim
Department of Microbial Ecology, University of Groningen, Centre for Ecological and Evolutionary Studies, Groningen, The Netherlands
Plant Research International, Wageningen, The Netherlands

Yuzhou Luo, Frank Spurlock, Xin Deng, Sheryl Gill and Kean Goh
Department of Pesticide Regulation, California Environmental Protection Agency, Sacramento, California, United States of America

Yunfeng Peng, Xuexian Li and Chunjian Li
Key Laboratory of Plant-Soil Interactions, Ministry of Education, Department of Plant Nutrition, China Agricultural University, Beijing, China

Guanpeng Gao, Danhan Yin, Shengju Chen, Fei Xia, Jie Yang, Qing Li and Wei Wang
State Key Laboratory of Bioreactor Engineering, East China University of Science and Technology, Shanghai, China

Andrey M. Yurkov and Dominik Begerow
Geobotany, Department of Evolution and Biodiversity of Plants, Faculty of Biology and Biotechnology, Ruhr-Universität Bochum, Bochum, Germany

Martin Kemler
Geobotany, Department of Evolution and Biodiversity of Plants, Faculty of Biology and Biotechnology, Ruhr-Universität Bochum, Bochum, Germany
Centre of Excellence in Tree Health Biotechnology, Forestry and Agricultural Biotechnology Institute (FABI), University of Pretoria, Pretoria, South Africa

Ana Fita, Fernando Nuez and Belén Picó,
Centro de Conservación y Mejora de la Agrodiversidad Valenciana, Universitat Politècnica de València, Valencia, Spain,

Helen C. Bowen, Rory M. Hayden and John P. Hammond
Warwick HRI, University of Warwick, Wellesbourne, Warwick, United Kingdom

Yuejian Mao
Energy Biosciences Institute, University of Illinois, Urbana, Illinois, United States of America
Institute for Genomic Biology, University of Illinois, Urbana, Illinois, United States of America

Anthony C. Yannarell
Energy Biosciences Institute, University of Illinois, Urbana, Illinois, United States of America
Institute for Genomic Biology, University of Illinois, Urbana, Illinois, United States of America
Department of Natural Resources and Environmental Sciences, University of Illinois, Urbana, Illinois, United States of America

Roderick I. Mackie
Energy Biosciences Institute, University of Illinois, Urbana, Illinois, United States of America
Institute for Genomic Biology, University of Illinois, Urbana, Illinois, United States of America
Department of Animal Sciences, University of Illinois, Urbana, Illinois, United States of America

Du Changwen and Zhou Jianmin
The State Key Laboratory of Soil and Sustainable Agriculture, Institute of Soil Science Chinese Academy of Sciences, Nanjing, People's Republic of China

Keith W. Goyne
Department of Soil, Environmental and Atmospheric Sciences, University of Missouri, Columbia, Missouri, United States of America

Shenzhong Tian, Tangyuan Ning, Hongxiang Zhao, Bingwen Wang, Na Li, Huifang Han and Zengjia Li
State Key Laboratory of Crop Biology, Shandong Key Laboratory of Crop Biology, Shandong Agricultural University, Taian, Shandong PR, China

Shuyun Chi
College of Mechanical and Electronic Engineering, Shandong Agricultural University, Taian, Shandong PR, China

Hong Wang, Brian McConkey and Ron DePauw
Semiarid Prairie Agricultural Research Centre, Agriculture and Agri-Food Canada, Swift Current, Saskatchewan, Canada

Yong He
Semiarid Prairie Agricultural Research Centre, Agriculture and Agri-Food Canada, Swift Current, Saskatchewan, Canada
Department of Soil and Water Sciences, China Agricultural University, Beijing, China

Budong Qian
Eastern Cereal and Oilseed Research Centre, Agriculture and Agri-Food Canada, Ottawa, Ontario, Canada

Wenyi Dong, Xinyu Zhang, Huimin Wang, Xiaoqin Dai, Xiaomin Sun and Fengting Yang
Key Laboratory of Ecosystem Network Observation and Modeling, Institute of Geographic Sciences and Natural Resources Research, Chinese Academy of Sciences, Beijing, People's Republic of China

Weiwen Qiu
The New Zealand Institute for Plant and Food Research Limited, Christchurch, New Zealand

Charlotte N. Legind, Arno Rein and Stefan Trapp
Department of Environmental Engineering, Technical University of Denmark, Lyngby, Denmark

Jeanne Serre and Violaine Brochier
Veolia Environnement – Research and Innovation, Rueil-Malmaison, France

Claire-Sophie Haudin, Philippe Cambier and Sabine Houot
INRA, UMR 1091 Environment and Arable Crop Research Unit, Thiverval-Grignon, France

Jinyang Wang, Xiaolin Zhang, Yinglie Liu, Xiaojian Pan, Zhaozhi Chen, Taiqing Huang and Zhengqin Xiong
Jiangsu Key Laboratory of Low Carbon Agriculture and GHGs Mitigation, College of Resources and Environmental Sciences, Nanjing Agricultural University, Nanjing, China

Pingli Liu
Jiangsu Key Laboratory of Low Carbon Agriculture and GHGs Mitigation, College of Resources and Environmental Sciences, Nanjing Agricultural University, Nanjing, China
Hebi Academy of Agricultural Sciences, Hebi, Henan, China

Jinman Wang and Zhongke Bai
College of Land Science and Technology of China University of Geosciences, Handian District, Beijing, People's Republic of China
Key Laboratory of Land Consolidation and Land Rehabilitation Ministry of Land and Resources, Beijing, People's Republic of China

Peiling Yang
College of Hydraulic and Civil Engineering, China Agricultural University, Handian District, Beijing, People's Republic of China

Xingang Zhou and Fengzhi Wu
Department of Horticulture, Northeast Agricultural University, Xiangfang, Harbin, People's Republic of China

Index